SOLVING APPLIED MATHEMATICAL PROBLEMS WITH MATLAB®

SOLVING APPLIED MATHEMATICAL PROBLEMS WITH MATLAB®

DINGYÜ XUE

YANGQUAN CHEN

CRC Press is an imprint of the
Taylor & Francis Group an **informa** business
A CHAPMAN & HALL BOOK

Chapman & Hall/CRC
Taylor & Francis Group
6000 Broken Sound Parkway NW, Suite 300
Boca Raton, FL 33487-2742

Printed in the United States of America on acid-free paper
10 9 8 7 6 5 4 3 2 1

International Standard Book Number-13: 978-1-4200-8250-0 (Hardcover)

Library of Congress Cataloging-in-Publication Data

Xue, Dingyu.
Solving applied mathematical problems with MATLAB / Dingyu Xue, YangQuan Chen.
p. cm.
Includes bibliographical references and index.
ISBN-13: 978-1-4200-8250-0
ISBN-10: 1-4200-8250-7
1. Engineering mathematics--Data processing. 2. MATLAB. 3. Numerical analysis--Data processing. 4. Mathematical optimization--Data processing. I. Chen, YangQuan, 1966- II. Title.

TA331.X84 2009
510.285'5133--dc22 2008025953

Visit the Taylor & Francis Web site at
http://www.taylorandfrancis.com

and the CRC Press Web site at
http://www.crcpress.com

Contents

Preface

Computational Thinking,[1] coined and promoted by Jeannette Wing of Carnegie Mellon University, is getting more and more attention. "It represents a universally applicable attitude and skill set everyone, not just computer scientists, would be eager to learn and use" as acknowledged by Dr. Wing, "Computational Thinking draws on math as its foundations." The present book responds to "Computational Thinking" by offering the readers enhanced math problem solving ability and therefore, the readers can focus more on "Computational Thinking" instead of "Computational Doing."

The breadth and depth of one's mathematical knowledge might not match his or her ability to solve mathematical problems. In today's applied science and applied engineering, one usually needs to get the mathematical problems at hand solved efficiently in a timely manner without complete understanding of the numerical techniques involved in the solution process. Therefore, today, arguably, it is a trend to focus more on how to formulate the problem in a form suitable for computer solution and on the interpretation of the results generated from the computer. We further argue that, even without a complete preparation of mathematics, it is possible to solve some advanced mathematical problems using a computer. We hope this book is useful for those who frequently feel that their level of math preparation is not high enough because they still can get their math problems at hand solved with the encouragement gained from reading this book.

Using computers to solve mathematical problems today is ubiquitous. MATLAB®/Simulink is considered as the dominant software platform for applied math related topics. Sometimes, one simply does not know one's problem could be solved in a much simpler way in MATLAB or Simulink. From what Confucius wrote, "The craftsman who wishes to work well has first to sharpen his implements,"[2] it is clear that MATLAB is the right, already sharpened "implement." However, a bothering practical problem is this: MATLAB documentation only shows "this function performs this," and what a user with a mathematical problem at hand wants is, "Given this math problem, through what reformulation and then use of what functions will get the problem solved." Frequently, it is very easy for one to get lost in thousands of functions offered in MATLAB plus the same amount, if not more, of functions contributed by the MATLAB users community. Therefore,

[1] `http://www.cs.cmu.edu/afs/cs/usr/wing/www/Computational_Thinking.pdf`

[2] Confucius. `http://www.confucius.org/lunyu/ed1509.htm`.

the major contribution of this book is to bridge the gap between "problems" and "solutions" through well grouped topics and tightly yet smoothly glued MATLAB example scripts and reproducible MATLAB-generated plots.

A distinguishing feature of the book is the organization and presentation of the material. Based on our teaching, research and industrial experience, we have chosen to present the course materials following the sequence

- Computer Mathematics Languages — An Overview
- Fundamentals of MATLAB Programming
- Calculus Problems
- Linear Algebra Problems
- Integral Transforms and Complex Variable Functions
- Nonlinear Equations and Optimization Problems
- Differential Equations Problems
- Data Interpolation and Functional Approximation Problems
- Probability and Statistics Problems
- Nontraditional Methods

In particular, in the nontraditional mathematical problem solution methods, we choose to cover some interesting and practically important topics such as set theory and fuzzy inference system, neural networks, wavelet transform, evolutional optimization methods including genetic algorithms and particle swarm optimization methods, rough set based data analysis problems, fractional-order calculus (derivative or integral of non-integer order) problems, etc., all with extensive problem solution examples. A dedicated CAI (computer aided instruction) kit including more than 1,300 interactive PowerPoint slides has been developed for this book for both instruction and self-learning purposes.

We hope that readers will enjoy playing with the scripts and changing them as they wish for a better understanding and deeper exploration with reduced efforts. Additionally, each chapter comes with a set of problems to strengthen the understanding of the chapter contents. It appears that the book is presenting in certain depth some mathematical problems. However, the ultimate objective of this book is to help the readers, after understanding *roughly* the mathematical background, to avoid the tedious and complex technical details of mathematics and find the reliable and accurate solutions to the interested mathematical problems with the use of MATLAB computer mathematics language. There is no doubt that the readers' ability to tackle mathematical problems can be significantly enhanced after reading this book.

This book can be used as a reference text for almost all college students, both undergraduates and graduates, in almost all disciplines which require certain levels of applied mathematics. The coverage of topics is practically broad yet with a balanced depth. The authors also believe that this book will be a good desktop reference for many who have graduated from college and are still involved in solving mathematical problems in their jobs.

Apart from the standard MATLAB, some of the commercial toolboxes may be needed. For instance, the Symbolic Math Toolbox is used throughout the book to provide alternative analytical solutions to certain problems. Optimization Toolbox, Partial Differential Equation Toolbox, Spline Toolbox, Statistics Toolbox, Fuzzy Logic Toolbox, Neural Network Toolbox, Wavelet Toolbox, and Genetic Algorithm and Direct Search Toolbox may be required in corresponding chapters or sections. A lot of MATLAB functions designed by the authors, plus some third-party free toolboxes, are also presented in the book. For more information on MATLAB and related products, please contact

The MathWorks, Inc.
3 Apple Hill Drive
Natick, MA, 01760-2098, USA
Tel: 508-647-7000
Fax: 508-647-7101
E-mail: `info@mathworks.com`
Web: `http://www.mathworks.com`

The writing of this book started more than 5 years ago, when a Chinese version[3] was published in 2004. Many researchers, professors and students have provided useful feedback comments and inputs for the newly extended English version. In particular, we thank the following professors: Xinhe Xu, Fuli Wang of Northeastern University; Hengjun Zhu of Beijing Jiaotong University; Igor Podlubny of Technical University of Kosice, Slovakia; Shuzhi Sam Ge of National University of Singapore, Wen Chen of Hohai University, China. The writing of some parts of this book has been helped by Drs. Feng Pan, Daoxiang Gao, Chunna Zhao and Dali Chen, and some of the materials are motivated by the talks with colleagues at Northeastern University, especially Drs. Xuefeng Zhang and Haibin Shi. The computer aided instruction kit and solution manual were developed by our graduate students Wenbin Dong, Jun Peng, Yingying Liu, Dazhi E, Lingmin Zhang and Ying Luo.

Moreover, we are grateful to the Editors, LiMing Leong and Marsha Hecht, CRC Press, Taylor & Francis Group, for their creative suggestions and professional help. The "Book Program" from The MathWorks Inc., in particular, Hong Yang, MathWorks, Beijing, Courtney Esposito, Meg Vuliez and Dee Savageau, is acknowledged for the latest MATLAB software and technical problem supports.

The authors are grateful to the following free toolbox authors, to allow the inclusion of their contributions in the companion CD:

Dr. Brian K Birge, for particle swarm optimization toolbox (PSOt)
John D'Errico, for `fminsearchbnd` Toolbox
Mr. Koert Kuipers for his BNB Toolbox
Dr. Johan Löfberg, University of Linköping, Sweden for YALMIP

[3]Xue D and Chen Y Q, *Advanced applied mathematical problem solutions using MATLAB*, Beijing: Tsinghua University Press, 2004

Mr. Xuefeng Zhang, Northeastern University, China for RSDA Toolbox

Last but not least, Dingyü Xue would like to thank his wife Jun Yang and his daughter Yang Xue; YangQuan Chen would like to thank his wife Huifang Dou and his sons Duyun, David and Daniel, for their patience, understanding and complete support throughout this work.

Dingyü Xue
Northeastern University
Shenyang, China
xuedingyu@mail.neu.edu.cn

YangQuan Chen
Utah State University
Logan, Utah, USA
yqchen@ieee.org

Chapter 1

Computer Mathematics Languages — An Overview

1.1 Computer Solutions to Mathematics Problems

1.1.1 Why should we study computer mathematics language?

We all know that manual derivation of solutions to mathematical problems is a useful skill when the problems are not so complicated. However, for a great variety of mathematical problems, manual solutions are laborious or even not possible. Thus computers must be employed for solving these problems. There are basically two ways in solving these problems by computers. One is to verbally implement the existing numerical algorithms using general purpose computer languages such as Fortran or C. The other way is to use specific computer languages with a good reputation. These languages include MATLAB, Mathematica and Maple. In this book, they are referred to as the *computer mathematics languages*. The numerical algorithms can only be used to handle computation problems by numbers, while for problems like to find the solutions to the symbolic equation $x^3 + ax + c = d$, where a, c, d are not given numerical values but symbolic variables, the numerical algorithms cannot be used. The computer mathematics languages with symbolic computation capabilities should be used instead.

Before systematically introducing the contents of the book, the following examples are given such that the readers may understand and appreciate the necessity of using the computer mathematics languages.

Example 1.1 In calculus courses, the concepts and derivation methods are introduced with an emphasis on manual deduction and computation. If a function $f(x)$ is given by $f(x) = \dfrac{\sin x}{x^2 + 4x + 3}$, how could one derive $\dfrac{\mathrm{d}^4 f(x)}{\mathrm{d}x^4}$ manually? Of course, one can derive it using the methods taught in calculus courses. For instance, the first-order derivative $\mathrm{d}f(x)/\mathrm{d}x$ can be derived first, the second-order derivative, third-order derivative and finally fourth-order derivative of the function $f(x)$ can be evaluated in turn. In this way, even higher-order derivatives of the function can be derived manually, in theory. However, the procedure is more suitable to be carried out with computers. With suitable computer mathematics languages, the fourth-order derivative of the function $f(x)$ can be calculated using a single statement as follows:

$$\frac{\sin x}{x^2+4x+3}+4\frac{(2x+4)\cos x}{(x^2+4x+3)^2}-12\frac{(2x+4)^2\sin x}{(x^2+4x+3)^3}+12\frac{\sin x}{(x^2+4x+3)^2}-24\frac{(2x+4)^3\cos x}{(x^2+4x+3)^4}$$
$$+48\frac{(2x+4)\cos x}{(x^2+4x+3)^3}+24\frac{(2x+4)^4\sin x}{(x^2+4x+3)^5}-72\frac{(2x+4)^2\sin x}{(x^2+4x+3)^4}+24\frac{\sin x}{(x^2+4x+3)^3}.$$

Of course, with built-in symbolic expression simplification methods, an even simpler form can be automatically derived for $\dfrac{\mathrm{d}^4 f(x)}{\mathrm{d}x^4}$ as follows:

$$(x^8+16x^7+72x^6-32x^5-1094x^4-3120x^3-3120x^2+192x+1581)\frac{\sin x}{(x^2+4x+3)^5}$$
$$+8(x^5+10x^4+26x^3-4x^2-99x-102)\frac{\cos x}{(x^2+4x+3)^4}.$$

It is obvious that manual derivation could be a tedious and laborious work and it could be quite complicated. Wrong results may be obtained even with a slightly careless manipulation of formulae. Thus, even though the results can be obtained, the results may be suspicious and untrustworthy. If the computer mathematics languages are used, the tedious and unreliable work can be avoided. For example, by using MATLAB language, the accurate $\mathrm{d}^{100}f(x)/\mathrm{d}x^{100}$ can be obtained in a second!

Example 1.2 In many fields, the roots of polynomial equations are often needed. The well-known Abel-Ruffini Theorem states that there is no general solution in radicals to polynomial equations of degree five or higher. The problems can be solved numerically using the Lin-Bairstrow algorithm. Now consider a polynomial equation

$$s^6+9s^5+\frac{135}{4}s^4+\frac{135}{2}s^3+\frac{1215}{16}s^2+\frac{729}{16}s+\frac{729}{64}=0.$$

Applying the Lin-Bairstrow method, under double-precision, the roots can be found as

$$s_{1,2}=-1.5056\pm \mathrm{j}0.0032,\quad s_{3,4}=-1.5000\pm \mathrm{j}0.0065,\quad s_{5,6}=-1.4944\pm \mathrm{j}0.0032.$$

Substituting s_1 back to the original equation, the error can be found to be $-8.7041\times 10^{-14}-\mathrm{j}1.8353\times 10^{-15}$. In fact, all the roots to the above equation are exactly -1.5, if the symbolic facilities of the computer mathematics languages are used.

Example 1.3 In linear algebra courses, the determinant of a matrix is suggested to be evaluated by algebraic complements. For instance, for an $n\times n$ matrix, its determinant can be evaluated from determinants of n matrices of size $(n-1)\times(n-1)$. Similarly, the determinant of each $(n-1)\times(n-1)$ matrix can be obtained from determinants of $n-1$ matrices of size $(n-2)\times(n-2)$. In other words, the determinant of an $n\times n$ matrix can finally be obtained from determinants of 1×1 matrices, i.e., the scalar itself. Thus, it can be concluded that the analytical solutions to the determinant of any given matrix exists.

In fact, the above mathematical conclusion neglected the computability and feasibility issue. The computation load for such an evaluation task could be extremely tremendous, which requires $(n-1)(n+1)!+n$ operations. For instance, when $n=20$, the number of floating-point operations (flops) for the computation is 9.7073×10^{20}, which amounts to 300 years of computation on mainframe computers of a billion flops per second. Thus, the algebraic complement method, although elegant and instructive, is not practically feasible. In real applications, the determinants of even larger sized matrices are usually needed ($n \gg 20$), which is clearly not possible to directly apply the algebraic complement method mentioned above.

In numerical analysis courses, various algorithms have been devised. However, due to finite precision numerical computation, these algorithms may have numerical problems when the matrix is close to being singular. For example, consider the Hilbert matrix given by

$$\boldsymbol{H} = \begin{bmatrix} 1 & 1/2 & 1/3 & \cdots & 1/n \\ 1/2 & 1/3 & 1/4 & \cdots & 1/(n+1) \\ \vdots & \vdots & \vdots & \ddots & \vdots \\ 1/n & 1/(n+1) & 1/(n+2) & \cdots & 1/(2n-1) \end{bmatrix}. \tag{1.1}$$

For $n = 20$, an erroneous determinant $\det(\boldsymbol{H}) = 0$ could actually be obtained even if double-precision is used. On the other hand, if computer mathematics language MATLAB is used, the analytical solution below can be obtained within 0.4 seconds:

$$\det(\boldsymbol{H}) = \frac{1}{\underbrace{2377454\cdots 68000000000000000000000000000000000000}_{\text{225 digits, with some digits omitted}}} \approx 4.2062 \times 10^{-224}.$$

Example 1.4 Consider the well-known nonlinear Van der Pol equation

$$\ddot{y} + \mu(y^2 - 1)\dot{y} + y = 0$$

and when μ is large, i.e., $\mu = 1000$, the conventional numerical algorithms for solving differential equations such as the standard Runge-Kutta method may cause numerical problems. Specialized numerical algorithms for stiff ordinary differential equations (ODEs) should be used instead, rather than the standard Runge-Kutta methods in numerical analysis courses.

As another example, the first-order delay differential equation (DDE)

$$\frac{\mathrm{d}y(t)}{\mathrm{d}t} = -0.1y(t) + 0.2\frac{y(t-30)}{1 + y^{10}(t-30)}$$

cannot be solved using the commonly taught algorithms in numerical analysis courses. The MATLAB function `dde23()` or block diagram modeling tool Simulink can be used instead. The details of the methods will be given later in the book (Section 7.3).

Example 1.5 Consider the linear programming problem given below

$$\min_{\boldsymbol{x} \text{ s.t. } \begin{cases} 2x_2+x_3+4x_4+2x_5 \leqslant 54 \\ 3x_1+4x_2+5x_3-x_4-x_5 \leqslant 62 \\ x_1, x_2 \geqslant 0, x_3 \geqslant 3.32, x_4 \geqslant 0.678, x_5 \geqslant 2.57 \end{cases}} (-2x_1 - x_2 - 4x_3 - 3x_4 - x_5).$$

Since the original problem is a linear constrained optimization problem, the analytical unconstrained method, i.e., setting the derivatives of the objective function with respect to each decision variable x_i to zeros, cannot be used. With linear programming tools in MATLAB, the numerical solutions can be found easily as $x_1 = 19.7850$, $x_2 = 0$, $x_3 = 3.3200$, $x_4 = 11.3850$, $x_5 = 2.5700$.

Applying algorithms in numerical analysis or optimization courses, conventional constrained optimization problems can be solved. However, if other special constraints are introduced, for instance, the decision variables are constrained to be integers, the integer programming must be used. There are not so many books introducing softwares that can tackle the integer and mixed-integer programming problems. If we use MATLAB, the solutions to this example problem are easily found as $x_1 = 19, x_2 = 0, x_3 = 4, x_4 = 10, x_5 = 5$.

Example 1.6 In many other courses of applied mathematics branches, such as integral transform, complex variable functions, partial differential equations, data interpolation and fitting, probability and statistics, can you still remember how to solve the problems after the final exams?

In many subjects, such as electric circuits, electronics, power electronics, motor drive, automatic control theory, more sophisticated examples and problems are usually skipped due to the lack of introduction of high-level computer software tools. If computer mathematics languages are introduced routinely in the above courses, complicated practical problems can be solved and innovative solutions to the problems can be explored.

1.1.2 Analytical solutions versus numerical solutions

The development of modern sciences and engineering depends heavily on mathematics. However, the research interests of pure mathematicians are different from other scientists and engineers. Mathematicians are often more interested in finding the analytical or closed-form solutions to mathematical problems. They are in particular interested in proving the existence and uniqueness of the solutions, and do not usually care much about what the solutions are. Engineers and scientists are more interested in finding the exact or approximate solutions to the problems at hand and usually do not care too much about the details on how the results are obtained, as long as the results are reliable and meaningful. The most widely used approaches for finding the approximate solutions are the numerical techniques.

It is quite common to find that analytical solutions do not exist in reality in many different mathematics branches. For instance, it is well-known that the definite integral $\dfrac{2}{\sqrt{\pi}}\displaystyle\int_0^a \mathrm{e}^{-x^2}\mathrm{d}x$ has no analytical solution. To solve the problem, mathematicians introduce a special function $\mathrm{erf}(a)$ to denote it and do not care what in particular the numerical value is. In order to find an approximate value, scientists and engineers have to use numerical approaches.

Another example is that the irrational number π has no closed-form solution. The ancient Chinese astronomer and scientist Zu Chongzhi, also known as *Tsu Ch'ung-chih*, found that the value is between 3.1415926 and 3.1415927, in about A.D. 480. This value is accurate enough in most science and engineering practice. Even with the imprecise value 3.14 found by Archimedes in about B.C 250 (?), the solutions to most engineering problems are often acceptable.

The above discussions hint that an approximate numerical solution is ubiquitous. In many cases, only showing existence and uniqueness of solutions is not enough. We need to compute the solution using computers. The breadth and depth of one's mathematical knowledge might not match one's ability of getting mathematical problems solved. In today's applied science and applied engineering, one usually needs to get the mathematical problems at hand solved efficiently in a timely manner without complete understanding of the numerical techniques involved even in the solution process. Therefore, today, arguably, it is a trend to focus more on how to formulate the problem in a form suitable for computer solution and on the interpretation of the results generated from the computer. Numerical techniques have already been used in many scientific and engineering areas. For instance, in mechanics, finite element methods (FEM) have been used in solving partial differential equations. In aerospace and control, numerical linear algebra and numerical solutions to ordinary differential equations have successfully been used for decades. For simulation experiments in engineering and non-engineering areas, numerical solutions to difference and differential equations are the core problems. In hi-tech developments, digital signal processing based on fast Fourier transform (FFT) has been regarded as a routine task. There is no doubt that if one masters one or more practical computation tools, significant enhancement of mathematical problem solving capability can be expected.

1.1.3 Mathematics software packages: an overview

The emerging digital computers fueled the developments of numerical as well as symbolic computation techniques. In the early stages of the development of numerical computation techniques, some well established packages, such as the eigenvalue-based package EISPACK[1, 2], linear algebra package LINPACK[3] in the USA, the NAG package by the Numerical Algorithm Group in the UK, and the package in the well accepted book *Numerical Recipes*[4], appeared and were widely used with good user feedback.

The famous EISPACK and LINPACK packages are both specific packages for numerical linear algebra applications. Originally developed in the USA, EISPACK and LINPACK packages were written in Fortran. To have a flavor of how to use the packages, let us consider eigenvalues ($\boldsymbol{W}_{\rm R}$, $\boldsymbol{W}_{\rm I}$ for their real and imaginary parts) and eigenvectors $\boldsymbol{Z}$ of an $N \times N$ real matrix $\boldsymbol{A}$. As suggested by EISPACK, the standard solution method is by sequentially calling relevant subroutines provided in EISPACK as follows:

```
CALL BALANC(NM,N,A,IS1,IS2,FV1)
CALL ELMHES(NM,N,IS1,IS2,A,IV1)
CALL ELTRAN(NM,N,IS1,IS2,A,IV1,Z)
CALL HQR2(NM,N,IS1,IS2,A,WR,WI,Z,IERR)
IF (IERR.EQ.0) GOTO 99999
```

```
      CALL BALBAK(NM,N,IS1,IS2,FV1,N,Z)
```

Apart from the main body of the program, the user should also write a few lines to input or initialize the matrix $\boldsymbol{A}$ to the above program and return or display the results obtained by adding some display or printing statements. Then, the whole program should be compiled and linked with the EISPACK library to generate an executable program. It can be seen that the procedure is quite complicated. Moreover, if another matrix is to be solved, the whole procedure might be repeated, which makes the solution process even more complicated.

It is good news that the mathematical software packages are continuously developing, implementing the leading-edge numerical algorithms, providing more efficient, more reliable, faster and more stable packages. For instance, in the area of numerical algebra, a new LaPACK is becoming the leading package. Unlike the original purposes of EISPACK or LINPACK, the objectives of LaPACK have been changed. LaPACK is no longer aiming at providing libraries or facilities for direct user applications. Instead, LaPACK provides support to mathematical software and languages. For example, MATLAB and a freeware Scilab have abandoned the packages of LINPACK and EISPACK, and adopted LaPACK as their low-level library support.

1.2 Summary of Computer Mathematics Languages

1.2.1 A brief historic review of MATLAB

In the late 1970's, Professor Cleve Moler, the Chairman of the Department of Computer Science at the University of New Mexico found that the solutions to linear algebraic problems using the most advanced EISPACK and LINPACK packages are too complicated. MATLAB (MATrix LABoratory) was then conceived and developed. The first release of MATLAB was freely distributed in 1980. Cleve Moler and Jack Little co-founded The MathWorks Inc. in 1984 to develop the MATLAB language. At that time, state-space-based control theory was rapidly developing and a significant amount of numerical algebra problems needed to be solved. The appearance of MATLAB and its Control Systems Toolbox soon attracted the attention of the control community. More and more control oriented toolboxes were written by distinguished experts in different control disciplines, which added higher reputations to MATLAB. It is true that MATLAB was initiated by numerical mathematicians but its impacts and innovations were first built by the control community. Soon it became the general purpose language of control scientists and engineers. With more and more new toolboxes in many other engineering disciplines, MATLAB is becoming the *de facto* standard language of science and engineering.

1.2.2 Three widely used computer mathematics languages

There are three leading computer mathematics languages in the world with high reputations. They are MATLAB of The MathWorks, Mathematica of Wolfram Research, and Maple of Waterloo Maple. They each have their own distinguishing merits, for instance, MATLAB is good at numerical computation and easy in programming, while Mathematica and Maple are powerful in pure mathematics problems involving symbolics and derivations.

The numerical computation capability of MATLAB is much stronger than the other two languages. Besides, various nice toolboxes by experts can be used to tackle the problems with high efficiency. In addition, the symbolic computation engine in Maple can be used to solve symbolic computation problems. Thus, the symbolic computation capability of MATLAB is essentially as good as Mathematica and Maple for most mathematical problems. When the readers have mastered such a computer mathematics language like MATLAB, the ability of handling mathematical problems could be enhanced significantly.

1.2.3 Introduction to free scientific open-source softwares

Although many extremely powerful scientific computation facilities have been provided in the computer mathematics languages such as MATLAB, Maple and Mathematica, there are certain limitations in their applications in research and education, for example, they are expensive commercial softwares. Moreover, some of the core source code are not accessible to the users. Thus the open-source softwares are welcome in scientific computation as well. Some influential softwares include:

(i) **Scilab** Scilab is developed and maintained by INRIA, France. The syntaxes are very similar with MATLAB. It is a free open-source software which concentrates in particular on control and signal processing. The Scicos in Scilab is a block diagram simulation environment similar to Simulink. The web-page of Scilab is `http://www.scilab.org/`.

(ii) **Octave** Octave was conceived in 1988 and first released in 1993. It is a promising open-source software for numerical computation, initiated from numerical linear algebra. The earlier objective of the software was to provide support in education. The web-page of Octave is `http://www.gnu.org/software/octave/`.

(iii) **Others** Some other small-scale numerical matrix computation softwares such as Freemat and SpeQ are all attractive free softwares. The web-pages are respectively `http://freemat.sourceforge.net/wiki/index.php/Main_Page` and `http://www.speqmath.com/index.php?id=1`.

1.3 Outline of the Book

The book can be used as a reference text or even a textbook of a new course on scientific computation. The applications of all branches of college mathematics can be taught in such a course with broad coverage, which enables the students view mathematics from a different angle. This will significantly increase the ability of the students for mathematical problem solutions. The book can also be used as a reference book for actual mathematical problem solutions.

The contents of the book are summarized below:

Chapter 1, the current chapter, gives an overview of the development of MATLAB and other computer mathematics languages.

Chapter 2, "Fundamentals of MATLAB Programming," introduces briefly the programming essentials of MATLAB, including data structure, flow control structures and M-function programming. Two-dimensional and three-dimensional graphics are also presented. This chapter is the basis for the materials in the book.

Chapter 3, "Calculus Problems," covers the problems in college calculus, from a different viewpoint. The subjects introduced in the chapter include limits, derivatives and integrals of single-variable and multivariable functions. Series expansion problems such as Taylor series and Fourier series expansions as well as series sums and products are covered. Numerical differentiation and integration or quadrature, are also introduced. Finally MATLAB solutions to path, line and surface integrals are illustrated.

Chapter 4, "Linear Algebra Problems," studies linear algebra problems using both analytical and numerical methods. Special matrices in MATLAB are first discussed followed by basic matrix analysis, matrix transformation and matrix decomposition problems. Matrix equation solutions, including linear equations, Lyapunov equation and Riccati equations, are introduced. How to evaluate matrix functions is introduced for both the exponential function and the functions of arbitrary forms.

Chapter 5, "Integral Transforms and Complex Variable Functions," includes the solutions to Laplace transform problems and their inverse, Fourier transforms and their variations, Z, Mellin and Hankel transforms. The analysis of complex variable functions are also introduced, including poles, residues, partial fraction expansion and closed-path integral problems, all with many illustrative solution examples.

Chapter 6, "Nonlinear Equations and Optimization Problems," explores the search methods for linear equations, nonlinear equations and nonlinear matrix equations. The unconstrained optimization, constrained optimization and mixed integer programming problems are demonstrated. Linear matrix inequality (LMIs) problems are also covered in the chapter.

Chapter 7, "Differential Equations Problems," mainly covers analytical as well as numerical solutions to ordinary differential equations. Different types of ordinary differential equations, including stiff equations, implicit equations, differential algebraic equations, delay differential equations and the boundary valued equations are illustrated. An introduction to partial differential equations is also given briefly through examples.

Chapter 8, "Data Interpolation and Functional Approximation Problems," studies the interpolation problems such as simple interpolation, cubic spline and B-spline problems. We show that numerical differentiation and integration problems can be solved with splines. Polynomial fitting, continued fraction expansion and Padé approximation as well as least squares curve fitting methods are all covered and illustrated. Fast Fourier transform, signal filtering and de-noising problems are also studied briefly in this chapter.

Chapter 9, "Probability and Mathematical Statistics Problems," studies the probability distributions and pseudo-random number generators first. Statistical analysis to the measured random data is then illustrated. Hypothesis tests for a few common applications are presented, and the analysis of variance method is also demonstrated briefly.

Chapter 10, "Nontraditional Solution Methods," covers a wide variety of interesting topics, such as traditional set theory, rough set theory, fuzzy set theory and fuzzy inference system, neural networks, wavelet transform, evolutional optimization methods including genetic algorithms and particle swarm optimization methods. Most interestingly, fractional-order calculus (derivative or integral of non-integer order) problems are introduced with basic numerical computational techniques and examples.

It appears that the book is presenting in certain depth some mathematical problems. However, the ultimate objective of this book is to help the readers, after understanding roughly the mathematical background, to avoid the tedious and complex technical details of mathematics and find the reliable and accurate solutions to the interested mathematical problems with the use of MATLAB computer mathematics language. There is no doubt that the readers' ability to tackle mathematical problems can be significantly enhanced after reading this book.

Exercises

1. Install MATLAB on your machine, and issue the command `demo`. From the dialog boxes and menu items of the demonstration program, experience the powerful functions provided in MATLAB.
2. Type the command `doc symbolic/diff`, and see whether it is possible, by reading the relevant help information, to solve the problem given in Example 1.1. If the solutions can be obtained, compare the solutions with the results in the example.
3. Solve the following Lyapunov equation by starting the command

```
lookfor lyapunov
```

and see whether there is any function related to the keyword `lyapunov`. If there is one, say, the `lyap` function is found, type `doc lyap` and see whether there is a way to solve this Lyapunov equation. Check the accuracy of the solution by back substitution.

$$\begin{bmatrix} 8 & 1 & 6 \\ 3 & 5 & 7 \\ 4 & 9 & 2 \end{bmatrix} \boldsymbol{X} + \boldsymbol{X} \begin{bmatrix} 16 & 4 & 1 \\ 9 & 3 & 1 \\ 4 & 2 & 1 \end{bmatrix} = \begin{bmatrix} 1 & 2 & 3 \\ 4 & 5 & 6 \\ 7 & 8 & 0 \end{bmatrix}.$$

4. Write a simple subroutine which can be used to perform matrix multiplications in other languages such as the C language. Try to make the code as general purpose as possible.
5. Write a piece of code in C language which can generate the Fibonacci sequence defined as $a_1 = a_2 = 1$, $a_{k+2} = a_k + a_{k+1}$, for $k = 1, 2, \cdots$. Generate the sequence with 50 terms. Observe whether the results are feasible. If there are serious abnormal problems, are there any possible solutions in C?

Chapter 2

Fundamentals of MATLAB Programming

MATLAB language is becoming a widely accepted scientific language, especially in the field of automatic control. In other engineering and non-engineering disciplines such as economics and even biology, MATLAB is also an attractive and promising computer mathematics language. In this book, we shall concentrate on the introduction to MATLAB with its applications in solving applied mathematics problems. A good working knowledge of MATLAB language will enable one not only understand in depth the concepts and algorithms in research but also increase the ability to do creative research work and apply MATLAB to actively tackle the problems in other related courses.

As a programming language, MATLAB has the following advantages:

(i) **Clarity and high efficiency** MATLAB language is a highly integrated language. A few MATLAB sentences may do the work of hundreds of lines of source code of other languages. Thus the MATLAB program is more reliable and easy to maintain.

(ii) **Scientific computation** The basic element in MATLAB is a complex matrix of double-precision. Matrix manipulations can be carried out directly. Numerical computation functions provided in MATLAB, such as the ones for solving optimization problems or other mathematical problems, can be used directly. Also symbolic computation facilities are provided in MATLAB's Symbolic Math Toolbox to support formula derivation.

(iii) **Graphics facilities** MATLAB language can be used to visualize the experimental data in an easy manner. Moreover, the graphical user interface is also supported in MATLAB.

(iv) **Comprehensive toolboxes and blocksets** There are a huge amount of MATLAB toolboxes and Simulink blocksets contributed by experienced programmers and researchers.

(v) **Powerful simulation facilities** The powerful block diagram-based modeling technique provided in Simulink can be used to analyze systems with almost any complexity. In particular, under Simulink, the control blocks, electronic blocks and mechanical blocks can be modeled together under the same framework, which is currently not possible in other computer mathematics languages.

In Section 2.1, the fundamental information about MATLAB programming, such as the data types, statement structures, colon expressions and sub-matrix extraction is introduced. In Section 2.2, the basic operations, including algebraic, logic and relationship operations, and simplification of symbolic formulae, and introduction to number theory are presented. The flow control such as loop structures, conditional structures, switches and trial structures are introduced in Section 2.3. In Section 2.4, the most important programming structure — the M-function — is illustrated with useful hints on high-level programming. In Section 2.5, two-dimensional graphics facilities are presented, where two-dimensional sketching and implicit function expressions are illustrated in particular. Three-dimensional graphics are presented in Section 2.6, where mesh and surface plots can be drawn and the viewpoint setting facilities are introduced.

2.1 Fundamentals of MATLAB Programming

2.1.1 Variables and constants in MATLAB

MATLAB variable names consist of a letter, followed by any number of letters, digits, or underscores. For instance, `MYvar12`, `MY_Var12` and `MyVar12_` are valid variable names, while `12MyVar` and `_MyVar12` are invalid ones, since the first character is not a letter. The variable names are case-sensitive, i.e., the variables `Abc` and `ABc` are different variables.

In MATLAB, some of the names are reserved for the constants. They can however be assigned to other values. It is suggested that these names should not be assigned to other values whenever possible.

- `eps` — error tolerance for floating-point operation. The default value is `eps`= 2.2204×10^{-16}, and if the absolute value of a quantity is smaller than `eps`, it can be regarded as 0.
- `i` and `j` — If `i` or `j` is not overwritten, they both represent $\sqrt{-1}$. If they are overwritten, they can be restored with the `i=sqrt(-1)` command.
- `Inf` — the MATLAB representation of infinity quantity $+\infty$. It can also be written as `inf`. Similarly $-\infty$ can be written as `-Inf`. When 0 is used in denominator, the value `Inf` can be generated, with a warning. This agrees with the IEEE standard. For mathematical computation, this definition has its advantages over C language.
- `NaN` — not a number, which is often returned by the operations 0/0, `Inf/Inf` and others. It should also be noted that `NaN` times `Inf` will return `NaN`.
- `pi` — double-precision representation of the circumference ratio π.

- `lasterr` — returns the error message received last time. It can be a string variable, with empty string for no error message generated.
- `lastwarn` — returns the last obtained warning message.

2.1.2 Data structure

Double-precision data type

Numerical computation is the most widely used computation form in MATLAB. To ensure high-precision computations, double-precision floating-point data type is used, which is 8 bytes (64 bits). According to the IEEE standard, it is composed of 11 exponential bits, 53 number bits and a sign bit, representing the data range of $\pm 1.7\times 10^{308}$. The MATLAB function for defining this data type is `double()`. In other special applications, i.e., in image processing, unsigned 8 bit integer can be used, whose function is `uint8()`, representing the value in $(0, 255)$. Thus significant memory space is saved. Also the data types such as `int8()`, `int16()`, `int32()`, `uint16()` and `uint32()` can be used.

Symbolic data type

"Symbolic" variables are also defined in MATLAB in contrast to the numerical variables. They can be used in formula derivation and analytical solutions of mathematical problems. Before finding analytical solutions, the related variables should be declared as `symbolic`, with the `syms` statement `syms` *var_list var_props*, where *var_list* is the list of variables to be declared, separated by spaces. If necessary, the types of the properties of the variables can be assigned by *var_props*, such as `real` or `positive`. For instance, if one wants to assume that `a,b` are symbolic variables, the statement `syms a b` can be used. Also the statement `syms a nonzero` can be used to say that `a` is a nonzero variable.

The variable precision arithmetic function `vpa()` can be used to display the symbolic variables in any precision. The syntax of the function is `vpa(`A`,`n`)` or `vpa(`A`)`, where A is the variable to be displayed, and n is the number of digits expected, with the default value of 32 decimal digits.

Example 2.1 Display the first 300 digits of π.

Solution The following statement can be used directly to display the exact value of π

```
>> vpa(pi,300)
```

and the result is shown as 3.1415926535897932384626433832795028841971693993751
058209749445923078164062862089986280348253421170679821480865132823066470
9384460955058223172535940812848111745028410270193852110555964462294895 49
3038196442881097566593344612847564823378678316527120190914564856692346 03

486104543266482133936072602491412 7.

One may also require large number of digits to be displayed. Also the result obtained with the statement `vpa(pi)` is 3.1415926535897932384626433832795.

Other data types

Apart from the commonly used numerical data types in MATLAB, the following data types are also provided such that

(i) **Strings** String variables are used to store messages. The syntax of string is slightly different from that in C; single quotation marks are used in MATLAB.

(ii) **Multi-dimensional arrays** Three-dimensional arrays are the direct extension of matrices. Multi-dimensional arrays are also provided in MATLAB.

(iii) **Cell arrays** Cells are extension of matrices, whose elements are no longer values. The element, referred to as *cells*, of cell arrays can be of any data type. For instance, $A\{i,j\}$ can be used to represent the (i,j)th term of cell array A.

(iv) **Classes and objects** MATLAB allows the use of classes in the programming. For instance, the transfer function class in control can be used to represent a transfer function of a system in one single variable. An example of the creation and overload function programming of an object is given in Section 10.6.

2.1.3 Basic structure of MATLAB

Two types of MATLAB statements can be used:

(i) **Direct assignment** The basic structure of this type of statement is `variable = expression`,
and *expression* can be evaluated and assigned to the variable defined in the left-hand-side, and established in MATLAB workspace. If there is a semicolon used at the end of the statement, the result is not displayed. Thus the semicolon can be used to suppress the display of intermediate results. If the left-hand-side variable is not given, the expression will be assigned to the reserved variable `ans`. Thus, the reserved variable `ans` always stores the latest statements without a left-hand-side variable.

Example 2.2 Specify the matrix $\boldsymbol{A} = \begin{bmatrix} 1 & 2 & 3 \\ 4 & 5 & 6 \\ 7 & 8 & 0 \end{bmatrix}$ into MATLAB workspace.

Solution The matrix $\boldsymbol{A}$ can easily be entered into MATLAB workspace, with the following statement

```
>> A=[1,2,3; 4 5,6; 7,8 0]
```

where >> is the MATLAB prompt, which is given automatically in MATLAB. Under the prompt, various MATLAB commands can be specified. For matrices, square brackets should be used to describe matrices, with the elements in the same row separated by commas, and the rows are separated by semicolons. The double matrix variable $\boldsymbol{A}$ can then be established in MATLAB workspace. The matrix $\boldsymbol{A}$ can be displayed in MATLAB command window

```
A =

     1     2     3
     4     5     6
     7     8     0
```

A semicolon at the end of the statement suppresses the display of such a matrix. The size of a matrix can be expanded or reduced dynamically, with the following statements.

```
>> A=[1,2,3; 4 5,6; 7,8 0];        % assignment is made, however no display
   A=[[A; [1 2 3]], [1;2;3;4]]; % dynamically define the size of matrix
```

Example 2.3 Enter complex matrix $\boldsymbol{B} = \begin{bmatrix} 1+\mathrm{j}9 & 2+\mathrm{j}8 & 3+\mathrm{j}7 \\ 4+\mathrm{j}6 & 5+\mathrm{j}5 & 6+\mathrm{j}4 \\ 7+\mathrm{j}3 & 8+\mathrm{j}2 & 0+\mathrm{j}1 \end{bmatrix}$ into MATLAB.

Solution Specifying a complex matrix in MATLAB is as simple as with the case for real matrices. The notations `i` and `j` can be used to describe the imaginary unit. Thus the following statement can be used to enter the complex matrix $\boldsymbol{B}$

```
>> B=[1+9i,2+8i,3+7j; 4+6j 5+5i,6+4i; 7+3i,8+2j 1i]
```

(ii) **Function call statement** The basic statement structure of function call is

[returned_arguments]=function_name(input_arguments)

where, the regulation for function names are the same as in variable names. Generally the function names are the file names in the MATLAB path. For instance, the function name `my_fun` corresponds to the file my_fun.m. Of course, some of the functions are built-in functions in MATLAB kernel, such as the `inv()` function.

More than one input arguments and returned arguments are allowed, in which case, commas should be used to separate the arguments. For instance, the function call `[U S V]=svd(X)` performs singular value decomposition to a given matrix $\boldsymbol{X}$, and the three arguments $\boldsymbol{U}, \boldsymbol{S}, \boldsymbol{V}$ will be returned.

2.1.4 Colon expressions and sub-matrices extraction

Colon expression is an effective way in defining row vectors. It is useful in generating vectors, and in extracting sub-matrices. The typical form of colon expression is $\boldsymbol{v}$=s_1:s_2:s_3. Thus a row vector $\boldsymbol{v}$ can be established in MATLAB workspace, with the initial value s_1, the increment s_2 and the final value s_3. If the term s_2 is omitted, a default increment of 1 is used instead. The examples given below illustrate the use of colon expressions.

Example 2.4 For different increments, establish vectors for $t \in [0, \pi]$.

Solution One may select an increment 0.2. The following statement can be used to establish a row vector such that

```
>> v1=0: 0.2: pi
```

and the row vector is then established such that $\boldsymbol{v}_1 = [0, 0.2, 0.4, 0.6, 0.8, 1, 1.2, 1.4, 1.6, 1.8, 2, 2.2, 2.4, 2.6, 2.8, 3]$. It is noted that the last term in $\boldsymbol{v}_1$ is 3, rather than π.

The following statements can be used to establish row vectors using colon expressions

```
>> v2=0: -0.1: pi  % negative step here means no vector generated
   v3=0:pi         % with the default step size of 1
   v4=pi:-1:0      % the new vector in the reversed order
```

thus $\boldsymbol{v}_2$ is a 1×0 (empty) matrix, $\boldsymbol{v}_3 = [0, 1, 2, 3]$, while $\boldsymbol{v}_4 = [3.142, 2.142, 1.142, 0.142]$.

The sub-matrix of a given matrix $\boldsymbol{A}$ can be extracted with the MATLAB statement, and the matrix can be extracted with `B=A(v1,v2)`, where $\boldsymbol{v}_1$ vector contains the numbers in the rows, and $\boldsymbol{v}_2$ contains the numbers of columns. Thus the relevant columns and rows can be extracted from matrix $\boldsymbol{A}$. The sub-matrix can be returned in matrix $\boldsymbol{B}$. If $\boldsymbol{v}_1$ is assigned to `:`, all the columns can be extracted. The keyword `end` can be used to indicate the last row or column.

Example 2.5 With the following statements, different sub-matrices can be extracted from the given matrix $\boldsymbol{A}$, such that

```
>> A=[1,2,3; 4,5,6; 7,8,0];
   B1=A(1:2:end, :)        % extract all the odd rows of matrix A
   B2=A([3,2,1],[1 1 1]) % copy the reversed first column to all columns
   B3=A(:,end:-1:1)        % flip left-right the given matrix A
```

and the sub-matrices extracted with the above statements are

$$\boldsymbol{B}_1 = \begin{bmatrix} 1 & 2 & 3 \\ 7 & 8 & 0 \end{bmatrix}, \ \boldsymbol{B}_2 = \begin{bmatrix} 7 & 7 & 7 \\ 4 & 4 & 4 \\ 1 & 1 & 1 \end{bmatrix}, \ \boldsymbol{B}_3 = \begin{bmatrix} 3 & 2 & 1 \\ 6 & 5 & 4 \\ 0 & 8 & 7 \end{bmatrix}.$$

2.2 Fundamental Mathematical Calculations

2.2.1 Algebraic operations of matrices

Suppose matrix $\boldsymbol{A}$ has n rows and m columns, it is then referred to as an $n \times m$ *matrix*. If $n = m$, then matrix $\boldsymbol{A}$ is also referred to as a *square matrix*. The following algebraic operations can be defined:

(i) **Matrix transpose** In mathematics textbooks, the transpose of matrices is often denoted as $\boldsymbol{A}^{\mathrm{T}}$. For an $n \times m$ matrix $\boldsymbol{A}$, the transpose matrix $\boldsymbol{B}$ can be defined as $b_{ji} = a_{ij}$, $i = 1, \cdots, n$, $j = 1, \cdots, m$, thus $\boldsymbol{B}$ is an $m \times n$ matrix. If matrix $\boldsymbol{A}$ contains complex elements, a special transpose can also be defined as $b_{ji} = a_{ij}^*$, $i = 1, \cdots, n$, $j = 1, \cdots, m$, i.e., the complex conjugate transpose matrix $\boldsymbol{B}$ is defined. This kind of transpose is referred to as the *Hermitian transpose*, denoted as $\boldsymbol{B} = \boldsymbol{A}^{\mathrm{H}}$. In MATLAB, `A'` can be used to evaluate the Hermitian matrix of $\boldsymbol{A}$. The simple transpose can be obtained with `A.'`. For a real matrix $\boldsymbol{A}$, `A'` is the same as `A.'`.

(ii) **Addition and subtraction** Assume that there are two matrices $\boldsymbol{A}$ and $\boldsymbol{B}$ in MATLAB workspace, the statements `C = A + B` and `C = A - B` can be used respectively to evaluate the addition and subtraction of these two matrices. If the matrices $\boldsymbol{A}$ and $\boldsymbol{B}$ are with the same size, the relevant results can be obtained. If one of the matrices is a scalar, it can be added to or subtracted from the other matrix. If the sizes of the two matrices are different, error messages can be displayed.

(iii) **Matrix multiplication** Assume that matrix $\boldsymbol{A}$ of size $n \times m$ and matrix $\boldsymbol{B}$ of size $m \times r$ are two variables in MATLAB workspace, and the columns of $\boldsymbol{A}$ equal the rows of $\boldsymbol{B}$, the two matrices are referred to as *compatible*. The product can be obtained from $c_{ij} = \sum_{k=1}^{m} a_{ik} b_{kj}$, where $i = 1, 2, \cdots, n$, $j = 1, 2, \cdots, r$. If one of the matrices is a scalar, the product can also be obtained. In MATLAB, the multiplication of the two matrices can be obtained with `C=A*B`. If the two matrices are not compatible, an error message will be given.

(iv) **Matrix left division** The left division of the matrices `A\B` can be used to solve the linear equations $\boldsymbol{AX} = \boldsymbol{B}$. If matrix $\boldsymbol{A}$ is non-singular, then $\boldsymbol{X} = \boldsymbol{A}^{-1}\boldsymbol{B}$. If $\boldsymbol{A}$ is not a square matrix, `A\B` can also be used to find the least squares solution to the equations $\boldsymbol{AX} = \boldsymbol{B}$.

(v) **Matrix right division** The statement `B/A` can be used to solve the linear equations $\boldsymbol{XA} = \boldsymbol{B}$. More precisely, `B/A=(A'\B')'`.

(vi) **Matrix flip and rotation** The left-right flip and up-down flip of a given matrix $\boldsymbol{A}$ can be obtained with `B=fliplr(A)` and `C=flipud(A)` respectively, such that $b_{ij} = a_{i,n+1-j}$ and $c_{ij} = a_{m+1-i,j}$. The command `D=rot90(A)` rotates matrix $\boldsymbol{A}$ counterclockwise by 90°, such that $d_{ij} = a_{j,n+1-i}$.

(vii) **Matrix power** $\boldsymbol{A}^x$ computes the matrix $\boldsymbol{A}$ to the power x when matrix $\boldsymbol{A}$ is square. In MATLAB, the power can be evaluated with `F=A^x`.

(viii) **Dot operation** A class of special operation is defined in MATLAB. The statement `C=A.*B` can be used to obtain element-by-element product of matrices $\boldsymbol{A}$ and $\boldsymbol{B}$, such that $c_{ij} = a_{ij}b_{ij}$. The dot product is also referred to as the *Hadamard product.*

Dot operation plays an important role in scientific computation. For instance, if a vector $\boldsymbol{x}$ is given, then the vector $[x_i^5]$ cannot be obtained with $\boldsymbol{x}$`^5`. Instead, the command $\boldsymbol{x}$`.^5` should be used. In fact, some of the functions such as `sin()` can also be used in element-by-element operation.

Dot operation can be used to deal with other problems, for instance, the statement $\boldsymbol{A}$`.^`$\boldsymbol{A}$ can be used, with the (i,j)th element then defined as $a_{ij}^{a_{ij}}$. Thus the matrix can be obtained

$$\begin{bmatrix} 1^1 & 2^2 & 3^3 \\ 4^4 & 5^5 & 6^6 \\ 7^7 & 8^8 & 0^0 \end{bmatrix} = \begin{bmatrix} 1 & 4 & 27 \\ 256 & 3125 & 46656 \\ 823543 & 16777216 & 1 \end{bmatrix}.$$

Example 2.6 Consider again the matrix $\boldsymbol{A}$ in Example 2.2. Find all the cubic roots of such a matrix and verify the results.

Solution The cubic root of the matrix $\boldsymbol{A}$ can easily be found that

```
>> A=[1,2,3; 4,5,6; 7,8,0]; C=A^(1/3), e=norm(A-C^3)
```

and it can be found, with an error of $e = 8.2375 \times 10^{-15}$, that

$$\boldsymbol{C} = \begin{bmatrix} 0.77179+\mathrm{j}0.6538 & 0.48688-\mathrm{j}0.015916 & 0.17642-\mathrm{j}0.2887 \\ 0.88854-\mathrm{j}0.072574 & 1.4473+\mathrm{j}0.47937 & 0.52327-\mathrm{j}0.49591 \\ 0.46846-\mathrm{j}0.64647 & 0.66929-\mathrm{j}0.6748 & 1.3379+\mathrm{j}1.0488 \end{bmatrix}.$$

In fact, the cubic root of matrix $\boldsymbol{A}$ should have three solutions. The other two roots can be rotated as $\boldsymbol{C}\mathrm{e}^{\mathrm{j}2\pi/3}$ and $\boldsymbol{C}\mathrm{e}^{\mathrm{j}4\pi/3}$, with the following statements

```
>> j1=exp(sqrt(-1)*2*pi/3); A1=C*j1, A2=C*j1^2,
   norm(A-A1^3), norm(A-A1^3)
```

and the other two roots, through verification, are

$$\boldsymbol{A}_1 = \begin{bmatrix} -0.9521+\mathrm{j}0.34149 & -0.22966+\mathrm{j}0.42961 & 0.16181+\mathrm{j}0.29713 \\ -0.38142+\mathrm{j}0.80579 & -1.1388+\mathrm{j}1.0137 & 0.16784+\mathrm{j}0.70112 \\ 0.32563+\mathrm{j}0.72893 & 0.24974+\mathrm{j}0.91702 & -1.5772+\mathrm{j}0.63425 \end{bmatrix}$$

and

$$\boldsymbol{A}_2 = \begin{bmatrix} 0.18031-\mathrm{j}0.99529 & -0.25722-\mathrm{j}0.41369 & -0.33823-\mathrm{j}0.008436 \\ -0.50712-\mathrm{j}0.73321 & -0.3085-\mathrm{j}1.4931 & -0.69111-\mathrm{j}0.20521 \\ -0.79409-\mathrm{j}0.082464 & -0.91904-\mathrm{j}0.24222 & 0.23934-\mathrm{j}1.6831 \end{bmatrix}.$$

2.2.2 Logic operations of matrices

Logical data was not implemented in earlier versions of MATLAB. The non-zero value is regarded as logic 1, while a zero value is defined as logic 0. In new versions of MATLAB, logical variables are defined and the above rules also apply.

Assume that the matrices $\boldsymbol{A}$ and $\boldsymbol{B}$ are both $n \times m$ matrices, the following logical operations are defined:

(i) **"And" operation** In MATLAB, the operator `&` is used to define element-by-element "and" operation. The statement $\boldsymbol{A}$ `&` $\boldsymbol{B}$ can then be defined.

(ii) **"Or" operation** In MATLAB, the operator | is used to define element-by-element "or" operation. The statement $\boldsymbol{A} \mid \boldsymbol{B}$ can then be defined.

(iii) **"Not" operation** In MATLAB, the operator ˜ can be used to define the "not" operation such that $\boldsymbol{B}$ =˜$\boldsymbol{A}$.

(iv) **Exclusive or** The exclusive or operation of two matrices $\boldsymbol{A}$ and $\boldsymbol{B}$ can be evaluated from `xor(A, B)`.

2.2.3 Relationship operations of matrices

Various relationship operators are provided in MATLAB. For example, $\boldsymbol{C}$=$\boldsymbol{A} > \boldsymbol{B}$ will perform element-by-element comparison, with the element $c_{ij} = 1$ for $a_{ij} > b_{ij}$, and $c_{ij} = 0$ otherwise. The equality relationship can be tested with `==` operator, while the other operators `>=`, `~=` can also be used.

The special functions such as `find()` and `all()` can also be used to perform relationship operations. For instance, the index of the elements in $\boldsymbol{C}$ equal to 1 can be obtained from `find(C==1)`. The following commands can be used:

```
>> A=[1,2,3; 4 5,6; 7,8 0]; % enter a matrix
   i=find(A>=5)' % find all the indices in A whose value is larger than 5
```

and the indices can be found as $i = 3, 5, 6, 8$. It can be seen that the function arranges first the original matrix $\boldsymbol{A}$ in a single column, on a columnwise basis. The indices can then be returned.

The functions `all()` and `any()` can also be used to check the values in the given matrices. For instance

```
>> a1=all(A>=5) % check each column whether all larger than 5
   a2=any(A>=5) % check each column whether any larger than 5
```

and it can be found that $\boldsymbol{a}_1 = [0, 0, 0]$, $\boldsymbol{a}_2 = [1, 1, 1]$.

2.2.4 Simplifications and presentations of analytical results

The Symbolic Math Toolbox can be used to derive mathematical formulas. The results however are often not presented in their simplest form. The results should then be simplified. The easiest way of simplification is by the use of `simple()` function, where different simplification methods are tested automatically until the simplest result can be obtained, with the syntaxes

```
s1=simple(s), % try various simplification methods and find the simplest
[s1,how]=simple(s), % return the simplest form and the method used
```

where s is the original expression, and s_1 is the simplified result. The string argument `how` will return the method of simplification. Apart from the easy-to-use `simple()` function, the function `collect()` can be used to collect the coefficients, and function `expand()` can be used to expand a polynomial. The function `factor()` can be used to perform factorization of a polynomial. The function `numden()` can be used to extract the numerator and denominator from a given expression.

Example 2.7 If a polynomial $P(s)$ is given by $P(s) = (s+3)^2(s^2+3s+2)(s^3+12s^2+48s+64)$, process it with various functions and understand the results converted.

Solution A symbolic variable s should be declared first, then the full polynomial can be expressed easily and the polynomial can then be established in MATLAB workspace. With the polynomial, one can first simplify it with the `simple()` function

```
>> syms s; P=(s+3)^2*(s^2+3*s+2)*(s^3+12*s^2+48*s+64)
   [P1,m]=simple(P) % a series of simplications made, find the simplest
```

and one finds that $P_1 = (s+3)^2(s+2)(s+1)(s+4)^3$, with the method m=`factor`, which means that factorization method is used to reach the conclusion. Also the `expand()` function can be tested

```
>> expand(P)            % expand the polynomial
```

and the expanded polynomial is $s^7+21s^6+185s^5+883s^4+2454s^3+3944s^2+3360s+1152$.

The function `subs()` provided in the Symbolic Math Toolbox can be used to perform variable substitution, and the syntaxes are

`f1=subs(f,x1,x1*)` or `f1=subs(f,{x1,x2,···,xn},{x1*,x2*,···,xn*})`

where f is the original expression. With the statement, the variable x_1 in the original function can be substituted with a new variable or expression x_1^*. The result is given in the variable f_1. The latter syntax can be used to substitute many variables simultaneously.

The function `latex()` can be used to convert a symbolic expression into a LaTeX-readable string, which can be embedded into a LaTeX document.

Example 2.8 For a given function $f(t) = \cos(at+b) + \sin(ct)\sin(dt)$, evaluate its Taylor expression with the function `taylor()` and convert the results in LaTeX.

Solution A full description on Taylor series expansion will be given in Section 3.2. Here the function `taylor()` can be used straightforwardly to get the results. Applying the function `latex()` to the results, the LaTeX can be obtained.

```
>> syms a b c d t;                     % declare symbolic variables
   f=cos(a*t+b)+sin(c*t)*sin(d*t);  % define the function f(t) with taylor()
   f1=taylor(f,5);                  % find first 5 terms in Taylor series
   latex(f1)                        % can be converted to a LaTeX string
```

The results can be embedded into a LaTeX document, and through compilation, the following results can be obtained

$$f(x) \approx \cos b - at \sin b + \left(-\frac{a^2 \cos b}{2} + cd\right) t^2 + \frac{a^3 \sin b}{6} t^3 + \left(\frac{a^4 \cos b}{24} - \frac{cd^3}{6} - \frac{c^3 d}{6}\right) t^4.$$

Unfortunately, there are no directly usable converters to other word processing programs such as Microsoft Word.

2.2.5 Basic number theory computations

Basic data transformation and number theory functions are provided in MATLAB, as shown in Table 2.1. The following examples are used to illustrate the functions. Through the example, the readers can observe the results.

TABLE 2.1: Functions for data transformations

function	syntax	function description
floor()	$\boldsymbol{n}$=floor($\boldsymbol{x}$)	round towards $-\infty$ for each value in variable $\boldsymbol{x}$, mathematically denoted as $\boldsymbol{n} = [\boldsymbol{x}]$
ceil()	$\boldsymbol{n}$=ceil($\boldsymbol{x}$)	round towards $+\infty$ for $\boldsymbol{x}$
round()	$\boldsymbol{n}$=round($\boldsymbol{x}$)	round to nearest integer for $\boldsymbol{x}$
fix()	$\boldsymbol{n}$=fix($\boldsymbol{x}$)	round towards zero for variable $\boldsymbol{x}$
rat()	[$\boldsymbol{n}$,$\boldsymbol{d}$]=rat($\boldsymbol{x}$)	find rational approximation for variable $\boldsymbol{x}$, and the numerator and denominator are returned respectively in $\boldsymbol{n}$ and $\boldsymbol{d}$
rem()	$\boldsymbol{B}$=rem($\boldsymbol{A}$,$\boldsymbol{C}$)	find the reminder after division to variable $\boldsymbol{A}$
gcd()	k=gcd(n,m)	compute the greatest common divisor for n and m
lcm()	k=lcm(n,m)	compute the least common multiplier for n and m
factor()	factor(n)	prime factorization
isprime()	$\boldsymbol{v}_1$=isprime($\boldsymbol{v}$)	check whether each component in $\boldsymbol{v}$ is prime or not. Set the corresponding value in $\boldsymbol{v}_1$ to 1 for prime numbers, otherwise set to 0

Example 2.9 For a given data set $-0.2765, 0.5772, 1.4597, 2.1091, 1.191, -1.6187$, observe the integers obtained using different rounding functions.

Solution The following statements can be used to round the original vector such that

```
>> A=[-0.2765,0.5772,1.4597,2.1091,1.191,-1.6187];
   v1=floor(A), v2=ceil(A) % round towards −∞ and +∞ respectively
   v3=round(A), v4=fix(A)  % round towards 0 and nearest integers
```

and the integer vectors obtained are $\boldsymbol{v}_1 = [-1, 0, 1, 2, 1, -2]$, $\boldsymbol{v}_2 = [0, 1, 2, 3, 2, -1]$, $\boldsymbol{v}_3 = [0, 1, 1, 2, 1, -2]$, $\boldsymbol{v}_4 = [0, 0, 1, 2, 1, -1]$.

Example 2.10 Assume that a 3×3 Hilbert matrix can be specified with the statement $\boldsymbol{A}$=hilb(3), perform the rational transformation.

Solution The following statements can be used to find the rational approximation

```
>> A=hilb(3); [n,d]=rat(A)
```

and the integer matrices obtained are $\boldsymbol{n} = \begin{bmatrix} 1 & 1 & 1 \\ 1 & 1 & 1 \\ 1 & 1 & 1 \end{bmatrix}$, $\boldsymbol{d} = \begin{bmatrix} 1 & 2 & 3 \\ 2 & 3 & 4 \\ 3 & 4 & 5 \end{bmatrix}$.

Example 2.11 Find the greatest common divisor and least common multiplier to the numbers 1856120 and 1483720, and find the prime factorization to the least common multiplier obtained.

Solution Since the values are very large, one should not use the double-precision representations. The symbolic representations must be used instead. The following statements can be used

```
>> m=sym(1856120); n=sym(1483720);
   gcd(m,n), lcm(m,n), factor(lcm(n,m))
```

which yield the greatest common divisor of 1960 and the greatest common multiplier of 1405082840, whose prime factorization is $(2)^3(5)(7)^2(757)(947)$.

Here the functions `gcd()` and `lcm()` can only be used to deal with two variables. If more than two variables are expected, nested calls are allowed such that `gcd(gcd(`m`,`n`),`k`)`.

Example 2.12 List all the prime numbers in the interval $[1, 1000]$.

Solution The prime numbers can easily be recognized by the function `isprime(`$\boldsymbol{A}$`)`. All the prime numbers less than 1000 can be extracted as shown in Table 2.2.

```
>> A=1:1000; B=A(isprime(A))
```

TABLE 2.2: The prime numbers less than 1000

2	29	67	107	157	199	257	311	367	421	467	541	599	647	709	769	829	887	967
3	31	71	109	163	211	263	313	373	431	479	547	601	653	719	773	839	907	971
5	37	73	113	167	223	269	317	379	433	487	557	607	659	727	787	853	911	977
7	41	79	127	173	227	271	331	383	439	491	563	613	661	733	797	857	919	983
11	43	83	131	179	229	277	337	389	443	499	569	617	673	739	809	859	929	991
13	47	89	137	181	233	281	347	397	449	503	571	619	677	743	811	863	937	997
17	53	97	139	191	239	283	349	401	457	509	577	631	683	751	821	877	941	
19	59	101	149	193	241	293	353	409	461	521	587	641	691	757	823	881	947	
23	61	103	151	197	251	307	359	419	463	523	593	643	701	761	827	883	953	

2.3 Flow Control Structures of MATLAB Language

As a programming language, the loop control structures, conditional control structures, switch structures and trial structures are provided in MATLAB. These structures are illustrated in this section.

2.3.1 Loop control structures

The loop structures can be introduced by the keywords `for` or `while`, and ended with the `end` command. The two kinds of loop structures are shown in Figures 2.1 (a) and (b), respectively.

(i) **The `for` loop structures** The syntax of the structure is

`for` $i = \boldsymbol{v}$, *loop programs body*, `end`

When using the `for` loop structure, a component in vector $\boldsymbol{v}$ is extracted and assigned to variable i each time, the loop body can be executed. Then go back to the `for` statement, until all the components in $\boldsymbol{v}$ are used.

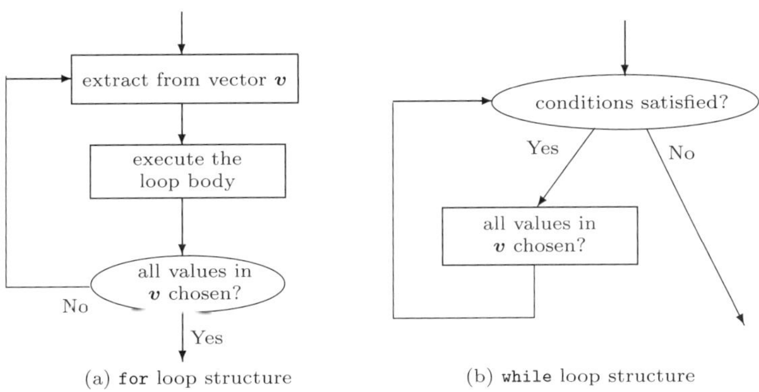

FIGURE 2.1: Illustration of loop structures

(ii) **The `while` loop structures** The syntax of the structure is

`while` (*condition*), *loop structure body*, `end`

The *condition* expression is crucial in the `while` loop structure. If it is true, the loop will be executed, and returned back to the `while` command. The loop structure will be executed, until *condition* becomes false.

There are differences between the two functions. Examples will be given below to show the advantages and disadvantages of these structures.

Example 2.13 Compute the sum of $\sum_{i=1}^{100} i$ using loop structures.

Solution The `for` and `while` loop structures can be used with the following statements, and the same results can be obtained

```
>> s1=0; for i=1:100, s1=s1+i; end
   s2=0; i=1; while (i<=100), s2=s2+i; i=i+1; end; s1, s2
```

where it can be seen that the `for` loop structure is simpler. In fact, the simplest statement for this problem is `sum(1:100)`. In the function call, the built-in function `sum()` can be used to solve the problem.

Example 2.14 Find the minimum value of m such that $\sum_{i=1}^{m} i > 10000$.

Solution It can be seen that it is not possible to solve such a problem with the `for` loop structure. However, the structure of `while` can be used easily to find the value of m

```
>> s=0; m=0;
   while (s<=10000), m=m+1; s=s+m; end, [s,m] % the value of m is expected
```

with $m = 141$ and the sum is $s = 10011$.

The loop statements can be used in nested format. The statement `break` can be used to terminate the loop structure of the current level.

The speed of loop is slow in MATLAB, compared with other programming languages. Thus the loops should be avoided, and vectorized programming techniques should be used instead.

Example 2.15 Evaluate the sum of the series[1] $S = \sum_{i=1}^{100000} \left(\frac{1}{2^i} + \frac{1}{3^i} \right)$.

Solution The execution time can be measured with the statements `tic` and `toc`. The time needed in vectorization is about 0.116 seconds, and the one needed in loops is 0.443 seconds. Thus the vectorization method is normally faster.

```
>> tic, s=0; for i=1:100000, s=s+1/2^i+1/3^i; end; toc
   tic, i=1:100000; s=sum(1./2.^i+1./3.^i); toc
```

2.3.2 Conditional control structures

Conditional control structures are the most widely used control structures. In MATLAB, `if ··· end` structure, as well as the complicated ones with `else` and `elseif` can be used. The structures can be shown in Figure 2.2.

```
if (condition 1) % If condition 1 is satisfied, statement group 1 is executed.
      statement group 1   % other sub-level if can be nested
elseif (condition 2) % Otherwise, if condition 2 is met, group 2 is executed.
      statement group 2
         :    :         % more conditional control statements
else    % if none of the above conditions are satisfied, define defaults.
      statement group n + 1
end
```

[1]In order to demonstrate the performance of vectorization, the number of terms are exaggerated. Normally 20~30 terms will be adequate for the exact solutions. The performance of loops was speeded up in new versions of MATLAB. And in version 7.x, the speed of loop execution is close to the vectorization method.

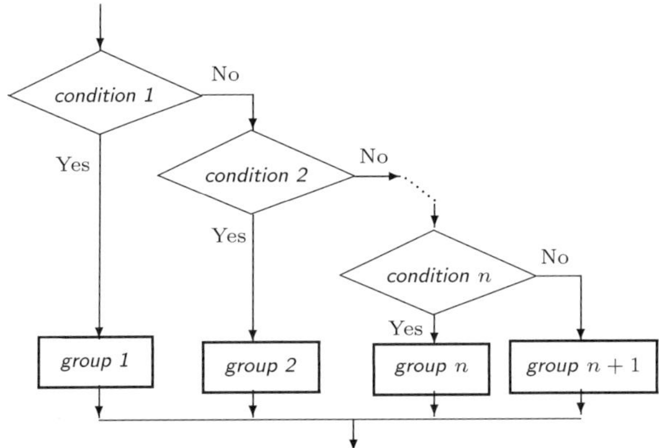

FIGURE 2.2: Illustrations of conditional control structures

Example 2.16 Solve the problem in Example 2.14 again using `for` and `if` statements.

Solution It has been shown in Example 2.14 that the `for` loop structure is not suitable for finding the minimum m such that the sum is greater than 10000. The `for` loop can be used with `if` structure to solve the problem.

```
>> s=0;
   for i=1:10000, s=s+i; if s>10000, break; end, end, s
```

Thus the structure of the program is more complicated than that of the `while` structure.

2.3.3 Switch structure

The switch structure is illustrated in Figure 2.3, and the fundamental structure is

```
switch switch expression
case expression 1, statements 1
case {expression 2, expression 3,···, expression m}, statements 2
      ⋮
otherwise, statements n
end
```

where the crucial part in switch structure is the evaluation of *switch expressions*. If it matches a value in a case statement, the statements after the case statement should be executed. Once completed, the switch structure is terminated.

There exist differences between the switch statements in MATLAB and in C languages. The following tips should be noted in programming with MATLAB:

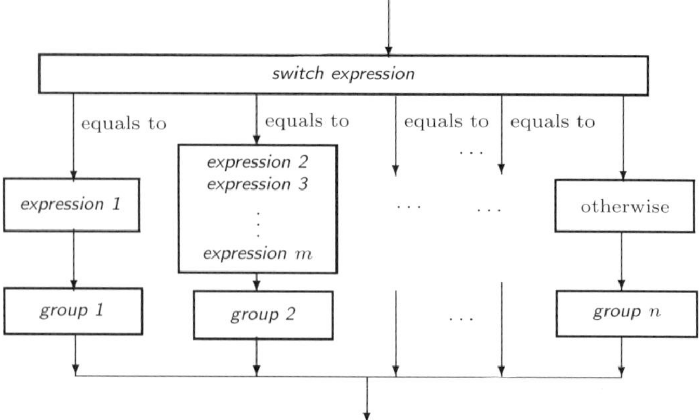

FIGURE 2.3: Illustrations of switch structures

(i) When the value of the switch expression equals *expression 1*, the *statement group 1* should be executed. After execution, the structure is completed. There is no need to introduce a `break` statement before the next `case`.

(ii) If one is checking whether one of several expressions is satisfied, the expressions must be given in cell format.

(iii) If none of the expressions are satisfied, the paragraph in `otherwise` should be executed. It is similar to the `default` statement in C language.

(iv) The execution results are independent of the orders of the `case` statement. When there exist two or more `case` statements having the same expressions, those listed behind may never be executed.

2.3.4 Trial structure

A brand new trial structure is provided in MATLAB, whose syntax is

```
try,    statement group 1,
catch,    statement group 2,
end
```

Normally, only the *statement group 1* is executed. However, if an error occurs during execution of any of the statements, the error is captured into `lasterror`, and the *statement group 2* is executed. The new structure is not available in languages such as C. The trial structure is useful in practical programming.

2.4 Writing and Debugging MATLAB Functions

Two types of source programs are supported in MATLAB, both in ASCII format. One of the code is the M-script program, which is a series of MATLAB statements to be evaluated in sequence, just as the batch files in DOS. The execution of this type of program is simple, one can simply key in the file name under the >> prompt. M-scripts process the data in MATLAB workspace, and the results are returned back to MATLAB workspace. M-scripts are suitable for dealing with small-scale computations.

Example 2.17 Consider again the problem in Example 2.14. The program can be used to find the smallest m such that the summation is greater than 10000. If one wants to find such m's for the summation greater than 20000 or 30000, the original program should be modified. This method is quite complicated and inconvenient. If a mechanism can be established such that the user may define 20000 or 30000 externally, without modifying the original program, the mechanism is quite reasonable. This kind of mechanism is often referred to as the *functions*.

M-function is the major structure in MATLAB programming. In practical programming, M-script programming is not recommended. In this section, MATLAB functions and some tricks in programming are given.

2.4.1 Basic structure of MATLAB functions

MATLAB functions are led by the statement of `function`, whose basic structure is

```
function [return argument list] =funname(input argument list)
    comments led by % sign
    input and output variables check
    main body of the function
```

The actual numbers of input and returned arguments can be extracted respectively by `nargin` and `nargout`. In the function call, the two variables are generated automatically.

If more than one input or returned arguments are needed, they should be separated with commas in the lists. The comments led by % will not be executed. The messages in the leading comments can be displayed by the `help` command.

From the system view points, the MATLAB functions can be regarded as a variable processing unit, which receives variables from the calling function. Once the variables are processed, the results will be returned back to the calling function. Apart from the input and returned arguments, the other variables within the function are local variables, which will be lost after function calls. Examples will be given to demonstrate the programming techniques.

Example 2.18 Consider the requests in Example 2.17. One may choose the input argument as k, and returned arguments of m and s, where s is the sum of first m terms. The function can then be written as

```
function [m,s]=findsum(k)
s=0; m=0; while (s<=k), m=m+1; s=s+m; end
```

The previous function can be saved as a function `findsum.m`. One can then call such a function for different values of k, without modifying the function. For instance, if the targeted summation is 145323, the following statements can be used to find the smallest value of m, which returns $m = 539$, $s_1 = 145530$.

```
>> [m1,s1]=findsum(145323)
```

It can be seen that the calling format is quite flexible, and we may find the needed results without modifying the original program. Thus this kind of method is recommended in programming.

Example 2.19 Assume that a MATLAB function is needed in obtaining an $n \times m$ Hilbert matrix[2], whose (i, j)th element is $h_{i,j} = 1/(i + j - 1)$. The following additional requests are also to be implemented:

(i) If only one input argument n is given in the calling command, a square matrix should be generated, such that $m = n$.

(ii) Certain help information to this function is required.

(iii) Check the formats of input and returned arguments.

Solution In actual programming, it is better to write adequate comments, which are beneficial to the programmer as well as to the maintainer of the program. The required MATLAB function `myhilb()` can be written and stored as `myhilb.m` in the default MATLAB path.

```
function A=myhilb(n, m)
%MYHILB  The function is used to illustrate MATLAB functions.
%   A=MYHILB(N, M) generates an NxM Hilbert matrix A;
%   A=MYHILB(N) generates an NxN square Hilbert matrix A;
%
%See also: HILB.

% Designed by Professor Dingyu XUE, Northeastern University, PRC
%     5 April, 1995, Last modified by DYX at 30 July, 2001
if nargout>1, error('Too many output arguments.'); end
if nargin==1, m=n;   % if one input argument used, square matrix
elseif nargin==0 | nargin>2
   error('Wrong number of iutput arguments.');
end
```

[2] A function `hilb()` is provided in MATLAB to create an $n \times n$ square Hilbert matrix.

```
for i=1:n, for j=1:m, A(i,j)=1/(i+j-1); end, end
```

In the program, the comments are led by % sign. To implement the requirement in item (i), one should check whether the number of input argument is 1, i.e., whether `nargin` is 1. If so, the column number m is set to n, the row number, thus a square matrix can be generated. If the numbers of input or returned arguments are not correct, the error messages can be given. The double `for` loops will generate the required Hilbert matrix.

The on-line help command `help myhilb` will display the following information

```
MYHILB  The function is used to illustrate MATLAB functions.
   A=MYHILB(N, M) generates an NxM Hilbert matrix A;
   A=MYHILB(N) generates an NxN square Hilbert matrix A;
See also: HILB.
```

It should be noted that only the first few lines of information are displayed, while the author information is not displayed. This is because there is a blank line before the author information.

The following commands can be used to generate Hilbert matrices

```
>> A1=myhilb(4,3)   % two input arguments yield a rectangular matrix
   A2=myhilb(4)     % while one input argument yields a square matrix
```

and the two matrices can then be established as

$$A_1 = \begin{bmatrix} 1 & 0.5 & 0.33333 \\ 0.5 & 0.33333 & 0.25 \\ 0.33333 & 0.25 & 0.2 \\ 0.25 & 0.2 & 0.16667 \end{bmatrix}, \quad A_2 = \begin{bmatrix} 1 & 0.5 & 0.33333 & 0.25 \\ 0.5 & 0.33333 & 0.25 & 0.2 \\ 0.33333 & 0.25 & 0.2 & 0.16667 \\ 0.25 & 0.2 & 0.16667 & 0.14286 \end{bmatrix}.$$

Example 2.20 MATLAB functions can be called recursively, i.e., a function may call itself. Please write a recursive function to evaluate the factorial $n!$.

Solution Consider the factorial $n!$. From the definition $n! = n(n-1)!$, it can be seen that the factorial of n can be evaluated from the factorial of $n-1$, while $n-1$ can be evaluated from $n-2$, and so on. The exits of the function call should be $1! = 0! = 1$. Thus the recursive function can be written as follows, with the comments omitted.

```
function k=my_fact(n)
if nargin~=1, error('Error: Only one input variable accepted'); end
if abs(n-floor(n))>eps | n<0 % judge whether n is a non-negative integer
   error('n should be a non-negative integer');
end
if n>1        % if n > 1, recursive calls are used
   k=n*my_fact(n-1);
elseif any([0 1]==n) % 0! = 1! = 1, the exit of the function
   k=1;
end
```

It can be seen that, in the function, the judgement whether n is a non-negative integer is made. If not, an error message will be declared. If it is, the recursive function calls will be used such that when $n = 1$ or 0, the result is 1, which can be used as an exit to the function. For instance, 11! can be evaluated with `my_fact(11)`, and the result obtained is 39916800.

In fact, the factorial for any non-negative integer can be evaluated directly with function `factorial(n)`, and the kernel of such a function is `prod(1:n)`.

Example 2.21 Compare the advantages and disadvantages of recursive algorithm with loop structure in constructing the Fibonacci arrays.

Solution It is for sure that the recursive algorithm is an effective method for a class of problems. However, this method should not be misused. A counter-example is shown in this example. Consider the Fibonacci array, where $a_1 = a_2 = 1$, and the kth term can be evaluated from $a_k = a_{k-1} + a_{k-2}$ for $k = 3, 4, \cdots$. A MATLAB function can be written for the problem

```
function a=my_fibo(k)
if k==1 | k==2, a=1; else, a=my_fibo(k-1)+my_fibo(k-2); end
```

and for $k = 1, 2$, the exit can be made such that it returns 1. If the 25th term is expected, the following statements can be used and the time required is 7.6 seconds.

```
>> tic, my_fibo(25), toc
```

If one is expecting the term $k = 30$, several hours of time might be required. If the loop structure is used, within 0.02 second, the whole array can be obtained for $k = 100$.

```
>> tic, a=[1,1]; for k=3:100, a(k)=a(k-1)+a(k-2); end, toc
```

It can be seen that the ordinary loop structure only requires a very short execution time. Thus the recursive function call should not be misused.

2.4.2 Programming of functions with variable inputs/outputs

In the following presentation, the variable number of input and returned arguments is introduced, based on the cell data type. It should be mentioned that most of the MATLAB functions are implemented in this format.

Example 2.22 The product of two polynomials can be evaluated from the `conv()` function, based on the algorithm of finding the convolution of two arrays. Write a function to evaluate directly the multiplications of arbitrary number of polynomials.

Solution Cell data type can be used to write the function `convs()`, which can be used to evaluate the multiplication of arbitrary number of polynomials.

```
function a=convs(varargin)
a=1; for i=1:length(varargin), a=conv(a,varargin{i}); end
```

The input argument list is passed to the function through the cell variable `varargin`. Consequently, the returned arguments can be specified in `varargout`, if necessary. Under such a function, the multiplication of arbitrary number of polynomials can be obtained. The following statements can be used to call the function

```
>> P=[1 2 4 0 5]; Q=[1 2]; F=[1 2 3]; D=convs(P,Q,F)
   E=conv(conv(P,Q),F)  % nested calls are to be used with conv() function
   G=convs(P,Q,F,[1,1],[1,3],[1,1])
```

where the obtained vectors are respectively

$\boldsymbol{E} = [1, 6, 19, 36, 45, 44, 35, 30]^{\mathrm{T}}$, $\boldsymbol{G} = [1, 11, 56, 176, 376, 578, 678, 648, 527, 315, 90]^{\mathrm{T}}$.

2.4.3 Inline functions and anonymous functions

In order to describe simply the mathematics functions, *inline functions* can be used. The functions are equivalent to the M-functions. However, with inline function, it may no longer be necessary to save files. The format of inline function is `fun=inline(function expression, list of variables)`, where the *function expression* is the actual contents of the function to be expressed, and the *list of variables* contains all the input variables, with each variable given as a string. The inline function is useful in the descriptions in differential equations and objective function given later. The function type accept only one returned variable. The mathematical function $f(x, y) = \sin(x^2 + y^2)$ can be expressed as `f=inline('sin(x.^2+y.^2)','x','y')`.

Anonymous function provided in MATLAB 7.x is a brand new type of function definition, the structure of the function is similar to the inline function, but it is more concise and easy to use. The syntax of the function is

`f=@(list of variables) function_contents`, e.g., `f=@(x,y)sin(x.^2+y.^2)`

Note that the variable currently existing in MATLAB workspace can be used directly in the function. For instance, the variables a and b in MATLAB workspace can be used in the anonymous function

```
f=@(x,y)a*x.^2+b*y.^2
```

to describe the mathematical function $f(x, y) = ax^2 + by^2$. If such a function has been defined, while the variables a, b change after that, the values of those in the anonymous function will not change, unless it is defined again.

2.5 Two-Dimensional Graphics

Graphics and visualization are the most significant advantages of MATLAB. A series of straightforward and simple functions are provided in MATLAB for two-dimensional and three-dimensional graphics. Experimental and simulation results can be easily interpreted in graphical form. In this section, the two-dimensional graphics functions will be illustrated.

2.5.1 Basic statements of two-dimensional plotting

Assume that a sequence of experimental data is acquired. For instance, at time instances $t = t_1, t_2, \cdots, t_n$, the function values are $y = y_1, y_2, \cdots, y_n$. The data can be entered to MATLAB workspace such that $\boldsymbol{t} = [t_1, t_2, \cdots, t_n]$ and $\boldsymbol{y} = [y_1, y_2, \cdots, y_n]$. The command `plot(t,y)` can be used to draw the curve for the data. The "curve" is in fact represented by poly-lines, joining the sample points.

It can be seen that the syntax of the function is quite straightforward. In actual applications, the `plot()` function can also be called in other extended ways.

(i) $\boldsymbol{t}$ is still a vector and $\boldsymbol{y}$ can be expressed by a matrix such that

$$\boldsymbol{y} = \begin{bmatrix} y_{11} & y_{12} & \cdots & y_{1n} \\ y_{21} & y_{22} & \cdots & y_{2n} \\ \vdots & \vdots & \ddots & \vdots \\ y_{m1} & y_{m2} & \cdots & y_{mn} \end{bmatrix}.$$

The same function can also be used to draw m curves, with each row of matrix $\boldsymbol{y}$ corresponding to a curve.

(ii) $\boldsymbol{t}$ and $\boldsymbol{y}$ are both matrices, and the sizes of the two matrices are the same. The plots between each row of $\boldsymbol{t}$ and $\boldsymbol{y}$ can be drawn.

(iii) Assume that there are many pairs of such vectors or matrices, $(\boldsymbol{t}_1, \boldsymbol{y}_1)$, $(\boldsymbol{t}_2, \boldsymbol{y}_2)$, $\cdots$, $(\boldsymbol{t}_m, \boldsymbol{y}_m)$, the following statement can be used directly to draw the corresponding curves.

`plot(`$\boldsymbol{t}_1, \boldsymbol{y}_1, \boldsymbol{t}_2, \boldsymbol{y}_2, \cdots, \boldsymbol{t}_m, \boldsymbol{y}_m$`)`

(iv) The line types, line width and color information of the curves can separately be specified with the command

`plot(`$\boldsymbol{t}_1, \boldsymbol{y}_1$, *option 1*, $\boldsymbol{t}_2, \boldsymbol{y}_2$, *option 2*, $\cdots$, $\boldsymbol{t}_m, \boldsymbol{y}_m$, *option m*`)`

where the available *options* are shown in Table 2.3. The combinations of the options are also allowed. For instance, the combination `'r-.pentagram'` indicates the red dash dot curve, with the sampling points marked by pentagrams.

After the curves are drawn, the command `grid on` can be used to add grids to the curves, while the `grid off` command may remove the grids. Also `hold on` command can reserve the current current axis. Other `plot()` function can be used to superimpose curves on top of the existing ones. `hold off` command may remove the holding status.

Example 2.23 Draw the curve of $y = \sin(\tan x) - \tan(\sin x)$ in the interval $x \in [-\pi, \pi]$.

Solution The curve of $f(x)$ can be drawn easily with the following statements

```
>> x=[-pi : 0.05: pi];          % specify the vector with a step-size of 0.05
```

TABLE 2.3: Options in MATLAB plotting commands

line type		line color				markers			
opts	meaning	opts	meaning	opts	meaning	opts	meaning	opts	meaning
'-'	solid	'b'	blue	'c'	cyan	'*'	$*$	'pentagram'	☆
'--'	dash	'g'	green	'k'	black	'.'	dotted	'o'	○
':'	dotted	'm'	magenta	'r'	red	'x'	$\times$	'square'	□
'-.'	dash-dot	'w'	white	'y'	yellow	'v'	∇	'diamond'	◇
'none'	none					'^'	$\triangle$	'hexagram'	✡
						'>'	$\triangleright$	'<'	$\triangleleft$

```
y=sin(tan(x))-tan(sin(x)); % evaluate the function values
plot(x,y)                  % draw the curve
```

and the curve in Figure 2.4 (a) can be obtained.

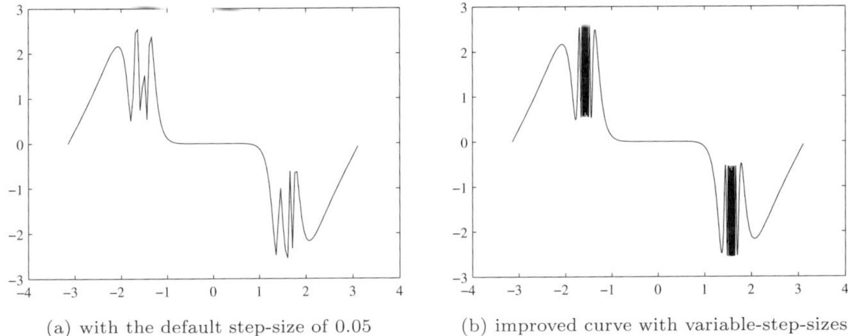

(a) with the default step-size of 0.05 (b) improved curve with variable-step-sizes

FIGURE 2.4: Two dimension curve for the given function

It can be seen from the curve that it is rather sluggish over the intervals $x \in (-1.8, -1.2)$ and $x \in (1.2, 1.8)$, since the step-size 0.05 is too large for these intervals. The step-size for these intervals should be selected smaller such that

```
>> x=[-pi:0.05:-1.8,-1.801:.001:-1.2, -1.2:0.05:1.2,...
      1.201:0.001:1.8, 1.81:0.05:pi]; % with variable-step-size
   y=sin(tan(x))-tan(sin(x));          % evaluate the function
   plot(x,y)                           % draw the curve
```

The modified curve of the function is given in Figure 2.4 (b). It can be seen that the curve is significantly improved in the new plot. Alternatively, for the whole interval, a fixed step-size of 0.001 can be selected.

Example 2.24 Please draw the saturation function $y = \begin{cases} 1.1\text{sign}(x), & |x| > 1.1 \\ x, & |x| \leqslant 1.1 \end{cases}$.

Solution It is obvious that one can create a vector of $\boldsymbol{x}$, then for each point, construct an `if` clause to calculate the value of $\boldsymbol{y}$. An alternative way is to use vectorized format to evaluate the function values. With the following statements, the segmented function can be drawn as shown in Figure 2.5.

```
>> x=[-2:0.02:2];  % generate the x vector
   y=1.1*sign(x).*(abs(x)>1.1) + x.*(abs(x)<=1.1); plot(x,y)
```

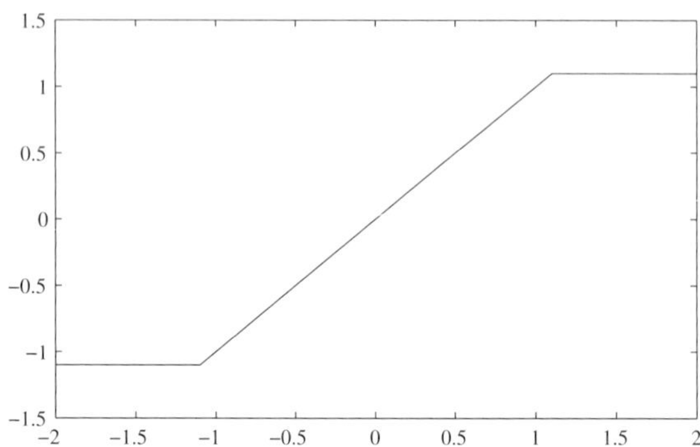

FIGURE 2.5: Segmented saturation function

Even more simply, the command `plot([-3,-1.1,1.1,3],[-1.1 -1.1 1.1 1.1])` can be used to draw the saturation poly-lines.

In MATLAB graphics, each curve or the axis is an object, and the window is another object. The properties of the objects can be assigned by `set()` function, or extracted by `get()` function. The syntaxes of the two functions are

```
set(handle,'p_name 1',p_value 1,'p_name 2',p_value 2, ···)
v=get(object, 'p_name')
```

where *p_name* and *p_value* are respectively the names and values of the corresponding properties. These two functions are very useful in graphical user interface programming.

2.5.2 Other two-dimensional plotting statements

Apart from the standard Descartes coordinate curves, MATLAB also provides other special two-dimensional graphical functions, and the common syntaxes of the functions are given in Table 2.4. In the functions, parameters $\boldsymbol{x}$, $\boldsymbol{y}$ are respectively the horizontal and vertical axis data, and `c` the color options. The parameters $\boldsymbol{y}_{\rm m}$, $\boldsymbol{y}_{\rm M}$ are the vectors of lower- and upper-boundaries in error plots. The functions are demonstrated through the following examples.

Example 2.25 Draw the polar plots for functions $\rho = 5\sin(4\theta/3)$ and $\rho = 5\sin(\theta/3)$.

Solution A vector $\boldsymbol{\theta}$ can be constructed first, over the interval $\theta \in (0, 6\pi)$, then

TABLE 2.4: Other two-dimensional plotting functions

general syntax	explanation	general syntax	explanation
bar($\boldsymbol{x}$,$\boldsymbol{y}$)	two-dimensional bar chart	comet($\boldsymbol{x}$,$\boldsymbol{y}$)	comet trajectory
compass($\boldsymbol{x}$,$\boldsymbol{y}$)	compass plot	errorbar($\boldsymbol{x}$,$\boldsymbol{y}$,$\boldsymbol{y}_{\rm m}$,$\boldsymbol{y}_{\rm M}$)	errorbar plot
feather($\boldsymbol{x}$,$\boldsymbol{y}$)	feather plot	fill($\boldsymbol{x}$,$\boldsymbol{y}$,c)	filled plot
hist($\boldsymbol{y}$,n)	histogram	loglog($\boldsymbol{x}$,$\boldsymbol{y}$)	logarithmic plot
polar($\boldsymbol{x}$,$\boldsymbol{y}$)	polar plot	quiver($\boldsymbol{x}$,$\boldsymbol{y}$)	quiver graph
stairs($\boldsymbol{x}$,$\boldsymbol{y}$)	stairs plot	stem($\boldsymbol{x}$,$\boldsymbol{y}$)	stem plot
semilogx($\boldsymbol{x}$,$\boldsymbol{y}$)	x-semi-logarithmic plot	semilogy($\boldsymbol{x}$,$\boldsymbol{y}$)	y-semi-logarithmic plot

the function value vector $\boldsymbol{\rho}$ can be calculated. With the `polar()` function, the polar plots can be drawn as shown in Figures 2.6 (a) and (b).

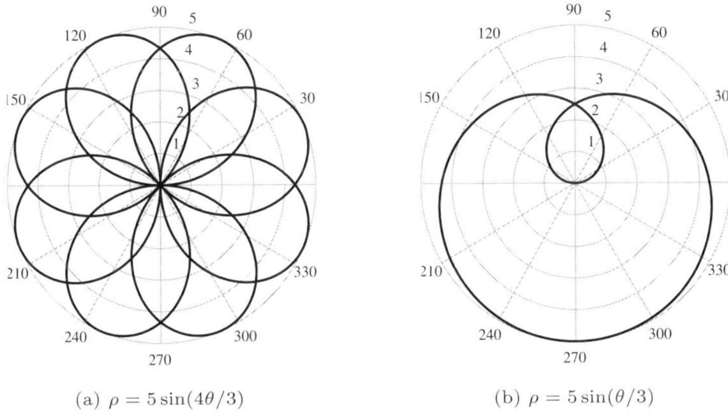

(a) $\rho = 5\sin(4\theta/3)$ (b) $\rho = 5\sin(\theta/3)$

FIGURE 2.6: Polar plots

```
>> theta=0:0.01:6*pi; rho=5*sin(4*theta/3); polar(theta,rho)
   figure; rho=5*sin(theta/3); polar(theta,rho)
```

Example 2.26 Draw the sinusoidal curve with different functions in different areas of the graphics window.

Solution The following commands can be used to draw the expected curves as shown in Figure 2.7, where function `subplot(`n`,`m`,`k`)` can be used to divide the graphics window into several parts, with n, m respectively the total numbers of rows and columns, and k indicates the serial of the area.

```
>> t=0:.2:2*pi; y=sin(t);           % generate the data for plots
   subplot(2,2,1), stairs(t,y)      % partition the graphics window
   subplot(2,2,2), stem(t,y)        % stem plot in upper-right portion
   subplot(2,2,3), bar(t,y)         % bar chart in lower-left portion
   subplot(2,2,4), semilogx(t,y)    % semilogx in lower-right portion
```

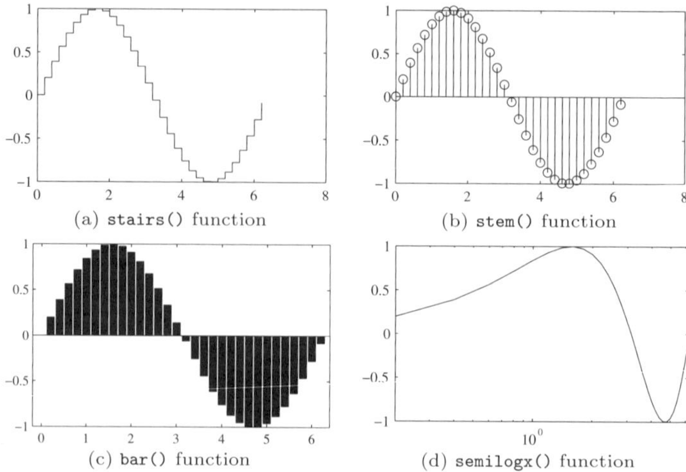

FIGURE 2.7: Different representations of the same function

2.5.3 Implicit function plotting and applications

For an implicit equation $f(x, y) = 0$, the relationship between x and y cannot be explicitly formulated. Thus the conventional `plot()` function cannot be used. The MATLAB function `ezplot()` can be used to draw the implicit function curve

`ezplot(`*implicit function expression*`)`

The following example is used to demonstrate the use of the function.

Example 2.27 Draw the curve of the implicit function

$$f(x, y) = x^2 \sin(x + y^2) + y^2 \mathrm{e}^{x+y} + 5\cos(x^2 + y) = 0.$$

Solution From the given function, it can be seen that the analytical explicit solution of x-y relationship cannot be found. Thus the `plot()` function cannot be used for such a function. The following MATLAB statements can be used to draw the implicit function as shown in Figure 2.8 (a).

```
>> ezplot('x^2 *sin(x+y^2) +y^2*exp(x+y)+5*cos(x^2+y)')
```

The above functions selected automatically the x range, the range can be enlarged with the following statements, with the implicit curve shown in Figure 2.8 (b).

```
>> ezplot('x^2 *sin(x+y^2) +y^2*exp(x+y)+5*cos(x^2+y)',[-10 10])
```

2.5.4 Graphics decorations

The graphics window with editing tools is shown in Figure 2.9. The user may choose to apply text and arrows to the plots. Local zooming and 3D view point settings are also provided in the plots. For instance, a subset of LaTeX commands can be used to add mathematical formula to the plots.

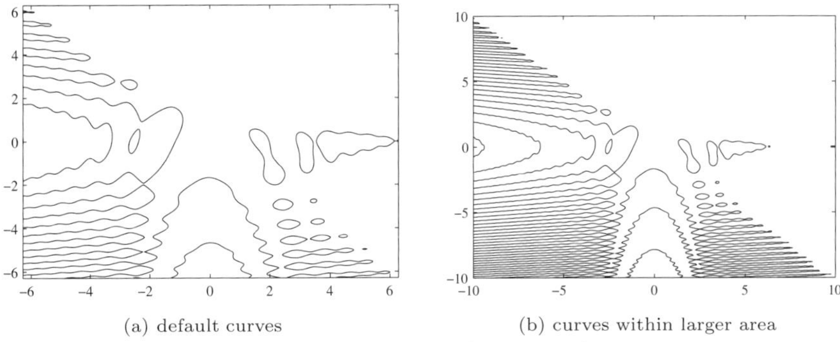

(a) default curves (b) curves within larger area

FIGURE 2.8: Curves of implicit functions

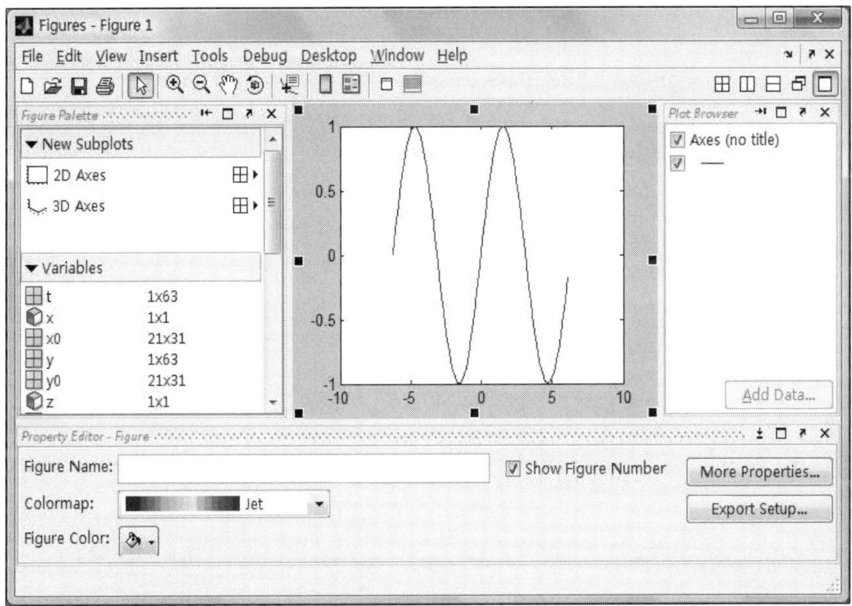

FIGURE 2.9: MATLAB graphics window with editing tools

In the graphics editing interface, there are three parts, with the left part corresponding to the View → Figure Palette menu item, where arrows and text can be added to the curve. 2D and 3D axes can also be added to the curve. The bottom part of the window corresponds to the Property Editor menu item, which allows the selections of color, line styles or fonts to the selected objects. The right part of the window corresponds to the View → Plot Browser menu item, which allows the user to add new data or superimpose new curves.

An example of a typical graphics display under the view-point change is shown in Figure 2.10, where 2D curve is displayed under a 3D framework.

LaTeX is a well established scientific type-setting system, and a subset of its mathematical symbols are supported in MATLAB. One may use them to

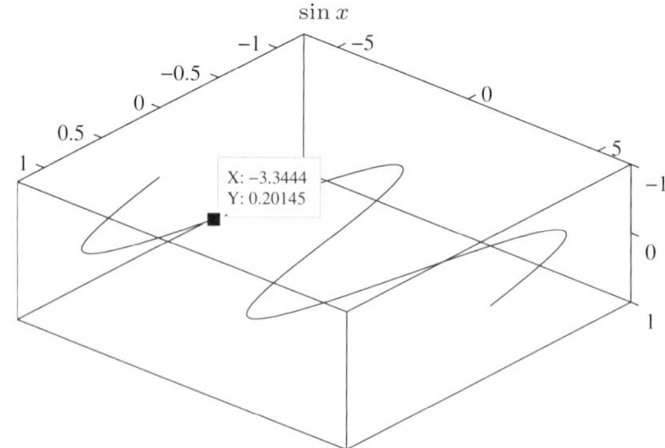

FIGURE 2.10: 3D representations of 2D curves

superimpose formula to the plots.

(i) The symbols are led by the backslash signs \, and the available symbols are listed in Table 2.5.

TABLE 2.5: TEX compatible commands in MATLAB

	c	TEX	c	TEX	c	TEX	c	TEX
	α	`\alpha`	β	`\beta`	γ	`\gamma`	δ	`\delta`
	ϵ	`\epsilon`	ε	`\varepsilon`	ζ	`\zeta`	η	`\eta`
lower-	θ	`\theta`	ϑ	`\vartheta`	ι	`\iota`	κ	`\kappa`
case	λ	`\lambda`	μ	`\mu`	ν	`\nu`	ξ	`\xi`
Greeks	o	`o`	π	`\pi`	ϖ	`\varpi`	ρ	`\rho`
	ι	`\iota`	κ	`\kappa`	ϱ	`\varrho`	σ	`\sigma`
	ς	`\varsigma`	τ	`\tau`	υ	`\upsilon`	ϕ	`\phi`
	φ	`\varphi`	χ	`\chi`	ψ	`\psi`	ω	`\omega`
upper-	Γ	`\Gamma`	Δ	`\Delta`	Θ	`\Theta`	Λ	`\Lambda`
case	Ξ	`\Xi`	Π	`\Pi`	Σ	`\Sigma`	Υ	`\Upsilon`
Greeks	Φ	`\Phi`	Ψ	`\Psi`	Ω	`\Omega`		
	$\aleph$	`\aleph`	$\prime$	`\prime`	$\forall$	`\forall`	$\exists$	`\exists`
common	$\wp$	`\wp`	$\Re$	`\Re`	$\Im$	`\Im`	∂	`\partial`
maths	∞	`\infty`	∇	`\nabla`	$\surd$	`\surd`	$\angle$	`\angle`
symbols	$\neg$	`\neg`	$\int$	`\int`	$\clubsuit$	`\clubsuit`	$\diamondsuit$	`\diamondsuit`
	$\heartsuit$	`\heartsuit`	$\spadesuit$	`\spadesuit`				
binary	$\pm$	`\pm`	$\cdot$	`\cdot`	$\times$	`\times`	$\div$	`\div`
maths	$\circ$	`\circ`	$\bullet$	`\bullet`	$\cup$	`\cup`	$\cap$	`\cap`
symbols	$\vee$	`\vee`	$\wedge$	`\wedge`	$\otimes$	`\otimes`	$\oplus$	`\oplus`
relat-	$\leqslant$	`\leq`	$\geqslant$	`\geq`	$\equiv$	`\equiv`	$\sim$	`\sim`
ional	$\subset$	`\subset`	$\supset$	`\supset`	$\approx$	`\approx`	$\subseteq$	`\subseteq`
maths	$\supseteq$	`\supseteq`	$\in$	`\in`	$\ni$	`\ni`	$\propto$	`\propto`
symbols	$\mid$	`\mid`	$\perp$	`\perp`				
	$\leftarrow$	`\leftarrow`	$\uparrow$	`\uparrow`	$\Leftarrow$	`\Leftarrow`	$\Uparrow$	`\Uparrow`
arrows	$\rightarrow$	`\rightarrow`	$\downarrow$	`\downarrow`	$\Rightarrow$	`\Rightarrow`	$\Downarrow$	`\Downarrow`
	$\leftrightarrow$	`\leftrightarrow`	$\updownarrow$	`\updownarrow`				

(ii) Superscripts and subscripts are represented by ^ and _ respectively. For instance, `a_2^2+b_2^2=c_2^2` represents $a_2^2 + b_2^2 = c_2^2$. If more than one symbol is used in the superscript, they should be written within the { and } signs. For instance `a^Abc` gives a^Abc, while `a^{Abc}` gives a^{Abc}.

LaTeX scientific type-setting system is widely used in the academic world. Interested readers may further refer to Reference [5].

2.6 Three-Dimensional Graphics

2.6.1 Plotting of three-dimensional curves

The two-dimensional function `plot()` can be extended to a three-dimensional (3D) curve drawing with the new `plot3()` function, whose syntaxes are

```
plot3(x,y,z)
plot3(x1,y1,z1,option 1,x2,y2,z2,option 2,··· ,xm,ym,zm,option m)
```

where the *options* are the same as shown in Table 2.3.

Similar to other 2D curve drawing functions, the functions `stem3()`, `fill3()` and `bar3()` can also be applied to 3D curves.

Example 2.28 Draw the curve of the parametric equations

$$x(t) = t^3 \sin(3t)e^{-t}, y(t) = t^3 \cos(3t)e^{-t}, z = t^2, \quad \text{where } t \in [0, 2\pi].$$

Solution A time vector $\boldsymbol{t}$ can be established first, then the vectors $\boldsymbol{x}, \boldsymbol{y}, \boldsymbol{z}$ can be computed. The 3D curve can be drawn with the `plot3()` function, as shown in Figure 2.11 (a). It should be noted that dot operations are used in the evaluations.

```
>> t=0:.1:2*pi;    % establish the t vector, with dot operation
   x=t.^3.*sin(3*t).*exp(-t); y=t.^3.*cos(3*t).*exp(-t); z=t.^2;
   plot3(x,y,z), grid  % 3D curve drawing
```

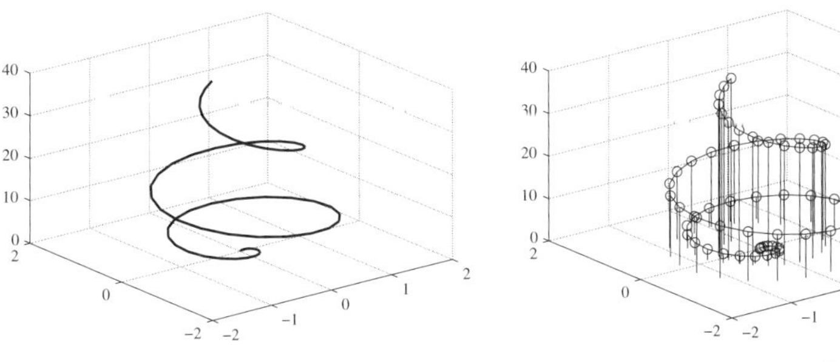

(a) 3D curve plots (b) superimposed with the plot with `stem3()`

FIGURE 2.11: Three-dimensional plots

The `stem3()` function can be used to obtained the plot in Figure 2.11 (b), superimposed by the 3D curve.

```
>> stem3(x,y,z); hold on; plot3(x,y,z), grid
```

2.6.2 Plotting of three-dimensional surfaces

If function $z = f(x, y)$ is given, the 3D surface of the function can be drawn. One can generate mesh grid data in the x-y plane, with the `meshgrid()` function. The function values $\boldsymbol{z}$ can be obtained. The functions `mesh()` and `surf()` can be used to draw the 3D mesh plots and surface plots. The syntaxes of the functions are

```
[x,y]=meshgrid(v1, v2)              % mesh grid generation
z= ..., for instance z=x.*y         % z matrix computation
surf(x,y,z)   or   mesh(x,y,z)      % mesh and surface plots
```

where $\boldsymbol{v}_1$ and $\boldsymbol{v}_2$ are the scales in the $\boldsymbol{x}$ and $\boldsymbol{y}$ axes. The 3D surface can also be drawn with the `surfc()`, `surfl()` and `waterfall()` functions. Also the `contour()` and `contour3()` functions can be used to draw 2D and 3D contour plots.

Example 2.29 Consider the function $z = f(x, y) = (x^2 - 2x)\mathrm{e}^{-x^2-y^2-xy}$. Select in the x-y plane an area and draw the 3D plots.

Solution One may use the `meshgrid()` function to specify the mesh grids on the x-y plane. The values of the function can be evaluated directly for the matrix $\boldsymbol{z}$. The mesh plot can be drawn as shown in Figure 2.12 (a).

```
>> [x,y]=meshgrid(-3:0.1:3,-2:0.1:2);
   z=(x.^2-2*x).*exp(-x.^2-y.^2-x.*y); mesh(x,y,z)
```

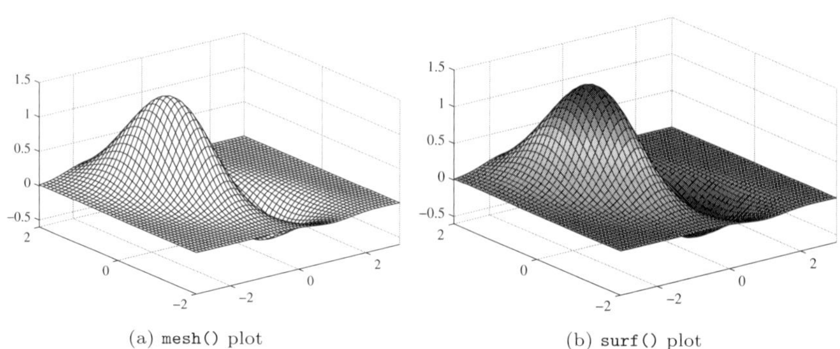

(a) `mesh()` plot (b) `surf()` plot

FIGURE 2.12: Mesh and surface plots of a given function

If one uses `surf()` function to replace the `mesh()` function, the corresponding surface plot can be obtained as shown in Figure 2.12 (b).

```
>> surf(x,y,z)   % surface plot
```

3D surface plots can be decorated by `shading` command, and the options `flat` and `interp` can be used. The decorations are shown in Figures 2.13 (a) and (b) respectively.

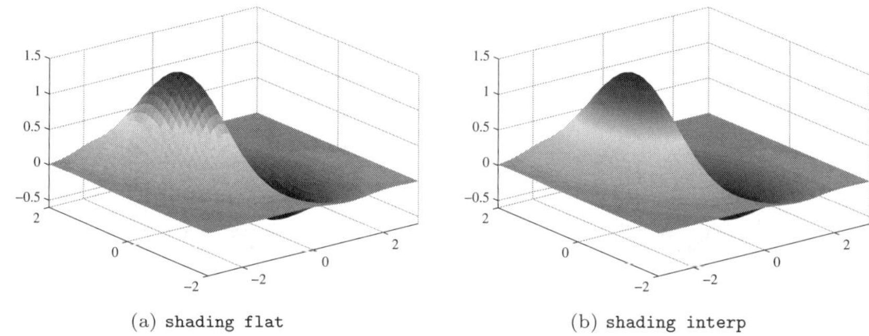

(a) `shading flat` (b) `shading interp`

FIGURE 2.13: 3D surfaces decorated by the `shading` command

Other functions, such as `waterfall(`x, y, z`)` and `contour3(`x, y, z`,30)` can be used to draw 3D plots as shown in Figures 2.14 (a) and (b).

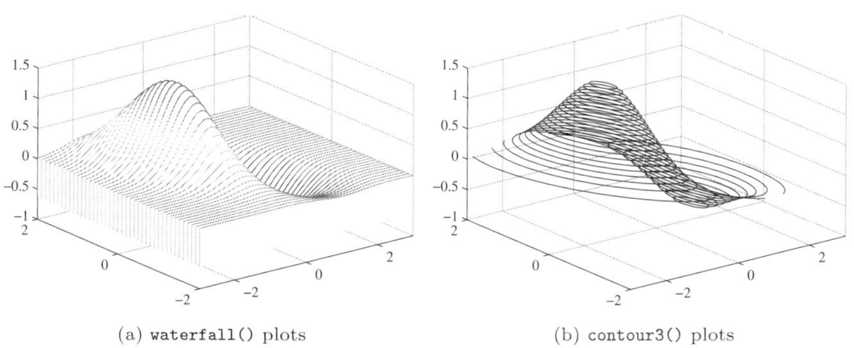

(a) `waterfall()` plots (b) `contour3()` plots

FIGURE 2.14: Other 3D representations

Example 2.30 Display graphically $z = f(x,y) = \dfrac{1}{\sqrt{(1-x)^2+y^2}} + \dfrac{1}{\sqrt{(1+x)^2+y^2}}$.

Solution The following statements can be used to draw the 3D surface of the function, as shown in Figure 2.15 (a).

```
>> [x,y]=meshgrid(-2:.1:2);
   z=1./(sqrt((1-x).^2+y.^2))+1./(sqrt((1+x).^2+y.^2));
   surf(x,y,z), shading flat
```

In fact, there are problems around the $(\pm 1, 0)$ points, where the function values tend to infinity. Thus variable-step-size mesh grids can be constructed, and the new 3D surface can be obtained as shown in Figure 2.15 (b).

```
>> xx=[-2:.1:-1.2,-1.1:0.02:-0.9,-0.8:0.1:0.8,0.9:0.02:1.1,1.2:0.1:2];
   yy=[-1:0.1:-0.2, -0.1:0.02:0.1, 0.2:.1:1];
   [x,y]=meshgrid(xx,yy);
```

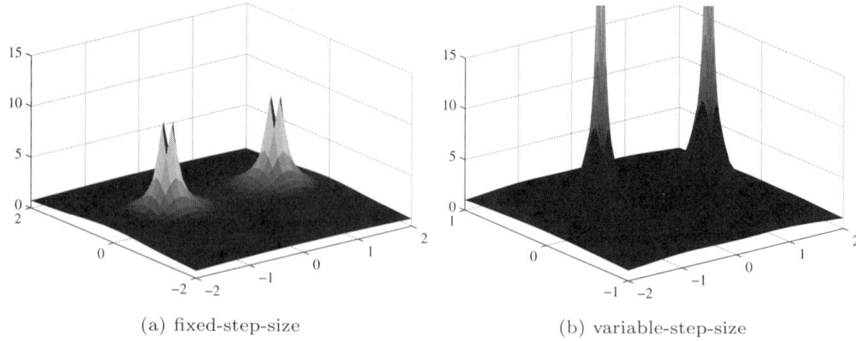

FIGURE 2.15: Three-dimensional surfaces under different grids

```
z=1./(sqrt((1-x).^2+y.^2))+1./(sqrt((1+x).^2+y.^2));
surf(x,y,z), shading flat; zlim([0,15])
```

Example 2.31 Assume that a piecewise function is described below[6]

$$p(x_1,x_2)=\begin{cases}0.5457\exp(-0.75x_2^2-3.75x_1^2-1.5x_1), & x_1+x_2>1\\ 0.7575\exp(-x_2^2-6x_1^2), & -1<x_1+x_2\leqslant 1\\ 0.5457\exp(-0.75x_2^2-3.75x_1^2+1.5x_1), & x_1+x_2\leqslant -1.\end{cases}$$

Show the function in a three-dimensional surface.

Solution Selecting $x = x_1$ and $y = x_2$, the function value can be evaluated with the `if` statements, however the process could be very complicated. Thus the piecewise function configuration statements based on relational operations can be used to evaluate the functions as follows

```
>> [x,y]=meshgrid(-1.5:.1:1.5,-2:.1:2);
   z= 0.5457*exp(-0.75*y.^2-3.75*x.^2-1.5*x).*(x+y>1)+...
      0.7575*exp(-y.^2-6*x.^2).*((x+y>-1) & (x+y<=1))+...
      0.5457*exp(-0.75*y.^2-3.75*x.^2+1.5*x).*(x+y<=-1);
   surf(x,y,z), xlim([-1.5 1.5]); shading flat
```

and the three-dimensional surface can be shown in Figure 2.16.

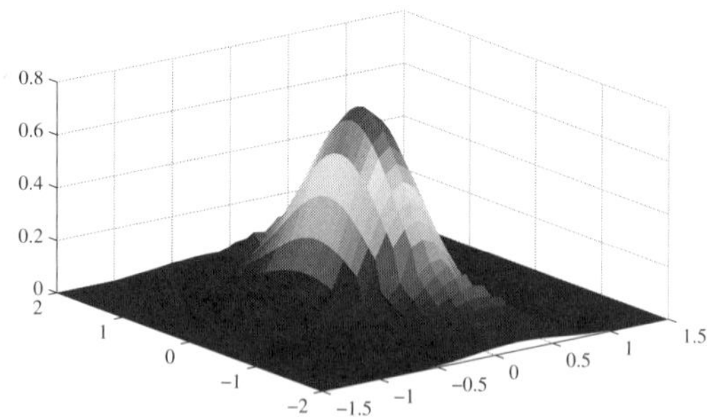

FIGURE 2.16: Surface of a piecewise function with two variables

2.6.3 Viewpoint setting in 3D graphs

In the MATLAB 3D graphics facilities, viewpoint setting functions are provided, which allows the user to view the plot from any angle. Two ways are provided: one is the toolbar facility in the figure window, and the other is the `view()` function.

An illustration to the definition of the viewpoint is given in Figure 2.17 (a), where the two angles α and β can be used to define uniquely the viewpoint. The azimuth α is defined as the angle between the projection line in x-y plane with the negative y-axis, with a default value of $\alpha = -37.5^\circ$. The elevation β is defined as the angle with the x-y plane, with a default value of $\beta = 30^\circ$.

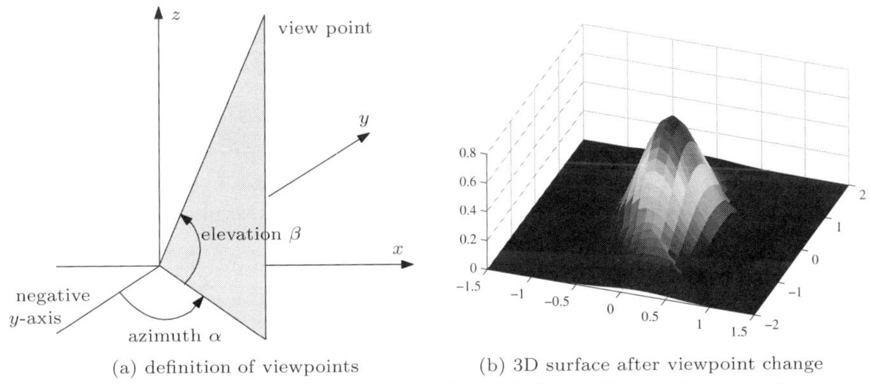

(a) definition of viewpoints (b) 3D surface after viewpoint change

FIGURE 2.17: Viewpoint settings of three-dimensional surfaces

The function `view(`α`,`β`)` can be used to set the viewpoint, where the angles α and β are the azimuth and elevation angles respectively. For instance, the setting `view(0,90)` shows the planform, while `view(0,0)` and `view(90,0)` show the front view and the side elevation respectively.

For instance, one may change the viewpoint in the three-dimensional surface display shown in Figure 2.16. One may set $\alpha = 20^\circ$, and $\beta = 50^\circ$, the following statements can be used and the results shown in Figure 2.17 (b) can be obtained.

```
>> view(20,50), xlim([-1.5 1.5]) % set the range of x-axis
```

Example 2.32 Consider again the surface plot in Example 2.29. View the surface from different angles.

Solution The surface plots from different viewpoints can be obtained using the following statements, as shown in Figure 2.18.

```
>> [x,y] = meshgrid(-3:0.1:3,-2:0.1:2);
   z=(x.^2-2*x).*exp(-x.^2-y.^2-x.*y);
   subplot(221), surf(x,y,z), view(0,90); % planform
```

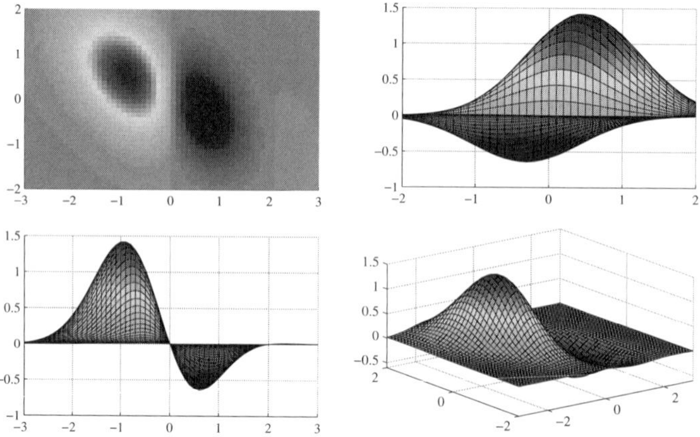

FIGURE 2.18: Surface view from different angles

```
subplot(222), surf(x,y,z), view(90,0); % side elevation
subplot(223), surf(x,y,z), view(0,0);  % front view
subplot(224), surf(x,y,z),             % 3D surface plot
```

Exercises

1. In MATLAB environment, the following statements can be given
`tic, A=rand(500); B=inv(A); norm(A*B-eye(500)), toc`
Run the statements and observe results. If you are not sure with the commands, just use the on-line `help` facilities to display information on the related functions. Then explain in detail the statement and the results.
2. Suppose that a polynomial can be expressed by $f(x) = x^5 + 3x^4 + 4x^3 + 2x^2 + 3x + 6$. If one wants to substitute x by $\dfrac{s-1}{s+1}$, the function $f(x)$ can be changed into a function of s. Use the Symbolic Math Toolbox to do the substitution and get the simplest result.
3. Input the matrices $\boldsymbol{A}$ and $\boldsymbol{B}$ into MATLAB workspace where
$$\boldsymbol{A} = \begin{bmatrix} 1 & 2 & 3 & 4 \\ 4 & 3 & 2 & 1 \\ 2 & 3 & 4 & 1 \\ 3 & 2 & 4 & 1 \end{bmatrix}, \quad \boldsymbol{B} = \begin{bmatrix} 1+\mathrm{j}4 & 2+\mathrm{j}3 & 3+\mathrm{j}2 & 4+\mathrm{j}1 \\ 4+\mathrm{j}1 & 3+\mathrm{j}2 & 2+\mathrm{j}3 & 1+\mathrm{j}4 \\ 2+\mathrm{j}3 & 3+\mathrm{j}2 & 4+\mathrm{j}1 & 1+\mathrm{j}4 \\ 3+\mathrm{j}2 & 2+\mathrm{j}3 & 4+\mathrm{j}1 & 1+\mathrm{j}4 \end{bmatrix}.$$
It is seen that $\boldsymbol{A}$ is a 4×4 matrix. If a command $\boldsymbol{A}(5,6) = 5$ is given, what will happen?
4. For a matrix $\boldsymbol{A}$, if one wants to extract all the even rows to form matrix $\boldsymbol{B}$, what command should be used? Suppose that matrix $\boldsymbol{A}$ is defined by $\boldsymbol{A}$ =`magic(8)`, establish matrix $\boldsymbol{B}$ with suitable statements and see whether the results are correct.

5. Implement the following piecewise function where $\boldsymbol{x}$ can be given by scalar, vectors, matrices or even other multi-dimensional arrays, the returned argument $\boldsymbol{y}$ should be the same size as that of $\boldsymbol{x}$. The parameters h and D are scalars.
$$\boldsymbol{y}=f(\boldsymbol{x})=\begin{cases} h, & \boldsymbol{x}>D \\ h/D\boldsymbol{x}, & |\,\boldsymbol{x}\,|\leqslant D \\ -h, & \boldsymbol{x}<-D \end{cases}$$
6. Evaluate using numerical method the sum $S=1+2+4+\cdots+2^{62}+2^{63}=\sum_{i=0}^{63}2^i$, the use of vectorized form is suggested. Check whether accurate solutions can be found and why. Find the accurate sum using the symbolic computation methods.
7. Write an M-function `mat_add()` with the syntax
   $\boldsymbol{A}$=`mat_add(`$\boldsymbol{A}_1$`,`$\boldsymbol{A}_2$`,`$\boldsymbol{A}_3$`,`$\cdots$`)`
   In the function, it is required that arbitrary number of input arguments $\boldsymbol{A}_i$ are allowed.
8. A MATLAB function can be written whose syntax is
   $\boldsymbol{v}$=`[`$h_1, h_2, h_m, h_{m+1}, \cdots, h_{2m-1}$`]` and $\boldsymbol{H}$=`myhankel(`$\boldsymbol{v}$`)`
   where the vector $\boldsymbol{v}$ is defined, and out of it, the output argument should be an $m\times m$ Hankel matrix.
9. From matrix theory, it is known that if a matrix $\boldsymbol{M}$ is expressed as $\boldsymbol{M}=\boldsymbol{A}+\boldsymbol{BCB}^{\mathrm{T}}$, where $\boldsymbol{A}$, $\boldsymbol{B}$ and $\boldsymbol{C}$ are the matrices of relevant sizes, the inverse of $\boldsymbol{M}$ can be calculated by the following algorithm
$$\boldsymbol{M}^{-1}=\left(\boldsymbol{A}+\boldsymbol{BCB}^{\mathrm{T}}\right)^{-1}=\boldsymbol{A}^{-1}-\boldsymbol{A}^{-1}\boldsymbol{B}\left(\boldsymbol{C}^{-1}+\boldsymbol{B}^{\mathrm{T}}\boldsymbol{A}^{-1}\boldsymbol{B}\right)^{-1}\boldsymbol{B}^{\mathrm{T}}\boldsymbol{A}^{-1}$$
   The matrix inversion can be carried out using the formula easily. Suppose that there is a 5×5 matrix $\boldsymbol{M}$, from which the three other matrices can be found.
$$\boldsymbol{M}=\begin{bmatrix} -1 & -1 & -1 & 1 & 0 \\ -2 & 0 & 0 & -1 & 0 \\ -6 & -4 & -1 & -1 & -2 \\ -1 & -1 & 0 & 2 & 0 \\ -4 & -3 & -3 & -1 & 3 \end{bmatrix},\quad \boldsymbol{A}=\begin{bmatrix} 1 & 0 & 0 & 0 & 0 \\ 0 & 3 & 0 & 0 & 0 \\ 0 & 0 & 4 & 0 & 0 \\ 0 & 0 & 0 & 2 & 0 \\ 0 & 0 & 0 & 0 & 4 \end{bmatrix}$$
$$\boldsymbol{B}=\begin{bmatrix} 0 & 1 & 1 & 1 & 1 \\ 0 & 2 & 1 & 0 & 1 \\ 1 & 1 & 1 & 2 & 1 \\ 0 & 1 & 0 & 0 & 1 \\ 1 & 1 & 1 & 1 & 1 \end{bmatrix},\quad \boldsymbol{C}=\begin{bmatrix} 1 & -1 & 1 & -1 & -1 \\ 1 & -1 & 0 & 0 & -1 \\ 0 & 0 & 0 & 0 & 1 \\ 1 & 0 & -1 & -1 & 0 \\ 0 & 1 & -1 & 0 & 1 \end{bmatrix}.$$
   Write the statement to evaluate the inverse matrix. Check the accuracy of the inversion. Compare the accuracy of the inversion method and the direct inversion method with `inv()` function.
10. Consider the following iterative model
$$\begin{cases} x_{k+1}=1+y_k-1.4x_k^2 \\ y_{k+1}=0.3x_k \end{cases}$$
   with initial conditions $x_0=0$, $y_0=0$. Write an M-function to evaluate the sequence x_i, y_i. 30000 points can be obtained by the function to construct the $\boldsymbol{x}$ and $\boldsymbol{y}$ vectors. The points can be expressed by a dot, rather than lines. In

this case, the so-called Hénon attractor can be drawn.

11. A regular triangle can be drawn by MATLAB statements easily. Use the loop structure to design an M-function that, in the same coordinates, a sequence of regular triangles can be drawn, each by rotating a small angle from the previous one.
12. Select suitable step-sizes and draw the function curve for $\sin(1/t)$, where $t \in (-1, 1)$.
13. For suitably assigned ranges of θ, draw polar plots for the following functions.
(i) $\rho = 1.0013\theta^2$, (ii) $\rho = \cos(7\theta/2)$,
(iii) $\rho = \sin(\theta)/\theta$, (iv) $\rho = 1 - \cos^3(7\theta)$
14. Find the solutions to the following equations using graphical methods and verify the solutions.
$$\begin{cases} x^2 + y^2 = 3xy^2 \\ x^3 - x^2 = y^2 - y \end{cases}$$
15. Draw the 3D surface plots for the functions xy and $\sin(xy)$ respectively. Also draw the contours of the functions. View the 3D surface plot from different angles.
16. In graphics command, there is a trick in hiding certain parts of the plot. If the function values are assigned to `NaN`s, the point on the curve or the surface will not be shown. Draw first the surface plot of the function $z = \sin xy$. Then cut off the region that satisfies $x^2 + y^2 \leqslant 0.5^2$.

Chapter 3

Calculus Problems

The calculus established by Isaac Newton and Gottfried Wilhelm Leibniz is fundamental to many branches of sciences and engineering. In traditional calculus courses, limits, differentiations, integrals, series expansions such as Taylor series and Fourier series expansions for single-variable and multivariable functions are the main topics discussed. The analytical solutions to these problems can be obtained by the direct use of the corresponding functions provided by the Symbolic Math Toolbox of MATLAB which will be discussed in Section 3.1. The Taylor series expansions for single- and multivariable functions as well as the Fourier series expansions are discussed in Section 3.2. Moreover, the series summation and product problems are discussed. Sections 3.5 and 3.6 present methods for path integrals, line integrals and surface integrals. Most of the materials presented in this chapter are symbolic-based, which cannot be solved using conventional computer programming languages such as C for average users. Computer mathematics languages such as MATLAB should be used instead.

In many scientific and engineering researches, the analytical solutions to calculus problems may face difficulties, since the original functions may not be given explicitly. For problems with measured data, numerical differentiations and integrals should be applied accordingly. They are illustrated in Sections 3.3 and 3.4, respectively. Alternative solutions to the same numerical calculus problems using spline interpolation are given in Chapter 8.

As an extension to the traditional (integer-order) calculus, non-integer-order or fractional-order calculus, will be discussed in Chapter 10.

For readers who wish to check the detailed explanations of calculus, we recommend the free textbooks [7, 8].

3.1 Analytical Solutions to Calculus Problems

The Symbolic Math Toolbox of MATLAB can be used directly in solving the limit problems, the differentiation problems, and the integral problems. Using the methods presented in this section, the readers will be equipped with the ability in solving ordinary calculus problems directly by computers.

3.1.1 Analytical solutions to limit problems

Limits of single-variable functions

Assume that the function to be analyzed is $f(x)$, the limit is defined as

$$L = \lim_{x \to x_0} f(x) \tag{3.1}$$

where x_0 can be either a given value or infinity. For certain functions, the left or right limit can be defined as

$$L_1 = \lim_{x \to x_0^-} f(x), \quad \text{or} \quad L_2 = \lim_{x \to x_0^+} f(x) \tag{3.2}$$

where the former means to approach the point x_0 from the left-hand side which is referred to as the *left limit* problem. The latter is referred to as the *right limit* problem. The limit problems summarized above can be solved by the use of the `limit()` function, where

L=`limit(`*fun*,x,x_0`)`	% calculate the limit
L=`limit(`*fun*,x,x_0`,'left'` or `'right')`	% the one-sided limit

To use the functions in Symbolic Math Toolbox, symbolic variables such as x should be declared first. Then, the limit function *fun* can be expressed. If x_0 is ∞, one can assign it to `inf`. If the one-sided limit is required, the `'left'` or `'right'` option should be specified. The following examples are used to demonstrate the use of the `limit()` function in MATLAB.

Example 3.1 Solve the limit problem $\lim_{x \to \infty} x\left(1+\dfrac{a}{x}\right)^x \sin\dfrac{b}{x}$.

Solution For this problem, one should first declare the variables a, b and x as symbolic variables. Then the function can be defined and the `limit()` function can be called directly to solve the problem, which returns $L = \mathrm{e}^a b$.

```
>> syms x a b; f=x*(1+a/x)^x*sin(b/x); L=limit(f,x,inf)
```

Example 3.2 Solve the one-sided limit problem $\lim_{x \to 0^+} \dfrac{\mathrm{e}^{x^3}-1}{1-\cos\sqrt{x-\sin x}}$.

Solution With the `limit()` function, the one-sided limit can easily be solved, with the limit of 12.

```
>> syms x; limit((exp(x^3)-1)/(1-cos(sqrt(x-sin(x)))),x,0,'right')
```

One can further verify the above problem graphically over a proper range of interest. For instance, if the interval $(-0.1, 0.1)$ is considered, the function over the interval can be drawn in Figure 3.1.

```
>> x=-0.1:0.001:0.1; y=(exp(x.^3)-1)./(1-cos(sqrt(x-sin(x))));
   plot(x,y,'-',[0],[12],'o')
```

It can be seen that the limit of the original problem is also 12.

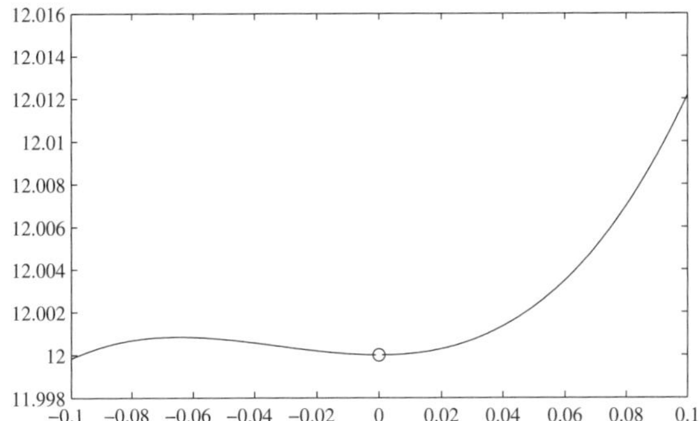

FIGURE 3.1: The curve of the function around $x = 0$

```
>> syms x; limit((exp(x^3)-1)/(1-cos(sqrt(x-sin(x)))),x,0)
```

Consider again the original problem. The aim of the original one-sided limit requirement ensures that the expression under the square root is positive. In fact, for imaginary variables, one can still find from the Euler's formula that $\cos \mathrm{j}\alpha = (\mathrm{e}^{\alpha} + \mathrm{e}^{-\alpha})/2$. Thus the one-sided limits for the function are the same for this example, which further verifies that the original function is continuous around $x = 0$ as also seen from Figure 3.1.

Limits of multivariable functions

The limit problems for multivariable functions can also be solved with the MATLAB function `limit()`. For instance, the limit to the function $f(x, y)$

$$L = \lim_{\substack{x \to x_0 \\ y \to y_0}} f(x, y) \tag{3.3}$$

can be solved by the nested use of the `limit()` function. For example, `L1=limit(limit(fun,x,x0),y,y0)` or `L1=limit(limit(fun,y,y0),x,x0)` where x_0 and y_0 can be either constants or functions of another variable, for instance $x \to g(y)$. In the latter case, the order of the function call cannot be changed.

Example 3.3 Solve the limit problem $\displaystyle\lim_{\substack{x \to 1/\sqrt{y} \\ y \to \infty}} \mathrm{e}^{-1/(y^2+x^2)} \frac{\sin^2 x}{x^2} \left(1 + \frac{1}{y^2}\right)^{x+a^2y^2}$.

Solution The problem can easily be solved with the following MATLAB scripts

```
>> syms x y a; f=exp(-1/(y^2+x^2))*sin(x)^2/x^2*(1+1/y^2)^(x+a^2*y^2);
   L=limit(limit(f,x,1/sqrt(y)),y,inf)
```

which yields $L = \mathrm{e}^{a^2}$.

3.1.2 Analytical solutions to derivative problems

Derivative and high-order derivatives

If the function is known, the function `diff()` can be used to calculate its derivatives. The syntaxes of the `diff()` function are

```
y=diff(fun,x)      % find the derivative
y=diff(fun,x,n)    % evaluate the nth order derivative
```

where *fun* is the symbolic expression of a given function; x is the symbolic independent variable; n is the order of the derivative to be taken.

Example 3.4 Compute $\dfrac{\mathrm{d}^4 f(x)}{\mathrm{d}x^4}$ for a given function $f(x) = \dfrac{\sin x}{x^2+4x+3}$.

Solution It should be noted that this is the first example given at the beginning of the book. The derivatives can easily be obtained with the following MATLAB functions. The variable x should be declared as a symbolic variable first, then the function `diff()` can be called to find the first-order derivative.

```
>> syms x; f=sin(x)/(x^2+4*x+3); f1=diff(f)
```

The readability of the results directly obtained may not be very high. It is suggested that the results should be converted with the use of `pretty()` function, or by `latex()` function. The latter can be used to convert the result into the form in the well-known LaTeX string, the best scientific documentation system. Under LaTeX, the result can be better displayed as $\dfrac{\cos x}{x^2+4x+3} - \dfrac{\sin x\,(2x+4)}{(x^2+4x+3)^2}$. It can be seen that the quality of LaTeX display is far better than the one obtained in MATLAB. In the later description, the LaTeX display will be extensively used to increase the readability.

The original function and the first-order derivative function can easily be obtained and their respective curves are shown in Figure 3.2.

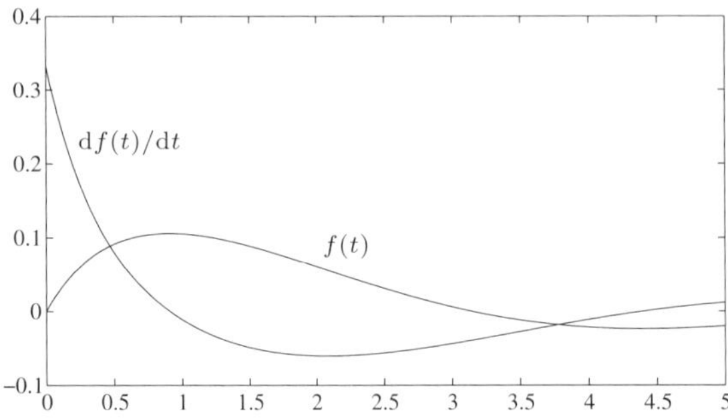

FIGURE 3.2: The curves of the original function and its derivative

```
>> x1=0:.01:5; y=subs(f,x,x1); y1=subs(f1,x,x1); plot(x1,y,x1,y1,':')
```

The fourth-order derivative can be simply calculated from

```
>> f4=diff(f,x,4)
```

and the result is displayed in LaTeX

$$\frac{\sin x}{x^2+4x+3}+4\frac{(2x+4)\cos x}{(x^2+4x+3)^2}-12\frac{(2x+4)^2\sin x}{(x^2+4x+3)^3}+12\frac{\sin x}{(x^2+4x+3)^2}-24\frac{(2x+4)^3\cos x}{(x^2+4x+3)^4}$$
$$+48\frac{(2x+4)\cos x}{(x^2+4x+3)^3}+24\frac{(2x+4)^4\sin x}{(x^2+4x+3)^5}-72\frac{(2x+4)^2\sin x}{(x^2+4x+3)^4}+24\frac{\sin x}{(x^2+4x+3)^3}.$$

From the above simplified results, it is clear that the direct use of the function simple() is not sufficient for this example. For the given example, it can immediately be found that one may extract the terms $\sin x$ and $\cos x$ from the results and the coefficients for these terms can be simplified separately such that

```
>> collect(simple(f4),sin(x)), collect(simple(f4),cos(x))
```

The even more concise results can be obtained shown as follows:

$$\frac{\mathrm{d}^4 f(x)}{\mathrm{d}x^4}=8(x^5+10x^4+26x^3-4x^2-99x-102)\frac{\cos x}{(x^2+4x+3)^4}+$$
$$(x^8+16x^7+72x^6-32x^5-1094x^4-3120x^3-3120x^2+192x+1581)\frac{\sin x}{(x^2+4x+3)^5}.$$

The differentiation function diff() can easily be used to find high-order derivatives. For instance, the 100th order derivative of the same function can be found within one second.

```
>> tic, diff(f,x,100); toc
```

Partial derivatives of multivariable functions

There is no direct function which can be used in finding the partial derivatives in MATLAB. The function diff() can actually be used instead. For instance, if a function $f(x,y)$ with two variables is defined, the partial derivative $\partial^{m+n}f/(\partial x^m \partial y^n)$ can be evaluated by the nested use of the diff() function as follows:

`f=diff(diff(fun,x,m),y,n)` , or `f=diff(diff(fun,y,n),x,m)`

Example 3.5 Find the partial derivatives of $z=f(x,y)=(x^2-2x)\mathrm{e}^{-x^2-y^2-xy}$ function and investigate the function further using graphical method.

Solution The partial derivatives $\partial z/\partial x$ and $\partial z/\partial y$ can be evaluated easily using

```
>> syms x y; z=(x^2-2*x)*exp(-x^2-y^2-x*y);
   zx=simple(diff(z,x)), zy=diff(z,y)
```

and the mathematical representations of the derivatives are

$$\frac{\partial z(x,y)}{\partial x}=-\mathrm{e}^{-x^2-y^2-xy}(-2x+2+2x^3+x^2y-4x^2-2xy)$$
$$\frac{\partial z(x,y)}{\partial y}=-x(x-2)(2y+x)\mathrm{e}^{-x^2-y^2-xy}.$$

Within the rectangular region where $x\in(-3,3), y\in(-2,2)$, mesh grids can be defined and the partial derivatives can be obtained numerically over the mesh grids. The three-dimensional surface of the original function is shown in Figure 3.3 (a)

```
>> [x0,y0]=meshgrid(-3:.2:3,-2:.2:2);
   z0=subs(z,{x,y},{x0,y0}); % substituting the two variables
   surf(x0,y0,z0), axis([-3 3 -2 2 -0.7 1.5]) % three-dimensional surface
```

From the partial derivatives obtained, the numerical solutions at the mesh grids can be evaluated. The function `quiver()` can then be used to draw attractive curves, and the curves can be superimposed over the contour of the original function with the following statements, as shown in Figure 3.3 (b).

```
>> contour(x0,y0,z0,30), hold on    % contours of the function
   zx0=subs(zx,{x,y},{x0,y0}); zy0=subs(zy,{x,y},{x0,y0});
   quiver(x0,y0,zx0,zy0)            % draw the attractive curves
```

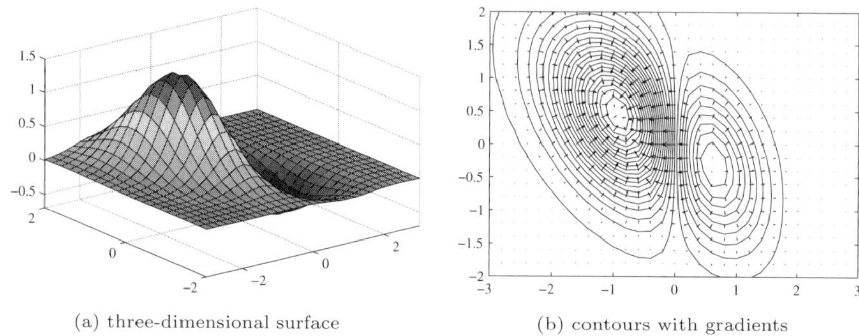

(a) three-dimensional surface (b) contours with gradients

FIGURE 3.3: Graphical interpretation of the functions with two variables

Example 3.6 For a given function with three independent variables x, y and z such that $f(x,y,z) = \sin(x^2 y)\mathrm{e}^{-x^2 y - z^2}$, find the partial derivative $\partial^4 f(x,y,z)/(\partial x^2 \partial y \partial z)$.

Solution The following MATLAB statements can be given to solve this problem

```
>> syms x y z; f=sin(x^2*y)*exp(-x^2*y-z^2);
   df=diff(diff(diff(f,x,2),y),z); df=simple(df)
```

the results can be obtained as

$$-4z\mathrm{e}^{-x^2 y - z^2}\left[\cos x^2 y - 10yx^2 \cos x^2 y + 4x^4 y^2 \sin x^2 y + 4x^4 y^2 \cos x^2 y - \sin x^2 y\right].$$

Jacobian matrix of multivariable functions

Assume that there are n independent variables, and m functions defined as

$$\begin{cases} y_1 = f_1(x_1, x_2, \cdots, x_n) \\ y_2 = f_2(x_1, x_2, \cdots, x_n) \\ \quad\vdots \qquad \vdots \\ y_m = f_m(x_1, x_2, \cdots, x_n). \end{cases} \tag{3.4}$$

The partial derivative $\partial y_i/\partial x_j$ for each combination of i and j can be represented in the matrix form as

$$\boldsymbol{J} = \begin{bmatrix} \partial y_1/\partial x_1 & \partial y_1/\partial x_2 & \cdots & \partial y_1/\partial x_n \\ \partial y_2/\partial x_1 & \partial y_2/\partial x_2 & \cdots & \partial y_2/\partial x_n \\ \vdots & \vdots & \ddots & \vdots \\ \partial y_m/\partial x_1 & \partial y_m/\partial x_2 & \cdots & \partial y_m/\partial x_n \end{bmatrix} \tag{3.5}$$

and such a matrix is referred to as the *Jacobian matrix*. Jacobian matrices are quite useful in many research areas, such as robotics and image processing. Jacobian matrix can be obtained using the `jacobian()` function of the Symbolic Math Toolbox directly. The syntax of the function is `J=jacobian(y,x)`, where $\boldsymbol{x}$ is the vector of independent variables, and $\boldsymbol{y}$ is the vector of multivariable functions.

Example 3.7 Consider that the functions for coordinate transformation are defined as $x = r\sin\theta\cos\phi$, $y = r\sin\theta\sin\phi$, and $z = r\cos\theta$. Find the Jacobian matrix of these functions.

Solution Three independent variables can be declared and the three functions can then be expressed. The following statements can be used to find the Jacobian matrix

```
>> syms r theta phi; x=r*sin(theta)*cos(phi);
   y=r*sin(theta)*sin(phi); z=r*cos(theta);
   J=jacobian([x; y; z],[r theta phi])
```

The Jacobian matrix is obtained as

$$\boldsymbol{J} = \begin{bmatrix} \sin\theta\cos\phi & r\cos\theta\cos\phi & -r\sin\theta\sin\phi \\ \sin\theta\sin\phi & r\cos\theta\sin\phi & r\sin\theta\cos\phi \\ \cos\theta & -r\sin\theta & 0 \end{bmatrix}.$$

Partial derivatives of implicit functions

Assume that an implicit function is defined as $f(x_1, x_2, \cdots, x_n) = 0$. The partial derivative $\partial x_i/\partial x_j$ among the independent variables can be obtained using the following formula

$$\frac{\partial x_i}{\partial x_j} = -\frac{\dfrac{\partial}{\partial x_j} f(x_1, x_2, \cdots, x_n)}{\dfrac{\partial}{\partial x_i} f(x_1, x_2, \cdots, x_n)}. \tag{3.6}$$

Since the derivatives of f with respect to x_i and x_j can easily be obtained separately with the function `diff()`, the partial derivative of $\partial x_i/\partial x_j$ can be obtained directly using the MATLAB functions `F=-diff(f,x_j)/diff(f,x_i)`.

Example 3.8 Consider again the implicit function $f(x, y) = (x^2-2x)\mathrm{e}^{-x^2-y^2-xy} = 0$. Evaluate $\partial y/\partial x$.

Solution It can be found from (3.6) that the partial derivative $\partial y/\partial x$ can be obtained with the following statements

```
>> syms x y; f=(x^2-2*x)*exp(-x^2-y^2-x*y);
   simple(-diff(f,x)/diff(f,y))
```

The result is $\dfrac{2x-2-2x^3-x^2y+4x^2+2xy}{x\,(x-2)\,(2y+x)}$.

Derivatives of parametric equations

When the function $y(x)$ is given as parametric equations $y = f(t), x = g(t)$, the kth order derivative of the function $\dfrac{\mathrm{d}^k y}{\mathrm{d}x^k}$ can be calculated using the following formula

$$\begin{aligned}
&\frac{\mathrm{d}y}{\mathrm{d}x} = \frac{f'(t)}{g'(t)} \\
&\frac{\mathrm{d}^2y}{\mathrm{d}x^2} = \frac{\mathrm{d}}{\mathrm{d}t}\left(\frac{f'(t)}{g'(t)}\right)\frac{1}{g'(t)} = \frac{\mathrm{d}}{\mathrm{d}t}\left(\frac{\mathrm{d}y}{\mathrm{d}x}\right)\frac{1}{g'(t)} \\
&\vdots \\
&\frac{\mathrm{d}^ny}{\mathrm{d}x^n} = \frac{\mathrm{d}}{\mathrm{d}t}\left(\frac{\mathrm{d}^{n-1}y}{\mathrm{d}x^{n-1}}\right)\frac{1}{g'(t)}.
\end{aligned} \tag{3.7}$$

Using the recursive calling structure, the following MATLAB function can be written to implement the above algorithm and the function should be placed in the `@sym` directory

```
function result=paradiff(y,x,t,n)
if mod(n,1)~=0, error('n should positive integer, please correct')
else
   if  n==1, result=diff(y,t)/diff(x,t);
   else, result=diff(paradiff(y,x,t,n-1),t)/diff(x,t);
end, end
```

Example 3.9 For the parametric equations $y = \dfrac{\sin t}{(t+1)^3}$, $x = \dfrac{\cos t}{(t+1)^3}$, find $\dfrac{\mathrm{d}^3y}{\mathrm{d}x^3}$.

Solution From the above parametric equations, the derivative can be found by

```
>> syms t; y=sin(t)/(t+1)^3; x=cos(t)/(t+1)^3;
   f=paradiff(y,x,t,3); [n,d]=numden(f); F=simple(n)/simple(d)
```

The results can be simplified into the following form:

$$\frac{\mathrm{d}^3y}{\mathrm{d}x^3} = \frac{-3(t+1)^7[(t^4+4t^3+6t^2+4t-23)\cos t-(4t^3+12t^2+32t+24)\sin t]}{(t\sin t+\sin t+3\cos t)^5}.$$

3.1.3 Analytical solutions to integral problems

In calculus, integral problems are often described mathematically as

$$\int f(x)\,\mathrm{d}x,\ \int_a^b f(x)\,\mathrm{d}x,\ \int\cdots\int f(x_1,x_2,\cdots,x_n)\,\mathrm{d}x_n\cdots\mathrm{d}x_2\mathrm{d}x_1 \tag{3.8}$$

where function $f(\cdot)$ is referred to as the *integrand.* The first integral is referred to as the *indefinite integral,* while $F(x)$ is referred to as the *primitive function.* The other two integrals are respectively referred to as the *definite integral* and *multiple integrals.* To solve the integral problems, according to calculus courses, one has to select, largely by experience, the integration methods, such as integration by substitution, or integration by parts, or others. Thus, solving integral problems could be a tedious task.

Indefinite integrals

The `int()` function provided in the Symbolic Math Toolbox of MATLAB can be used to evaluate the indefinite integrals to given functions. The syntax of the function is `F=int(fun,x)`, where the integrand can be described by *fun*. If only one variable appears in the integrand, the argument x can be omitted. The returned argument is the primitive $F(x)$. In fact, the general solution to the indefinite integral problem is $F(x)+C$, with C an arbitrary constant.

For any integrable functions, the use of the function `int()` can reduce the complicated work such that the primitive function can be obtained directly. However, for symbolically non-integrable functions, the `int()` function may not give useful results. In this case, numerical methods have to be used instead.

Example 3.10 Consider the function given in Example 3.4. The `diff()` function can be used to find the derivative of $f(x)$. If the indefinite integral is made upon the results, check whether the original function can be restored.

Solution The original function can be defined and the integral can be taken on the first-order derivative such that

```
>> syms x; y=sin(x)/(x^2+4*x+3); y1=diff(y); y0=int(y1)
```

the result is then $\dfrac{\sin x}{2(x+1)}-\dfrac{\sin x}{2(x+3)}$. It can be seen that the result restores the original function.

Now consider taking the fourth-order derivative to the original function by applying `int()` four times in a nested way as follows:

```
>> y4=diff(y,4); y0=int(int(int(int(y4)))); simple(y0)
```

and the result is $\dfrac{\sin x}{(x+1)(x+3)}$, which is still the same as the original function.

Example 3.11 Show that

$$\int x^3\cos^2 ax\,\mathrm{d}x=\frac{x^4}{8}+\left(\frac{x^3}{4a}-\frac{3x}{8a^3}\right)\sin 2ax+\left(\frac{3x^2}{8a^2}-\frac{3}{16a^4}\right)\cos 2ax+C.$$

Solution The following MATLAB statements can be used:

```
>> syms a x; f=simple(int(x^3*cos(a*x)^2,x))
```

and the simplified results can be obtained as

$$\frac{1}{16a^4}\left[4a^3x^3\sin(2ax)+2a^4x^4+6a^2x^2\cos(2ax)-6\,ax\sin(2ax)+3-3\cos(2ax)\right].$$

It can be seen that the result is not the same as the one on the right-hand side. Let us check the difference. Using the following scripts

```
>> f1=x^4/8+(x^3/(4*a)-3*x/(8*a^3))*sin(2*a*x)+...
      (3*x^2/(8*a^2)-3/(16*a^4))*cos(2*a*x);
   simple(f-f1)  % difference is taken and simplified
```

after simplification, the difference is $-3/(16a^4)$ which is not zero. However, fortunately, since the difference between the two primitive functions is a constant, it can be included into the final constant C. Thus the original equation can be proved.

Example 3.12 Consider the two integrands

$$f(x)=\mathrm{e}^{-x^2/2}, \quad \text{and} \quad g(x)=x\sin(ax^4)\mathrm{e}^{x^2/2}.$$

They are all known to be not integrable. Compute the integral to the two functions.

Solution Let us consider the integral to the first integrand $f(x)=\mathrm{e}^{-x^2/2}$. The following MATLAB functions can be used

```
>> syms x; int(exp(-x^2/2))
```

and the result obtained is $\mathrm{erf}(\sqrt{2})/\sqrt{2\pi}$. Since the original integrand is not integrable, a function $\mathrm{erf}(x)=\dfrac{2}{\sqrt{\pi}}\displaystyle\int_0^x \mathrm{e}^{-t^2}\mathrm{d}t$ is introduced. Thus the "analytical" solution to the original problem can be obtained.

The second integrand can be tested under the `int()` function. The following MATLAB statements

```
>> syms a x; int(x*sin(a*x^4)*exp(x^2/2))
```

result in the following returned warning message, which means that the explicit solutions cannot be obtained.

```
Warning: Explicit integral could not be found.
> In sym.int at 58
ans =
   int(x*sin(a*x^4)*exp(1/2*x^2),x)
```

Computing definite and infinite integrals

The definite integrals and infinite integrals are also part of calculus. For instance, although the function $\mathrm{erf}(x)$ is defined previously, the integral of a particular value of x cannot be obtained analytically. In this case, definite integrals, in cooperation with numerical methods, can be obtained. The function `int()` can be used to evaluate the definite and infinite integrals. The syntax of the function is `I=int(fun,x,a,b)`, where x is the independent variable, (a, b) is the integral interval. For infinite integrals, the arguments a and b can be assigned to `-Inf` or `Inf`. Also if no exact value can be obtained directly, the `vpa()` function can be used to evaluate the solutions numerically.

Example 3.13 Consider the integrands given previously in Example 3.12. When $a = 0$, $b = 1.5$ (or ∞), evaluate the values of the integral.

Solution The following statements can be used in solving the definite and infinite integral problems

```
>> syms x; I1=int(exp(-x^2/2),x,0,1.5)
   vpa(I1,70), I2=int(exp(-x^2/2),x,0,inf)
```

where $I_1 = \sqrt{\pi/2}\,\mathrm{erf}[3/(2\sqrt{2})]$, and the high-precision numerical solution to the definite integral is $I_1 = 1.085853317666016569702419076542265042534236293532156326729917229308 53$. The analytical solution to the infinite integral is $I_2 = \sqrt{\pi/2}$.

Example 3.14 Solve the definite integral problems for functional boundaries

$$I(t) = \int_{\cos t}^{\mathrm{e}^{-2t}} \frac{-2x^2+1}{(2x^2-3x+1)^2}\,\mathrm{d}x.$$

Solution The function `int()` can be used in solving definite integrals and the following statements can be used

```
>> syms x t; f=(-2*x^2+1)/(2*x^2-3*x+1)^2;
   I=simple(int(f,x,cos(t),exp(-2*t))),
```

and the results can be expressed as

$$I(t) = -\frac{(2\mathrm{e}^{-2t}\cos t - 1)(\mathrm{e}^{-2t} - \cos t)}{(\mathrm{e}^{-2t} - 1)(2\mathrm{e}^{-2t} - 1)(\cos t - 1)(2\cos t - 1)}.$$

Computing multiple integrals

Multiple integral problems can also be solved by using the same MATLAB function `int()`. Generally speaking, usually the inner integrals should be carried out first and then outer integrals. However, the sequence of integrals should be observed. In each integration step, the `int()` function can be used. Thus sometimes in certain integration steps, the inner integral may not yield a primitive function, which results in no analytical solution to the overall integral problem. If the sequence of integrals can be changed, analytical solutions may be obtained. Numerical solutions to multiple integral problems will be presented in Section 3.4.3.

Example 3.15 Compute the multiple integrals $\int\cdots\int F(x,y,z)\,\mathrm{d}x^2\mathrm{d}y\mathrm{d}z$ where the integrand $F(x,y,z)$ is defined as

$$-4z\mathrm{e}^{-x^2y-z^2}\left[\cos x^2y - 10yx^2\cos x^2y + 4x^4y^2\sin x^2y + 4x^4y^2\cos x^2y - \sin x^2y\right].$$

Solution In fact, the above $F(x,y,z)$ function was obtained by taking partial derivatives to the function $f(x,y,z)$ defined in Example 3.6. Thus taking inverse operations in this example should restore the same primitive function.

One may integrate once with respect to z, once to y and twice to x. The following results can be obtained through simplification

```
>> syms x y z;
   f0=-4*z*exp(-x^2*y-z^2)*(cos(x^2*y)-10*cos(x^2*y)*y*x^2+...
      4*sin(x^2*y)*x^4*y^2+4*cos(x^2*y)*x^4*y^2-sin(x^2*y));
   f1=int(f0,z); f1=int(f1,y); f1=int(f1,x); f1=simple(int(f1,x))
```

with the primitive function $\sin(x^2y)\mathrm{e}^{-x^2y-z^2}$, which is exactly the same as the function defined in Example 3.6. Now if one alters the order of integration, i.e., change the order to $z \to x \to x \to y$

```
>> f2=int(f0,z); f2=int(f2,x); f2=int(f2,x); f2=simple(int(f2,y))
```

the result becomes $2\dfrac{\mathrm{e}^{-x^2y-z^2}\tan\left(x^2y/2\right)}{1+\tan^2\left(x^2y/2\right)}$. The primitive function obtained does not look the same as the original function in Example 3.6. If one simplifies the difference between the two functions, i.e., `simple(f1-f2)`, it can be seen that the difference is 0, which means that the two functions are identical.

Example 3.16 Compute the definite integral $\displaystyle\int_0^2\int_0^\pi\int_0^\pi 4xz\mathrm{e}^{-x^2y-z^2}\mathrm{d}z\mathrm{d}y\mathrm{d}x$.

Solution The following statements can be given to calculate the triple definite integral

```
>> syms x y z
   int(int(int(4*x*z*exp(-x^2*y-z^2),x,0,2),y,0,pi),z,0,pi)
```

and the results obtained are

```
pi*Ei(1,4*pi)*(1/pi-1/pi*exp(-pi^2))+pi*log(pi)*(1/pi-1/pi*exp(-pi^2))+
pi*eulergamma*(1/pi-1/pi*exp(-pi^2))+2*pi*log(2)*(1/pi-1/pi*exp(-pi^2))
```

where `eulergamma` is the Euler constant γ, $\mathrm{Ei}(n,z)=\displaystyle\int_1^\infty \mathrm{e}^{-zt}t^{-n}\,dt$ is an exponential integral. The integrand is not integrable analytically. However, numerical solutions can be found. Thus the accurate numerical solution to the original problem can be found from `vpa(ans)` command and the integral value is 3.10807940208541272.

3.2 Series Expansions and Series Evaluations

Taylor series expansions to functions with a single variable and multiple variables will be discussed in this section. The Fourier series expansion to given functions are also to be discussed. Summations and products of series are illustrated.

3.2.1 Taylor series expansion

Taylor series expansion of single-variable functions

The Taylor series expansion about the point $x = 0$ can be written as

$$f(x) = a_1 + a_2 x + a_3 x^2 + \cdots + a_k x^{k-1} + o(x^k) \tag{3.9}$$

where the coefficients a_i can be obtained from

$$a_i = \frac{1}{i!} \lim_{x\to 0} \frac{\mathrm{d}^{i-1}}{\mathrm{d}x^{i-1}} f(x), \quad i = 1, 2, 3, \cdots. \tag{3.10}$$

The expansion is also referred to as the *Maclaurin series*. If the Taylor series expansion is made about the $x = a$ point, the series can then be written as

$$f(x) = b_1 + b_2(x-a) + b_3(x-a)^2 + \cdots + b_k(x-a)^{k-1} + o[(x-a)^k] \tag{3.11}$$

where the b_i coefficients can be obtained from

$$b_i = \frac{1}{i!} \lim_{x\to a} \frac{\mathrm{d}^{i-1}}{\mathrm{d}x^{i-1}} f(x), \quad i = 1, 2, 3, \cdots. \tag{3.12}$$

Taylor series expansion can be obtained by the use of the `taylor()` function, provided in the Symbolic Math Toolbox. The syntaxes of the function are

```
taylor(fun,x,k)      % Taylor series about x = 0 point
taylor(fun,x,k,a)    % Taylor series expansion about the x = a point
```

where *fun* is a symbolic expression of the original function and x is the independent variable. If there is only one independent variable in *fun*, x can be omitted. The argument k is the number of terms required in the expansion, with a default number of terms of 6. If an extra argument a is given, the expansion is then made about the $x = a$ point. The Taylor series expansion solutions are demonstrated in the following examples.

Example 3.17 Consider again the function $f(x) = \sin x/(x^2 + 4x + 3)$ given in Example 3.4. Find the first 9 terms of Taylor series expansion about $x = 0$ point. Consider also the series expansions about points $x = 2$ and $x = a$.

Solution The following statements can be used to specify the given function. The first 9 terms of Taylor series expansion can be obtained using the following statements

```
>> syms x; f=sin(x)/(x^2+4*x+3); y1=taylor(f,x,9)
```

and the result is

$$f(x) = \frac{1}{3}x - \frac{4}{9}x^2 + \frac{23}{54}x^3 - \frac{34}{81}x^4 + \frac{4087}{9720}x^5 - \frac{3067}{7290}x^6 + \frac{515273}{1224720}x^7 - \frac{386459}{918540}x^8 + \cdots.$$

In classical calculus courses, no analysis had been made upon the fitting quality of the finite number of terms approximation for a given function, since there were no ready tools available. With the use of MATLAB, the original function as well as the finite term Taylor series approximation can be comapred graphically as shown in Figure 3.4 (a).

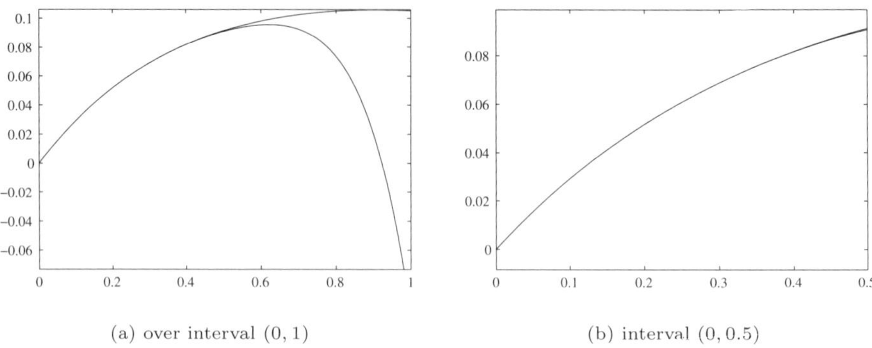

(a) over interval (0, 1) (b) interval (0, 0.5)

FIGURE 3.4: Finite term Taylor series approximation

```
>> ezplot(f,[0,5]), hold on; ezplot(y1,[0,5])
```

It can be seen that the fitting over the specified interval (0,1) is not satisfactory in the sense that when t is close to 1, the fitting is very poor. Thus 9 terms of Taylor series expansion for the original function are not enough. If the interval is changed to $(0, 0.5)$, the fitting quality is shown in Figure 3.4 (b) which is good enough. Thus with the graphical facilities in MATLAB, the fitting qualities can be examined easily.

Now consider the Taylor series expansion about the point $x = 2$. The series can be derived using the following statement:

```
>> taylor(f,x,9,2)
```

Since the expansion is lengthy, only first five terms are shown here

$$f(x) \approx \frac{\sin 2}{15} + \left(\frac{\cos 2}{15} - \frac{8\sin 2}{225}\right)(x-2) - \left(\frac{127\sin 2}{6750} + \frac{8\cos 2}{225}\right)(x-2)^2$$
$$+ \left(\frac{23\cos 2}{6750} + \frac{628\sin 2}{50625}\right)(x-2)^3 + \left(-\frac{15697}{6075000}\sin(2) + \frac{28}{50625}\cos(2)\right)(x-2)^4.$$

If one wants to find the series expansion about the $x = a$ point, Taylor series expansion can still be derived using similar statements

```
>> syms a; taylor(f,x,9,a)
```

Here only the first three terms are shown

$$\frac{\sin a}{a^2+3+4a} + \left[\frac{\cos a}{a^2+3+4a} - \frac{(4+2a)\sin a}{(a^2+3+4a)^2}\right](x-a) + \left[-\frac{\sin a}{(a^2+3+4a)^2}\right.$$
$$\left. - \frac{\sin a}{2(a^2+3+4a)} - \frac{\left(a^2\cos a + 3\cos a + 4a\cos a - 4\sin a - 2a\sin a\right)(4+2a)}{(a^2+3+4a)^3}\right](x-a)^2.$$

Example 3.18 Expand the sinusoidal function $y = \sin x$ into Taylor series, and compare the approximation quality for different terms.

Solution In order to find out the relationship between the fitting quality and the number of terms used, the loop structure should be used. The following statements can be issued to solve the problem, where the fitting curves shown in Figure 3.5 can be obtained.

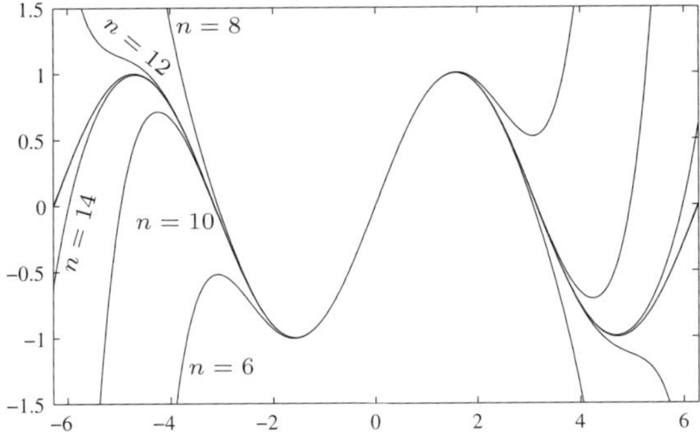

FIGURE 3.5: Taylor series approximation to given sinusoidal functions

```
>> x0=-2*pi:0.01:2*pi; y0=sin(x0); syms x; y=sin(x);
   plot(x0,y0), axis([-2*pi,2*pi,-1.5,1.5]);
   for n=[8:2:20]
      p=taylor(y,x,n), y1=subs(p,x,x0); line(x0,y1)
   end
```

For fewer terms, the satisfactory fitting interval is small. If the number of terms is increased, the satisfactory fitting interval will also increase. For instance, if one selects $n = 16$, the fitting is satisfactory over the interval $(-2\pi, 2\pi)$. The first 20 terms in the Taylor series expansion are obtained as

$$\sin x \approx x - \frac{1}{6}x^3 + \frac{1}{120}x^5 - \frac{1}{5040}x^7 + \frac{1}{362880}x^9 - \frac{1}{39916800}x^{11} + \frac{1}{6227020800}x^{13} - \frac{1}{1307674368000}x^{15} + \frac{1}{355687428096000}x^{17} - \frac{1}{121645100408832000}x^{19}.$$

Taylor series expansion of multivariable functions

The Taylor series expansion of a multivariable function $f(x_1, x_2, \cdots, x_n)$ is

$$\begin{aligned} f(x_1,\cdots,x_n) &= f(a_1,\cdots,a_n)+ \\ &\left[(x_1-a_1)\frac{\partial}{\partial x_1}+\cdots+(x_n-a_n)\frac{\partial}{\partial x_n}\right]f(a_1,\cdots,a_n)+ \\ &\frac{1}{2!}\left[(x_1-a_1)\frac{\partial}{\partial x_1}+\cdots+(x_n-a_n)\frac{\partial}{\partial x_n}\right]^2 f(a_1,\cdots,a_n)+\cdots+ \\ &\frac{1}{k!}\left[(x_1-a_1)\frac{\partial}{\partial x_1}+\cdots+(x_n-a_n)\frac{\partial}{\partial x_n}\right]^k f(a_1,\cdots,a_n)+\cdots, \end{aligned} \tag{3.13}$$

where $(a_1, \cdots, a_n)$ is the center point of Taylor series expansion. In order to avoid misunderstanding, the terms can be regarded as the derivatives of function *fun*. Then the function evaluation can be made to the point $(a_1, a_2, \cdots, a_n)$. There is no existing function provided in the Symbolic Math Toolbox of MATLAB. However, the `mtaylor()` function in Maple can be used instead. The Taylor series expansion to multivariable functions can be obtained from

```
F=maple('mtaylor',fun,'[x_1,···,x_n]',k)              % about the origin
F=maple('mtaylor',fun,'[x_1=a_1,···,x_n=a_n]',k)      % about (a_1,···,a_n)
```

where $k-1$ is the highest degree in the expansion, and *fun* is the multivariable function. It should be noted that the quotation marks cannot be omitted since the information within the quotation marks will be passed to the Maple function directly.

Example 3.19 Consider again the function $z = f(x, y) = (x^2 - 2x)\mathrm{e}^{-x^2-y^2-xy}$ shown in Example 3.5. Find its Taylor series expansion.

Solution The following statements can be used to get the Taylor series expansion about the origin

```
>> syms x y; f=(x^2-2*x)*exp(-x^2-y^2-x*y);
   F=maple('mtaylor',f,'[x,y]',9); collect(F,x) % collect the polynomial
```

whose mathematical representation is

$$f(x,y) = -\frac{1}{6}x^8 + \left(-\frac{1}{2}y + \frac{1}{3}\right)x^7 + \left(-y^2 + y + \frac{1}{2}\right)x^6 + \left(-\frac{7}{6}y^3 + 2y^2 - 1 + y\right)x^5$$
$$+ \left(-y^4 - 1 - 2y + \frac{7}{3}y^3 + \frac{3}{2}y^2\right)x^4 + \left(2 + 2y^4 - y - \frac{1}{2}y^5 + y^3 - 3y^2\right)x^3$$
$$+ \left(2y + 1 + \frac{1}{2}y^4 - \frac{1}{6}y^6 - 2y^3 - y^2 + y^5\right)x^2 + \left(-2 - y^4 + 2y^2 + \frac{1}{3}y^6\right)x.$$

If one wants to expand the original function about $x = 1, y = a$ point, the following statements can be used

```
>> syms a; F=maple('mtaylor',f,'[x=1,y=a]',5);
```

and the expansion can be found as

$$f(x,y) = -\mathrm{e}^{-1-a-a^2} - \mathrm{e}^{-1-a-a^2}(-2-a)(x-1) - \mathrm{e}^{-1-a-a^2}(-2a-1)(y-a) +$$
$$\left[-\mathrm{e}^{-1-a-a^2}\left(1 + 2a + \frac{a^2}{2}\right) + \mathrm{e}^{-1-a-a^2}\right](x-1)^2 -$$
$$\mathrm{e}^{-1-a-a^2}(5a+1+2a^2)(y-a)(x-1) - \mathrm{e}^{-1-a-a^2}\left(-\frac{1}{2}+2a+2a^2\right)(y-a)^2 + \cdots.$$

In fact, the Maple function `mtaylor()` can also be used in evaluating the Taylor series expansion for single variable functions. The function call is almost as simple as `taylor()` function

```
>> F=maple('mtaylor',f,'[x=a]',5);
```

and the result obtained is

$$f(x,y) = (a^2-2a)\mathrm{e}^{-a^2-y^2-ay} + \left[(a^2-2a)(-2a-y) + (2a-2)\right]\mathrm{e}^{-a^2-y^2-ay}(x-a) +$$
$$\left[(a^2-2a)(-1+2a^2+2ay+y^2/2) + 1 + (2a-2)(-2a-y)\right]\mathrm{e}^{-a^2-y^2-ay}(x-a)^2 + \cdots.$$

3.2.2 Fourier series expansion

Consider a periodic function $f(x)$ defined over the interval $x \in [-L, L]$. The function is with a period of $T = 2L$. For the function defined on other intervals, it can be extended to periodic functions such that $f(x) = f(kT+x)$, where k is an arbitrary integer. A given function $f(x)$ can be expressed as an infinite series such that

$$f(x) = \frac{a_0}{2} + \sum_{n=1}^{\infty} \left(a_n \cos \frac{n\pi}{L} x + b_n \sin \frac{n\pi}{L} x \right) \tag{3.14}$$

where

$$\begin{cases} a_n = \dfrac{1}{L} \displaystyle\int_{-L}^{L} f(x) \cos \frac{n\pi x}{L} \, \mathrm{d}x, \quad n = 0, 1, 2, \cdots \\ b_n = \dfrac{1}{L} \displaystyle\int_{-L}^{L} f(x) \sin \frac{n\pi x}{L} \, \mathrm{d}x, \quad n = 1, 2, 3, \cdots . \end{cases} \tag{3.15}$$

Such a series is referred to as the *Fourier series* and a_n, b_n are referred to as *Fourier coefficients*. If the function is defined over $x \in (a, b)$, it can be found that $L = (b - a)/2$. One may introduce a new variable $\hat{x}$ such that $x = \hat{x} + L + a$, the function $f(\hat{x})$ can be mapped into the symmetrical interval $(-L, L)$. Fourier series can be established for the new transformed function. Then the variable substitution $\hat{x} = x - L - a$ can be used to map the series back to the function of x.

There is no existing function for Fourier series expansion provided in MATLAB and Maple. Thus based on the above formula, the following MATLAB function can be prepared and placed in the `@sym` directory

```
function [A,B,F]=fseries(f,x,p,a,b)
if nargin==3, a=-pi; b=pi; end
L=(b-a)/2; if a+b, f=subs(f,x,x+L+a); end
A=int(f,x,-L,L)/L; B=[]; F=A/2;
for n=1:p
   an=int(f*cos(n*pi*x/L),x,-L,L)/L;
   bn=int(f*sin(n*pi*x/L),x,-L,L)/L; A=[A, an]; B=[B,bn];
   F=F+an*cos(n*pi*x/L)+bn*sin(n*pi*x/L);
end
if a+b, F=subs(F,x,x-L-a); end
```

The syntax of the function is `[A,B,F]=fseries(f,x,p,a,b)`, where f is the given function; x is the independent variable; p is number of the terms required in the expansion and (a, b) is the interval for x. If a, b arguments are omitted, the default interval $[-\pi, \pi]$ will be used. The returned arguments $\boldsymbol{A}$, $\boldsymbol{B}$ contain the Fourier coefficients, F is the Fourier series expansion obtained. Similar to the analytical function `fseries()`, the numerical version can also be written easily.

Example 3.20 Find the Fourier series expansion to the function $y = x(x - \pi)(x - 2\pi)$, where $x \in (0, 2\pi)$.

Solution The Fourier series for the given function can easily be expressed

```
>> syms x; f=x*(x-pi)*(x-2*pi); [A,B,F]=fseries(f,x,12,0,2*pi)
```

where the first 12 terms in the Fourier series are as follows:

$$f(x) = 12\sin x + \frac{3\sin 2x}{2} + \frac{4\sin 3x}{9} + \frac{3\sin 4x}{16} + \frac{12\sin 5x}{125} + \frac{\sin 6x}{18} + \frac{12\sin 7x}{343} + \frac{3\sin 8x}{128} + \frac{4\sin 9x}{243} + \frac{3\sin 10x}{250} + \frac{12\sin 11x}{1331} + \frac{\sin 12x}{144}.$$

From these results, the analytical form can be summarized as $f(x) = \sum_{n=1}^{\infty} \frac{12}{n^3}\sin nx$.

The first 12 terms in the Fourier series expansion and the original function can be graphically compared as shown in Figure 3.6 (a) with the following statements

```
>> ezplot(f,[0,2*pi]), hold on, ezplot(F,[0,2*pi])
```

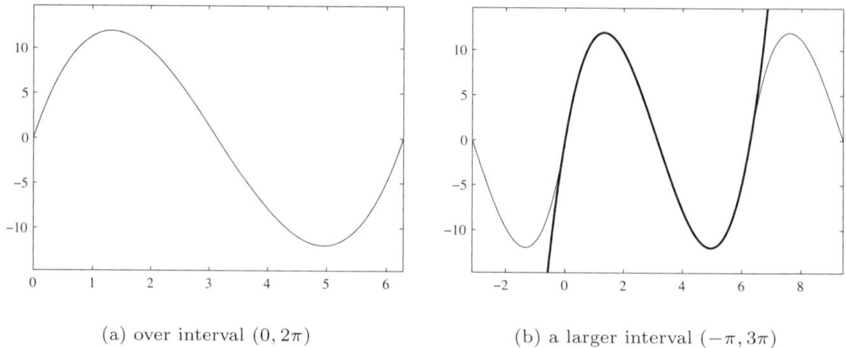

(a) over interval $(0, 2\pi)$ (b) a larger interval $(-\pi, 3\pi)$

FIGURE 3.6: Accuracy of finite term Fourier series approximation

If one wants to further examine the approximation over a larger interval $x \in (-\pi, 3\pi)$, the following statements should be used

```
>> ezplot(f,[-pi,3*pi]), hold on, ezplot(F,[-pi,3*pi])
```

and the curves are shown in Figure 3.6 (b). It can be seen that over the $(0, 2\pi)$ interval the fitting is quite good. In other regions, since the Fourier series is made upon the assumption that it is periodically extended, thus it cannot approximate the original function in other intervals at all.

Example 3.21 Now consider a square wave defined over the interval $(-\pi, \pi)$, where $y = 1$ when $x \geqslant 0$, and $y = -1$ otherwise. Expand the function using Fourier series and observe how many terms in the function may give good approximation.

Solution Since in symbolic expressions inequality cannot be used, the square wave can be expressed as $f(x) = |x|/x$. In this way, the numerical and analytical expressions in Fourier series can be obtained for different terms in the expression. The curves can be obtained as shown in Figure 3.7 (a).

```
>> syms x; f=abs(x)/x;  % square wave definition
   xx=[-pi:pi/200:pi]; xx=xx(xx~=0); xx=sort([xx,-eps,eps]); % remove 0
   yy=subs(f,x,xx); plot(xx,yy), hold on  % draw the original function
```

```
for n=1:20
   [a,b,f1]=fseries(f,x,n); y1=subs(f1,x,xx); plot(xx,y1)
end
```

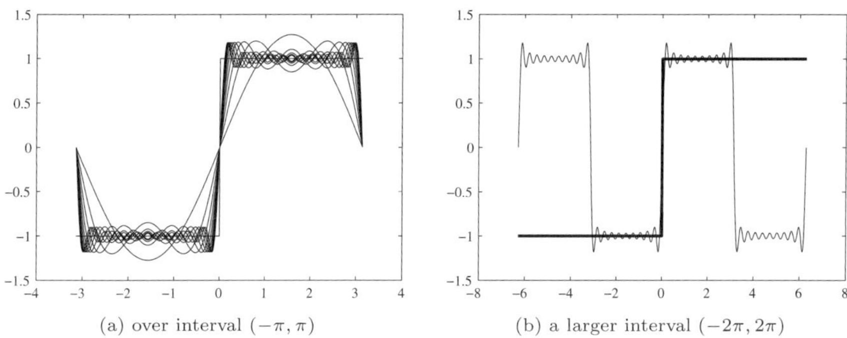

(a) over interval $(-\pi, \pi)$ (b) a larger interval $(-2\pi, 2\pi)$

FIGURE 3.7: Approximation of square wave by Fourier series

It can be seen that when 10 terms are used, the approximation is satisfactory. Even if the number of terms increases, the fitting accuracy may not be improved significantly. A finite Fourier series of the original function can be obtained by

```
>> [a,b,f1]=fseries(f,x,14); f1
```

and the expansion can be written as

$$f(x) \approx 4\frac{\sin x}{\pi} + \frac{4\sin 3x}{3\pi} + \frac{4\sin 5x}{5\pi} + \frac{4\sin 7x}{7\pi} + \frac{4\sin 9x}{9\pi} + \frac{4\sin 11x}{11\pi} + \frac{4\sin 13x}{13\pi}$$

which can further be summarized as $f(x) = \dfrac{4}{\pi}\sum_{k=1}^{\infty}\dfrac{\sin(2k-1)x}{2k-1}$.

Again the Fourier series expansion is established upon the assumption that it is periodically extended over the original function, thus the fitting in other intervals may be incorrect, as shown in Figure 3.7 (b).

```
>> xx=[-2*pi:pi/200:2*pi]; xx=xx(xx~=0); xx=sort([xx,-eps,eps]);
   yy=subs(f,x,xx); plot(xx,yy), y1=subs(f1,x,xx); line(xx,y1)
```

3.2.3 Series

The function `symsum()` provided in the Symbolic Math Toolbox can be used to evaluate the finite and infinite series with known general terms. The syntax of the function is `S=symsum(`f_k`,`k`,`k_0`,`k_n`)`, where f_k is the general term of the series, k is the independent term, and k_0 and k_n are the initial and final terms of the series, respectively. They can be set to `inf` for infinite series. The series can be written as

$$S = \sum_{k=k_0}^{k_n} f_k. \tag{3.16}$$

If there is only one independent variable defined in f_k, the variable k can be omitted in the function call.

Example 3.22 Compute the finite sum $S = 2^0 + 2^1 + 2^2 + \cdots + 2^{62} + 2^{63} = \sum_{i=0}^{63} 2^i$.

Solution Numerical solution to the problem can be found from

```
>> format long; s=sum(2.^[0:63])
```

with $s = 1.844674407370955 \times 10^{19}$. Since the data type of `double` is used, only 16 digits can be reserved. Thus the exact result cannot be obtained under double-precision scheme. The function `symsum()` can be used to solve the problem

```
>> syms k; symsum(2^k,0,63)
```

where $s_1 = 18446744073709551615$ can be obtained. The problem can even be solved with a simpler command `sum(sym(2).^[0:63])`, and the same result can be obtained. The method can be extended to calculate for more terms, for instance, it is possible to calculate the sum to 201 terms

```
>> s2=symsum(2^k,0,200)
```

and $s_2 = 3213876088517980551083924184682325205044405987565585670602751$. The exact solution cannot possibly be obtained using the double-precision data type.

Example 3.23 Compute the infinite series

$$S = \frac{1}{1 \times 4} + \frac{1}{4 \times 7} + \frac{1}{7 \times 10} + \cdots + \frac{1}{(3n-2)(3n+1)} + \cdots.$$

Solution With the use of the symbolic function

```
>> syms n; s=symsum(1/((3*n-2)*(3*n+1)),n,1,inf)
```

the sum result $s = 1/3$ can be obtained. The same problem can be tried using numerical method with `double` data type. For instance, if 10,000,000 terms are selected to be added up, the following statements can be used directly

```
>> m=1:10000000; s1=sum(1./((3*m-2).*(3*m+1))); format long; s1
```

and the sum is $s_1 = 0.33333332222165$. It can be seen that although a very large number of terms are selected with much long time consumed, there still exists unavoidable difference and the error reaches 10^{-6} level. It can be seen that when $m = 10^7$, the value of the general term is around 10^{-15}, thus it seems that the additional error in the summation may not be very large. In fact, since double-precision data type is used, some of the terms may not be added to the S variable. Thus even though more terms are used in the summation, the accuracy cannot be further increased.

Example 3.24 Evaluate the infinite series with an extra variable x.

$$J = 2\sum_{n=0}^{\infty} \frac{1}{(2n+1)(2x+1)^{2n+1}}.$$

Solution In the examples studied earlier, numerical methods can be used to find the approximate solutions. If in the general term, extra independent variables are involved, numerical methods can no longer be used. Symbolic method has to be used to solve the problem. For instance, the sum can be evaluated with

```
>> syms n x; s1=symsum(2/((2*n+1)*(2*x+1)^(2*n+1)),n,0,inf);
   simple(s1)
```

and the infinite sum is $s_1 = \ln[(x+1)/x]$.

Example 3.25 Solve the limit problem with the series

$$\lim_{n\to\infty}\left[\left(1+\frac{1}{2}+\frac{1}{3}+\frac{1}{4}+\cdots+\frac{1}{n}\right)-\ln n\right].$$

Solution So far, the series and limit problems have been discussed and illustrated separately. For this mixed problem, the following MATLAB statements can be used to solve it, where the finite sum should be made first using `symsum(1/m,m,1,n)`

```
>> syms m n; limit(symsum(1/m,m,1,n)-log(n),n,inf)
```

and `eulergamma` can be obtained, i.e., the Euler constant γ can be obtained whose value can be evaluated with `vpa(ans)` such that $\gamma = 0.57721566490153286060651209$.

It should be noted that in the computation, one should not evaluate the infinite sum before limit. Otherwise the original problem cannot be correctly solved.

3.2.4 Sequence product

The computation of sequence product $\prod_{n=a}^{b} f(n)$ is not directly supported in MATLAB. We can call Maple function `product()` from MATLAB to solve the product problem. The syntaxes of the function are

`maple('product(fun, n=a..b)')` or `maple('product',fun,'n=a..b')`

Example 3.26 Calculate the sequence product $\displaystyle\prod_{i=2}^{\infty}\left(1-\frac{2}{i(i+1)}\right)$.

Solution The sequence product is simply 1/3 .

```
>> syms i; maple('product',1-2/i/(i+1),'i=2..Inf')
```

3.3 Numerical Differentiation

If the original function is symbolically given, the analytical solutions to the differentiation problem can be obtained directly with the MATLAB built-in function `diff()`. The 100th order derivative can be obtained within seconds. However, in some applications where the original function is not known, only experimental data are given, the analytical or symbolic methods cannot be used. In this case, numerical methods must be used to get the derivatives from the experimental data. There is no dedicated function available in solving numerical differentiation problems in MATLAB. Thus, simple numerical algorithms are presented in this section with detailed implementation of the algorithms together with examples on how to solve the numerical differentiation problems in MATLAB.

3.3.1 Numerical differentiation algorithms

Assume that there is a set of measured data (t_i, y_i) with evenly distributed time instances $t_i = i\Delta t$, $i = 1, \cdots, N$, and the sampling period is Δt. The approximate derivative of the function can be defined as

$$y_i' \approx \frac{\Delta y_i}{\Delta t}; \quad y_i' = \frac{y_{i+1} - y_i}{\Delta t} + o(\Delta t). \tag{3.17}$$

This formula is also referred to as the *forward difference algorithm.*

Similarly, *backward difference formula* is defined as

$$y_i' \approx \frac{\Delta y_i}{\Delta t}; \quad y_i' = \frac{y_i - y_{i-1}}{\Delta t} + o(\Delta t). \tag{3.18}$$

From calculus, it is known that when $\Delta t \to 0$, the forward and backward formula can be used to solve analytically the differentiation problem. However, unfortunately, in practical applications, the condition $\Delta t \to 0$ cannot be satisfied. When the value of Δt is large, the accuracy of the differentiation cannot be guaranteed. So other improved numerical differentiation algorithms should be considered. For instance, the *central-point algorithm* can be used. The first-order derivative can also be defined as

$$y_i' \approx \frac{\Delta y_i}{\Delta t}; \quad y_i' = \frac{y_{i+1} - y_{i-1}}{2\Delta t}. \tag{3.19}$$

Denote $\widetilde{f}'(x) = \dfrac{f(x+\Delta t) - f(x-\Delta t)}{2\Delta t}$. From Taylor series expansion, the above method can further be written as

$$\begin{aligned} \widetilde{f}'(x) = & \frac{f(x) + \Delta t f'(x) + \Delta t^2 f''(x)/2! + \Delta t^3 f''(\xi)/3! + o(\Delta t^4)}{2\Delta t} - \\ & \frac{f(x) - \Delta t f'(x) + \Delta t^2 f''(x)/2! - \Delta t^3 f''(\xi)/3! + o(\Delta t^4)}{2\Delta t} \\ = & f'(x) + \frac{\Delta t^3}{3!} f''(\xi). \end{aligned} \tag{3.20}$$

It can be shown that the precision of the numerical approximation algorithm of first order differentiation is $o(\Delta t^2)$. High-order differentiation formulae can be similarly derived as follows:

$$\begin{aligned} y_i'' &\approx \frac{y_{i+1} - 2y_i + y_{i-1}}{\Delta t^2} \\ y_i''' &\approx \frac{y_{i+2} - 2y_{i+1} + 2y_{i-1} - y_{i-2}}{2\Delta t^3} \\ y_i^{(4)} &\approx \frac{y_{i+2} - 4y_{i+1} + 6y_i - 4y_{i-1} + y_{i-2}}{\Delta t^4}. \end{aligned} \tag{3.21}$$

There is yet another set of central-point difference algorithms with even higher accuracy of $o(\Delta t^4)$, defined as follows:

$$\begin{aligned} y_i' &\approx \frac{-y_{i+2} + 8y_{i+1} - 8y_{i-1} + y_{i-2}}{12\Delta t} \\ y_i'' &\approx \frac{-y_{i+2} + 16y_{i+1} - 30y_i + 16y_{i-1} - y_{i-2}}{12\Delta t^2} \\ y_i''' &\approx \frac{-y_{i+3} + 8y_{i+2} - 13y_{i+1} + 13y_{i-1} - 8y_{i-2} + y_{i-3}}{8\Delta t^3} \\ y_i^{(4)} &\approx \frac{-y_{i+3} + 12y_{i+2} - 39y_{i+1} + 56y_i - 39y_{i-1} + 12y_{i-2} - y_{i-3}}{6\Delta t^4} \end{aligned} \tag{3.22}$$

3.3.2 Central-point difference algorithm with MATLAB implementation

The numerical differentiation algorithm given in (3.22) has the error level of $o(\Delta t^4)$ which can be used to solve numerical differentiation problems with higher numerical accuracy. Even when Δt is not too small, good approximation can still be expected due to its error level. Based on the algorithm, a MATLAB function is prepared as follows:

```
function [dy,dx]=diff_ctr(y,Dt,n)
yx1=[y 0 0 0 0 0]; yx2=[0 y 0 0 0 0]; yx3=[0 0 y 0 0 0];
yx4=[0 0 0 y 0 0]; yx5=[0 0 0 0 y 0]; yx6=[0 0 0 0 0 y];
switch n
case 1
   dy = (-diff(yx1)+7*diff(yx2)+7*diff(yx3)-diff(yx4))/(12*Dt); L0=3;
case 2
   dy=(-diff(yx1)+15*diff(yx2)-15*diff(yx3)+diff(yx4))/(12*Dt^2);L0=3;
case 3
   dy=(-diff(yx1)+7*diff(yx2)-6*diff(yx3)-6*diff(yx4)+...
         7*diff(yx5)-diff(yx6))/(8*Dt^3); L0=5;
case 4
   dy = (-diff(yx1)+11*diff(yx2)-28*diff(yx3)+28*diff(yx4)-...
         11*diff(yx5)+diff(yx6))/(6*Dt^4);L0=5;
end
dy=dy(L0+1:end-L0); dx=([1:length(dy)]+L0-2-(n>2))*Dt;
```

The syntax of the function is `[`d_y`,`d_x`]=diff_ctr(`y`,`Δt`,`n`)`, where $\boldsymbol{y}$ is the vector containing measured data for evenly distributed points, and Δt is the sampling period. The argument n specifies the order of derivatives. The returned arguments $\boldsymbol{d}_y$ is the derivative vector computed, while the argument $\boldsymbol{d}_x$ is the corresponding vector of independent variables. It should be noted that the two vectors are a few points shorter than the original $\boldsymbol{y}$ vector.

Example 3.27 The function defined in Example 3.4 is still used in the demonstration of the algorithm. Since the original function is known, the analytical solution can be obtained for comparison. Sample data of the function can be generated from the function, and with the help of the data, the derivatives of the first- up to the fourth-order can be calculated and the results can be compared with the analytical solutions.

Solution An evenly spaced vector $\boldsymbol{x}$ is generated first. Since the original function is known, the analytical solutions to derivatives can be obtained. Then, if one substitutes the vector $\boldsymbol{x}$ into the obtained analytical functions, the theoretical derivative vectors can be obtained for comparison.

```
>> h=0.05; x=0:h:pi; syms x1; y=sin(x1)/(x1^2+4*x1+3);
   yy1=diff(y); f1=subs(yy1,x1,x); % get the contrast data analytically
   yy2=diff(yy1); f2=subs(yy2,x1,x); yy3=diff(yy2); f3=subs(yy3,x1,x);
   yy4=diff(yy3); f4=subs(yy4,x1,x);
```

From the data points y_i generated above, the first-order up to the fourth-order derivatives from the data can be calculated easily with the function `diff_ctr()` and the results are shown in Figure 3.8, together with the exact solutions. It can be seen that one may not observe the difference.

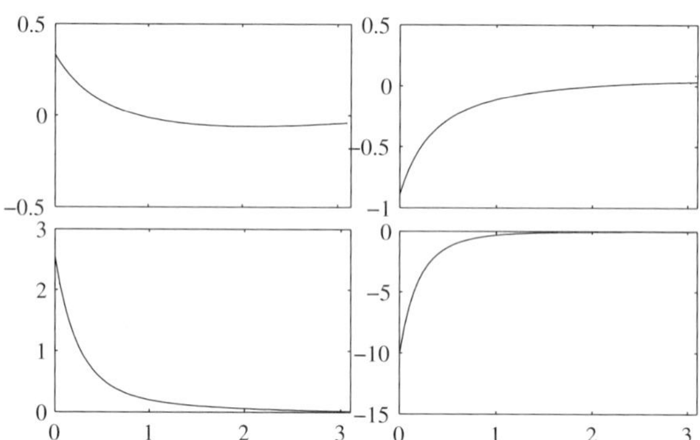

FIGURE 3.8: Comparisons of derivatives of different orders

```
>> y=sin(x)./(x.^2+4*x+3);     % generate the data to be used
   [y1,dx1]=diff_ctr(y,h,1); subplot(221), plot(x,f1,dx1,y1,':');
   [y2,dx2]=diff_ctr(y,h,2); subplot(222), plot(x,f2,dx2,y2,':')
   [y3,dx3]=diff_ctr(y,h,3); subplot(223), plot(x,f3,dx3,y3,':');
   [y4,dx4]=diff_ctr(y,h,4); subplot(224), plot(x,f4,dx4,y4,':')
```

Quantitative studies for the fourth-order derivative show that the maximum error between the exact results and the calculated results is as small as 3.5025×10^{-4}.

```
>> norm((y4-f4(4:60))./f4(4:60))
```

3.3.3 Gradient computations of functions with two variables

Consider the function $z(x,y)$ with two variables representing a 3D surface. The function `gradient()` can be used to calculate the gradients for the function. The syntax of the function is `[`f_x`,`f_y`]=gradient(`z`)`, where the "gradients" f_x and f_y thus calculated are not the actual gradients, since the coordinates x and y are not considered. If the matrix z is obtained, the gradients can be obtained using the following statements f_x`=`f_x`/`Δx, f_y`=`f_y`/`Δy, where Δx and Δy are respectively the step-sizes for x and y.

Example 3.28 Consider the function given in Example 3.5. Assume that the mesh grid data can be generated. Compute the gradients of the original function and analyze the error.

Solution The data can be generated using the following statements. The gradients here are obtained from the data rather than from the analytical function. The 3D attractive curves can also be drawn as shown in Figure 3.9 and it should be the same as the one in Figure 3.3 (b).

```
>> syms x y; z=(x^2-2*x)*exp(-x^2-y^2-x*y);
   [x0,y0]=meshgrid(-3:.2:3,-2:.2:2); z0=subs(z,{x,y},{x0,y0});
   [fx,fy]=gradient(z0); fx=fx/0.2; fy=fy/0.2;
   contour(x0,y0,z0,30); hold on; quiver(x0,y0,fx,fy)
```

The error surface is shown in Figure 3.9 where it can be seen that in most regions, the errors are relatively small. In other areas, the errors are large. This means that the spacing in the grid is too large to provide accurate gradient information. In order to reduce the error, the step-size should be reduced.

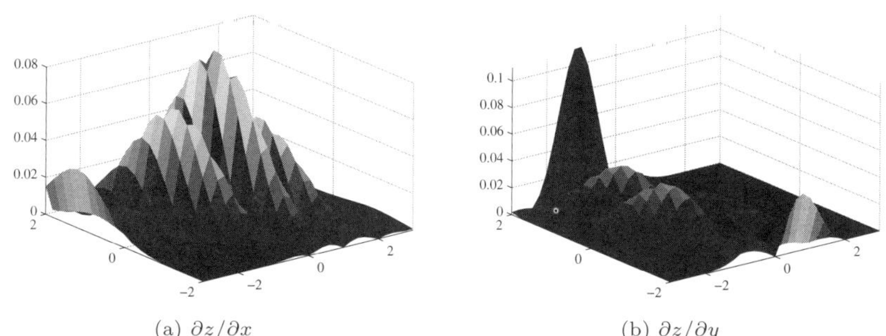

(a) $\partial z/\partial x$ (b) $\partial z/\partial y$

FIGURE 3.9: Error surface of the gradient of function with two variables

```
>> zx=diff(z,x); zx0=subs(zx,{x,y},{x0,y0});
   zy=diff(z,y); zy0=subs(zy,{x,y},{x0,y0});
   surf(x0,y0,abs(fx-zx0)); axis([-3 3 -2 2 0,0.08])
   figure; surf(x0,y0,abs(fy-zy0)); axis([-3 3 -2 2 0,0.11])
```

If the spacing in grids is reduced both by half, the following statements can be used and the new error surface can be calculated again as shown in Figure 3.10. It can be observed that the error is also reduced compared to Figure 3.9.

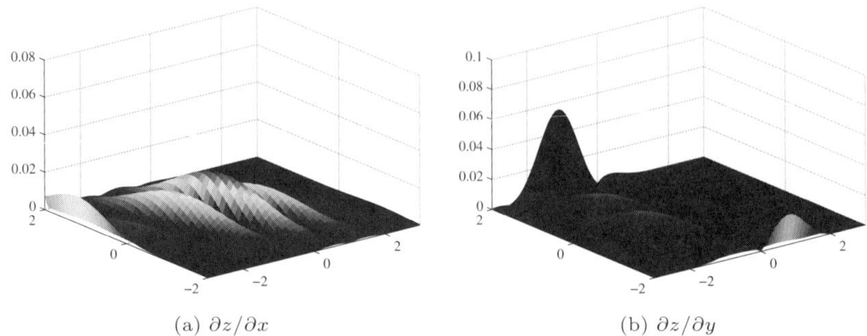

(a) $\partial z/\partial x$ (b) $\partial z/\partial y$

FIGURE 3.10: The error surface with reduced spacing in mesh grids

```
>> [x1,y1]=meshgrid(-3:.1:3,-2:.1:2); z1=subs(z,{x,y},{x1,y1});
   [fx,fy]=gradient(z1); fx=fx/0.1; fy=fy/0.1;
   zx1=subs(zx,{x,y},{x1,y1}); zy1=subs(zy,{x,y},{x1,y1});
   surf(x1,y1,abs(fx-zx1)); axis([-3 3 -2 2 0,0.08])
   figure; surf(x1,y1,abs(fy-zy1)); axis([-3 3 -2 2 0,0.1])
```

3.4 Numerical Integration Problems

3.4.1 Numerical integration from given data using trapezoidal method

Definite integral of the function with a single variable is defined as

$$I = \int_a^b f(x)\,\mathrm{d}x. \tag{3.23}$$

It is known that if the integrand $f(x)$ is theoretically not integrable, even with the powerful computer program, the analytical solutions to the problem cannot be obtained. Thus numerical solutions to the problems should be pursued instead. Numerical computation of an integral of single-variable function is also known as *quadrature*. There are various numerical quadrature algorithms to solve the integration problem. The widely used algorithms include the trapezoidal method, the Simpson's algorithm, the Romberg's algorithm, etc. The basic idea of the algorithms is to divide the whole interval $[a, b]$ into several sub-intervals $[x_i, x_{i+1}]$, $i = 1, 2, \cdots, N$, where $x_1 = a$ and $x_{N+1} = b$. Then the integration problem can be converted to the summation problem as follows:

$$\int_a^b f(x)\,\mathrm{d}x = \sum_{i=1}^{N} \int_{x_i}^{x_{i+1}} f(x)\,\mathrm{d}x = \sum_{i=1}^{N} \Delta f_i. \tag{3.24}$$

The easiest method is to use trapezoidal approximation to each sub-interval. The numerical integration can be obtained by the use of `trapz()` function, whose syntax is `S=trapz(x,y)`, where $\boldsymbol{x}$ is a vector, and the number of rows of matrix $\boldsymbol{y}$ equals the number of the elements in vector $\boldsymbol{x}$. If the variable $\boldsymbol{y}$ is given as a multi-column matrix, the numerical integration to several functions can be evaluated simultaneously.

Example 3.29 Compute the definite integrals to the functions $\sin x, \cos x, \sin x/2$ within the interval $x \in (0, \pi)$ using the trapezoidal algorithm.

Solution The vector for horizontal axis is generated first and from it, the values of different functions can be evaluated such that the numerical integration can be obtained

```
>> x1=[0:pi/30:pi]'; y=[sin(x1) cos(x1) sin(x1/2)]; S=trapz(x1,y)
```

and the results are $S = [1.99817196134365, 0, 1.99954305299081]$.

Since the step-size is selected as $h = \pi/30 \approx 0.1$ which is considered as quite large, there exist errors in the results. In Section 8.1.2, the algorithm will be used with interpolation method to improve the quality of numerical integration results.

Example 3.30 Compute $\displaystyle\int_0^{3\pi/2} \cos 15x\,\mathrm{d}x$ with various step-sizes.

Solution Before solving the problem, the following statements can be used to draw the curves of the integrand as shown in Figure 3.11. It can be seen that there exists strong oscillation in the integrand.

```
>> x=[0:0.01:3*pi/2, 3*pi/2]; % the vector is assigned to ensure that
   y=cos(15*x); plot(x,y)       % the 3π/2 point is included
```

The theoretical solution to the problem is $1/15$. For different step-sizes, $h =$ 0.1, 0.01, 0.001, 0.0001, 0.00001, 0.000001, the following statements can be used in solving approximately the integrals. The relevant results are given in Table 3.1.

```
>> syms x, A=int(cos(15*x),0,3*pi/2)
   h0=[0.1,0.01,0.001,0.0001,0.00001,0.000001]; v=[]; H=3*pi/2;
   for h=h0,
      x=[0:h:H,H]; y=cos(15*x); I=trapz(x,y); v=[v; h,I,1/15-I];
   end
```

It can be seen that when the step-size h reduces, so does the integral accuracy. For instance, if the step-size is selected as $h = 10^{-6}$, 11 digits can be preserved in

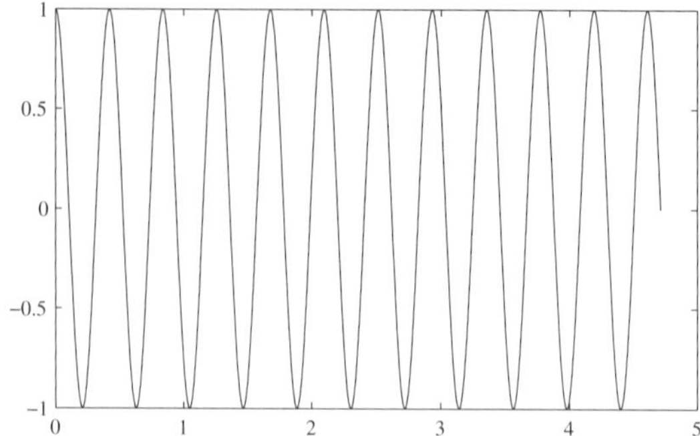

FIGURE 3.11: The plot of the integrand $f(x) = \cos 15x$

TABLE 3.1: Step-size selection and computation results

step	integral	error	time (s)	step	integral	error	time (s)
0.1	0.053891752	0.0127749	0.000	0.0001	0.066666654	1.25×10^{-8}	0.008
0.01	0.0665417	0.000125	0.005	10^{-5}	0.06666667	1.25×10^{-10}	0.033
0.001	0.066665417	1.25×10^{-6}	0.241	10^{-6}	0.066666667	1.25×10^{-12}	8.357

the result. Thus for this example, it takes as long as eight seconds for computation. If the step-size is further reduced, the computational effort demanded will be too high to be accepted.

3.4.2 Numerical integration of single variable functions

In traditional numerical analysis courses, several other numerical algorithms are usually explored for single variable functions. For instance, the approximate solutions Δf_i to the numerical integration problem can be solved with the Simpson's algorithm within the $[x_i, x_{i+1}]$ interval

$$\Delta f_i \approx \frac{h_i}{12}\left[f(x_i)+4f\left(x_i+\frac{h_i}{4}\right)+2f\left(x_i+\frac{h_i}{2}\right)+4f\left(x_i+\frac{3h_i}{4}\right)+f(x_i+h_i)\right] \tag{3.25}$$

where $h_i = x_{i+1} - x_i$. Based on the algorithm, a function `quad()` is provided in MATLAB to implement the variable-step-size Simpson's algorithm. The syntax of the function is

```
[y,k]=quad(fun,a,b)     % evaluate definite integral
[y,k]=quad(fun,a,b,ϵ)   % ibid with user-specified error
```

where *fun* can be used to specify the integrand. It can either be an M-file saved in `fun.m` file, or an anonymous function or an inline function. The syntax of such a function should be `y=fun(x)`. The arguments a and b are the lower- and upper-bounds in the definite integral, respectively. The argument ϵ is the user specified error tolerance, with a default value of 10^{-6}. The argument k returns the number of integrand function calls. With the information provided, the function `quad()` can be used directly to solve the numerical integration problem.

Example 3.31 For the integral $\operatorname{erf}(x) = \dfrac{2}{\sqrt{\pi}}\displaystyle\int_0^x \mathrm{e}^{-t^2}\mathrm{d}t$, which was shown not integrable, compute the integral using numerical methods.

Solution Before finding the numerical integration of a given function, the integrand should be specified first. There are three ways for specifying the integrand.

(i) **M-function** The first method is to express the integrand using a MATLAB function, where the input argument is the variable x. Since many x values need to be processed simultaneously, $\boldsymbol{x}$ vector can finally be used as the input argument, and the computation within the function should be expressed in dot operations. An example for expressing such a function is shown as

```
function y=c3ffun(x)
y=2/sqrt(pi)*exp(-x.^2);
```

The function can be saved as `c3ffun.m` file.

(ii) **Inline function** The integrand can also be described by the inline function, where the input argument $\boldsymbol{x}$ should be appended after the integrand expression. The integrand in this example can be written as

```
>> f=inline('2/sqrt(pi)*exp(-x.^2)','x');
```

(iii) **Anonymous function** Anonymous function expression is an effective way for describing the integrand. The format of the function is even more straightforward than the inline expression. The integrand can be expressed by the anonymous function as follows:

```
>> f=@(x)2/sqrt(pi)*exp(-x.^2);
```

It should be pointed out that the anonymous function expression is the fastest among the three. The drawbacks of the representation are that it can only return one argument, and function evaluations with intermediate computations are not allowed. Thus the anonymous function is used throughout the book whenever possible. If anonymous function cannot be used, the M-function description will be used.

When the integrand has been declared by any of the above three methods, the `quad()` function can be used to solve the definite integral problem

```
>> f=@(x)2/sqrt(pi)*exp(-x.^2); % anonymous function expression
   [I1,k1]=quad(f,0,1.5)         % evaluate the integration
   [I2,k2]=quad(@c3ffun,0,1.5)   % Alternatively M-function can be used
```

and $I_1 = I_2 = 0.96610518623173$, with 25 function calls. In fact, the high-precision solution to the same problem can be obtained with the use of Symbolic Math Toolbox

```
>> syms x, y0=vpa(int(2/sqrt(pi)*exp(-x^2),0,1.5),60)
```

where $y_0 = 0.966105146475310713936933729949905794996224943257461473285 75$.

Comparing the results obtained above, it can be found that the accuracy of the numerical method is not very high. This is due to the default setting of the error tolerance ϵ. One may reduce the value of ϵ to find solutions with higher accuracy. However, over-demanding in expected accuracy may lead to the failure of computation due to possible singularity problems

```
>> [y,k2]=quad(f,0,1.5,1e-20)    % high-precision is expected but failed
```

and a warning message is given, with an unreliable result $y = 0.96606$, $k_2 = 10009$.

```
Warning: Maximum function count exceeded; singularity likely.
  > In quad at 100
```

A new function quadl() is provided in MATLAB. The syntax of the function is exactly the same as the quad() function. The effective Lobatto algorithm is implemented in the function which is much more accurate than the quad() function.

Example 3.32 Consider the above example. Let us try to use the quadl() function to solve numerically the same problem and observe how the precision can be increased.

Solution Using quadl() function the following results can be obtained. Compared with the analytical solution, it can be found that the accuracy may reach 10^{-16} level. Although the pre-specified 10^{-20} error tolerance cannot be reached with double-precision computation, the solution is accurate enough for most applications

```
>> [y,k3]=quadl(f,0,1.5,1e-20), e=abs(y-y0)
```

and it can be found that $y = 0.96610514647531$, $e = 6.4\times 10^{-17}$, $k_3 = 1098$.

Example 3.33 Compute the integral of a piecewise function

$$I = \int_0^4 f(x)\,\mathrm{d}x, \text{ where } f(x) = \begin{cases} \mathrm{e}^{x^2}, & 0 \leqslant x \leqslant 2 \\ \dfrac{80}{4-\sin(16\pi x)}, & 2 < x \leqslant 4. \end{cases}$$

Solution The piecewise function is displayed in filled curve in Figure 3.12. It can be seen that the curve is not continuous at $x = 2$ point.

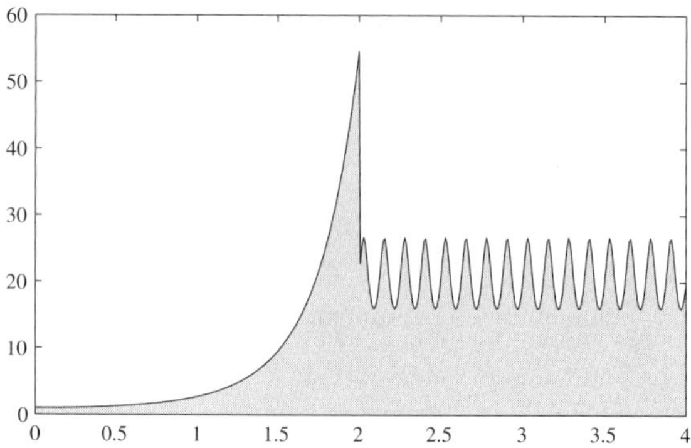

FIGURE 3.12: Filled plot of the integrand

```
>> x=[0:0.01:2, 2+eps:0.01:4,4];
   y=exp(x.^2).*(x<=2)+80./(4-sin(16*pi*x)).*(x>2);
   y(end)=0; x=[eps, x]; y=[0,y]; fill(x,y,'g')
```

With the use of relationship expressions, the integrand can be described and the functions quad() and quadl() can be used respectively to solve the original problem

```
>> f=@(x)exp(x.^2).*(x<=2)+80*(x>2)./(4-sin(16*pi*x));
   I1=quad(f,0,4), I2=quadl(f,0,4)
```

and it is found that $I_1 = 57.76435412500863$, $I_2 = 57.76445016946768$.

It can be seen from the obtained two results that there is a significant difference from the two methods. In fact, the original problem can also be divided into $\int_0^2 + \int_2^4$. The analytical solution function int() can then be used to find the analytical solutions to the original problem

```
>> syms x; I=vpa(int(exp(x^2),0,2)+int(80/(4-sin(16*pi*x)),2,4))
```

with $I = 57.764450125053010333315235385182$.

Compared with the analytical solutions, the results obtained by the quad() function may have large errors. If one divides the interval into two sub-intervals, there still exist errors. It can be concluded that quad() is not a good method to use. Let us try the quadl() function. Furthermore, the control options can be assigned such that even more accurate solutions can be obtained

```
>> f1=@(x)exp(x.^2); f2=@(x)80./(4-sin(16*pi*x));
   I1=quad(f1,0,2)+quad(f2,2,4), I2=quadl(f1,0,2)+quadl(f2,2,4)
   I3=quadl(f1,0,2,1e-11)+quadl(f2,2,4,1e-11)  % with error tolerance
```

the results are $I_1 = 57.764442889$, $I_2 = 57.76445012538$, and $I_3 = 57.764450125053$.

Example 3.34 Compute again the integral defined in Example 3.30.

Solution From the fixed-step algorithm demonstrated in Example 3.30, it can be found that only when the step-size is selected to be a very small value, the high accuracy can be achieved. However, with the help of variable-step algorithms, the original problem can be solved within a much shorter time and with much higher accuracy.

```
>> f=@(x)cos(15*x); tic, S=quadl(f,0,3*pi/2,1e-15), toc
```

It can be found that $S = 0.06666666666667$, and the elapsed time is 0.43 seconds, much faster than the fixed-step method. If quad() function is used, with error tolerance specified, the same warning and erroneous results are obtained.

Thus it can be concluded that the variable-step Lobatto algorithm of integrals has much more advantages over the fixed-step method taught in numerical analysis courses.

3.4.3 Numerical solutions to double integrals

Now consider the double integrals defined over a rectangular region

$$I = \int_{y_\mathrm{m}}^{y_\mathrm{M}} \int_{x_\mathrm{m}}^{x_\mathrm{M}} f(x, y)\, \mathrm{d}x\mathrm{d}y \tag{3.26}$$

and the function dblquad() can be used to solve this type of problem, with the syntaxes

```
y=dblquad(fun,x_m,x_M,y_m,y_M)      % double integral
y=dblquad(fun,x_m,x_M,y_m,y_M,ε)    % with given error tolerance
```

It should be noted that the number of calls to the integrand is not returned in this function. The users may set a global counter to check it when necessary.

Example 3.35 Compute the double definite integral

$$J = \int_{-1}^{1} \int_{-2}^{2} e^{-x^2/2} \sin(x^2 + y) \, dx dy.$$

Solution With the anonymous function to describe the integrand, the double integral can be evaluated numerically from the following statements

```
>> f=@(x,y)exp(-x.^2/2).*sin(x.^2+y);
   y=dblquad(f,-2,2,-1,1),
```

and the result is $y = 1.57456866245358$.

Unfortunately, the MATLAB function cannot be used in solving the double integral problem defined over a non-rectangular region as

$$I = \int_{x_m}^{x_M} \int_{y_m(x)}^{y_M(x)} f(x, y) \, dy dx. \tag{3.27}$$

A free toolbox, the Numerical Integration Toolbox (NIT) developed by Howard Wilson and Bryce Gardner can be downloaded from MathTools's website[1]. The function gquad2dggen() in the toolbox can be used to solve the numerical integration problem defined in (3.27). The syntaxes of the function are

```
J=quad2dggen(fun,F_lower,F_upper,x_m,x_M)      % double integral
J=quad2dggen(fun,F_lower,F_upper,x_m,x_M,ε)    % integral with error control
```

where ϵ is the error tolerance. This value can control the error in computation. However, if it is selected too small, significant amount of computational efforts will be required. In the function, three other MATLAB functions, i.e., the integrand and the inner upper- and lower-bound functions, are required to solve the problem.

It should be noted that the order of the integration is made with respect to x first then to y. If the user wants to integrate with respect to y first, the order x, y in the function should be changed first, before integration can be carried out. An illustrative example will be given to demonstrate this phenomenon.

Example 3.36 Compute the double definite integral

$$J = \int_{-1/2}^{1} \int_{-\sqrt{1-x^2/2}}^{\sqrt{1-x^2/2}} e^{-x^2/2} \sin(x^2 + y) \, dy dx.$$

[1] Download address: http://www.mathtools.net/files/net/nittbx.zip

Solution One can first integrate with respect to y, then to x, and then the inner bounds $y_{\rm M}(x)$ and $y_{\rm m}(x)$ can be defined. The following statements can then be used. Please note that the order of integration should be swaped first

```
>> fh=@(x)sqrt(1-x.^2/2);  fl=@(x)-sqrt(1-x.^2/2); % inner bounds
   f=@(y,x)exp(-x.^2/2).*sin(x.^2+y);              % order swaped
   y=quad2dggen(f,fl,fh,-1/2,1,eps),
```

and the value of integration can be found as $y = 0.41192954617630$. Now consider the analytical method

```
>> syms x y
   i1=int(exp(-x^2/2)*sin(x^2+y),y,-sqrt(1-x^2/2),sqrt(1-x^2/2));
   int(i1,x,-1/2,1)    % warning message given
```

and the following warning is displayed, after some time

```
Warning: Explicit integral could not be found.
> In sym.int at 58
ans =
int(2*exp(-1/2*x^2)*sin(x^2)*sin(1/2*(4-2*x^2)^(1/2)),x = -1/2..1)
```

and it can be seen that no explicit solution exists for this problem. One can get the high-precision numerical method with `vpa(ans,70)`, the result can be written as .4119295461762951196517599401760134672761827128252391627415959533602675.

It can be seen that the numerical solutions are very accurate.

If the original integral problem is changed to

$$J = \int_{-1}^{1}\int_{-\sqrt{1-y^2}}^{\sqrt{1-y^2}} \mathrm{e}^{-x^2/2}\sin(x^2+y)\,\mathrm{d}x\mathrm{d}y$$

the analytical problem cannot be used to find the results, even with the help of `vpa()` function. However, if the numerical method is used

```
>> fh=@(y)sqrt(1-y.^2); fl=@(y)-sqrt(1-y.^2); % inner bounds
   f=@(x,y)exp(-x.^2/2).*sin(x.^2+y);         % integrand
   I=quad2dggen(f,fl,fh,-1,1,eps),
```

which yields $I = 0.53686038269795$, which is different. For numerical methods, the numerical integration results will not be affected by whether the integrand is theoretically integrable or not.

3.4.4 Numerical solutions to triple integrals

The triple definite integral over the 3D rectangular region is given by

$$I = \int_{x_{\rm m}}^{x_{\rm M}}\int_{y_{\rm m}}^{y_{\rm M}}\int_{z_{\rm m}}^{z_{\rm M}} f(x,y,z)\,\mathrm{d}z\mathrm{d}y\mathrm{d}x; \tag{3.28}$$

the problem can be solved with the `triplequad()` function whose syntax is

I=`triplequad(`*fun*`,`$x_{\rm m}$`,`$x_{\rm M}$`,`$y_{\rm m}$`,`$y_{\rm M}$`,`$z_{\rm m}$`,`$z_{\rm M}$`,`ϵ`,@quadl)`

where *fun* describes the integrand. The argument ϵ can still be used in controlling the accuracy of the integration, with a default value of 10^{-6}. In order to increase the accuracy, smaller error tolerance can be assigned.

The extra function @quadl can be used to implement the integration for single variable functions. It can also be assigned to @quad or any other user functions.

Example 3.37 Compute the triple integral in Example 3.16

$$\int_0^2\int_0^\pi\int_0^\pi 4xz\mathrm{e}^{-x^2y-z^2}\mathrm{d}z\mathrm{d}y\mathrm{d}x.$$

Solution The anonymous function is used to specify the integrand. Thus the following statements can be used to compute the triple integral

```
>> f=@(x,y,z)4*x.*z.*exp(-x.*x.*y-z.*z);
   I=triplequad(f, 0,2, 0,pi, 0,pi, 1e-10,@quadl)
```

and $I = 3.108079402072966$.

NIT Toolbox can be used to solve multiple integral problems with other hyper-rectangular regions. For instance, the quadndg() function can be used for these problems. However, if the integration regions are not hyper rectangular regions, there are no existing implemented MATLAB functions available for numerical triple integrals.

3.5 Path Integrals and Line Integrals

Surprisingly, path integrals and line integrals cannot be solved by the existing MATLAB or Maple functions. In this section, the concepts and integration method for path and line integrals are summarized first and then solutions to these problems will be demonstrated through examples.

3.5.1 Path integrals

Path integrals are originated from the evaluation of the total mass of a spatial wire with unevenly distributed density. Assume that the density of a path l is $f(x, y, z)$. Then the total mass of the wire can be evaluated from the following equation

$$I_1 = \int_l f(x, y, z)\,\mathrm{d}s \tag{3.29}$$

where $\mathrm{d}s$ is the arc length at a certain point. Thus this kind of integral is also known as the *integral with respect to arc*. If $f(x, y, z) \equiv 1$, i.e., the density is evenly distributed and equals unity, the total length of the wire is calculated.

If the variables x, y and z are given respectively by parametric equations $x = x(t)$, $y = y(t)$, $z = z(t)$, they can be substituted into the $f(\cdot)$ function, and the differentiation of the arc can be written as

$$\mathrm{d}s=\sqrt{\left(\frac{\mathrm{d}x}{\mathrm{d}t}\right)^2+\left(\frac{\mathrm{d}y}{\mathrm{d}t}\right)^2+\left(\frac{\mathrm{d}z}{\mathrm{d}t}\right)^2}\,\mathrm{d}t,\text{ or simply }\mathrm{d}s=\sqrt{x_t^2+y_t^2+z_t^2}\,\mathrm{d}t. \tag{3.30}$$

Then, the path integral can be converted into an ordinary integral with respect to t

$$I=\int_{t_\mathrm{m}}^{t_\mathrm{M}} f[x(t),y(t),z(t)]\sqrt{x_t^2+y_t^2+z_t^2}\,\mathrm{d}t. \tag{3.31}$$

For the integrand with two variables, $f(x,y)$, it can also be converted into ordinary integrals. Therefore, the solutions to the path integral problem can be solved with MATLAB using the previously described procedures.

Example 3.38 Compute $\displaystyle\int_l \frac{z^2}{x^2+y^2}\,\mathrm{d}s$, where the path l is defined as $x = a\cos t, y = a\sin t, z = at$, with $0 \leqslant t \leqslant 2\pi$ and $a > 0$.

Solution The following statements can be used for this path integral problem:

```
>> syms t; syms a positive; x=a*cos(t); y=a*sin(t); z=a*t;
   dx=diff(x,t); dy=diff(y,t); dz=diff(z,t);
   I=int(z^2/(x^2+y^2)*sqrt(dx^2+dy^2+dz^2),t,0,2*pi)
```

and the result is $I = \dfrac{8\sqrt{2}}{3}\pi^3 a$.

Example 3.39 Compute $\displaystyle\int_l (x^2+y^2)\,\mathrm{d}s$ where path l is defined as the positive direction curve encircled by the paths $y = x$ and $y = x^2$.

Solution The following statements can be used to draw the two paths shown in Figure 3.13.

```
>> x=0:.001:1.2; y1=x; y2=x.^2; plot(x,y1,x,y2)
```

It can be seen that the original integration problem can be divided into two sub-integration problems. Thus the following statements can be used to add the two sub-integrations up to get the final solutions

```
>> syms x; y1=x; y2=x^2; I1=int((x^2+y2^2)*sqrt(1+diff(y2,x)^2),x,0,1);
   I2=int((x^2+y1^2)*sqrt(1+diff(y1,x)^2),x,1,0); I=I2+I1
```

and $I = -\dfrac{2}{3}\sqrt{2}+\dfrac{349}{768}\sqrt{5}+\dfrac{7}{512}\ln\left(-2+\sqrt{5}\right)$.

3.5.2 Line integrals

Line integral problems are originated from physics, where the total work is done by the force $\vec{f}(x,y,z)$ along a spatial curve l. This kind of integral problem can be expressed as

$$I_2=\int_l \vec{\boldsymbol{f}}(x,y,z)\cdot\mathrm{d}\vec{s} \tag{3.32}$$

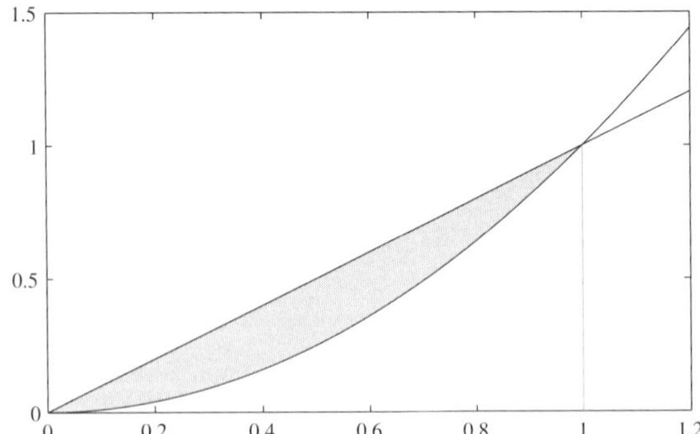

FIGURE 3.13: Illustration of the integration paths

where $\vec{\boldsymbol{f}}(x, y, z) = [P(x, y, z), Q(x, y, z), R(x, y, z)]$ is a row vector. The differentiation of the line $\mathrm{d}\vec{\boldsymbol{s}}$ is a column vector. If the line can be described by a parametric equation of t such as $x(t), y(t), z(t)$, with $t \in (a, b)$, the vector $\mathrm{d}\vec{\boldsymbol{s}}$ can then be written as

$$\mathrm{d}\vec{\boldsymbol{s}} = \left[\frac{\mathrm{d}x}{\mathrm{d}t}, \frac{\mathrm{d}y}{\mathrm{d}t}, \frac{\mathrm{d}z}{\mathrm{d}t}\right]^{\mathrm{T}} \mathrm{d}t. \tag{3.33}$$

The dot product of two vectors can be carried out directly and the line integrals can be re-defined as an ordinary integral as follows:

$$I_2 = \int_a^b [P(x, y, z), Q(x, y, z), R(x, y, z)] \left[\frac{\mathrm{d}x}{\mathrm{d}t}, \frac{\mathrm{d}y}{\mathrm{d}t}, \frac{\mathrm{d}z}{\mathrm{d}t}\right]^{\mathrm{T}} \mathrm{d}t \tag{3.34}$$

which can be solved by using MATLAB.

Example 3.40 Compute the integral $\displaystyle\int_l \frac{x+y}{x^2+y^2}\,\mathrm{d}x - \frac{x-y}{x^2+y^2}\,\mathrm{d}y$, where the line l is defined as the positive circle given by $x^2 + y^2 = a^2$, $a > 0$.

Solution If one wants to evaluate the line integral, the circle can be interpreted as the parametric equations $x = a\cos(t)$, $y = a\sin(t)$ for $0 \leqslant t \leqslant 2\pi$. Thus the following statements can be used to calculate the line integral, with the result $I = 2\pi$.

```
>> syms t; syms a positive; x=a*cos(t); y=a*sin(t);
   F=[(x+y)/(x^2+y^2),-(x-y)/(x^2+y^2)]; ds=[diff(x,t);diff(y,t)];
   I=int(F*ds,t,2*pi,0)  % positive circle, clockwise, t from 2π to 0
```

Example 3.41 Compute the line integral $\displaystyle\int_l (x^2 - 2xy)\,\mathrm{d}x + (y^2 - 2xy)\,\mathrm{d}y$, where the line l is defined as the parabolic curve $y = x^2$ $(-1 \leqslant x \leqslant 1)$.

Solution In fact, the equations given are already the parametric equations of x. The derivative of x with respective to x is 1. The following statements can be used to solve the line integral problem, with the result $I = -14/15$.

```
>> syms x; y=x^2; F=[x^2-2*x*y,y^2-2*x*y]; ds=[1; diff(y,x)];
   I=int(F*ds,x,-1,1)
```

3.6 Surface Integrals

Two types of surface integrals are considered in this section, the scalar type and the vector type. The definitions and solutions to the problems will be summarized first followed by the detailed solution procedures with MATLAB script-based examples.

3.6.1 Scalar surface integrals

The scalar-type surface integrals are defined as

$$I = \iint_S \phi(x, y, z)\,\mathrm{d}S \tag{3.35}$$

where $\mathrm{d}S$ is the differentiated area. Thus this kind of integral is also referred to as the *surface integrals with respect to area.* If $\phi(x, y, z) \equiv 1$, the area of the surface can be computed.

Let the surface S be defined by $z = f(x, y)$. The original surface integral can be converted into a double integral over the x-y plane, such that

$$I = \iint_{\sigma_{xy}} \phi[x, y, f(x, y)]\sqrt{1 + f_x^2 + f_y^2}\,\mathrm{d}x\mathrm{d}y \tag{3.36}$$

where σ_{xy} is the integration region, which is an ordinary double integral problem.

Example 3.42 Compute $\displaystyle\iint_S xyz\mathrm{d}S$, where the integration surface S is defined as the region enclosed by the four planes $x = 0$, $y = 0$, $z = 0$, and $x + y + z = a$, where $a > 0$.

Solution Denote the four planes by S_1, S_2, S_3, and S_4. The original surface integral can be calculated using $\displaystyle\iint_S = \iint_{S_1} + \iint_{S_2} + \iint_{S_3} + \iint_{S_4}$. Considering the planes S_1, S_2, S_3, since the integrands are all 0, only the integral on the S_4 should be considered. The plane S_4 can mathematically be described as $z = a - x - y$. Then, the following statements can be used to evaluate the surface integral

```
>> syms x y; syms a positive; z=a-x-y;
   I=int(int(x*y*z*sqrt(1+diff(z,x)^2+diff(z,y)^2),y,0,a-x),x,0,a)
```

which gives $I = \sqrt{3}a^5/120$.

If the parametric equations for the surface are given by

$$x = x(u,v),\ y = y(u,v),\ z = z(u,v), \tag{3.37}$$

the surface integral can then be obtained using the following formula

$$I = \iint_{\Sigma} \phi[x(u,v), y(u,v), z(u,v)]\sqrt{EG - F^2}\, \mathrm{d}u\mathrm{d}v \tag{3.38}$$

where

$$E = x_u^2 + y_u^2 + z_u^2,\quad F = x_u x_v + y_u y_v + z_u z_v,\quad G = x_v^2 + y_v^2 + z_v^2. \tag{3.39}$$

Example 3.43 Compute the surface integral $\iint (x^2y+zy^2)\,\mathrm{d}S$, where the surface S is defined as the surfaces composed of $x = u\cos v, y = u\sin v, z = v,\ 0 \leqslant u \leqslant a, 0 \leqslant v \leqslant 2\pi$.

Solution The following statements can be used to calculate the integrals

```
>> syms u v; syms a positive;
   x=u*cos(v); y=u*sin(v); z=v; f=x^2*y+z*y^2;
   E=simple(diff(x,u)^2+diff(y,u)^2+diff(z,u)^2);
   F=diff(x,u)*diff(x,v)+diff(y,u)*diff(y,v)+diff(z,u)*diff(z,v);
   G=simple(diff(x,v)^2+diff(y,v)^2+diff(z,v)^2);
   I=int(int(f*sqrt(E*G-F^2),u,0,a),v,0,2*pi)
```

and the result is $I = \dfrac{1}{8}\pi^2\left[2a\left(a^2+1\right)^{3/2} - a\sqrt{a^2+1} - \operatorname{arcsinh} a\right]$.

3.6.2 Vector surface integrals

The second category of surface integral is also referred to as the *surface integrals in vector fields*. Suppose the integrand is given by a row vector $\vec{\boldsymbol{\Gamma}} = [P, Q, R]$, while $\mathrm{d}\vec{\boldsymbol{V}}$ is given by a column vector $\mathrm{d}\vec{\boldsymbol{V}} = [\mathrm{d}y\mathrm{d}z, \mathrm{d}x\mathrm{d}z, \mathrm{d}x\mathrm{d}y]^{\mathrm{T}}$, the mathematical description to the problem is

$$I = \iint_{S^+} \vec{\boldsymbol{\Gamma}}\cdot \mathrm{d}\vec{\boldsymbol{V}} = \iint_{S^+} P(x,y,z)\,\mathrm{d}y\mathrm{d}z + Q(x,y,z)\,\mathrm{d}x\mathrm{d}z + R(x,y,z)\,\mathrm{d}x\mathrm{d}y, \tag{3.40}$$

where the positive surface S^+ is defined with $z = f(x,y)$. The surface integral problem can then be converted into the scalar surface integral problem

$$I = \iint_{S^+} [P(x,y,z)\cos\alpha + Q(x,y,z)\cos\beta + R(x,y,z)\cos\gamma]\ \mathrm{d}\boldsymbol{S} \tag{3.41}$$

where z is replaced by $f(x,y)$, and

$$\cos\alpha = \frac{-f_x}{\sqrt{1+f_x^2+f_y^2}},\qquad \cos\beta = \frac{-f_y}{\sqrt{1+f_x^2+f_y^2}},\qquad \cos\gamma = \frac{1}{\sqrt{1+f_x^2+f_y^2}}. \tag{3.42}$$

Thus, the $\sqrt{1+f_x^2+f_y^2}$ term may cancel the relevant term in (3.36), and the surface integral can be written as

$$I=\iint_{\sigma_{xy}} -Pf_x\,\mathrm{d}x\mathrm{d}y-Qf_y\,\mathrm{d}x\mathrm{d}z+R\,\mathrm{d}y\mathrm{d}z. \tag{3.43}$$

If the surface is described by the parametric equations in (3.37), the following equations can be obtained

$$\cos\alpha=\frac{A}{\sqrt{A^2+B^2+C^2}},\quad \cos\beta=\frac{B}{\sqrt{A^2+B^2+C^2}},\quad \cos\gamma=\frac{C}{\sqrt{A^2+B^2+C^2}} \tag{3.44}$$

where $A=y_uz_v-z_uy_v$, $B=z_ux_v-x_uz_v$, $C=x_uy_v-y_ux_v$. Then from the converted scalar surface integral (3.41), it can be found that the denominator in (3.44) cancels the $\sqrt{EG-F^2}$ term. Thus the vector surface integral can be simplified as the following standard double integral

$$I=\int_{v_\mathrm{m}}^{v_\mathrm{M}}\int_{u_\mathrm{m}(v)}^{u_\mathrm{M}(v)}\left[AP(u,v)+BQ(u,v)+CR(u,v)\right]\mathrm{d}u\mathrm{d}v. \tag{3.45}$$

Example 3.44 Compute the surface integral $\iint x^3\mathrm{d}y\mathrm{d}z$, where the surface S is defined as the positive side of the ellipsoid surface $\dfrac{x^2}{a^2}+\dfrac{y^2}{b^2}+\dfrac{z^2}{c^2}=1$.

Solution The parametric equations can be introduced such that $x=a\sin u\cos v$, $y=b\sin u\sin v$, $z=c\cos u$, and $\left(0\leqslant u\leqslant\dfrac{\pi}{2}\right)$, $(0\leqslant v\leqslant 2\pi)$. The following statements can be used to compute the surface integral, with the result $I=2\pi a^3cb/5$.

```
>> syms u v; syms a b c positive;
   x=a*sin(u)*cos(v); y=b*sin(u)*sin(v); z=c*cos(u);
   A=diff(y,u)*diff(z,v)-diff(z,u)*diff(y,v);
   I=int(int(x^3*A,u,0,pi/2),v,0,2*pi)
```

Exercises

1. Compute the following limit problems:

   (i) $\lim\limits_{x\to\infty}\left(3^x+9^x\right)^{\frac{1}{x}}$, (ii) $\lim\limits_{x\to\infty}\dfrac{(x+2)^{x+2}(x+3)^{x+3}}{(x+5)^{2x+5}}$

2. Compute the following double limit problems:

   (i) $\lim\limits_{\substack{x\to-1\\y\to2}}\dfrac{x^2y+xy^3}{(x+y)^3}$, (ii) $\lim\limits_{\substack{x\to0\\y\to0}}\dfrac{xy}{\sqrt{xy+1}-1}$, (iii) $\lim\limits_{\substack{x\to0\\y\to0}}\dfrac{1-\cos\left(x^2+y^2\right)}{\left(x^2+y^2\right)\mathrm{e}^{x^2+y^2}}$

3. Compute the derivatives of the following functions:

(i) $y(x)=\sqrt{x\sin x\sqrt{1-\mathrm{e}^x}}$, (ii) $y(t)=\sqrt{\dfrac{(x-1)(x-2)}{(x-3)(x-4)}}$

(iii) $\operatorname{atan}\dfrac{y}{x}=\ln(x^2+y^2)$, (iv) $y(x)=-\dfrac{1}{na}\ln\dfrac{x^n+a}{x^n}$, $n>0$

4. Compute the 10th order derivative of the function $y=\dfrac{1-\sqrt{\cos ax}}{x\left(1-\cos\sqrt{ax}\right)}$.

5. In calculus courses, when the limit of a ratio is required, where both the numerator and the denominator tend to 0 or ∞, simultaneously, L'Hôpital's law can be used, i.e., to evaluate the limits of derivatives of numerator and denominator. Verify the $\lim\limits_{x\to 0}\dfrac{\ln(1+x)\ln(1-x)-\ln(1-x^2)}{x^4}$ by the consecutive use of L'Hôpital's law, and compare with the results directly obtained.

6. For parametric equation $\begin{cases} x=\ln\cos t \\ y=\cos t-t\sin t\end{cases}$, compute $\dfrac{\mathrm{d}y}{\mathrm{d}x}$ and $\left.\dfrac{\mathrm{d}^2y}{\mathrm{d}x^2}\right|_{t=\pi/3}$.

7. Assume that $u=\cos^{-1}\sqrt{\dfrac{x}{y}}$. Verify that $\dfrac{\partial^2 u}{\partial x\partial y}=\dfrac{\partial^2 u}{\partial y\partial x}$.

8. For a given function $\begin{cases} xu+yv=0 \\ yu+xv=1\end{cases}$, compute $\dfrac{\partial^2 u}{\partial x\partial y}$.

9. Assume that $f(x,y)=\displaystyle\int_0^{xy}\mathrm{e}^{-t^2}\mathrm{d}t$. Compute $\dfrac{x}{y}\dfrac{\partial^2 f}{\partial x^2}-2\dfrac{\partial^2 f}{\partial x\partial y}+\dfrac{\partial^2 f}{\partial y^2}$.

10. Given a matrix $\boldsymbol{f}(x,y,z)=\begin{bmatrix} 3x+\mathrm{e}^y z \\ x^3+y^2\sin z\end{bmatrix}$, compute its Jacobian matrix.

11. Compute the following infinite integrals:

(i) $I(x)=-\displaystyle\int\frac{3x^2+a}{x^2\left(x^2+a\right)^2}\,\mathrm{d}x$, (ii) $I(x)=\displaystyle\int\frac{\sqrt{x(x+1)}}{\sqrt{x}+\sqrt{1+x}}\,\mathrm{d}x$

(iii) $I(x)=\displaystyle\int x\mathrm{e}^{ax}\cos bx\,\mathrm{d}x$, (iv) $I(t)=\displaystyle\int\mathrm{e}^{ax}\sin bx\sin cx\,\mathrm{d}x$

12. Compute the definite integrals and infinite integrals

(i) $I=\displaystyle\int_0^\infty\frac{\cos x}{\sqrt{x}}\,\mathrm{d}x$, (ii) $I=\displaystyle\int_0^1\frac{1+x^2}{1+x^4}\,\mathrm{d}x$

13. For the function $f(x)=\mathrm{e}^{-5x}\sin(3x+\pi/3)$, compute $\displaystyle\int_0^t f(x)f(t+x)\,\mathrm{d}x$.

14. For different values of a, compute the integral $I=\displaystyle\int_0^\infty\frac{\cos ax}{1+x^2}\,\mathrm{d}x$.

15. Show that for any function $f(t)$, $\displaystyle\int_a^b f(t)\,\mathrm{d}t=-\int_b^a f(t)\,\mathrm{d}t$.

16. Solve the following multiple integral problems:

(i) $\displaystyle\int_0^2\int_0^1\sqrt{4-x^2-y^2}\,\mathrm{d}y\mathrm{d}x$, (ii) $\displaystyle\int_0^3\int_0^{3-x}\int_0^{3-x-y}xyz\,\mathrm{d}z\mathrm{d}y\mathrm{d}x$

(iii) $\displaystyle\int_0^2\int_0^{\sqrt{4-x^2}}\int_0^{\sqrt{4-x^2-y^2}}z(x^2+y^2)\,\mathrm{d}z\mathrm{d}y\mathrm{d}x$

(iv) $\displaystyle\int_0^{7/10}\int_0^{4/5}\int_0^{9/10}\int_0^1\int_0^{11/10}\sqrt{6-x^2-y^2-z^2-w^2-u^2}\,\mathrm{d}w\mathrm{d}u\mathrm{d}z\mathrm{d}y\mathrm{d}x$

17. Compute the Fourier series expansions for the following functions, and compare

graphically the approximation and exact results, using finite numbers of terms:

(i) $f(x) = (\pi - |x|)\sin x, \ -\pi \leqslant x < \pi$, (ii) $f(x) = \mathrm{e}^{|x|}, \ -\pi \leqslant x < \pi$,

(iii) $f(x) = \begin{cases} 2x/l, & 0 < x < l/2 \\ 2(l-x)/l, & l/2 < x < l \end{cases}$, where $l = \pi$.

18. Obtain the Taylor series expansions for the following functions, and compare graphically the approximation and exact results with finite numbers of terms:

    (i) $\displaystyle\int_0^x \frac{\sin t}{t}\,\mathrm{d}t$, (ii) $\ln\left(\dfrac{1+x}{1-x}\right)$, (iii) $\ln\left(x+\sqrt{1+x^2}\right)$, (iv) $(1+4.2x^2)^{0.2}$,

    (v) $\mathrm{e}^{-5x}\sin(3x+\pi/3)$ expansions about $x=0$ and $x=a$ points respectively.

    (vi) $f(x,y) = \dfrac{1-\cos\left(x^2+y^2\right)}{\left(x^2+y^2\right)\mathrm{e}^{x^2+y^2}}$ expansion about $x=1, y=0$ point.

19. Compute the first n term finite sums and infinite sums.

    (i) $\dfrac{1}{1\times 6} + \dfrac{1}{6\times 11} + \cdots + \dfrac{1}{(5n-4)(5n+1)} + \cdots$

    (ii) $\left(\dfrac{1}{2}+\dfrac{1}{3}\right) + \left(\dfrac{1}{2^2}+\dfrac{1}{3^2}\right) + \cdots + \left(\dfrac{1}{2^n}+\dfrac{1}{3^n}\right) + \cdots$

20. Compute the following limits:

    (i) $\displaystyle\lim_{n\to\infty}\left[\frac{1}{2^2-1}+\frac{1}{4^2-1}+\frac{1}{6^2-1}+\cdots+\frac{1}{(2n)^2-1}\right]$,

    (ii) $\displaystyle\lim_{n\to\infty} n\left(\frac{1}{n^2+\pi}+\frac{1}{n^2+2\pi}+\frac{1}{n^2+3\pi}+\cdots+\frac{1}{n^2+n\pi}\right)$

21. Show that $\cos\theta + \cos 2\theta + \cdots + \cos n\theta = \dfrac{\sin(n\theta/2)\cos[(n+1)\theta/2]}{\sin\theta/2}$.

22. For the following tabulated measured data, evaluate numerically its derivatives and definite integral.

x_i	0	0.1	0.2	0.3	0.4	0.5	0.6	0.7	0.8	0.9	1	1.1	1.2
y_i	0	2.2077	3.2058	3.4435	3.241	2.8164	2.311	1.8101	1.3602	0.9817	0.6791	0.4473	0.2768

23. Evaluate the definite integral $\displaystyle\int_0^\pi (\pi-t)^{\frac{1}{4}} f(t)\mathrm{d}t$, $f(t)=\mathrm{e}^{-t}\sin(3t+1)$ numerically.

    Also evaluate the integration function $F(t) = \displaystyle\int_0^t (t-\tau)^{\frac{1}{4}} f(\tau)\,\mathrm{d}\tau$ numerically for different sample points of t, such that $t = 0.1, 0.2, \cdots, \pi$, and draw the $F(t)$ plot.

24. Evaluate numerically the following multiple integral problems. It should be noted that there are no analytical solutions to these problems. Therefore, the obtained numerical results should be double-checked by varying step sizes or default accuracies.

    (i) $\displaystyle\int_0^2 \int_0^{\mathrm{e}^{-x^2/2}} \sqrt{4-x^2-y^2}\ \mathrm{e}^{-x^2-y^2}\,\mathrm{d}y\mathrm{d}x$

    (ii) $\displaystyle\int_0^2 \int_0^{\sqrt{4-x^2}} \int_0^{\sqrt{4-x^2-y^2}} z(x^2+y^2)\,\mathrm{e}^{-x^2-y^2-z^2-xz}\,\mathrm{d}z\mathrm{d}y\mathrm{d}x$

(iii) $\int_0^1 \int_0^2 \int_0^{xy} \int_0^u e^{6-x^2-y^2-z^2-u^2} \, du dz dy dx$

25. Compute the gradient of the measured data for a function of two variables. Assume that the data were generated by the function $f(x, y) = 4 - x^2 - y^2$. Generate the data and verify the results of gradient with theoretical results.

0	0	0.2	0.4	0.6	0.8	1	1.2	1.4	1.6	1.8	2
0	4	3.96	3.84	3.64	3.36	3	2.56	2.04	1.44	0.76	0
0.2	3.96	3.92	3.8	3.6	3.32	2.96	2.52	2	1.4	0.72	−0.04
0.4	3.84	3.8	3.68	3.48	3.2	2.84	2.4	1.88	1.28	0.6	−0.16
0.6	3.64	3.6	3.48	3.28	3	2.64	2.2	1.68	1.08	0.4	−0.36
0.8	3.36	3.32	3.2	3	2.72	2.36	1.92	1.4	0.8	0.12	−0.64
1	3	2.96	2.84	2.64	2.36	2	1.56	1.04	0.44	−0.24	−1
1.2	2.56	2.52	2.4	2.2	1.92	1.56	1.12	0.6	0	−0.68	−1.44
1.4	2.04	2	1.88	1.68	1.4	1.04	0.6	0.08	−0.52	−1.2	−1.96
1.6	1.44	1.4	1.28	1.08	0.8	0.44	0	−0.52	−1.12	−1.8	−2.56
1.8	0.76	0.72	0.6	0.4	0.12	−0.24	−0.68	−1.2	−1.8	−2.48	−3.24
2	0	−0.04	−0.16	−0.36	−0.64	−1	−1.44	−1.96	−2.56	−3.24	−4

26. Compute the following path and line integrals:

(i) $\int_l (x^2 + y^2) \, ds$, l: $x = a(\cos t + t \sin t), y = a(\sin t - t \cos t)$, for $0 \leqslant t \leqslant 2\pi$

(ii) $\int_l (yx^3 + e^y) \, dx + (xy^3 + xe^y - 2y) \, dy$, where l is given by the upper-semi-ellipsis of $a^2x^2 + b^2y^2 = c^2$.

(iii) $\int_l y \, dx - x \, dy + (x^2 + y^2) \, dz$, l: $x = e^t, y = e^{-t}, z = at$, $0 \leqslant t \leqslant 1$, for $a > 0$.

(iv) $\int_l (e^x \sin y - my) dx + (e^x \cos y - m) \, dy$, where l is defined as the closed path from $(a, 0)$ to $(0, 0)$, then with the upper-semi-circle $x^2 + y^2 = ax$.

27. Compute the surface integrals, where S is the bottom side of the semi-sphere $z = \sqrt{R^2 - x^2 - y^2}$.

(i) $\int_S xyz^3 \, ds$, (ii) $\int_S (x + yz^3) \, dxdy$.

Chapter 4

Linear Algebra Problems

Linear algebra deals with vectors, vector spaces or linear spaces, linear maps or linear transformations, and systems of linear equations. It is ubiquitous in modern applied mathematics. Although nonlinear models are true models of the real world systems, in the natural sciences and the social sciences nonlinear models are usually approximated by linear ones for initial effective characterization. The importance of linear algebra and the ability to solve linear algebra problems is obvious.

Although many readers might not be able to quickly compute the determinant of a given 3×3 or even 4×4 matrix by hand, one should not feel bad about this inability. We leave this low level computation to computer mathematics languages, such as MATLAB. In fact, many computer mathematics languages, such as MATLAB, originated from the early research of numerical linear algebra. For instance, the well-known EISPACK package[1] focused on the computation of eigen-systems of matrices. Another well-known package, LINPACK[3], was developed to solve general linear algebra problems using numerical algorithms. With the development of computer science, matrix computations are now no longer restricted to numerical computations. Analytical solutions can also be found for many linear algebra problems. Successful computer mathematics languages such as Mathematica, Maple and the Symbolic Math Toolbox of MATLAB can be used to analytically solve certain problems in linear algebra.

In Section 4.1, as a warming up, how to input some special matrices, such as identity matrix, companion matrix and Hankel matrix, and symbolic matrices is presented. In Section 4.2, basic concepts of matrix analysis are presented and illustrative MATLAB scripts are given for solving matrix determinant, trace, rank, norm, inverse matrix and eigen-system problems. Matrix decomposition methods such as similarity transformation, orthogonal decomposition, triangular decomposition and singular value decomposition are explained and demonstrated in Section 4.3. These decomposition methods can be used to simplify matrix analysis problems among many other potential benefits. Section 4.4 presents solutions to various matrix equations, such as linear algebraic equations, Lyapunov equations, Sylvester equations as well as Riccati equations. Both analytical solution algorithms and their MATLAB implementations are given. In Section 4.5, evaluations of matrix functions such as exponential functions and trigonometry functions will be discussed.

In particular, this chapter ends with an introduction of a general method for computing matrix functions of arbitrary forms by using detailed illustrative examples with the respective MATLAB scripts.

For readers who wish to check the detailed explanations of linear algebra, we recommend the free textbooks [9, 10, 11].

4.1 Inputting Special Matrices

4.1.1 Numerical matrix input

Although all the matrices can be entered into MATLAB workspace using the low-level statement discussed earlier, it might be complicated for some matrices with special structures. For instance, if one wants to enter an identity matrix, one should use the existing function `eye()` instead. In this section, the specifications of some special matrices will be presented.

Matrices of zeros, ones and identity matrices

In matrix theory, a matrix with all its elements 0 is referred to as a *zero matrix*, while a matrix with all its elements 1 is referred to as a *matrix of ones*. If the diagonal elements are 1 with the rest of the elements 0, the matrix is referred to as an *identity matrix*. This concept can be extended to $m \times n$ matrices. Matrix of zeros, matrix of ones and identity matrix can be entered into MATLAB using the following statements

```
A=zeros(n), B=ones(n), C=eye(n)   % n × n square matrix
A=zeros(m,n); B=ones(m,n); C=eye(m,n)   % m × n rectangular matrix
A=zeros(size(B))   % with the same size of B
```

Example 4.1 A 3×8 zero matrix $\boldsymbol{A}$ and an extended identity matrix $\boldsymbol{B}$, with the same size of $\boldsymbol{A}$ can be entered into MATLAB environment using the following statements

```
>> A=zeros(3,8), B=eye(size(A))
```

The two matrices can be established in MATLAB workspace

$$\boldsymbol{A} = \begin{bmatrix} 0 & 0 & 0 & 0 & 0 & 0 & 0 & 0 \\ 0 & 0 & 0 & 0 & 0 & 0 & 0 & 0 \\ 0 & 0 & 0 & 0 & 0 & 0 & 0 & 0 \end{bmatrix}, \quad \boldsymbol{B} = \begin{bmatrix} 1 & 0 & 0 & 0 & 0 & 0 & 0 & 0 \\ 0 & 1 & 0 & 0 & 0 & 0 & 0 & 0 \\ 0 & 0 & 1 & 0 & 0 & 0 & 0 & 0 \end{bmatrix}.$$

Functions `zeros()` and `ones()` can also be used to define multi-dimensional arrays. For instance, `zeros(3,4,5)` can be used to define a $3\times4\times5$ zero array.

Matrices with random elements

If all the elements in a matrix satisfy uniform distribution within the $[0, 1]$ interval, it can be defined using MATLAB function `rand()`. The syntax of such a function is

```
A=rand(n)     % generates an n × n uniformly distributed random matrix
A=rand(n,m)   % generates an n × m random matrix
```

Function `rand()` can also be used to define multi-dimensional random arrays. Another function `randn()` can be used to define standard normal distributed random matrices. The statement `rand(size(A))` can be used to declare a random matrix of size $\boldsymbol{A}$.

Here, the random number is in fact pseudo-random numbers, and can be generated mathematically. It is easy to generate random numbers satisfying certain predefined distributions. Another advantages of pseudo-random numbers are that they can be generated repeatedly.

If one wants to obtain a uniformly distributed random number over (a, b) interval, one may generate uniformly distributed pseudo-random matrix $\boldsymbol{V}$=`rand(n,m)` over $(0, 1)$ interval, then the expected matrix can be generated by $\boldsymbol{V}_1 = a + (b - a)*\boldsymbol{V}$ command.

Diagonal matrices

Mathematical description to a diagonal matrix is

$$\mathrm{diag}(\alpha_1, \alpha_2, \cdots, \alpha_n) = \begin{bmatrix} \alpha_1 & & & \\ & \alpha_2 & & \\ & & \ddots & \\ & & & \alpha_n \end{bmatrix} \tag{4.1}$$

where all the non-diagonal elements are 0. A MATLAB function `diag()` can be used to deal with diagonal matrix related problems

```
A=diag(V)     % define a matrix from given vector
V=diag(A)     % extract diagonal vector from a given matrix
A=diag(V,k)   % define the kth diagonal elements V
```

Example 4.2 MATLAB function `diag()` is an interesting function. Different syntax are allowed for different tasks. For instance, the following statements can be used to define different matrices

```
>> C=[1 2 3]; V=diag(C)
   V1=diag(V)     % diagonal elements extraction
   C=[1 2 3]; V2=diag(C,2), V3=diag(C,-1)
```

and they yield

$$\boldsymbol{V} = \begin{bmatrix} 1 & 0 & 0 \\ 0 & 2 & 0 \\ 0 & 0 & 3 \end{bmatrix}, \ \boldsymbol{V}_1 = \begin{bmatrix} 1 \\ 2 \\ 3 \end{bmatrix}, \ \boldsymbol{V}_2 = \begin{bmatrix} 0 & 0 & 1 & 0 & 0 \\ 0 & 0 & 0 & 2 & 0 \\ 0 & 0 & 0 & 0 & 3 \\ 0 & 0 & 0 & 0 & 0 \\ 0 & 0 & 0 & 0 & 0 \end{bmatrix}, \ \boldsymbol{V}_3 = \begin{bmatrix} 0 & 0 & 0 & 0 \\ 1 & 0 & 0 & 0 \\ 0 & 2 & 0 & 0 \\ 0 & 0 & 3 & 0 \end{bmatrix}.$$

In fact, k can be assigned to negative integers, indicating one wants to specify the kth lower-diagonal elements. With the use of such properties

```
>> V=diag([1 2 3 4])+diag([2 3 4],1)+diag([5 4 3],-1)
```

the tri-diagonal matrix $\boldsymbol{V}=\begin{bmatrix}1&2&0&0\\5&2&3&0\\0&4&3&4\\0&0&3&4\end{bmatrix}$ can then be established.

Assume that there exist the following matrices $\boldsymbol{A}_1, \boldsymbol{A}_2, \cdots, \boldsymbol{A}_n$, the following MATLAB function `diagm()` can be written to construct block diagonal matrices

```
function A=diagm(varargin)
A=[];
for i=1:length(varargin),
    A1=varargin{i}; [n,m]=size(A); [n1,m1]=size(A1);
    A(n+1:n+n1,m+1:m+m1)=A1;
end
```

such that

$$\boldsymbol{A}=\begin{bmatrix}\boldsymbol{A}_1 & & & \\ & \boldsymbol{A}_2 & & \\ & & \ddots & \\ & & & \boldsymbol{A}_n\end{bmatrix}. \tag{4.2}$$

Hankel matrices

The general form of a Hankel matrix is given below, with all the elements in each back-diagonal the same.

$$\boldsymbol{H}=\begin{bmatrix}c_1 & c_2 & \cdots & c_m\\ c_2 & c_3 & \cdots & c_{m+1}\\ \vdots & \vdots & \ddots & \vdots\\ c_n & c_{n+1} & \cdots & c_{n+m-1}\end{bmatrix}. \tag{4.3}$$

In MATLAB, the Hankel matrix based on two given vectors $\boldsymbol{c}$ and $\boldsymbol{r}$ can be constructed by `H=hankel(c,r)`, and the first column of matrix $\boldsymbol{H}$ can be assigned to vector $\boldsymbol{c}$, with the last row assigned to $\boldsymbol{r}$. Thus using the properties of a Hankel matrix, the full Hankel matrix can then be established.

If there is only one vector $\boldsymbol{c}$ specified, the command `H=hankel(c)` can be used to construct an upper-triangular Hankel matrix.

Example 4.3 Establish the following Hankel matrices using MATLAB statements.

$$\boldsymbol{H}_1=\begin{bmatrix}1&2&3&4&5&6&7\\2&3&4&5&6&7&8\\3&4&5&6&7&8&9\end{bmatrix},\quad \boldsymbol{H}_2=\begin{bmatrix}1&2&3\\2&3&0\\3&0&0\end{bmatrix}.$$

Solution In order to construct the above Hankel matrix, the $\boldsymbol{c}$ and $\boldsymbol{r}$ vectors should be assigned to `c=[1,2,3]`, `r=[3 4 5 6 7 8 9]`. The Hankel matrices can then be established with the following statements.

```
>> c=[1 2 3]; r=[3 4 5 6 7 8 9];
   H1=hankel(c,r), H2=hankel(c)
```

Hilbert matrices and their inverses

Hilbert matrix is a special matrix whose (i,j)th element is defined as $h_{i,j} = 1/(i+j-1)$. An $n \times n$ square Hilbert can be written as

$$\boldsymbol{H} = \begin{bmatrix} 1 & 1/2 & 1/3 & \cdots & 1/n \\ 1/2 & 1/3 & 1/4 & \cdots & 1/(n+1) \\ \vdots & \vdots & \vdots & \ddots & \vdots \\ 1/n & 1/(n+1) & 1/(n+2) & \cdots & 1/(2n-1) \end{bmatrix}. \tag{4.4}$$

The syntax for generating the Hilbert matrix is `A=hilb(n)`.

Large-sized Hilbert matrices are bad-conditioned matrices. Overflow will often occur during inverting such a matrix. Thus a direct inverse Hilbert matrix can be obtained with function `B=invhilb(n)`.

Since Hilbert matrices are very close to singular matrices, one must be very careful in dealing with such matrices. It is suggested here that symbolic computation be used. If numerical methods are used, do validate the results.

Vandermonde matrices

For a given sequence $\boldsymbol{c} = \{c_1, c_2, \cdots, c_n\}$, a Vandermonde matrix can be established such that the (i,j)th element is defined as $v_{i,j} = c_i^{n-j}$, $i,j = 1,2,\cdots,n$.

$$\boldsymbol{V} = \begin{bmatrix} c_1^{n-1} & c_1^{n-2} & \cdots & c_1 & 1 \\ c_2^{n-1} & c_2^{n-2} & \cdots & c_2 & 1 \\ \vdots & \vdots & \ddots & \vdots & \vdots \\ c_n^{n-1} & c_n^{n-2} & \cdots & c_n & 1 \end{bmatrix}. \tag{4.5}$$

A Vandermonde matrix can be established with `V=vander(c)` function in MATLAB for a given vector $\boldsymbol{c}$.

Example 4.4 Establish a Vandermonde matrix $\boldsymbol{A} = \begin{bmatrix} 1 & 1 & 1 & 1 & 1 \\ 16 & 8 & 4 & 2 & 1 \\ 81 & 27 & 9 & 3 & 1 \\ 256 & 64 & 16 & 4 & 1 \\ 625 & 125 & 25 & 5 & 1 \end{bmatrix}$.

Solution For such a matrix, one should select `c=[1,2,3,4,5]`. Thus with the following statements, the corresponding Vandermonde can be constructed.

```
>> c=[1, 2, 3, 4, 5];  V=vander(c)
```

Companion matrices

Assume that there exists a monic polynomial (the highest order term with coefficient 1)

$$P(s) = s^n + a_1 s^{n-1} + a_2 s^{n-2} + \cdots + a_{n-1}s + a_n \tag{4.6}$$

a companion matrix can be established such that

$$\boldsymbol{A}_{\rm c} = \begin{bmatrix} -a_1 & -a_2 & \cdots & -a_{n-1} & -a_n \\ 1 & 0 & \cdots & 0 & 0 \\ 0 & 1 & \cdots & 0 & 0 \\ \vdots & \vdots & \ddots & \vdots & \vdots \\ 0 & 0 & \cdots & 1 & 0 \end{bmatrix}. \tag{4.7}$$

A companion matrix can be established using `B=compan(p)`, where $\boldsymbol{p}$ is a polynomial coefficient vector, and `compan()` function will automatically transform it into monic form.

Example 4.5 Consider a polynomial $P(s) = 2s^4+4s^2+5s+6$. Find its companion matrix.

Solution The characteristic polynomial can be entered first and then the companion matrix $\boldsymbol{A}$ can be established using the following statements

```
>> P=[2 0 4 5 6]; A=compan(P)
```

and the matrix $\boldsymbol{A} = \begin{bmatrix} 0 & -2 & -2.5 & -3 \\ 1 & 0 & 0 & 0 \\ 0 & 1 & 0 & 0 \\ 0 & 0 & 1 & 0 \end{bmatrix}$ can be obtained.

4.1.2 Defining symbolic matrices

For a given numerical matrix $\boldsymbol{A}$, one may transform it by `B=sym(A)` into a symbolic matrix. Thus all the numerical matrices can be transformed into symbolic matrices so as to achieve higher accuracy. In some cases, it is even possible to find analytical solutions.

For other special matrices, such as Vandermonde matrix, Hankel matrix and companion matrix, the functions described above do not support symbolic matrices. One may write special purpose functions as follows to set up such matrices from given vectors. These functions should be placed under the `@sym` directory.

For instance, the symbolic companion matrix can be established from the following function

```
function A=compan(c)
if min(size(c))>1, error('Input argument must be a vector.'), end
n=length(c);
if n<=1, A=[];
elseif n==2, A=-c(2)/c(1);
else, c=c(:).'; A=sym(diag(ones(1,n-2),-1)); A(1,:)=-c(2:n)./ c(1);
end
```

Example 4.6 Establish a companion matrix from the following polynomial

$$P(\lambda) = a_1\lambda^9 + a_2\lambda^8 + a_3\lambda^7 + \cdots + a_8\lambda^2 + a_9\lambda + a_{10}.$$

Solution With the use of the extended `compan()` function given above, the required matrix can be established using the following statements

```
>> syms a1 a2 a3 a4 a5 a6 a7 a8 a9 a10
   A=compan([a1 a2 a3 a4 a5 a6 a7 a8 a9 a10])
```

the following matrix can be generated

$$\begin{bmatrix} -a_2/a_1 & -a_3/a_1 & -a_4/a_1 & -a_5/a_1 & -a_6/a_1 & -a_7/a_1 & -a_8/a_1 & -a_9/a_1 & -a_{10}/a_1 \\ 1 & 0 & 0 & 0 & 0 & 0 & 0 & 0 & 0 \\ 0 & 1 & 0 & 0 & 0 & 0 & 0 & 0 & 0 \\ 0 & 0 & 1 & 0 & 0 & 0 & 0 & 0 & 0 \\ 0 & 0 & 0 & 1 & 0 & 0 & 0 & 0 & 0 \\ 0 & 0 & 0 & 0 & 1 & 0 & 0 & 0 & 0 \\ 0 & 0 & 0 & 0 & 0 & 1 & 0 & 0 & 0 \\ 0 & 0 & 0 & 0 & 0 & 0 & 1 & 0 & 0 \\ 0 & 0 & 0 & 0 & 0 & 0 & 0 & 1 & 0 \end{bmatrix}.$$

Similarly, the Hankel matrix can be established using the new function

```
function H=hankel(c,r)
c=c(:); nc=length(c); if nargin==1, r=zeros(size(c)); end
r=r(:); nr=length(r); x=[c; r((2:nr)')]; cidx=(1:nc)';
ridx=0:(nr-1); H1=cidx(:,ones(nr,1))+ridx(ones(nc,1),:); H=x(H1);
```

and the symbolic Vandermonde matrix can also be set up from using

```
function A=vander(v)
n=length(v); v=v(:); A=sym(ones(n));
for j=n-1:-1:1, A(:,j)=v.*A(:,j+1); end
```

4.2 Fundamental Matrix Operations

4.2.1 Basic concepts and properties of matrices

Determinant

The determinant of matrix $\boldsymbol{A} = \{a_{ij}\}$ is defined as

$$\boldsymbol{D} = |\, \boldsymbol{A} \,| = \det(\boldsymbol{A}) = \sum (-1)^k a_{1k_1} a_{2k_2} \cdots a_{nk_n} \tag{4.8}$$

where the sum is taken over all possible permutations on n elements and the sign is positive if the permutation is even and negative if the permutation is odd.

There are many algorithms which can be used to compute the determinant of a matrix. A built-in function `det()` provided in MATLAB can be used to calculate the determinant `d=det(A)`, and this function applies both to symbolic and numerical matrices.

Example 4.7 Compute the determinant of a given matrix $\boldsymbol{A} = \begin{bmatrix} 16 & 2 & 3 & 13 \\ 5 & 11 & 10 & 8 \\ 9 & 7 & 6 & 12 \\ 4 & 14 & 15 & 1 \end{bmatrix}$.

Solution The determinant of matrix $\boldsymbol{A}$ can be obtained and it equals 0, which means that matrix $\boldsymbol{A}$ is singular.

```
>> A=[16 2 3 13; 5 11 10 8; 9 7 6 12; 4 14 15 1]; det(A)
```

Example 4.8 From the example given in Chapter 1, it is known that large-sized Hilbert matrix is very close to singular. Calculate analytically the determinant of a 20×20 Hilbert matrix.

Solution The function `hilb()` can be used to declare a numerical 20×20 Hilbert matrix. It can then be transformed into a symbolic matrix. The `det()` function in MATLAB can be used to calculate analytically the determinant of the matrix.

```
>> A=sym(hilb(20)); det(A)
```

Thus the determinant can be obtained as

$$\det(\boldsymbol{H}) = \frac{1}{\underbrace{237745\cdots 6800000000000000000000000000000000000}_{\text{225 digits, some are omitted}}} \approx 4.206179\times 10^{-224}.$$

Trace

For a square matrix $\boldsymbol{A} = \{a_{ij}\}$, $i, j = 1, 2, \cdots, n$, the trace of the $\boldsymbol{A}$ is defined as

$$\mathrm{tr}(\boldsymbol{A}) = \sum_{i=1}^{n} a_{ii} \tag{4.9}$$

i.e., the trace of a matrix is defined as the sum of diagonal elements. From linear algebra theory, the trace of a matrix equals the sum of the eigenvalues. The trace of matrix $\boldsymbol{A}$ can be obtained using the MATLAB function `trace()`, such that `t=trace(A)`.

The trace of the matrix in Example 4.7 can be obtained directly from trace($\boldsymbol{A}$)=34.

Rank

If for a given $n \times m$ matrix, there exist maximum r_c linearly independent columns, the column rank of the matrix is r_c. If $r_\mathrm{c} = m$, the matrix is referred to as a *full column rank matrix*. Similarly, if there exist maximum r_r linearly independent rows, the row rank of the matrix is r_r. If $r_\mathrm{r} = n$, matrix $\boldsymbol{A}$ is referred to as a *full row rank matrix*. It can be shown that the column rank and row rank of the same matrix are identical, and they both can be called the *rank* of the matrix, i.e., $\mathrm{rank}(\boldsymbol{A}) = r_\mathrm{c} = r_\mathrm{r}$. The rank of matrix $\boldsymbol{A}$ is mathematically denoted as $\mathrm{rank}(\boldsymbol{A})$.

There are various algorithms for calculating the ranks of given matrices. Some of the algorithms may be numerically unstable. A built-in function `rank()` has been provided in MATLAB which applies to both numerical and symbolic matrices. The syntax of the function is

```
r=rank(A)     % symbolic or numerical
r=rank(A,ε)   % numerical rank with error tolerance of ε
```

where $\boldsymbol{A}$ is the given matrix, and ε is user specified error tolerance.

Example 4.9 Find the rank of matrix $\boldsymbol{A}$ in Example 4.7.

Solution The MATLAB function `rank(A)` can be used to calculate the rank of matrix $\boldsymbol{A}$ and $\text{rank}(\boldsymbol{A}) = 3$, which indicates that $\boldsymbol{A}$ is not a full rank matrix.

Example 4.10 Now consider the 20×20 Hilbert matrix in Example 4.8. Find the rank of the matrix in numerical and analytic methods respectively.

Solution The following commands can be used to find the rank in a numerical way

```
>> H=hilb(20); rank(H)
```

and it can be seen that the rank of the matrix is 13, which means that the matrix is not of full rank. Now let us try the analytic method

```
>> H=sym(hilb(20)); rank(H)
```

and it is concluded that the rank of the matrix is 20. So, if the numerical method is adopted, one should be very careful, since misleading results can sometimes be obtained.

Norms

The norms of a matrix can be considered as a measure of "size" of the matrix. Before introducing the concept of the norms of the matrices, the norms of the vectors are introduced first. For a vector $\boldsymbol{x}$ in linear space, if there exists a function $\rho(\boldsymbol{x})$ satisfying the following three conditions:

(i) $\rho(\boldsymbol{x}) \geqslant 0$ and $\rho(\boldsymbol{x}) = 0$ if and only if $\boldsymbol{x} = \boldsymbol{0}$;
(ii) $\rho(a\boldsymbol{x}) = |a|\rho(\boldsymbol{x})$, a is any given scalar;
(iii) for vectors $\boldsymbol{x}$ and $\boldsymbol{y}$, there exists $\rho(\boldsymbol{x}+\boldsymbol{y}) \leqslant \rho(\boldsymbol{x}) + \rho(\boldsymbol{y})$.

then $\rho(\boldsymbol{x})$ is referred to as the *norm* of the vector $\boldsymbol{x}$. There are various forms of the norms. It can be shown that a class of norms defined below satisfies all the above three conditions.

$$||\boldsymbol{x}||_p = \left(\sum_{i=1}^{n} | x_i |^p\right)^{1/p}, \ p = 1, 2, \cdots, \text{and } ||\boldsymbol{x}||_\infty = \max_{1\leqslant i\leqslant n} | x_i | \tag{4.10}$$

where the notation $||\boldsymbol{x}||_p$ is used to define the p-norm of the vector $\boldsymbol{x}$.

The definition of the norms of a matrix $\boldsymbol{A}$ is

$$||\boldsymbol{A}|| = \sup_{\boldsymbol{x}\neq 0} \frac{||\boldsymbol{A}\boldsymbol{x}||}{||\boldsymbol{x}||} \tag{4.11}$$

for any non-zero vector $\boldsymbol{x}$. Similar to vector norms, the commonly used norms of a matrix are the following three

$$||\boldsymbol{A}||_1 = \max_{1\leqslant j\leqslant n} \sum_{i=1}^{n} |a_{ij}|, \ ||\boldsymbol{A}||_2 = \sqrt{s_{\max}(\boldsymbol{A}^{\mathrm{T}}\boldsymbol{A})}, \ ||\boldsymbol{A}||_\infty = \max_{1\leqslant i\leqslant n} \sum_{j=1}^{n} |a_{ij}| \tag{4.12}$$

where $s(\boldsymbol{X})$ is the eigenvalue of matrix $\boldsymbol{X}$, while $s_{\max}(\boldsymbol{A}^{\mathrm{T}}\boldsymbol{A})$ is the maximum eigenvalue of matrix $\boldsymbol{A}^{\mathrm{T}}\boldsymbol{A}$. In fact, $||\boldsymbol{A}||_2$ equals the maximum singular value of matrix $\boldsymbol{A}$.

The norms of a matrix can be evaluated using the MATLAB function `norm()`. Note that the `norm()` function applies only to numerical matrices. The syntax of the function is

```
N=norm(A)          % default for ||A||_2
N=norm(A,opts)     % opts could be 1,2,inf, see Table 4.1
```

Thus different norms of matrix $\boldsymbol{A}$ in Example 4.7 can be evaluated using the following MATLAB statements

```
>> A=[16 2 3 13; 5 11 10 8; 9 7 6 12; 4 14 15 1];
   [norm(A), norm(A,2), norm(A,1), norm(A,Inf), norm(A,'fro')]
```

where $||\boldsymbol{A}||_1 = ||\boldsymbol{A}||_2 = ||\boldsymbol{A}||_\infty = 34$, $||\boldsymbol{A}||_F = 38.6782$. The matrix here is a special matrix such that the norms $||\boldsymbol{A}||_1 = ||\boldsymbol{A}||_2 = ||\boldsymbol{A}||_\infty$ but in general cases, the specific values of the norms should not be the same.

TABLE 4.1: The options for the `norm()` function

options	definitions and algorithms
none	the maximum singular value, i.e., $\|\|\boldsymbol{A}\|\|_2$
2	the same as defaults, i.e., $\|\|\boldsymbol{A}\|\|_2$
1	1-norm of the matrix, i.e., $\|\|\boldsymbol{A}\|\|_1$
`Inf` or `'inf'`	infinite norm, i.e., $\|\|\boldsymbol{A}\|\|_\infty$
`'fro'`	Frobenius norm of the matrix, i.e., $\|\|\boldsymbol{A}\|\|_F = \mathrm{tr}(\boldsymbol{A}^{\mathrm{T}}\boldsymbol{A})$
integer p	applies only to vectors. For matrices, only 1,2, `inf` and `'fro'` are allowed
`-inf`	For vectors only, where $\|\|\boldsymbol{A}\|\|_{-\infty} = \min(\|\sum a_i\|)$

It should be again noted that the function `norm()` is not applicable to symbolic matrices. One should convert the symbolic matrix to the numerical one using the function `double()`, then the function `norm()` can be used to evaluate the norm.

Characteristic polynomials

When a symbolic variable s is introduced, the matrix $s\boldsymbol{I} - \boldsymbol{A}$ can be constructed, and if one wants to calculate its determinant, a polynomial of s

can be obtained such that

$$C(s) = \det(s\boldsymbol{I} - \boldsymbol{A}) = s^n + c_1 s^{n-1} + \cdots + c_{n-1}s + c_n \tag{4.13}$$

and polynomial $C(s)$ is referred to as the *characteristic polynomial* of matrix $\boldsymbol{A}$. In the formula, the coefficients c_i, $i = 1, 2, \cdots, n$ are referred to as the *coefficients of the characteristic polynomial.*

A MATLAB function `c=poly(A)` can be applied to evaluate the coefficients of the characteristic polynomial of matrix $\boldsymbol{A}$, where the returned vector $\boldsymbol{c}$ contains the coefficients in descending order of s of the characteristic polynomial. If the input argument $\boldsymbol{A}$ is a vector, the `poly()` function will return the polynomial coefficients vector whose roots are given in $\boldsymbol{A}$.

It should also be noted that this function also applies to symbolic $\boldsymbol{A}$ matrix, and in this case, the returned variable is no longer the coefficient vector. The mathematical representation of a polynomial will be returned.

Example 4.11 Find the characteristic polynomial of matrix $\boldsymbol{A}$ given in Example 4.7.

Solution Using the `poly()` function, the following results can be obtained

```
>> A=[16 2 3 13; 5 11 10 8; 9 7 6 12; 4 14 15 1]; c1=poly(A)
```

the coefficient vector is $\boldsymbol{c}_1 = 10^3 \times [0.0010, -0.0340, -0.0800, 2.7200, -0.0000]^{\mathrm{T}}$, and it can be seen that there might exist minor errors in the results.

Using symbolic matrix $\boldsymbol{A}$, the function `poly()` will yield a characteristic polynomial such that $\phi(x) = x^4 - 34x^3 - 80x^2 + 2720x$.

```
>> A=sym(A); poly(A)
```

In practical applications, there are other algorithms which can be used to calculate the coefficients of the characteristic polynomial of a matrix. For instance, the Leverrier-Faddeev recursive algorithm is an accurate algorithm for solving such problems.

$$c_{k+1} = -\frac{1}{k}\mathrm{tr}(\boldsymbol{A}\boldsymbol{R}_k), \quad \boldsymbol{R}_{k+1} = \boldsymbol{A}\boldsymbol{R}_k + c_{k+1}\boldsymbol{I}, \quad k = 1, \cdots, n \tag{4.14}$$

where $\boldsymbol{R}_1 = \boldsymbol{I}, c_1 = 1$.

In the above algorithm, an identity matrix $\boldsymbol{I}$ is assigned to matrix $\boldsymbol{R}_1$. Then for each value of k, the matrix $\boldsymbol{R}_k$ is updated, and then the value of c_k can be found. Based on such an algorithm, the following MATLAB function can be written

```
function c=poly1(A)
[nr,nc]=size(A);
if nc==nr, I=eye(nc);  R=I; c=[1 zeros(1,nc)];
   for k=1:nc, c(k+1)=-1/k*trace(A*R); R=A*R+c(k+1)*I; end
elseif (nr==1 | nc==1), A=A(isfinite(A)); n=length(A);
   c = [1 zeros(1,n)];
```

```
    for j=1:n, c(2:(j+1))=c(2:(j+1))-A(j).*c(1:j); end
  else, error('Argument must be a vector or a square matrix.'); end
```

With the new `poly1()` function, accurate result can be achieved

```
>> poly1(A)
```

such that $\boldsymbol{c} = [1, -34, -80, 2720, 0]$, which is the accurate result.

Example 4.12 For a given vector $\boldsymbol{B} = [a_1, a_2, a_3, a_4, a_5]$, establish the corresponding Hankel matrix, then find the characteristic polynomial.

Solution The command **A**`=hankel(`**B**`)` can be used to construct a Hankel matrix. Thus the function `poly(`**A**`)` can then be used to calculate its characteristic polynomial.

```
>> syms a1 a2 a3 a4 a5 x; A=hankel([a1 a2 a3 a4 a5]);
   collect(poly(A),x) % polynomial collecting in terms of x
```

The mathematical representation of the characteristic equation can be written as

$$\begin{aligned}p(\boldsymbol{A}) = {} & x^5 + (-a_3 - a_1 - a_5)x^4 + (a_5a_1 + a_3a_1 + a_5a_3 - 2a_4^2 - 2a_5^2 - a_2^2 - a_3^2)x^3 \\ & + (a_4^2a_1 + a_5^2a_3 + a_4^2a_5 + a_2^2a_5 + a_4^2a_3 - a_3a_1a_5 - 2a_2a_5a_4a_5^2a_1 - 2a_2a_4a_3 + 2a_5^3 + a_3^3)x^2 \\ & + (-3a_4^2a_3a_5 + a_5^4 + a_4^2a_5^2 + a_4^4 - a_5^3a_1 + a_5^2a_3^2 + 2a_2a_5^2a_4 - a_5^3a + 3)x - a_5^5.\end{aligned}$$

There is no existing MATLAB function to extract the coefficients from a symbolic polynomial $p(x)$. If the polynomial can be expressed as $p(x) = p_1x^n + p_2x^{n-1} + p_3x^{n-2} + \cdots + p_nx + p_{n+1}$, then it is easily found that

$$p_{n+1} = p(0), \text{ and } p_i = \frac{1}{(n-i+1)!} \left. \frac{\mathrm{d}^{n-i+1}p(x)}{\mathrm{d}x^{n-i+1}} \right|_{t=0}, \quad i = 1, 2, \cdots, n. \tag{4.15}$$

The MATLAB implementation of the parameters can be written as

```
function pc=polycoef(p,x,n)
pc(n+1)=subs(p,x,0); p1=p;
for i=1:n, p1=diff(p1,x)/i; pc(n-i+1)=subs(p1,x,0); end
```

Example 4.13 Extract the coefficients of the characteristic polynomial of matrix **A** in Example 4.12.

Solution The following statements can be used to extract the coefficients of the characteristic polynomial

```
>> syms a1 a2 a3 a4 a5 x; A=hankel([a1 a2 a3 a4 a5]);
   p=poly(A); P=polycoef(p,x,5)
```

and it is found that

$$\begin{aligned}& p_1 = 1, \quad p_2 = -a_3 - a_5 - a_1, \quad p_3 = a_1a_5 - a_2^2 + a_3a_5 - 2a_4^2 + a_1a_3 - a_3^2 - 2a_5^2 \\ & p_4 = a_3^3 + a_4^2a_3 - 2a_2a_5a_4 + a_1a_5^2 - a_1a_3a_5 - 2a_2a_4a_3 + a_4^2a_1 + a_5a_4^2 + 2a_5^3 + a_2^2a_5 + a_3a_5^2 \\ & p_5 = 2a_2a_5^2a_4 + a_3^2a_5^2 + a_5^4 - 3a_3a_5a_4^2 - a_3a_5^3 + a_5^2a_4^2 - a_1a_5^3 + a_4^4, \quad p_6 = -a_5^5.\end{aligned}$$

Evaluation of polynomial matrices

Polynomial matrices take the following form

$$\boldsymbol{B} = a_1\boldsymbol{A}^n + a_2\boldsymbol{A}^{n-1} + \cdots + a_n\boldsymbol{A} + a_{n+1}\boldsymbol{I} \tag{4.16}$$

where $\boldsymbol{A}$ is a given matrix, $\boldsymbol{I}$ is an identity matrix whose size is the same as matrix $\boldsymbol{A}$. The matrix $\boldsymbol{B}$ is then the polynomial matrix. In MATLAB, the polynomial matrix can be evaluated using the function `polyvalm()`, such that `B=polyvalm(a,A)`, where $\boldsymbol{a}$=`[`$a_1, a_2, \cdots, a_n, a_{n+1}$`]` is the coefficients in descending order of s of the polynomial.

On the other hand, if polynomial operation is defined upon "dot operation" basis such that

$$\boldsymbol{C} = a_1\boldsymbol{x}\texttt{.\^{}}n + a_2\boldsymbol{x}\texttt{.\^{}}(n-1) + \cdots + a_{n+1} \tag{4.17}$$

the matrix $\boldsymbol{C}$ can be evaluated from `C=polyval(a,x)`.

If polynomial p is defined in symbolic form, the function `subs()` can be used to evaluate the polynomial upon "dot operation" basis, where `C=subs(p,s,x)`.

Example 4.14 Cayley-Hamilton Theorem is a very important theorem in linear algebra. The theorem states that if the characteristic polynomial of matrix $\boldsymbol{A}$ is

$$f(s) = \det(s\boldsymbol{I} - \boldsymbol{A}) = a_1 s^n + a_2 s^{n-1} + \cdots + a_n s + a_{n+1} \tag{4.18}$$

then $f(\boldsymbol{A}) = 0$, i.e.,

$$a_1\boldsymbol{A}^n + a_2\boldsymbol{A}^{n-1} + \cdots + a_n\boldsymbol{A} + a_{n+1}\boldsymbol{I} = \boldsymbol{0}. \tag{4.19}$$

Assume that matrix $\boldsymbol{A}$ is a Vandermonde matrix. Verify that it satisfies the Cayley-Hamilton Theorem.

Solution The following statements can be used to verify the Theorem.

```
>> A=vander([1 2 3 4 5 6 7])
   aa=poly(A); B=polyvalm(aa,A); norm(B)
```

The norm of the error matrix is 2.1887×10^6, which is too large because the `poly()` function is not accurate. Thus something strange happened in the verification.

It has been indicated that the function `poly()` may cause errors, and for this particular matrix, the error is so large that misleading results are obtained. So for this matrix, the new function `poly1()` should be used instead. And it can be seen that with the new function, exact solutions can easily be obtained.

```
>> aa1=poly1(A); B1=polyvalm(aa1,A); norm(B1)
```

It can be seen from the results that $\boldsymbol{B}$ matrix obtained is a matrix of zeros, which means that for the given matrix, the Cayley-Hamilton Theorem holds.

Conversion between symbolic and numerical polynomials

A polynomial can either be represented in a numerical way or in a symbolic way. In the former case, the polynomial can be expressed by coefficient vector such that `p=[`a_1`, `a_2`, ···, `a_{n+1}`]`. In the latter case, it can be expressed by symbolic polynomials. One may convert a numerical polynomial to a symbolic one using `poly2sym()` function, and the function `sym2poly()` can be used to convert in the other way. The syntaxes of the two functions are rather simple.

`f=poly2sym(p)` or `f=poly2sym(p,x)` , or `p=sym2poly(f)`

Example 4.15 Represent the $f = s^5 + 2s^4 + 3s^3 + 4s^2 + 5s + 6$ in both numerical and symbolic forms.

Solution For simplicity, a vector can be constructed from the coefficients of the polynomial and then the function `poly2sym()` can be used to find the symbolic representation

```
>> P=[1 2 3 4 5 6];  % coefficients of the polynomial in descending order
   f=poly2sym(P,'v') % v is used as the operator
```

and the symbolic polynomial can be obtained as $f = v^5 + 2v^4 + 3v^3 + 4v^2 + 5v + 6$.

The numerical function can be obtained with `P=sym2poly(f)`.

4.2.2 Matrix inversion and generalized inverse of a matrix

Inverse matrix

For an $n \times n$ non-singular square matrix $\boldsymbol{A}$, if there exists a matrix $\boldsymbol{C}$ of the same size satisfying

$$\boldsymbol{AC} = \boldsymbol{CA} = \boldsymbol{I} \tag{4.20}$$

where $\boldsymbol{I}$ is an identity matrix, then matrix $\boldsymbol{C}$ is referred to as the *inverse matrix* of $\boldsymbol{A}$, denoted as $\boldsymbol{C} = \boldsymbol{A}^{-1}$.

A MATLAB function `C=inv(A)` is provided to calculate the inverse matrix $\boldsymbol{C}$, and this function is applicable for both numerical and symbolic matrices.

Example 4.16 Compute the inverse matrix for the given Hilbert matrix.

Solution Let us consider first a 4×4 Hilbert matrix. The MATLAB function `inv()` can be used to find the inverse matrix

```
>> format long; H=hilb(4); H1=inv(H), norm(H*H1-eye(4))
```

and the equation obtained below with the error is around 1.3931×10^{-12}. The inverse matrix obtained is

$$\begin{bmatrix} 15.999999999999 & -119.99999999999 & 239.99999999998 & -139.99999999999 \\ -119.99999999999 & 1199.9999999999 & -2699.9999999997 & 1679.9999999998 \\ 239.99999999998 & -2699.9999999997 & 6479.9999999994 & -4199.9999999996 \\ -139.99999999999 & 1679.9999999998 & -4199.9999999996 & 2799.9999999997 \end{bmatrix}.$$

Since a large-sized Hilbert matrix is close to a singular matrix, the use of numerical function `inv()` is not recommended for Hilbert matrices. The function `invhilb()`

can be used instead to find the accurate inverse matrix. For the 4×4 Hilbert matrix, the inverse matrix can be obtained

```
>> H2=invhilb(4); norm(H*H2-eye(size(H)))
```

and the error is reduced to 5.6843×10^{-14}, which means that the function `invhilb()` improves significantly for inverse matrices. Now consider a 10×10 Hilbert matrix, the inverse matrices by `inv()` and `invhilb()` can be obtained

```
>> H=hilb(10); H1=inv(H); norm(H*H1-eye(size(H)))
   H2=invhilb(10); norm(H*H2-eye(size(H)))
```

and the errors by these approaches are respectively 0.0032 and 2.5249×10^{-5}. The accuracy is very low. If the size of the matrix is further increased to 13, then the commands

```
>> H=hilb(13); H1=inv(H); norm(H*H1-eye(size(H)))
   H2=invhilb(13); norm(H*H2-eye(size(H)))
```

will detect the error norms by using the above two methods respectively as 81.1898, 11.7781. They are too high to be practically used.

Fortunately, the function `inv()` is also provided in the Symbolic Math Toolbox, which can be used to evaluate the inverse matrix for symbolic matrices. Even for large-sized non-singular matrices, the use of such a function can return error-free solutions. Using the following commands, the inverse matrix of a 7×7 Hilbert can be obtained and displayed

```
>> H=sym(hilb(7)); inv(H)
```

and the exact inverse can then be found

$$\begin{bmatrix} 49 & -1176 & 8820 & -29400 & 48510 & -38808 & 12012 \\ -1176 & 37632 & -317520 & 1128960 & -1940400 & 1596672 & -504504 \\ 8820 & -317520 & 2857680 & -10584000 & 18711000 & -15717240 & 5045040 \\ -29400 & 1128960 & -10584000 & 40320000 & -72765000 & 62092800 & -20180160 \\ 48510 & -1940400 & 18711000 & -72765000 & 133402500 & -115259760 & 37837800 \\ -38808 & 1596672 & -15717240 & 62092800 & -115259760 & 100590336 & -33297264 \\ 12012 & -504504 & 5045040 & -20180160 & 37837800 & -33297264 & 11099088 \end{bmatrix}.$$

In fact, even for a 30×30 Hilbert matrix, the exact inverse matrix can be obtained and it can be shown that there is zero error norm in the results

```
>> H=sym(hilb(30)); norm(double(H*inv(H)-eye(size(H))))
```

Example 4.17 Compute the inverse matrix for the matrix $\boldsymbol{A}$ in Example 4.7 and observe the differences using numerical and analytical methods.

Solution One can enter the matrix and then use `inv()` function to find its numerical inverse

```
>> A=[16 2 3 13; 5 11 10 8; 9 7 6 12; 4 14 15 1];
   B=inv(A), A*B
```

and it prompted that

```
Warning: Matrix is close to singular or badly scaled.
         Results may be inaccurate. RCOND = 1.306145e-017.
```

and the matrix $\boldsymbol{A}$ and $\boldsymbol{AB}$ obtained are respectively

$$\boldsymbol{B}=\begin{bmatrix}-2.6495 & -7.9484 & 7.9484 & 2.6495\\ -7.9484 & -23.845 & 23.845 & 7.9484\\ 7.9484 & 2.38455 & -23.845 & -7.9484\\ 2.6495 & 7.9484 & -7.9484 & -2.6495\end{bmatrix}\times 10^{14},\ \boldsymbol{AB}=\begin{bmatrix}1 & 0 & -1 & -0.25\\ -0.25 & 0 & 0 & 0.875\\ 0.25 & 0.5 & 0 & 0.25\\ 0.15625 & 0.125 & 0 & 1.7344\end{bmatrix}.$$

It can be seen that a warning message is displayed claiming that $\boldsymbol{A}$ matrix is close to a singular matrix, thus the results are useless. The product $\boldsymbol{AB}$ is no longer an identity matrix. The command `norm(A*B-eye(size(A)))` indicates that the norm of the error matrix is as big as 1.6408.

In fact, for singular matrices, there is no inverse matrix satisfying (4.20). For the same problem, from the Symbolic Math Toolbox function `inv(sym(A))`, it is clearly indicated that $\boldsymbol{A}$ is a singular matrix.

```
??? Error using ==> sym/inv
Error, (in inverse) singular matrix
```

Example 4.18 The symbolic function `inv()` can be applied also to matrices with symbolic variables. Find the inverse of Hankel matrix with variables.

Solution For instance, the inverse matrix of a given Hankel matrix can easily be obtained with the direct use of `inv()`

```
>> syms a1 a2 a3 a4; H=hankel([a1 a2 a3 a4]); inv(H)
```

the inverse matrix is

$$\boldsymbol{H}^{-1}=\begin{bmatrix}0 & 0 & 0 & 1/a_4\\ 0 & 0 & 1/a_4 & -1/a_4^2a_3\\ 0 & 1/a_4 & -1/a_4^2a_3 & -1/a_4^3(a_2a_4-a_3^2)\\ 1/a_4 & -1/a_4^2a_3 & -1/a_4^3(a_2a_4-a_3^2) & -(a_1a_4^2-2a_2a_3a_4+a_3^3)/a_4^4\end{bmatrix}.$$

Generalized matrix inverse

It can be seen that even with the Symbolic Math Toolbox, the inverse problems to a singular matrix cannot be handled. In fact, in a strict sense, no inverse matrix exists at all for singular matrices. In practical applications, one may need an "inverse" for singular or even rectangular matrices. Thus a generalized matrix should be defined. For a given matrix $\boldsymbol{A}$, if there exists another matrix $\boldsymbol{N}$ such that

$$\boldsymbol{ANA}=\boldsymbol{A} \tag{4.21}$$

the matrix $\boldsymbol{N}$ is referred to as the *generalized inverse matrix* of matrix $\boldsymbol{A}$, denoted by $\boldsymbol{N}=\boldsymbol{A}^{-}$. For an $n\times m$ rectangular matrix $\boldsymbol{A}$, the generalized inverse matrix $\boldsymbol{N}$ is an $m\times n$ matrix. It can be shown that there are an infinite number of $\boldsymbol{N}$ satisfying such a condition.

One may introduce a norm criterion such that

$$\min_{\boldsymbol{x}} ||\boldsymbol{A}\boldsymbol{x} - \boldsymbol{B}|| \tag{4.22}$$

is minimized and it can be shown that for any given matrix $\boldsymbol{A}$, there exists a unique matrix $\boldsymbol{M}$ such that the three conditions below are satisfied

(i) $\boldsymbol{AMA} = \boldsymbol{A}$

(ii) $\boldsymbol{MAM} = \boldsymbol{M}$

(iii) $\boldsymbol{AM}$ and $\boldsymbol{MA}$ are all Hermitian symmetrical matrices

Such a matrix $\boldsymbol{M}$ is referred to as the *Moore-Penrose pseudo-inverse matrix* of $\boldsymbol{A}$, denoted by $\boldsymbol{M} = \boldsymbol{A}^+$.

A MATLAB function `pinv()` can be used to find the Moore-Penrose pseudo-inverse of a given matrix. The syntax of the function is

```
M=pinv(A)      % evaluate the Moore-Penrose pseudo-inverse
M=pinv(A,ε)    % evaluate the inverse numerically with precision of ε
```

where the variable ϵ is used to judge whether a value is zero or not. The returned matrix $\boldsymbol{M}$ is the Moore-Penrose pseudo-inverse of the original matrix $\boldsymbol{A}$. If $\boldsymbol{A}$ is a non-singular square matrix, the resulted pseudo-inverse is in fact the inverse of the original matrix. However, the speed of the `pinv()` function is significantly lower than that of the `inv()` function.

Example 4.19 Find the pseudo-inverse of the singular matrix $\boldsymbol{A}$ in Example 4.7.

Solution For the singular matrix, the Moore-Penrose pseudo-inverse of the matrix should be established instead, using the following statements.

```
>> A=[16 2 3 13; 5 11 10 8; 9 7 6 12; 4 14 15 1]; B=pinv(A), A*B
```

The Moore-Penrose inverse of the matrix $\boldsymbol{B}$ and $\boldsymbol{AB}$ are

$$\boldsymbol{B} = \begin{bmatrix} 0.1011 & -0.0739 & -0.0614 & 0.0636 \\ -0.0364 & 0.0386 & 0.0261 & 0.001103 \\ 0.0136 & -0.0114 & -0.0239 & 0.0511 \\ -0.0489 & 0.0761 & 0.0886 & -0.0864 \end{bmatrix}, \quad \boldsymbol{AB} = \begin{bmatrix} 0.95 & -0.15 & 0.15 & 0.05 \\ -0.15 & 0.55 & 0.45 & 0.15 \\ 0.15 & 0.45 & 0.55 & -0.15 \\ 0.05 & 0.15 & -0.15 & 0.95 \end{bmatrix}.$$

From these results obtained, it can be observed that the values in $\boldsymbol{A}^+$ are meaningful. Note that $\boldsymbol{AA}^+$ is no longer an identify matrix. Now the three conditions for the Moore-Penrose pseudo-inverse matrix can be checked using the following statements:

```
>> [norm(A*B*A-A), norm(B*A*B-B), norm(A*B-(A*B)'), norm(B*A-(B*A)')]
```

with errors respectively $2.2383\times10^{-14}, 7.6889\times10^{-17}, 1.0753\times10^{-15}, 9.3653\times10^{-16}$. It can be seen that the above obtained matrix $\boldsymbol{B}$ is the Moore-Penrose inverse of the original matrix. Performing Moore-Penrose inverse to matrix $\boldsymbol{B}$, it can be seen that $(\boldsymbol{A}^+)^+ = \boldsymbol{A}$, with error 1.9278×10^{-14}.

```
>> pinv(B), norm(ans-A)
```

Example 4.20 For a rectangular matrix $A = \begin{bmatrix} 6 & 1 & 4 & 2 & 1 \\ 3 & 0 & 1 & 4 & 2 \\ -3 & -2 & -5 & 8 & 4 \end{bmatrix}$, find its rank and Moore-Penrose inverse and check whether the pseudo-inverse matrix obtained is correct or not.

Solution The following commands can be given to get the rank of the matrix. It can be seen that the rank is 2, rather than 3, which means that the matrix is not a full-rank matrix.

```
>> A=[6,1,4,2,1; 3,0,1,4,2; -3,-2,-5,8,4]; rank(A)
```

Since matrix A is a singular rectangular matrix, the function `pinv()` can be used to evaluate the Moore-Penrose pseudo-inverse of the matrix. Then each of the three conditions for Moore-Penrose pseudo-inverse can be verified and it can be shown that the resulted matrix is correct.

```
>> iA=pinv(A) % pseudo-inverse of a non-full-rank matrix
   [norm(A*iA*A-A), norm(iA*A-A'*iA'), ...
      norm(iA*A-A'*iA'), norm(A*iA-iA'*A')]
```

The pseudo-inverse is

$$A^+ = \begin{bmatrix} 0.073024740622506 & 0.041300877893057 & -0.022146847565842 \\ 0.010774142059058 & 0.0019952114924182 & -0.015562649640862 \\ 0.045889864325619 & 0.017757382282522 & -0.038507581803671 \\ 0.032721468475658 & 0.043096568236233 & 0.063846767757382 \\ 0.016360734237829 & 0.021548284118117 & 0.031923383878691 \end{bmatrix}$$

and $||A^+AA^+ - A^+|| = 1.0263\times10^{-16}$, $||AA^+A - A|| = 8.1145\times10^{-15}$, $||A^+A - (A)^{\rm H}(A^+)^{\rm H}|| = 3.9098\times10^{-16}$, $||A^+A - (A)^{\rm H}(A^+)^{\rm H}|| = 1.6653\times10^{-16}$.

4.2.3 Matrix eigenvalue problems

Eigenvalues and eigenvectors of a matrix

For the given matrix A, if there exists a non-zero vector x and a scalar λ satisfying

$$Ax = \lambda x \tag{4.23}$$

then λ is referred to as an *eigenvalue* of matrix A, while the vector x is referred to as the *eigenvector* of A corresponding to the eigenvalue λ. Strictly speaking, the eigenvector x should be referred to as the *right eigenvector*. If the eigenvalues are distinct, the eigenvectors are linearly independent. Thus a non-singular square eigen-matrix can be constructed. A diagonal matrix can be obtained if such a matrix is used to perform similar transformation. The eigenvalues and eigenvectors can easily be obtained using the `eig()` function. The syntax of the function is

```
d=eig(A)        % only eigenvalues are required
[V, D]=eig(A)   % if both eigenvalues and eigenvectors are expected
```

where d is a vector containing all the eigenvalues, while D is a diagonal matrix whose diagonal elements are the eigenvalues of the matrix, and each column in matrix $\boldsymbol{V}$ contains the eigenvector to the corresponding eigenvalues. The following relationship $\boldsymbol{AV} = \boldsymbol{VD}$ is satisfied. This function applies also to complex $\boldsymbol{A}$ matrix as well as symbolic ones.

The definition of the roots of the characteristic polynomial discussed earlier is exactly the same as the eigenvalues. If the characteristic polynomial can be exactly known, the function `roots()` can also be used in evaluating the eigenvalues of the matrix.

Example 4.21 Compute the eigenvalues and eigenvectors of matrix $\boldsymbol{A}$ in Example 4.7.

Solution Using numerical method, the eigenvalues can be obtained by the direct use of the `eig()` function such that

```
>> A=[16 2 3 13; 5 11 10 8; 9 7 6 12; 4 14 15 1]; eig(A)
```

and it can be seen that they are $34, \pm 8.9442719, -2.234826\times 10^{-15}$.

The `eig()` function provided in the Symbolic Math Toolbox can also be used to evaluate the eigenvalues and eigenvectors of a given matrix. Even for large-sized matrices, results with very high accuracy can be obtained.

```
>> eig(sym(A))
```

Thus the exact eigenvalues of the matrix are $0, 34, \pm 4\sqrt{5}$.

For the same matrix $\boldsymbol{A}$, the eigenvalues and eigenvectors can be solved numerically such that

```
>> [v,d]=eig(A)
```

the numerical solutions to the eigenvector and eigenvalues are

$$\boldsymbol{v} = \begin{bmatrix} -0.5 & -0.8236 & 0.3764 & -0.2236 \\ -0.5 & 0.4236 & 0.02361 & -0.6708 \\ -0.5 & 0.02361 & 0.4236 & 0.6708 \\ -0.5 & 0.3764 & -0.8236 & 0.2236 \end{bmatrix}, \quad \boldsymbol{d} = \begin{bmatrix} 34 & 0 & 0 & 0 \\ 0 & 8.9443 & 0 & 0 \\ 0 & 0 & -8.9443 & 0 \\ 0 & 0 & 0 & 9.416\times 10^{-16} \end{bmatrix}.$$

If the Symbolic Math Toolbox is used, the eigenvalues and eigenvectors can easily be found such that

```
>> [v,d]=eig(sym(A))
```

and it can be found

$$\boldsymbol{v} = \begin{bmatrix} -1 & 1 & -8\sqrt{5}-17 & 8\sqrt{5}-17 \\ -3 & 1 & 4\sqrt{5}+9 & -4\sqrt{5}+9 \\ 3 & 1 & 1 & 1 \\ 1 & 1 & 4\sqrt{5}+7 & -4\sqrt{5}+7 \end{bmatrix}, \quad \boldsymbol{d} = \begin{bmatrix} 0 & 0 & 0 & 0 \\ 0 & 34 & 0 & 0 \\ 0 & 0 & 4\sqrt{5} & 0 \\ 0 & 0 & 0 & -4\sqrt{5} \end{bmatrix}.$$

If matrix $\boldsymbol{A}$ has repeated eigenvalues, the eigenvector matrix will be a singular matrix.

Generalized eigenvalues and eigenvectors

Assume that there exists a scalar λ and a non-zero vector $\boldsymbol{x}$ such that

$$\boldsymbol{A}\boldsymbol{x} = \lambda\boldsymbol{B}\boldsymbol{x} \tag{4.24}$$

where $\boldsymbol{B}$ is a symmetrical positive-definite matrix, λ is referred to as the *generalized eigenvalue*, while $\boldsymbol{x}$ is the *generalized eigenvector*. In fact, the ordinary eigenvalue problem is a special case of the generalized eigenvalue problem when $\boldsymbol{B} = \boldsymbol{I}$ is assumed.

If matrix $\boldsymbol{B}$ is a non-singular square matrix, the generalized eigenvalue problem can be converted to the eigenvalue problem for matrix $\boldsymbol{B}^{-1}\boldsymbol{A}$.

$$\boldsymbol{B}^{-1}\boldsymbol{A}\boldsymbol{x} = \lambda\boldsymbol{x} \tag{4.25}$$

i.e., λ and $\boldsymbol{x}$ are respectively the eigenvalues and eigenvectors of matrix $\boldsymbol{B}^{-1}\boldsymbol{A}$. In MATLAB, the function `eig()` can be used to compute directly the generalized eigenvalues and eigenvectors such that

```
d=eig(A,B)          % generalized eigenvalue evaluation
[V,D]=eig(A,B)      % generalized eigenvalues and eigenvectors
```

With the `eig()` function, the generalized eigenvalues and eigenvectors can be obtained in matrices $\boldsymbol{D}$ and $\boldsymbol{V}$, where $\boldsymbol{AV} = \boldsymbol{BVD}$. It should be noted that the matrix $\boldsymbol{B}$ is no longer restricted to positive-definite matrices.

Example 4.22 Now consider the matrices

$$\boldsymbol{A} = \begin{bmatrix} 5 & 7 & 6 & 5 \\ 7 & 10 & 8 & 7 \\ 6 & 8 & 10 & 9 \\ 5 & 7 & 9 & 10 \end{bmatrix}, \quad \boldsymbol{B} = \begin{bmatrix} 2 & 6 & -1 & -2 \\ 5 & -1 & 2 & 3 \\ -3 & -4 & 1 & 10 \\ 5 & -2 & -3 & 8 \end{bmatrix}.$$

Compute the generalized eigenvalues and eigenvector matrices for the $(\boldsymbol{A}, \boldsymbol{B})$ pair.

Solution The following statements can be entered

```
>> A=[5,7,6,5; 7,10,8,7; 6,8,10,9; 5,7,9,10];
   B=[2,6,-1,-2; 5,-1,2,3; -3,-4,1,10; 5,-2,-3,8];
   [V,D]=eig(A,B), norm(A*V-B*V*D)
```

and the eigenvalues and eigenvector matrices can be obtained

$$\boldsymbol{V} = \begin{bmatrix} 0.3697 & -0.37409 + \mathrm{j}0.62591 & -0.37409 - \mathrm{j}0.62591 & 1 \\ 0.99484 & -0.067434 - \mathrm{j}0.25314 & -0.067434 + \mathrm{j}0.25314 & -0.60903 \\ 0.79792 & 0.92389 + \mathrm{j}0.026381 & 0.92389 - \mathrm{j}0.026381 & -0.23164 \\ 1 & -0.65986 - \mathrm{j}0.32628 & -0.65986 + \mathrm{j}0.32628 & 0.13186 \end{bmatrix}$$

$$\boldsymbol{D} = \begin{bmatrix} 4.7564 & 0 & 0 & 0 \\ 0 & 0.047055 + \mathrm{j}0.17497 & 0 & 0 \\ 0 & 0 & 0.047055 - \mathrm{j}0.17497 & 0 \\ 0 & 0 & 0 & -0.003689 \end{bmatrix}$$

and the norm of the error matrix is 1.5783×10^{-14}.

4.3 Fundamental Matrix Transformations

4.3.1 Similarity transformations and orthogonal matrices

For a square matrix $\boldsymbol{A}$, if there exists a non-singular matrix $\boldsymbol{B}$, then the original $\boldsymbol{A}$ matrix can be transformed into the following form

$$\boldsymbol{X} = \boldsymbol{B}^{-1}\boldsymbol{A}\boldsymbol{B} \tag{4.26}$$

and this transformation is referred to as *similarity transformation* and matrix $\boldsymbol{B}$ is referred to as the *similarity transformation matrix*. It can be shown that the determinant, rank, trace and eigenvalues of the transformed matrix will not change. Through properly chosen transformation matrix $\boldsymbol{B}$, one may transform the original matrix $\boldsymbol{A}$ to other forms without changing important properties of matrix $\boldsymbol{A}$.

For a class of special transformation matrices $\boldsymbol{T}$, if it satisfies $\boldsymbol{T}^{-1} = \boldsymbol{T}^{\mathrm{H}}$, where $\boldsymbol{T}^{\mathrm{H}}$ is the Hermitian conjugate transpose of matrix $\boldsymbol{T}$, matrix $\boldsymbol{T}$ is then referred to as an *orthogonal matrix*, and it can be denoted that $\boldsymbol{Q} = \boldsymbol{T}$. So it can be seen that the orthogonal matrix $\boldsymbol{Q}$ satisfies

$$\boldsymbol{Q}^{\mathrm{H}}\boldsymbol{Q} = \boldsymbol{I}, \quad \text{and} \quad \boldsymbol{Q}\boldsymbol{Q}^{\mathrm{H}} = \boldsymbol{I} \tag{4.27}$$

where $\boldsymbol{I}$ is an $n \times n$ identity matrix.

A MATLAB function `Q=orth(A)` can be used to construct the orthogonal basis of matrix $\boldsymbol{A}$. If matrix $\boldsymbol{A}$ is non-singular, the orthogonal basis matrix $\boldsymbol{Q}$ obtained satisfies the conditions in (4.27). If matrix $\boldsymbol{A}$ is singular, however, the columns in matrix $\boldsymbol{Q}$ equals the rank of matrix $\boldsymbol{A}$, and satisfies $\boldsymbol{Q}^{\mathrm{H}}\boldsymbol{Q} = \boldsymbol{I}$, other than $\boldsymbol{Q}\boldsymbol{Q}^{\mathrm{H}} = \boldsymbol{I}$.

Example 4.23 Compute the orthogonal basis for matrix $\boldsymbol{A} = \begin{bmatrix} 5 & 9 & 8 & 3 \\ 0 & 3 & 2 & 4 \\ 2 & 3 & 5 & 9 \\ 3 & 4 & 5 & 8 \end{bmatrix}$.

Solution The orthogonal basis of a given matrix $\boldsymbol{A}$ can be established directly with the `orth()` function, and the following statements can also be used to verify the properties of the orthogonal matrix obtained.

```
>> A=[5,9,8,3; 0,3,2,4; 2,3,5,9; 3,4,5,8]; Q=orth(A)
   [norm(Q'*Q-eye(4)), norm(Q*Q'-eye(4))]
```

The orthogonal basis of $\boldsymbol{A}$ can then be found as

$$\boldsymbol{Q} = \begin{bmatrix} -0.61967134 & 0.77381388 & -0.026187275 & -0.12858357 \\ -0.25484758 & -0.15505966 & 0.94903028 & 0.10173858 \\ -0.51978107 & -0.52982004 & -0.15628279 & -0.65168555 \\ -0.52998848 & -0.31057898 & -0.27245447 & 0.74055484 \end{bmatrix}$$

with calculation errors $||\boldsymbol{Q}^{\mathrm{H}}\boldsymbol{Q} - \boldsymbol{I}|| = 4.6395\times10^{-16}$, $||\boldsymbol{Q}\boldsymbol{Q}^{\mathrm{H}} - \boldsymbol{I}|| = 4.9270\times10^{-16}$.

Example 4.24 Consider the singular matrix $\boldsymbol{A}$ defined in Example 4.7. Compute the orthogonal basis matrix and then verify the its properties.

Solution The orthogonal basis of matrix $\boldsymbol{A}$ can be obtained easily by the direct use of the function `orth()`.

```
>> A=[16,2,3,13; 5,11,10,8; 9,7,6,12; 4,14,15,1];
   Q=orth(A), a=norm(Q'*Q-eye(3))
```

It can be seen that since $\boldsymbol{A}$ is a singular matrix, the orthogonal basis is rectangular such that

$$\boldsymbol{Q} = \begin{bmatrix} -0.5 & 0.67082039324994 & 0.5 \\ -0.5 & -0.22360679774998 & -0.5 \\ -0.5 & 0.22360679774998 & -0.5 \\ -0.5 & -0.67082039324994 & 0.5 \end{bmatrix}$$

and the error $||\boldsymbol{Q}^{\mathrm{H}}\boldsymbol{Q} - \boldsymbol{I}|| = 1.0140\times 10^{-15}$.

Example 4.25 For a given matrix $\boldsymbol{A}$, if there exists a column vector $\boldsymbol{x}$, such that $\boldsymbol{T} = [\boldsymbol{x}, \boldsymbol{A}\boldsymbol{x}, \cdots, \boldsymbol{A}^{n-1}\boldsymbol{x}]$ is non-singular, the matrix $\boldsymbol{A}$ can be transformed into a companion-like matrix. Please transform the matrix in Example 4.23 into a companion matrix.

Solution A column vector $\boldsymbol{x}$ can be generated randomly and the transformation matrix and the transformed matrix can then be obtained

```
>> A=[5,7,6,5; 7,10,8,7; 6,8,10,9; 5,7,9,10];
   while(1), x=floor(2*rand(4,1)); T=sym([x A*x A^2*x A^3*x]);
   if rank(T)==4, break; end, end
   T, A1=inv(T)*A*T
```

which yields

$$\boldsymbol{T} = \begin{bmatrix} 1 & 11 & 326 & 9853 \\ 0 & 15 & 453 & 13696 \\ 1 & 16 & 472 & 14296 \\ 0 & 14 & 444 & 13489 \end{bmatrix}, \quad \boldsymbol{A}_1 = \begin{bmatrix} 0 & 0 & 0 & -1 \\ 1 & 0 & 0 & 100 \\ 0 & 1 & 0 & -146 \\ 0 & 0 & 1 & 35 \end{bmatrix}.$$

It should be noted that the matrix $\boldsymbol{T}$ is not unique. The matrix $\boldsymbol{A}_1$ is quite similar to the companion matrix defined in (4.7). If one does need the standard companion form, the following statements can further be given

```
>> T1=inv(T*fliplr(eye(4)))', A2=inv(T1)*A*T1,
```

the transformation matrix and companion form are as follows:

$$\boldsymbol{T} = \frac{1}{14053}\begin{bmatrix} -318 & 10591 & -29493 & 19064 \\ -176 & 5243 & 3298 & -11368 \\ 318 & -10591 & 29493 & -5011 \\ 75 & -1835 & -13063 & 2928 \end{bmatrix}, \quad \boldsymbol{A}_2 = \begin{bmatrix} 35 & -146 & 100 & -1 \\ 1 & 0 & 0 & 0 \\ 0 & 1 & 0 & 0 \\ 0 & 0 & 1 & 0 \end{bmatrix}.$$

4.3.2 Triangular and Cholesky decompositions

Triangular decompositions

The triangular decomposition of a matrix is also known as the *LU decomposition*, where the original matrix can be decomposed into the product of a lower-triangular matrix $\boldsymbol{L}$ and an upper-triangular matrix $\boldsymbol{U}$, such that $\boldsymbol{A} = \boldsymbol{L}\boldsymbol{U}$, where $\boldsymbol{L}$ and $\boldsymbol{U}$ can respectively be written as

$$\boldsymbol{L} = \begin{bmatrix} 1 & & & \\ l_{21} & 1 & & \\ \vdots & \vdots & \ddots & \\ l_{n1} & l_{n2} & \cdots & 1 \end{bmatrix}, \quad \boldsymbol{U} = \begin{bmatrix} u_{11} & u_{12} & \cdots & u_{1n} \\ & u_{22} & \cdots & u_{2n} \\ & & \ddots & \vdots \\ & & & u_{nn} \end{bmatrix} \tag{4.28}$$

where it can easily be shown that l_{ij} and u_{ij} can be calculated recursively that

$$l_{ij} = \frac{a_{ij} - \sum_{k=1}^{j-1} l_{ik}u_{kj}}{u_{jj}}, \quad (j < i), \quad u_{ij} = a_{ij} - \sum_{k=1}^{i-1} l_{ik}u_{kj}, \quad (j \geqslant i) \tag{4.29}$$

with initial values defined as $u_{1i} = a_{1i}, \quad i = 1, 2, \cdots, n$.

It should be noted that since the pivot element was not selected in the above formula, the direct use of such an algorithm may not be numerically stable, since small values or even 0 might be used as denominators. In MATLAB, a pivot-based LU decomposition function `lu()` is provided such that

```
[L,U]=lu(A)      % LU decomposition A = LU
[L,U,P]=lu(A)    % P is the permutation matrix, A = P^(-1)LU
```

where $\boldsymbol{L}$ and $\boldsymbol{U}$ are transformed lower- and upper-triangular matrices. In MATLAB, the `lu()` function considers the selection of pivoting element, thus reliable results will be ensured. The actual matrix $\boldsymbol{L}$ is not necessarily lower-triangular.

A MATLAB function `lu()` is written for symbolic matrix $\boldsymbol{A}$, and the function should be placed under the `@sym` directory. There is no pivot selected in the function, thus the $\boldsymbol{L}$ is a lower-triangular matrix

```
function [L,U]=lu(A)
n=length(A); U=sym(zeros(size(A))); L=sym(eye(size(A)));
U(1,:)=A(1,:); L(:,1)=A(:,1)/U(1,1);
for i=2:n,
   for j=2:i-1, L(i,j)=(A(i,j)-L(i,1:j-1)*U(1:j-1,j))/U(j,j); end
   for j=i:n, U(i,j)=A(i,j)-L(i,1:i-1)*U(1:i-1,j); end
end
```

Example 4.26 Consider the LU decomposition problem to Example 4.7. Try to use the two calling syntaxes of `lu()` to compute the triangular decomposition and then compare the results.

Solution For matrix $\boldsymbol{A}$, the triangular decomposition is performed such that

```
>> A=[16 2 3 13; 5 11 10 8; 9 7 6 12; 4 14 15 1]; [L1,U1]=lu(A)
```

where

$$\boldsymbol{L}_1=\begin{bmatrix}1&0&0&0\\0.3125&0.76852&1&0\\0.5625&0.43519&1&1\\0.25&1&0&0\end{bmatrix},\quad \boldsymbol{U}_1=\begin{bmatrix}16&2&3&13\\0&13.5&14.25&-2.25\\0&0&-1.8889&5.6667\\0&0&0&3.5527\times10^{-15}\end{bmatrix}.$$

It can be seen that $\boldsymbol{L}_1$ matrix is not a lower-triangular matrix. It is the permutated triangular matrix. Now let us consider the other syntax to the `lu()` function

```
>> [L,U,P]=lu(A)
```

where

$$\boldsymbol{L}=\begin{bmatrix}1&0&0&0\\0.25&1&0&0\\0.3125&0.7685&1&0\\0.5625&0.4352&1&1\end{bmatrix},\ \boldsymbol{U}=\begin{bmatrix}16&2&3&13\\0&13.5&14.25&-2.25\\0&0&-1.8889&5.6667\\0&0&0&3.55\times10^{-15}\end{bmatrix},\ \boldsymbol{P}=\begin{bmatrix}1&0&0&0\\0&0&0&1\\0&1&0&0\\0&0&1&0\end{bmatrix}.$$

It should be noted that the matrix $\boldsymbol{P}$ is not an identity matrix, and it is just a permutation matrix of such a matrix. The matrix $\boldsymbol{A}$ can be transformed back if the statement `inv(P)*L*U` is used. If the symbolic `lu()` function is used

```
>> A=sym(A); [L2,U2]=lu(A)
```

the exact triangular matrices can be found as

$$\boldsymbol{L}_2=\begin{bmatrix}1&0&0&0\\5/16&1&0&0\\9/16&47/83&1&0\\1/4&108/83&-3&1\end{bmatrix},\quad \boldsymbol{U}_2=\begin{bmatrix}16&2&3&13\\0&83/8&145/16&63/16\\0&0&-68/83&204/83\\0&0&0&0\end{bmatrix}.$$

Cholesky decomposition of symmetrical matrices

If matrix $\boldsymbol{A}$ is a symmetrical matrix, it can be decomposited using LU decomposition algorithm such that

$$\boldsymbol{A}=\boldsymbol{L}\boldsymbol{L}^{\mathrm{T}}=\begin{bmatrix}l_{11}&&&\\l_{21}&l_{22}&&\\\vdots&\vdots&\ddots&\\l_{n1}&l_{n2}&\cdots&l_{nn}\end{bmatrix}\begin{bmatrix}l_{11}&l_{21}&\cdots&l_{n1}\\&l_{22}&\cdots&l_{n2}\\&&\ddots&\vdots\\&&&l_{nn}\end{bmatrix}\tag{4.30}$$

and the LU decomposition algorithm can be simplified such that

$$l_{ii}=\sqrt{a_{ii}-\sum_{k=1}^{i-1}l_{ik}^2},\quad l_{ji}=\frac{1}{l_{jj}}\left(a_{ij}-\sum_{k=1}^{j-1}l_{ik}l_{jk}\right),\quad j<i\tag{4.31}$$

and such an algorithm is referred to as the *Cholesky decomposition algorithm*. To start with the algorithm, one should have initially $l_{11}=\sqrt{a_{11}}, l_{j1}=a_{j1}/l_{11}$.

A MATLAB function `chol()` is provided to perform Cholesky decomposition such that an upper-triangular matrix $\boldsymbol{D}$ is returned `D=chol(A)`, where $\boldsymbol{D}=\boldsymbol{L}^{\mathrm{T}}$.

An overloaded function `chol()` for symbolic $\boldsymbol{A}$ is written

```
function L=chol(A)
n=length(A); L(1,1)=sqrt(A(1,1)); L(2:n,1)=A(2:n,1)/L(1,1);
for i=2:n, k=1:i-1; L(i,i)=sqrt(A(i,i)-sum(L(i,k).^2));
   for j=i+1:n, L(j,i)=(A(j,i)-sum(L(j,k).*L(i,k)))/L(i,i); end
end
```

Example 4.27 Consider a matrix $\boldsymbol{A}=\begin{bmatrix}9&3&4&2\\3&6&0&7\\4&0&6&0\\2&7&0&9\end{bmatrix}$, perform numerical and analytical Cholesky decompositions.

Solution The Cholesky decompositions can be obtained by the use of `chol()` function

```
>> A=[9,3,4,2; 3,6,0,7; 4,0,6,0; 2,7,0,9]; D=chol(A), L=chol(sym(A))
```

and the solutions are respectively

$$\boldsymbol{D}=\begin{bmatrix}3&1&1.3333&0.66667\\0&2.2361&-0.59628&2.8324\\0&0&1.9664&0.40684\\0&0&0&0.60648\end{bmatrix},\ \boldsymbol{L}=\begin{bmatrix}3&0&0&0\\1&\sqrt{5}&0&0\\\dfrac{4}{3}&-\dfrac{4\sqrt{5}}{15}&\dfrac{\sqrt{870}}{15}&0\\\dfrac{2}{3}&\dfrac{19\sqrt{5}}{15}&\dfrac{2\sqrt{870}}{145}&\dfrac{4\sqrt{174}}{87}\end{bmatrix}.$$

Positive-definiteness, regular matrix: definitions and tests

The concept of positive-definiteness of matrices is established upon symmetrical matrices. Before introducing the concept, the main sub-matrices of a given matrix is defined. Assume that a symmetrical matrix $\boldsymbol{A}$ is defined as

$$\boldsymbol{A}=\begin{bmatrix}a_{11}&a_{12}&a_{13}&\cdots&a_{1n}\\a_{12}&a_{22}&a_{23}&\cdots&a_{2n}\\a_{13}&a_{23}&a_{33}&\cdots&a_{3n}\\\vdots&\vdots&\vdots&\ddots&\vdots\\a_{1n}&a_{2n}&a_{3n}&\cdots&a_{nn}\end{bmatrix}\tag{4.32}$$

with the upper-left cornered sub-matrices defined as the *main sub-matrices*. The determinants of each such sub-matrices can be calculated directly. If all the determinants of the main sub-matrices have the same sign, the matrix is referred to as a *positive-definite matrix*. If they take alternative signs, the matrix is referred to as a *negative-definite matrix*. If all the determinants are non-negative, the matrix is referred to as a *semi-positive-definite matrix*.

The MATLAB function `[D,p]=chol(A)` can also be used to check whether a matrix is positive-definite or not, where for positive-definite matrix $\boldsymbol{A}$, $p=$

0 will be returned. Thus such a function can be used to check whether a symmetrical matrix is positive-definite or not. For matrices which are not positive-definite, a variable p will be returned, where $p-1$ is the size of the sub-matrix in $\boldsymbol{A}$ which is positive-definite, i.e., the size of matrix $\boldsymbol{D}$.

If a complex matrix $\boldsymbol{A}$ satisfies

$$\boldsymbol{A}^{\mathrm{H}}\boldsymbol{A} = \boldsymbol{A}\boldsymbol{A}^{\mathrm{H}} \tag{4.33}$$

where $\boldsymbol{A}^{\mathrm{H}}$ is the Hermitian transpose of matrix $\boldsymbol{A}$, then the matrix is referred to as a *regular matrix*. The judgement can be made using the following MATLAB statements `norm(A'*A-A*A')<` ϵ, if 1 is returned, then it can be concluded that $\boldsymbol{A}$ is a regular matrix.

Example 4.28 Judge whether matrix $\boldsymbol{A} = \begin{bmatrix} 7 & 5 & 5 & 8 \\ 5 & 6 & 9 & 7 \\ 5 & 9 & 9 & 0 \\ 8 & 7 & 0 & 1 \end{bmatrix}$ is a positive-definite matrix. Then perform Cholesky decomposition to the matrix.

Solution Cholesky decomposition to matrix $\boldsymbol{A}$ is performed

```
>> A=[7,5,5,8; 5,6,9,7; 5,9,9,0; 8,7,0,1]; [D,p]=chol(A)
```

The positive-definite part $\boldsymbol{D}$ of the matrix can be obtained. It can be seen that matrix $\boldsymbol{A}$ is not a positive-definite matrix, thus $p \neq 0$. It can be found that $\begin{bmatrix} 2.6457513 & 1.8898224 \\ 0 & 1.5583874 \end{bmatrix}$, and $p = 3$.

If one calls the `chol()` function to an asymmetrical matrix $\boldsymbol{A}$, the results are also obtained. However, the results are useless, since it erroneously forced the original matrix into the symmetrical one. Strictly speaking, Cholesky decomposition cannot be performed upon asymmetrical matrices.

4.3.3 Jordan transformations

The matrices containing repeated eigenvalues cannot be decomposed into diagonal matrices. Instead, Jordanian decomposition should be used.

Example 4.29 For a given matrix $\boldsymbol{A} = \begin{bmatrix} -71 & -65 & -81 & -46 \\ 75 & 89 & 117 & 50 \\ 0 & 4 & 8 & 4 \\ -67 & -121 & -173 & -58 \end{bmatrix}$, compute its eigenvalues and eigenvector matrix using both numerical and analytical methods.

Solution With the use of MATLAB, the numerical solutions to the eigenvalues can be found from the following statements

```
>> A=[-71,-65,-81,-46; 75,89,117,50; 0,4,8,4; -67,-121,-173,-58];
   D=eig(A)
```

where the eigenvalues of matrix $\boldsymbol{A}$ are $\lambda(\boldsymbol{A}) = -8.0045, -8\pm 8+\mathrm{j}0.004, -7.9955$. In fact, -8 is the quadruple eigenvalue of the matrix. Thus using numerical methods, there exist rather large errors. For this example, the numerical method is not satisfacory, and analytical methods should be used instead.

If the analytical method is used, one has

```
>> [v,d]=eig(sym(A))
```

It can be found that $\boldsymbol{v} = \begin{bmatrix} 17/8 \\ -13/8 \\ 1 \\ -19/8 \end{bmatrix}$, $\boldsymbol{d} = \begin{bmatrix} -8 & 0 & 0 & 0 \\ 0 & -8 & 0 & 0 \\ 0 & 0 & -8 & 0 \\ 0 & 0 & 0 & -8 \end{bmatrix}$, where -8 is a quadruple eigenvalue. Thus the eigenvector matrix is singular. Since all the four columns are the same, only one column is reserved.

Due to the restrictions of numerical packages and languages, including the numerical solutions provided by MATLAB language, the best data type is the double-precision one. Thus in computation with such languages, computation errors are unavoidable.

In order to solve this kind of problem, it is suggested that the symbolic data type be used instead. Using the `jordan()` function in the Symbolic Math Toolbox, the Jordanian matrix as well as the non-singular generalized eigenvector matrix can be obtained. The syntax of the function is

```
J=jordan(A)          % only the Jordanian matrix J returned
[V,J]=jordan(A)      % Jordanian J and generalized vector matrix V
```

When the generalized eigenvector matrix $\boldsymbol{V}$ is found, the Jordanian canonical form can be obtained from $\boldsymbol{J} = \boldsymbol{V}^{-1}\boldsymbol{A}\boldsymbol{V}$. It should be noted that the main diagonal elements in the Jordanian matrix are the eigenvalues, with the main sub-diagonal elements taking 1's.

Example 4.30 Perform Jordanian decompositions for the matrix in Example 4.29.

Solution The Jordanian decomposition for a given symbolic matrix can be obtained directly by the use of `jordan()` function such that

```
>> A=[-71,-65,-81,-46; 75,89,117,50; 0,4,8,4; -67,-121,-173,-58];
   [V,J]=jordan(sym(A))
```

and one has

$$\boldsymbol{V} = \begin{bmatrix} -18496 & 2176 & -63 & 1 \\ 14144 & -800 & 75 & 0 \\ -8704 & 32 & 0 & 0 \\ 20672 & -1504 & -67 & 0 \end{bmatrix}, \quad \boldsymbol{J} = \begin{bmatrix} -8 & 1 & 0 & 0 \\ 0 & -8 & 1 & 0 \\ 0 & 0 & -8 & 1 \\ 0 & 0 & 0 & -8 \end{bmatrix}.$$

The matrix $\boldsymbol{V}$ is now a full-rank matrix and is invertible. Thus it is possible to implement some special operations which are very hard to implement using numerical methods. The applications of such a function will be demonstrated in the examples later.

Example 4.31 For a given matrix $\boldsymbol{A}=\begin{bmatrix}-10 & -14 & -17 & -9\\ 10 & 16 & 18 & 10\\ 0 & -1 & 0 & 0\\ -9 & -15 & -17 & -10\end{bmatrix}$, perform Jordanian decompositions to the matrix.

Solution Matrix $\boldsymbol{A}$ defined here has complex eigenvalues. The Jordanian canonical forms of a complex matrix can also be obtained directly by the used of `jordan()` function, and the following statements can be used to find Jordanian matrix and its similarity transformation matrix

```
>> A=[-10,-14,-17,-9; 10,16,18,10; 0,-1,0,0; -9,-15,-17,-10];
   [V,J]=jordan(sym(A))
```

and it can be found that

$$\boldsymbol{V}=\begin{bmatrix}-11+\mathrm{j}2 & -11-\mathrm{j}2 & -5 & 23\\ 5-\mathrm{j}5 & 5+\mathrm{j}5 & 0 & -10\\ 5 & 5 & 0 & -10\\ -6+\mathrm{j}7 & -6-\mathrm{j}7 & 5 & 12\end{bmatrix},\quad \boldsymbol{J}=\left[\begin{array}{cc:cc}-1+\mathrm{j}1 & 0 & 0 & 0\\ 0 & -1-\mathrm{j}1 & 0 & 0\\ \hdashline 0 & 0 & -1 & 1\\ 0 & 0 & 0 & -1\end{array}\right].$$

If one prefers real matrices, the following statements can further be given

```
>> V(:,1:2)=[real(V(:,1)), imag(V(:,1))], J1=inv(V)*A*V
```

and the modified Jordanian form and its transformation matrix can be obtained as

$$\boldsymbol{J}_1=\left[\begin{array}{cc:cc}-1 & 1 & 0 & 0\\ -1 & -1 & 0 & 0\\ \hdashline 0 & 0 & -1 & 1\\ 0 & 0 & 0 & -1\end{array}\right],\quad \boldsymbol{V}=\begin{bmatrix}-11 & 2 & -5 & 23\\ 5 & -5 & 0 & -10\\ 5 & 0 & 0 & -10\\ -6 & 7 & 5 & 12\end{bmatrix}.$$

4.3.4 Singular value decompositions

Singular values of a matrix can be regarded as a measure of the matrix. For any given $n\times m$ matrix $\boldsymbol{A}$, one has

$$\boldsymbol{A}^{\mathrm{T}}\boldsymbol{A}\geqslant 0,\quad \boldsymbol{A}\boldsymbol{A}^{\mathrm{T}}\geqslant 0 \tag{4.34}$$

and in theory, it follows that

$$\mathrm{rank}(\boldsymbol{A}^{\mathrm{T}}\boldsymbol{A})=\mathrm{rank}(\boldsymbol{A}\boldsymbol{A}^{\mathrm{T}})=\mathrm{rank}(\boldsymbol{A}). \tag{4.35}$$

It can further be shown that the matrices $\boldsymbol{A}^{\mathrm{T}}\boldsymbol{A}$ and $\boldsymbol{A}\boldsymbol{A}^{\mathrm{T}}$ have the same non-negative eigenvalues λ_i. The square roots of these non-negative eigenvalues are referred to as the *singular values* of matrix $\boldsymbol{A}$, denoted as $\sigma_i(\boldsymbol{A})=\sqrt{\lambda_i(\boldsymbol{A}^{\mathrm{T}}\boldsymbol{A})}$.

Example 4.32 For a matrix $\boldsymbol{A}=\begin{bmatrix}1 & 1\\ \mu & 0\\ 0 & \mu\end{bmatrix}$, where $\mu=5$`eps`, find the rank of matrix $\boldsymbol{A}$ using (4.35).

Solution It is obvious that the rank of matrix $\boldsymbol{A}$ is 2. The same result can be obtained from the following MATLAB statement.

```
>> A=[1 1; 5*eps,0; 0,5*eps]; rank(A)
```

Now consider the method in (4.35) for calculating the rank of matrix $\boldsymbol{A}$. If $\boldsymbol{A}^{\mathrm{T}}\boldsymbol{A}$ is used for finding the rank of matrix $\boldsymbol{A}$, it can be seen that

$$\boldsymbol{A}^{\mathrm{T}}\boldsymbol{A} = \begin{bmatrix} 1+\mu^2 & 1 \\ 1 & 1+\mu^2 \end{bmatrix}$$

and under double-precision scheme, since μ^2 is around 10^{-30}, it can be completely neglected when added to 1. Thus matrix $\boldsymbol{A}^{\mathrm{T}}\boldsymbol{A}$ is reduced to a matrix of ones. It can be concluded that the rank of matrix $\boldsymbol{A}$ is 1, which is obviously wrong. Thus the concepts of singular values for matrix $\boldsymbol{A}$ should be introduced to provide a better characterization for the matrices.

If matrix $\boldsymbol{A}$ is an $n \times m$ matrix, it can be decomposed as

$$\boldsymbol{A} = \boldsymbol{L}\boldsymbol{A}_1\boldsymbol{M} \tag{4.36}$$

where matrices $\boldsymbol{L}$ and $\boldsymbol{M}$ are orthogonal matrices, and $\boldsymbol{A}_1 = \mathrm{diag}(\sigma_1, \cdots, \sigma_n)$ is a diagonal matrix, whose elements satisfy the inequality $\sigma_1 \geqslant \sigma_2 \geqslant \cdots \geqslant \sigma_n \geqslant 0$. If $\sigma_n = 0$, then matrix $\boldsymbol{A}$ is singular. The rank of matrix $\boldsymbol{A}$ is in fact the number of non-zero quantities in the diagonal elements of matrix $\boldsymbol{A}_1$. The transformation is referred to as *singular value decomposition (SVD)*.

A singular value decomposition function `svd()` is provided in MATLAB

```
S = svd(A)              % only singular value decomposition required
[L,A1,M]=svd(A)         % singular value decomposition
```

where $\boldsymbol{A}$ is the original matrix and the returned matrix $\boldsymbol{A}_1$ is a diagonal matrix, while the matrices $\boldsymbol{L}$ and $\boldsymbol{M}$ are orthogonal matrices, satisfying $\boldsymbol{A} = \boldsymbol{L}\boldsymbol{A}_1\boldsymbol{M}^{\mathrm{T}}$.

The singular values of a matrix often determine the properties of the matrix. When some of the singular values are large, while others are small, and they differ significantly, then very small perturbation to certain elements in the matrix may significantly affect the behavior of the matrix. This kind of matrices are often referred to as *ill-* or *bad-conditioned* matrices. If zeros exist in the singular values, then the matrix is a singular matrix. The ratio of maximum singular value $\sigma_{\max}$ and the minimum one $\sigma_{\min}$ is defined as the *condition number* of the matrix, denoted by $\mathrm{cond}(\boldsymbol{A})$, i.e., $\mathrm{cond}(\boldsymbol{A}) = \sigma_{\max}/\sigma_{\min}$. The larger the condition number of the matrix, the more sensitive the matrix is. The maximum and minimum singular values of the matrix may also be denoted by $\bar{\sigma}(\boldsymbol{A})$ and $\underline{\sigma}(\boldsymbol{A})$, respectively. A MATLAB function `cond(A)` is provided to calculate the condition number of the matrix $\boldsymbol{A}$, and for singular matrices, the condition number is infinity.

Example 4.33 Performing SVD to the matrix $\boldsymbol{A}$ in Example 4.7.

Solution If the MATLAB function `svd()` is used, the matrices $\boldsymbol{L}$, $\boldsymbol{A}_1$ and $\boldsymbol{M}$ can be obtained. The condition number can also be calculated by the singular values.

```
>> A=[16,2,3,13; 5,11,10,8; 9,7,6,12; 4,14,15,1]; [L,A1,M]=svd(A)
```

It can be found that the decomposed matrices are

$$\boldsymbol{L}=\begin{bmatrix}-0.5 & 0.67082 & 0.5 & -0.22361\\ -0.5 & -0.22361 & -0.5 & -0.67082\\ -0.5 & 0.22361 & -0.5 & 0.67082\\ -0.5 & -0.67082 & 0.5 & 0.22361\end{bmatrix},\ \boldsymbol{A}_1=\begin{bmatrix}34 & 0 & 0 & 0\\ 0 & 17.889 & 0 & 0\\ 0 & 0 & 4.4721 & 0\\ 0 & 0 & 0 & 0\end{bmatrix}$$

$$\boldsymbol{M}=\begin{bmatrix}-0.5 & 0.5 & 0.67082 & -0.22361\\ -0.5 & -0.5 & -0.22361 & -0.67082\\ -0.5 & -0.5 & 0.22361 & 0.67082\\ -0.5 & 0.5 & -0.67082 & 0.22361\end{bmatrix}.$$

It can be seen that since there exists a zero singular value, thus the original matrix is a singular matrix. The condition number of the matrix should be infinity; however, since numerical calculation is used, there might be minor errors. The commands `cond(A)` will result the condition number 3.2592×10^{16}.

One can convert matrix $\boldsymbol{A}$ into a symbolic variable, then with the function `svd()`, more accurate singular value decomposition can be obtained.

Example 4.34 For a rectangular matrix $\boldsymbol{A}=\begin{bmatrix}1 & 3 & 5 & 7\\ 2 & 4 & 6 & 8\end{bmatrix}$, perform SVD to matrix $\boldsymbol{A}$ and then verify the results.

Solution The following statements can be given such that

```
>> A=[1,3,5,7; 2,4,6,8]; [L,A1,M]=svd(A)
   A2=L*A1*M'; norm(A-A2)
```

The decomposited matrices are

$$\boldsymbol{L}=\begin{bmatrix}-0.64142 & -0.76719\\ -0.76719 & 0.64142\end{bmatrix},\ \boldsymbol{A}_1=\begin{bmatrix}14.269 & 0 & 0 & 0\\ 0 & 0.62683 & 0 & 0\end{bmatrix}$$

$$\boldsymbol{M}=\begin{bmatrix}-0.15248 & 0.82265 & -0.3945 & -0.37996\\ -0.34992 & 0.42138 & 0.2428 & 0.80066\\ -0.54735 & 0.020103 & 0.69791 & -0.46143\\ -0.74479 & -0.38117 & -0.54621 & 0.040738\end{bmatrix}$$

and it can be seen that the error $||\boldsymbol{L}\boldsymbol{A}_1\boldsymbol{V}^{\mathrm{T}}-\boldsymbol{A}||=9.7277\times 10^{-15}$ and for this example, $\boldsymbol{L}\boldsymbol{A}_1\boldsymbol{V}^{\mathrm{T}}$ will restore the original matrix $\boldsymbol{A}$, with very small error.

4.4 Solving Matrix Equations

4.4.1 Solutions to linear algebraic equations

Consider the linear algebraic equation

$$\boldsymbol{A}\boldsymbol{x}=\boldsymbol{B} \tag{4.37}$$

where $\boldsymbol{A}$ are $\boldsymbol{B}$ given matrices

$$\boldsymbol{A}=\begin{bmatrix} a_{11} & a_{12} & \cdots & a_{1n} \\ a_{21} & a_{22} & \cdots & a_{2n} \\ \vdots & \vdots & \ddots & \vdots \\ a_{m1} & a_{m2} & \cdots & a_{mn} \end{bmatrix}, \quad \boldsymbol{B}=\begin{bmatrix} b_{11} & b_{12} & \cdots & b_{1p} \\ b_{21} & b_{22} & \cdots & b_{2p} \\ \vdots & \vdots & \ddots & \vdots \\ b_{m1} & b_{m2} & \cdots & b_{mp} \end{bmatrix} \tag{4.38}$$

and the target is to find matrix $\boldsymbol{x}$ for the equation. According to linear algebra theory, in some cases, there are unique solutions, and in other cases, the equation may have an infinite number of solutions, or have no solution at all. Thus here the solutions to the linear equation can be considered thoroughly.

(i) **Unique solutions** If $m = n$ and $\text{rank}(\boldsymbol{A}) = n$, then the equations in (4.37) have unique solutions

$$\boldsymbol{x} = \boldsymbol{A}^{-1}\boldsymbol{B}. \tag{4.39}$$

The solution `x=inv(A)*B` can be obtained immediately using MATLAB to the original equation. It should be noted that if the numerical method is used, sometimes the `inv()` function may lead to erroneous results. For instance, if $\text{cond}(\boldsymbol{A})$ is very large, the results may be unreliable. If the Symbolic Math Toolbox is used, then this problem may be avoided.

Example 4.35 Solve the linear algebraic equations $\begin{bmatrix} 1 & 2 & 3 & 4 \\ 4 & 3 & 2 & 1 \\ 1 & 3 & 2 & 4 \\ 4 & 1 & 3 & 2 \end{bmatrix} \boldsymbol{X} = \begin{bmatrix} 5 & 1 \\ 4 & 2 \\ 3 & 3 \\ 2 & 4 \end{bmatrix}$.

Solution The analytical solutions to the given equations can be obtained using the following MATLAB statements.

```
>> A=[1 2 3 4; 4 3 2 1; 1 3 2 4; 4 1 3 2]; B=[5 1; 4 2; 3 3; 2 4];
   x=inv(sym(A))*B
```

Thus the solutions can be written as $\boldsymbol{x} = \begin{bmatrix} -9/5 & 12/5 \\ 28/15 & -19/15 \\ 58/15 & -49/15 \\ -32/15 & 41/15 \end{bmatrix}$.

Substituting the results back to the equations, one may find that there is no error. If in the above statement, the numerical statement is used instead, i.e., `x=inv(sym(A))*B` is replaced by `x=inv(A)*B`, an error level of 10^{-15} may be achieved.

(ii) **Equations with infinite number of solutions** One may construct first the judging matrix $\boldsymbol{C}$ out of $\boldsymbol{A}$, $\boldsymbol{B}$ matrices

$$\boldsymbol{C}=\begin{bmatrix} a_{11} & a_{12} & \cdots & a_{1n} & b_{11} & b_{12} & \cdots & b_{1p} \\ a_{21} & a_{22} & \cdots & a_{2n} & b_{21} & b_{22} & \cdots & b_{2p} \\ \vdots & \vdots & \ddots & \vdots & \vdots & \vdots & \ddots & \vdots \\ a_{m1} & a_{m2} & \cdots & a_{mn} & b_{m1} & b_{m2} & \cdots & b_{mp} \end{bmatrix}. \tag{4.40}$$

If rank($\boldsymbol{A}$) = rank($\boldsymbol{C}$) = $r < n$, the original equations (4.37) have an infinite number of solutions. One may find the $n - r$ basic set of solutions $\boldsymbol{x}_i$, $i = 1, 2, \cdots, n - r$ to the homogeneous equations $\boldsymbol{A}\boldsymbol{x} = 0$. From which, for any constant $\alpha_i, i = 1, 2, \cdots, n-r$, the general solutions to the homogeneous equations can be written as

$$\hat{\boldsymbol{x}} = \alpha_1 \boldsymbol{x}_1 + \alpha_2 \boldsymbol{x}_2 + \cdots + \alpha_{n-r} \boldsymbol{x}_{n-r}. \tag{4.41}$$

The solution basis can be found directly using MATLAB function `null()` such that `Z=null(sym(A))`, and the function `null()` can also be used in numerical cases, where in this case, the syntax `Z=null(A,'r')` should be used instead. The resulted $\boldsymbol{Z}$ matrix should have $n - r$ columns, to which each one is referred as a *basic set of solutions* for the given matrix $\boldsymbol{A}$.

Solving equations (4.37) is not a difficult task. If a special solution $\boldsymbol{x}_0$ to (4.37) can be found, the general solutions to the original equations can be constructed from $\boldsymbol{x} = \hat{\boldsymbol{x}} + \boldsymbol{x}_0$. In fact, the special solution can be found from `x0=sym(pinv(A)*B)`.

Example 4.36 Find the solutions to the equations $\begin{bmatrix} 1 & 2 & 3 & 4 \\ 2 & 2 & 1 & 1 \\ 2 & 4 & 6 & 8 \\ 4 & 4 & 2 & 2 \end{bmatrix} \boldsymbol{X} = \begin{bmatrix} 1 \\ 3 \\ 2 \\ 6 \end{bmatrix}$.

Solution One may enter matrices $\boldsymbol{A}$ and $\boldsymbol{B}$ first, and then establish $\boldsymbol{C}$ matrix from (4.40), then the solvability of the equations can be judged

```
>> A=[1 2 3 4; 2 2 1 1; 2 4 6 8; 4 4 2 2];  B=[1;3;2;6];
   C=[A B]; [rank(A), rank(C)]
```

Both the norms equal 2, smaller than that of the size of matrix $\boldsymbol{A}$. It can be concluded that there are an infinite number of solutions in the original equations. The general solutions to the homogeneous equations, and a special solution to the original equation, and finally the general solutions to the given solutions can all be obtained from

```
>> Z=null(sym(A)); x0=sym(pinv(A)*B); syms a b; x=a*Z(:,1)+b*Z(:,2)+x0
```

and the analytical solutions are $\boldsymbol{x} = a \begin{bmatrix} 2 \\ -5/2 \\ 1 \\ 0 \end{bmatrix} + b \begin{bmatrix} 3 \\ -7/2 \\ 0 \\ 1 \end{bmatrix} + \begin{bmatrix} 125/131 \\ 96/131 \\ -10/131 \\ -39/131 \end{bmatrix}$.

(iii) **Equations with no solutions** If rank($\boldsymbol{A}$) < rank($\boldsymbol{C}$), then the equations in (4.37) are conflict equations, and there are no solutions to such equations. Applying Moore-Penrose inverse `x=pinv(A)*B`, one may obtain the least squares solution to the original equations such that the norm of the error $||\boldsymbol{A}\boldsymbol{x} - \boldsymbol{B}||$ is minimized.

Example 4.37 If matrix $\boldsymbol{B}$ in the previous equation is changed to $\boldsymbol{B} = [1, 2, 3, 4]^{\mathrm{T}}$, it can be found that the rank of matrix $\boldsymbol{C}$ is 3, which is higher than that of matrix $\boldsymbol{A}$, indicating that there is no solution to the original equations

```
>> B=[1:4]'; C=[A B]; [rank(A), rank(C)]
```

One may use function `pinv()` to evaluate the Moore-Penrose inverse to matrix $\boldsymbol{A}$, then the least squares solutions to the conflict equations can be obtained as

```
>> x=pinv(A)*B, norm(A*x-B)
```

Using the above statements, one may find the solutions $\boldsymbol{x}^{\mathrm{T}} = [0.9542, 0.7328, -0.0763, -0.29771]$. Substituting the solution back to the original equations, one may expect the norm of the solution error of 0.4472, which is the smallest possible error level.

For linear algebraic equations of the form $\boldsymbol{xA} = \boldsymbol{B}$, one may perform transpose to both sides of the equation to transform the original equation to the following form

$$\boldsymbol{A}^{\mathrm{T}}\boldsymbol{z} = \boldsymbol{B}^{\mathrm{T}} \tag{4.42}$$

where $\boldsymbol{z} = \boldsymbol{x}^{\mathrm{T}}$, the new equations can be transformed into the form in (4.37), and the above methods can be applied directly to solve the original problems.

4.4.2 Solutions to Lyapunov equations

Continuous Lyapunov equations

A continuous Lyapunov equation can be expressed as

$$\boldsymbol{AX} + \boldsymbol{XA}^{\mathrm{T}} = -\boldsymbol{C} \tag{4.43}$$

and it is known that Lyapunov equations are originated from stability theory of differential equations, where one often expects $-\boldsymbol{C}$ to be symmetrical positive-definite $n \times n$ matrix. It follows that the solution $\boldsymbol{X}$ is also an $n \times n$ symmetrical matrix. Direct solutions to such equations were rather difficult; however, with the use of powerful computer mathematics languages, such a function can be solved easily by using the `lyap()` function provided in the Control Systems Toolbox. The syntax of the function is `X=lyap(A,C)`, and if one specifies matrices $\boldsymbol{A}$ and $\boldsymbol{C}$, the numerical solutions to the Lyapunov equations can be obtained immediately.

Example 4.38 Assume that in (4.43), the matrices $\boldsymbol{A}$ and $\boldsymbol{C}$ are given by

$$\boldsymbol{A} = \begin{bmatrix} 1 & 2 & 3 \\ 4 & 5 & 6 \\ 7 & 8 & 0 \end{bmatrix}, \quad \boldsymbol{C} = -\begin{bmatrix} 10 & 5 & 4 \\ 5 & 6 & 7 \\ 4 & 7 & 9 \end{bmatrix}$$

solve the corresponding Lyapunov equation and check the accuracy of the solutions.

Solution One may enter the matrices into MATLAB, then solve the equation using the following MATLAB statements

```
>> A=[1 2 3;4 5 6; 7 8 0];  C=-[10, 5, 4; 5, 6, 7; 4, 7, 9];
   X=lyap(A,C), norm(A*X+X*A'+C)
```

the solution obtained is

$$X = \begin{bmatrix} -3.9444444444444 & 3.8888888888889 & 0.38888888888889 \\ 3.8888888888889 & -2.7777777777778 & 0.22222222222222 \\ 0.38888888888889 & 0.22222222222222 & -0.11111111111111 \end{bmatrix}$$

and the norm of the error matrix is $||\boldsymbol{A}\boldsymbol{X} + \boldsymbol{X}\boldsymbol{A}^{\mathrm{T}} + \boldsymbol{C}|| = 2.64742\times 10^{-14}$. It can be seen that the accuracy in the solutions obtained is very high.

Analytical solutions to Lyapunov equations

For simplicity, the matrices in the Lyapunov equation can be rearranged such that

$$\boldsymbol{X} = \begin{bmatrix} x_1 & x_2 & \cdots & x_m \\ x_{m+1} & x_{m+2} & \cdots & x_{2m} \\ \vdots & \vdots & \ddots & \vdots \\ x_{(n-1)m+1} & x_{(n-1)m+2} & \cdots & x_{nm} \end{bmatrix}, \quad \boldsymbol{C} = \begin{bmatrix} c_1 & c_2 & \cdots & c_m \\ c_{m+1} & c_{m+2} & \cdots & c_{2m} \\ \vdots & \vdots & \ddots & \vdots \\ c_{(n-1)m+1} & c_{(n-1)m+2} & \cdots & c_{nm} \end{bmatrix}.$$

The Lyapunov equation can be rewritten as a simple linear equation

$$(\boldsymbol{A} \otimes \boldsymbol{I} + \boldsymbol{I} \otimes \boldsymbol{A})\boldsymbol{x} = -\boldsymbol{c} \tag{4.44}$$

where $\boldsymbol{A} \otimes \boldsymbol{B}$ denotes the Kronecker product of matrices $\boldsymbol{A}$ and $\boldsymbol{B}$ such that

$$\boldsymbol{A} \otimes \boldsymbol{B} = \begin{bmatrix} a_{11}\boldsymbol{B} & \cdots & a_{1m}\boldsymbol{B} \\ \vdots & \ddots & \vdots \\ a_{n1}\boldsymbol{B} & \cdots & a_{nm}\boldsymbol{B} \end{bmatrix} \tag{4.45}$$

and MATLAB evaluation of the product is `C =kron(A,B)`.

It can now be seen that the conditions when the equation has a unique solution is no longer that $-\boldsymbol{C}$ is a positive-definite symmetrical matrix. It requires that the matrix $(\boldsymbol{A} \otimes \boldsymbol{I} + \boldsymbol{I} \otimes \boldsymbol{A})$ is a non-singular square matrix.

Example 4.39 Consider again the Lyapunov equation in Example 4.38, and find the analytical solutions.

Solution The analytical solution of the Lyapunov equation can be obtained from the following statements, and if one substitutes the solutions back to the original equation, there is no error.

```
>> A0=sym(kron(A,eye(3))+kron(eye(3),A));
   c=reshape(C',9,1);  x0=-inv(A0)*c; x=reshape(x0,3,3)'
```

The solutions can be written as $\boldsymbol{x} = \begin{bmatrix} -71/18 & 35/9 & 7/18 \\ 35/9 & -25/9 & 2/9 \\ 7/18 & 2/9 & -1/9 \end{bmatrix}$.

Example 4.40 Traditionally in Lyapunov equations, one may assume that matrix $\boldsymbol{C}$ is a real and symmetrically positive-definite matrix. Find whether there are solutions to Lyapunov equations if $\boldsymbol{C}$ is not a symmetrical real matrix.

Solution Since the Lyapunov equation was originated from stability theory, it was usually assumed that matrix $-\boldsymbol{C}$ is a real and symmetrically positive-definite matrix. In fact, (4.46) will also have unique solutions even when the above conditions are not satisfied. For instance, if matrix $\boldsymbol{C}$ is changed into a complex asymmetrical matrix such that $\boldsymbol{C} = -\begin{bmatrix} 1+\mathrm{j}1 & 3+\mathrm{j}3 & 12+\mathrm{j}10 \\ 2+\mathrm{j}5 & 6 & 11+\mathrm{j}6 \\ 5+\mathrm{j}2 & 11+\mathrm{j}1 & 2+\mathrm{j}12 \end{bmatrix}$, while matrix $\boldsymbol{A}$ remains unchanged in Example 4.38, the complex solution matrix $\boldsymbol{X}$ will be immediately obtained.

```
>> A=[1 2 3;4 5 6; 7 8 0];
   C=-[1+1i, 3+3i, 12+10i; 2+5i, 6, 11+6i; 5+2i, 11+1i, 2+12i];
   A0=sym(kron(A,eye(3))+kron(eye(3),A));
   c=reshape(C',9,1);   x0=-inv(A0)*c; x=reshape(x0,3,3)'
   norm(double(A*x+x*A+C))
```

The solution is

$$\boldsymbol{x} = \begin{bmatrix} -5/102+\mathrm{j}1457/918 & 15/17-\mathrm{j}371/459 & -61/306+\mathrm{j}166/459 \\ 4/17-\mathrm{j}626/459 & -10/51+\mathrm{j}160/459 & 115/153+\mathrm{j}607/459 \\ -55/306+\mathrm{j}166/459 & -26/153-\mathrm{j}209/459 & 203/153+\mathrm{j}719/918 \end{bmatrix}$$

and it can be concluded that there is no error at all in the solution. Thus if one does not consider the energy concept in the physical model, matrix $\boldsymbol{C}$ can be generalized into any matrix in the Lyapunov matrix.

Discrete Lyapunov equations

A discrete Lyapunov equation can be expressed in the form

$$\boldsymbol{A}\boldsymbol{X}\boldsymbol{A}^{\mathrm{T}} - \boldsymbol{X} + \boldsymbol{Q} = 0 \tag{4.46}$$

which can be solved easily by using `dlyap()` function provided by Control Systems Toolbox of MATLAB. The syntax of the function is `X=dlyap(A,Q)`.

In fact, if matrix $\boldsymbol{A}$ is a non-singular matrix, then one can multiply both sides of the equation to the right by matrix $(\boldsymbol{A}^{\mathrm{T}})^{-1}$, and it can then be transformed into the Sylvester equation. In Section 4.4.3, the analytical solution to the equation can be obtained.

Example 4.41 Solve the discrete Lyapunov equation

$$\begin{bmatrix} 8 & 1 & 6 \\ 3 & 5 & 7 \\ 4 & 9 & 2 \end{bmatrix} \boldsymbol{X} \begin{bmatrix} 8 & 1 & 6 \\ 3 & 5 & 7 \\ 4 & 9 & 2 \end{bmatrix}^{\mathrm{T}} - \boldsymbol{X} + \begin{bmatrix} 16 & 4 & 1 \\ 9 & 3 & 1 \\ 4 & 2 & 1 \end{bmatrix} = 0.$$

Solution The numerical solution to the equation can easily be solved by the use of the function `dlyap()` such that

```
>> A=[8,1,6; 3,5,7; 4,9,2]; Q=[16,4,1; 9,3,1; 4,2,1];
   X=dlyap(A,Q), norm(A*X*A'-X+Q)    % accuracy check
```

with the error of 2.7778×10^{-14}, and $\boldsymbol{X} = \begin{bmatrix} -0.16474 & 0.06915 & -0.016785 \\ 0.052843 & -0.029785 & -0.0061542 \\ -0.10198 & 0.044959 & -0.030541 \end{bmatrix}$.

4.4.3 Solutions to Sylvester equations

A Sylvester equation takes the general form

$$\boldsymbol{AX} + \boldsymbol{XB} = -\boldsymbol{C} \tag{4.47}$$

where $\boldsymbol{A}$ is an $n \times n$ matrix, $\boldsymbol{B}$ is an $m \times m$ matrix, and $\boldsymbol{C}$ and $\boldsymbol{X}$ are $n \times m$ matrices. This equation is also known as the *generalized Lyapunov equation*. The function `lyap()` can still be used such that `X=lyap(A,B,C)`. Schur decomposition is an effective way in solving such an equation.

Similar to the above mentioned Lyapunov equation, the analytical solutions can also be found with the help of Kronecker products

$$(\boldsymbol{A} \otimes \boldsymbol{I}_m + \boldsymbol{I}_n \otimes \boldsymbol{B}^{\mathrm{T}})\boldsymbol{x} = \boldsymbol{c}. \tag{4.48}$$

If $(\boldsymbol{A} \otimes \boldsymbol{I}_m + \boldsymbol{I}_n \otimes \boldsymbol{B}^{\mathrm{T}})$ is a non-singular matrix, the Sylvester equation has a unique solution.

Combining the above mentioned analytical solution algorithms, a symbolic-based MATLAB function lyap.m for Sylvester equations can be written and should be placed under the `@sym` directly. The listing of the function is

```
function X=lyap(A,B,C)
if nargin==2, C=B; B=A'; end
[nr,nc]=size(C); A0=kron(A,eye(nc))+kron(eye(nr),B');
try
    C1=C'; x0=-inv(A0)*C1(:); X=reshape(x0,nc,nr)';
catch, error('singular matrix found.'), end
```

Considering the discrete Lyapunov equation shown in (4.46). If one multiplies both sides of the equation by $(\boldsymbol{A}^{\mathrm{T}})^{-1}$, then the original discrete Lyapunov equation can be rewritten as

$$\boldsymbol{AX} + \boldsymbol{X}[-(\boldsymbol{A}^{\mathrm{T}})^{-1}] = -\boldsymbol{Q}(\boldsymbol{A}^{\mathrm{T}})^{-1}$$

Let $\boldsymbol{B} = -(\boldsymbol{A}^{\mathrm{T}})^{-1}$, $\boldsymbol{C} = \boldsymbol{Q}(\boldsymbol{A}^{\mathrm{T}})^{-1}$, the equation can be transformed into a Sylvester equation defined in (4.47). Thus the new `lyap()` function can be used to solve such equations. The syntax of the function now is

```
X=lyap(sym(A),C)                      % continuous Lyapunov equation
X=lyap(sym(A),-inv(A'),Q*inv(A'))     % discrete Lyapunov equation
X=lyap(sym(A),B,C)                    % Sylvester equation
```

Example 4.42 Solve the following Sylvester equation

$$\begin{bmatrix} 8 & 1 & 6 \\ 3 & 5 & 7 \\ 4 & 9 & 2 \end{bmatrix} \boldsymbol{X} + \boldsymbol{X} \begin{bmatrix} 16 & 4 & 1 \\ 9 & 3 & 1 \\ 4 & 2 & 1 \end{bmatrix} = \begin{bmatrix} 1 & 2 & 3 \\ 4 & 5 & 6 \\ 7 & 8 & 0 \end{bmatrix}.$$

Solution The original equation can easily be solved by calling `lyap()` function

```
>> A=[8,1,6; 3,5,7; 4,9,2]; B=[16,4,1; 9,3,1; 4,2,1];
   C=-[1,2,3; 4,5,6; 7,8,0]; X=lyap(A,B,C), norm(A*X+X*B+C)
```

with the equation of $\boldsymbol{X}=\begin{bmatrix}0.074872 & 0.089913 & -0.43292\\ 0.0080716 & 0.48144 & -0.21603\\ 0.019577 & 0.18264 & 1.1579\end{bmatrix}$, and it can be found that the norm of the error matrix is 4.5315×10^{-15}. If one wants to have the analytical solutions, the following statements can be given

```
>> x=lyap(sym(A),B,C), norm(double(A*x+x*B+C))
```

the solution now is

$$x=\begin{bmatrix}1349214/18020305 & 648107/7208122 & -15602701/36040610\\ 290907/36040610 & 3470291/7208122 & -3892997/18020305\\ 70557/3604061 & 1316519/7208122 & 8346439/7208122\end{bmatrix}$$

and it can be seen that there is no longer any error in the solution.

Example 4.43 Consider again the discrete Lyapunov equation given in Example 4.41, now find the analytical solution of the equation.

Solution The analytical solutions to the original equation can be found

```
>> A=[8,1,6; 3,5,7; 4,9,2]; Q=[16,4,1; 9,3,1; 4,2,1];
   x=lyap(sym(A),-inv(A'),Q*inv(A')), norm(double(A*x*A'-x+Q))
```

the solution can then be found and it is with zero error.

$$x=\begin{bmatrix}-22912341/139078240 & 48086039/695391200 & -11672009/695391200\\ 36746487/695391200 & -20712201/695391200 & -4279561/695391200\\ -70914857/695391200 & 31264087/695391200 & -4247541/139078240\end{bmatrix}$$

Example 4.44 Solve the Sylvester equation with

$$\boldsymbol{A}=\begin{bmatrix}8&1&6\\3&5&7\\4&9&2\end{bmatrix},\quad \boldsymbol{B}=\begin{bmatrix}2&3\\4&5\end{bmatrix},\quad \boldsymbol{C}=\begin{bmatrix}1&2\\3&4\\5&6\end{bmatrix}.$$

Solution In Sylvester equations, the matrix $\boldsymbol{C}$ is not necessarily square. The analytical solutions to the equation can be found using the new `lyap()` function

```
>> A=[8,1,6; 3,5,7; 4,9,2]; B=[2,3; 4,5]; C=-[1,2; 3,4; 5,6];
   X=lyap(sym(A),B,C), norm(double(A*X+X*B+C))
```

and the solution is $\boldsymbol{X}=\begin{bmatrix}-2853/14186 & -11441/56744\\ -557/14186 & -8817/56744\\ 9119/14186 & 50879/56744\end{bmatrix}$, with zero error.

4.4.4 Solutions to Riccati equations

A Riccati equation is a quadratic matrix equation which can be written as

$$\boldsymbol{A}^{\mathrm{T}}\boldsymbol{X}+\boldsymbol{X}\boldsymbol{A}-\boldsymbol{X}\boldsymbol{B}\boldsymbol{X}+\boldsymbol{C}=0. \tag{4.49}$$

Due to the quadratic term of matrix $\boldsymbol{X}$, solving such an equation is more difficult than Lyapunov type equations. There are no analytical solutions to such equations. A numerical-based function `are()` is provided in the Control Systems Toolbox, such that `X=are(A,B,C)`.

Example 4.45 Consider the Riccati equation in (4.49), where

$$A=\begin{bmatrix}-2 & 1 & -3\\ -1 & 0 & -2\\ 0 & -1 & -2\end{bmatrix},\ B=\begin{bmatrix}2 & 2 & -2\\ -1 & 5 & -2\\ -1 & 1 & 2\end{bmatrix},\ C=\begin{bmatrix}5 & -4 & 4\\ 1 & 0 & 4\\ 1 & -1 & 5\end{bmatrix}$$

find the numerical solution and then validate the results.

Solution The following commands can be used in solving the Riccati equation

```
>> A=[-2,1,-3; -1,0,-2; 0,-1,-2]; B=[2,2,-2; -1 5 -2; -1 1 2];
   C=[5 -4 4; 1 0 4; 1 -1 5]; X=are(A,B,C), norm(A'*X+X*A-X*B*X+C)
```

and the solution obtained is $X=\begin{bmatrix}0.98739491 & -0.7983277 & 0.418869\\ 0.57740565 & -0.13079234 & 0.57754777\\ -0.284045 & -0.073036978 & 0.69241149\end{bmatrix}$.

It can also be found that the norm of the error matrix is 1.8605×10^{-14}, which is small enough. More solutions can be given in Chapter 6.

4.5 Nonlinear Functions and Matrix Function Evaluations

Two kinds of functions for matrices are provided in MATLAB. One is for element-by-element calculation of matrices, while the other is for matrix functions. In this section, these two types of functions are explored.

4.5.1 Element-by-element computations

A large number of functions has been provided in MATLAB to carry out element-by-element nonlinear function computation. For instance, the `sin(x)` function used in Chapter 2 calculates sinusoidal to each element of the matrix x. The element-by-element computation is useful especially in graphics. The "dot operation" is another example of element-by-element operation. The commonly used element-by-element nonlinear functions are summarized in Table 4.2. The syntax for this kind of operation is very simple, `B=funname(A)` for instance `B=sin(A)`.

Example 4.46 Consider the matrix A in Example 4.7. The following functions can be used and the element-by-element exponential and sinusoidal functions can be found

```
>> A=[16,2,3,13; 5,11,10,8; 9,7,6,12; 4,14,15,1]; exp(A), sin(A)
```

the results obtained are

TABLE 4.2: Commonly used element-by-element nonlinear functions

Function name	Meaning	Function name	Meaning
`abs()`	absolute value	`asin(),acos(),atan()`	inverse triangular functions
`sqrt()`	square roots	`log(),log10()`	logarithmic function
`exp()`	exponential function	`real(),imag(),conj()`	real, imaginary or conjugates
`sin(),cos(),tan()`	triangular functions	`round(),floor(),ceil()`	integer functions

$$\texttt{exp}(\boldsymbol{A}) = \begin{bmatrix} 8.8861\times 10^6 & 7.3891 & 20.086 & 4.4241\times 10^5 \\ 148.41 & 59874 & 22026 & 2981 \\ 8103.1 & 1096.6 & 403.43 & 1.6275\times 10^5 \\ 54.598 & 1.2026\times 10^6 & 3.269\times 10^6 & 2.7183 \end{bmatrix},$$

$$\texttt{sin}(\boldsymbol{A}) = \begin{bmatrix} -0.2879 & 0.9093 & 0.14112 & 0.42017 \\ -0.95892 & -0.99999 & -0.54402 & 0.98936 \\ 0.41212 & 0.65699 & -0.27942 & -0.53657 \\ -0.7568 & 0.99061 & 0.65029 & 0.84147 \end{bmatrix}.$$

4.5.2 Matrix function evaluations

Computations of matrix exponentials

Apart from element-by-element nonlinear computation of matrices, sometimes the nonlinear functions on the whole matrices, i.e., matrix functions, are expected. For instance, if one wants to have the exponential function to a matrix, special algorithms should be required. Nineteen different numerical algorithms have been summarized in Reference [12], and each algorithm has its own advantages. The built-in MATLAB function `expm()` can be used with $\boldsymbol{E}$=`expm(`$\boldsymbol{A}$`)`. Taylor series approximation can also be used to evaluate the exponential function from the formula

$$\mathrm{e}^{\boldsymbol{A}} = \sum_{i=0}^{\infty} \frac{1}{i!}\boldsymbol{A}^i = \boldsymbol{I} + \boldsymbol{A} + \frac{1}{2}\boldsymbol{A}^2 + \frac{1}{3!}\boldsymbol{A}^3 + \cdots + \frac{1}{m!}\boldsymbol{A}^m + \cdots. \tag{4.50}$$

Example 4.47 Consider matrix $\boldsymbol{A} = \begin{bmatrix} -2 & 1 & 0 & & \\ 0 & -2 & 1 & & \\ 0 & 0 & -2 & & \\ & & & -5 & 1 \\ & & & 0 & -5 \end{bmatrix}$. Compute the exponential function $\mathrm{e}^{\boldsymbol{A}}$, then calculate the logarithmic function and see whether the original matrix $\boldsymbol{A}$ can be restored. Also find analytically $\mathrm{e}^{\boldsymbol{A}t}$.

Solution The following MATLAB statements can be used to calculate the exponential matrix $\boldsymbol{B} = \mathrm{e}^{\boldsymbol{A}}$, then the use of `logm()` function can be used to calculate the logarithmic function to restore matrix $\boldsymbol{A}$

```
>> A=[[-2 1 0; 0 -2 1; 0 0 -2], zeros(3,2); zeros(2,3) [-5 1; 0 -5]];
   B=expm(A), C=logm(B), norm(C-A)
```

Thus one may find that

$$B = \begin{bmatrix} 0.13534 & 0.13534 & 0.067668 & 0 & 0 \\ 0 & 0.13534 & 0.13534 & 0 & 0 \\ 0 & 0 & 0.13534 & 0 & 0 \\ 0 & 0 & 0 & 0.0067379 & 0.0067379 \\ 0 & 0 & 0 & 0 & 0.0067379 \end{bmatrix}$$

and $\boldsymbol{C}$ is almost equal to $\boldsymbol{A}$, with the restoration error $||\boldsymbol{C}-\boldsymbol{A}|| = 3.9014\times 10^{-15}$. It can be seen that the accuracy of such algorithms is very high.

The `expm()` function is also applicable to symbolic matrices. For instance, the exponential function $\mathrm{e}^{\boldsymbol{A}t}$ can also be obtained by the direct use of such a function. This kind of problem cannot be solved using the numerical methods of course.

```
>> syms t; expm(A*t)
```

So the results can be written as $\mathrm{e}^{\boldsymbol{A}t} = \begin{bmatrix} \mathrm{e}^{-2t} & t\mathrm{e}^{-2t} & t^2\mathrm{e}^{-2t}/2 & 0 & 0 \\ 0 & \mathrm{e}^{-2t} & t\mathrm{e}^{-2t} & 0 & 0 \\ 0 & 0 & \mathrm{e}^{-2t} & 0 & 0 \\ 0 & 0 & 0 & \mathrm{e}^{-5t} & t\mathrm{e}^{-5t} \\ 0 & 0 & 0 & 0 & \mathrm{e}^{-5t} \end{bmatrix}$. In fact, since matrix $\boldsymbol{A}$ is a block Jordanian matrix, the results can easily be written out, even without the help of computers.

Example 4.48 For a given matrix $\boldsymbol{A} = \begin{bmatrix} -3 & -1 & -1 \\ 0 & -3 & -1 \\ 1 & 2 & 0 \end{bmatrix}$, calculate $\mathrm{e}^{\boldsymbol{A}t}$.

Solution The exponential function of $\boldsymbol{A}$ can be obtained directly using the `expm()` function such that

```
>> syms t; A=[-3,-1,-1; 0,-3,-1; 1,2,0]; simple(expm(A*t))
```

and the result is $\begin{bmatrix} -\mathrm{e}^{-2t}(-1+t) & -t\mathrm{e}^{-2t} & -t\mathrm{e}^{-2t} \\ -t^2\mathrm{e}^{-2t}/2 & -\mathrm{e}^{-2t}(-1+t+t^2/2) & -t\mathrm{e}^{-2t}(2+t/2) \\ t\mathrm{e}^{-2t}/2 & t\mathrm{e}^{-2t}(2+t/2) & \mathrm{e}^{-2t}(1+2t+t^2/2) \end{bmatrix}$.

Now consider a Jordanian transformation technique for solving such a problem. First, Jordanian transformation should be obtained such that

```
>> [V,J]=jordan(A)    % Jordanian transformation
```

the transformation can then be obtained

$$\boldsymbol{V} = \begin{bmatrix} 0 & -1 & 1 \\ -1 & 0 & 0 \\ 1 & 1 & 0 \end{bmatrix}, \quad \boldsymbol{J} = \begin{bmatrix} -2 & 1 & 0 \\ 0 & -2 & 1 \\ 0 & 0 & -2 \end{bmatrix}$$

where the matrix $\boldsymbol{V}$ obtained is no longer singular.

Since the exponential matrix of a Jordanian matrix is known, it can the be directly entered, and then the exponential function of the original matrix can be obtained from

```
>> J1=[exp(-2*t), t*exp(-2*t), 1/2*t^2*exp(-2*t);
            0,   exp(-2*t),          t*exp(-2*t);
            0,           0,            exp(-2*t)];
   A1=simple(V*J1*inv(V))
```

and the results are exactly the same as the one obtained above.

It should be noted that the use of Jordanian matrix to solve the exponential matrix function is not the best method, since it can be calculated directly otherwise. The use of the Jordanian function here indicates that this method can be extended to other applications. And in later subsections, the use of Jordanian matrices will be explored thoroughly.

Trigonometric functions of matrices

There are no MATLAB functions for trigonometric matrix operations. One may use the universal function `funm()` instead to find the numerical solution to such problems. This function is intended to be used for any nonlinear matrix functions, such that `A1=funm(A,fun)`, where the name of the function should be quoted using single quotation marks. For instance, if one wants to evaluate $\sin \boldsymbol{A}$, the statement `B=funm(A,'sin')` should be provided.

Example 4.49 Consider again the matrix given in Example 4.47. The function `funm()` can be used to calculate the sinusoidal matrix[1].

```
>> A=[[-2 1 0; 0 -2 1; 0 0 -2], zeros(3,2); zeros(2,3) [-5 1; 0 -5]];
   funm(A,'sin')
```

Thus the result can be obtained

$$\sin \boldsymbol{A} = \begin{bmatrix} -0.9093 & -0.41615 & 0.45465 & 0 & 0 \\ 0 & -0.9093 & -0.41615 & 0 & 0 \\ 0 & 0 & -0.9093 & 0 & 0 \\ 0 & 0 & 0 & 0.95892 & 0.28366 \\ 0 & 0 & 0 & 0 & 0.95892 \end{bmatrix}.$$

In fact, some certain matrix functions can be evaluated by the use of Taylor series expansion technique. For instance, sinusoidal function can be evaluated from

$$\sin \boldsymbol{A} = \sum_{i=0}^{\infty} (-1)^i \frac{\boldsymbol{A}^{2i+1}}{(2i+1)!} = \boldsymbol{A} - \frac{1}{3!}\boldsymbol{A}^3 + \frac{1}{5!}\boldsymbol{A}^5 + \cdots. \tag{4.51}$$

So the MATLAB function can be written to compute the sinusoidal matrix

```
function E=sinm1(A)
E=zeros(size(A)); F=A; k=1;
while norm(E+F-E,1)>0, E=E+F; F=-A^2*F/((k+2)*(k+1)); k=k+2; end
```

Example 4.50 From the above explanation, it can be seen that the seemingly complicated sinusoidal matrix functions can easily be solved using a few statements in MATLAB. With such a function, $\sin \boldsymbol{A}$ can easily be obtained

[1]It should be noted that in earlier versions of MATLAB, since the computation used the eigenvector-based algorithm, erroneous results may be obtained if $\boldsymbol{A}$ has repeated eigenvalues.

```
>> E=sinm1(A)
```

It can be measured that 39 terms have been added in the above computation, and the results are exactly the same as the one obtained from the previous example.

For sinusoidal and cosine functions, if a scalar a is given, the trigonometric functions satisfy the well-known Euler formula, such that $e^{ja} = \cos a + j\sin a$ and $e^{-ja} = \cos a - j\sin a$. From the above formula, one can find immediately that

$$\sin a = \frac{1}{j2}(e^{ja} - e^{-ja}), \quad \cos a = \frac{1}{2}(e^{ja} + e^{-ja}) \tag{4.52}$$

and the new formula holds also if the scalar a is replaced by a matrix $\boldsymbol{A}$. Some examples will be given to show such computations.

Example 4.51 Consider also the matrix in Example 4.47, evaluate $\sin \boldsymbol{A}$.

Solution The existing `expm()` function can be used to evaluate the sinusoidal function

```
>> A=[[-2 1 0; 0 -2 1; 0 0 -2], zeros(3,2); zeros(2,3) [-5 1; 0 -5]];
   j=sqrt(-1); A1=(expm(A*j)-expm(-A*j))/(2*j),
```

which yields

$$\sin \boldsymbol{A} = \begin{bmatrix} -0.9093 & -0.41615 & 0.45465 & 0 & 0 \\ 0 & -0.9093 & -0.41615 & 0 & 0 \\ 0 & 0 & -0.9093 & 0 & 0 \\ 0 & 0 & 0 & 0.95892 & 0.28366 \\ 0 & 0 & 0 & 0 & 0.95892 \end{bmatrix}$$

which agrees well with the results in Example 4.50, which means that the algorithm is correct.

Example 4.52 Consider a given matrix $\boldsymbol{A} = \begin{bmatrix} -7 & 2 & 0 & -1 \\ 1 & -4 & 2 & 1 \\ 2 & -1 & -6 & -1 \\ -1 & -1 & 0 & -4 \end{bmatrix}$. It is known that the matrix contains repeated eigenvalues. Compute matrix functions $\sin \boldsymbol{A}t$ and $\cos \boldsymbol{A}t$.

Solution From (4.52), the following statements can be obtained for the sinusoidal and cosine functions.

```
>> A=[-7,2,0,-1; 1,-4,2,1; 2,-1,-6,-1; -1,-1,0,-4];
   syms t; j=sym(sqrt(-1));
   A1=simple((expm(A*j*t)-expm(-A*j*t))/(2*j)),
   A2=simple((expm(A*j*t)+expm(-A*j*t))/2)
```

The results obtained are

$$\sin \boldsymbol{A}t = \left[\begin{matrix} -2/9\sin 3t + (t^2-7/9)\sin 6t - 5/3t\cos 6t & -1/3\sin 3t + 1/3\sin 6t + t\cos 6t \\ -2/9\sin 3t + (t^2+2/9)\sin 6t + 1/3t\cos 6t & -1/3\sin 3t - 2/3\sin 6t + t\cos 6t \\ -2/9\sin 3t + (-2t^2+2/9)\sin 6t + 4/3t\cos 6t & -1/3\sin 3t + 1/3\sin 6t - 2t\cos 6t \\ 4/9\sin 3t + (t^2-4/9)\sin 6t + 1/3t\cos 6t & 2/3\sin 3t - 2/3\sin 6t + t\cos 6t \end{matrix}\right.$$

$$\left.\begin{array}{ll}
-2/9\sin 3t+(2/9+t^2)\sin 6t-2/3t\cos 6t & 1/9\sin 3t+(-1/9+t^2)\sin 6t-2/3t\cos 6t \\
-2/9\sin 3t+(2/9+t^2)\sin 6t+4/3t\cos 6t & 1/9\sin 3t+(-1/9+t^2)\sin 6t+4/3t\cos 6t \\
-2/9\sin 3t-(7/9+2t^2)\sin 6t-2/3t\cos 6t & 1/9\sin 3t-(1/9+2t^2)\sin 6t-2/3t\cos 6t \\
4/9\sin 3t+(-4/9+t^2)\sin 6t+4/3t\cos 6t & -2/9\sin 3t+(-7/9+t^2)\sin 6t+4/3t\cos 6t
\end{array}\right]$$

$$\cos \boldsymbol{A}t=\left[\begin{array}{ll}
2/9\cos 3t+(-t^2+7/9)\cos 6t-5/3t\sin 6t & 1/3\cos 3t-1/3\cos 6t+t\sin 6t \\
2/9\cos 3t-(t^2+2/9)\cos 6t+1/3t\sin 6t & 1/3\cos 3t+2/3\cos 6t+t\sin 6t \\
2/9\cos 3t+(2t^2-2/9)\cos 6t+4/3t\sin 6t & 1/3\cos 3t-1/3\cos 6t-2t\sin 6t \\
-4/9\cos 3t+(-t^2+4/9)\cos 6t+1/3t\sin 6t & -2/3\cos 3t+2/3\cos 6t+t\sin 6t
\end{array}\right.$$

$$\left.\begin{array}{ll}
2/9\cos 3t-(2/9+t^2)\cos 6t-2/3t\sin 6t & -1/9\cos 3t+(1/9-t^2)\cos 6t-2/3t\sin 6t \\
2/9\cos 3t-(2/9+t^2)\cos 6t+4/3t\sin 6t & -1/9\cos 3t+(1/9-t^2)\cos 6t+4/3t\sin 6t \\
2/9\cos 3t+(7/9+2t^2)\cos 6t-2/3t\sin 6t & -1/9\cos 3t+(1/9+2t^2)\cos 6t-2/3t\sin 6t \\
-4/9\cos 3t+(4/9-t^2)\cos 6t+4/3t\sin 6t & 2/9\cos 3t+(7/9-t^2)\cos 6t+4/3t\sin 6t
\end{array}\right].$$

General matrix functions

Other MATLAB functions can be used for solving matrix function problems, for instance, `logm()`, `sqrtm()` and more universal `funm()`, etc. It can be seen that an `m` is appended to the function names, indicating the evaluations of matrix functions, rather than element-by-element computation.

Unfortunately, the existing function `funm()` has many limitations. For instance, it can only be used for certain functions such as $\cos \boldsymbol{A}$, $\sin \boldsymbol{A}$, but cannot be used for an arbitrary function $\psi(\boldsymbol{A})$. Also the matrix functions containing the time variable t cannot be evaluated. Here a Jordanian matrix-based algorithm is provided [13], with MATLAB implementations.

To solve such a problem, one can first write an $m_i \times m_i$ Jordanian block $\boldsymbol{J}_i$ as $\boldsymbol{J}_i = \lambda_i \boldsymbol{I} + \boldsymbol{H}_{m_i}$, where λ_i is a repeated eigenvalue of multiplicity m_i, and $\boldsymbol{H}_{m_i}$ is a nilpotent matrix, i.e., when $k \geqslant m_i$, $\boldsymbol{H}_{m_i}^k \equiv \boldsymbol{0}$. It can be shown that the matrix function $\psi(\boldsymbol{J}_i)$ can be obtained as follows:

$$\psi(\boldsymbol{J}_i) = \psi(\lambda_i)\boldsymbol{I}_{m_i} + \psi'(\lambda_i)\boldsymbol{H}_{m_i} + \cdots + \frac{\psi^{(m_i-1)}(\lambda_i)}{(m_i-1)!}\boldsymbol{H}_{m_i}^{m_i-1}. \tag{4.53}$$

With Jordan decomposition, the given matrix $\boldsymbol{A}$ can be transformed such that

$$\boldsymbol{A} = \boldsymbol{V}\begin{bmatrix} \boldsymbol{J}_1 & & & \\ & \boldsymbol{J}_2 & & \\ & & \ddots & \\ & & & \boldsymbol{J}_m \end{bmatrix}\boldsymbol{V}^{-1}. \tag{4.54}$$

Using Jordanian matrix decomposition, the arbitrary function $\psi(\boldsymbol{A})$ can be evaluated from

$$\psi(\boldsymbol{A}) = \boldsymbol{V}\begin{bmatrix} \psi(\boldsymbol{J}_1) & & & \\ & \psi(\boldsymbol{J}_2) & & \\ & & \ddots & \\ & & & \psi(\boldsymbol{J}_m) \end{bmatrix}\boldsymbol{V}^{-1}. \tag{4.55}$$

Based on the algorithm above, the following new MATLAB `funm()` function can be written and it should be placed under the `@sym` directory. This function can be used to find the analytical solution of any matrix function. The listing of the function is

```
function F=funm(A,fun,x)
[V,J]=jordan(A); v1=[0,diag(J,1)']; v2=[find(v1==0), length(v1)+1];
for i=1:length(v2)-1
   v_lambda(i)=J(v2(i),v2(i)); v_n(i)=v2(i+1)-v2(i);
end
m=length(v_lambda); F=sym([]);
for i=1:m
   J1=J(v2(i):v2(i)+v_n(i)-1,v2(i):v2(i)+v_n(i)-1);
   fJ=funJ(J1,fun,x); F=diagm(F,fJ);
end
F=V*F*inv(V);
function fJ=funJ(J,fun,x)
lam=J(1,1); f1=fun; fJ=subs(fun,x,lam)*eye(size(J));
H=diag(diag(J,1),1); H1=H;
for i=2:length(J)
   f1=diff(f1,x); a1=subs(f1,x,lam); fJ=fJ+a1*H1; H1=H1*H/i;
end
function A=diagm(A1,A2)
A=A1; [n,m]=size(A); [n1,m1]=size(A2); A(n+1:n+n1,m+1:m+m1)=A2;
```

The syntax of the function is $\boldsymbol{A}_1$=`funm(`$\boldsymbol{A}$`,`*funx*`,`x`)`, where x is a symbolic variable, *funx* is the function $\psi(\cdot)$ of variable x. For instance, if one wants $\mathrm{e}^{\boldsymbol{A}}$, function *funx* should be defined as `exp(x)`. In fact, the variable *funx* can be assigned to an arbitrarily complicated function, i.e., `exp(x*t)` means that one wants to have $\mathrm{e}^{\boldsymbol{A}t}$, and t could also be a symbolic variable. Composite functions such as `exp(x*cos(x*t))` can also be used here to indicate that one wants to calculate $\psi(\boldsymbol{A}) = \mathrm{e}^{\boldsymbol{A}\cos(\boldsymbol{A}t)}$.

Example 4.53 For matrix $\boldsymbol{A} = \begin{bmatrix} -7 & 2 & 0 & -1 \\ 1 & -4 & 2 & 1 \\ 2 & -1 & -6 & -1 \\ -1 & -1 & 0 & -4 \end{bmatrix}$, compute the matrix function $\psi(\boldsymbol{A}) = \mathrm{e}^{\boldsymbol{A}\cos \boldsymbol{A}t}$.

Solution If the matrix function $\psi(\boldsymbol{A}) = \mathrm{e}^{\boldsymbol{A}\cos \boldsymbol{A}t}$ is required, the following MATLAB statements should be entered

```
>> A=[-7,2,0,-1; 1,-4,2,1; 2,-1,-6,-1; -1,-1,0,-4];
   syms x t; A1=funm(sym(A),exp(x*cos(x*t)),x)
```

Since the results obtained are very lengthy, only the first term of the results is displayed as follows:

$$\psi_{1,1}(\boldsymbol{A}) = 2/9\mathrm{e}^{-3\cos 3t} + (2t\sin 6t + 6t^2\cos 6t)\mathrm{e}^{-6\cos 6t} + (\cos 6t - 6t\sin 6t)^2\mathrm{e}^{-6\cos 6t} - 5/3(\cos 6t - 6t\sin 6t)\mathrm{e}^{-6\cos 6t} + 7/9\mathrm{e}^{-6\cos 6t}.$$

It can be seen from the results that in $\psi_{1,1}(t)$, the term $\mathrm{e}^{-6\cos 6t}$ is repeatedly used, thus one may collect that term using the following statement to simplify the result

```
>> collect(A1(1,1),exp(-6*cos(6*t)))
```

which leads to the following simplified result of $\psi_{1,1}(\boldsymbol{A})$

$$\left[12t\sin 6t + 6t^2\cos 6t + (\cos 6t - 6t\sin 6t)^2 - \tfrac{5}{3}\cos 6t + \tfrac{7}{9}\right]\mathrm{e}^{-6\cos 6t} + \tfrac{2}{9}\mathrm{e}^{-3\cos 3t}.$$

Further, if one assumes $t = 1$, the matrix $\mathrm{e}^{\boldsymbol{A}\cos\boldsymbol{A}}$ can be found such that

```
>> subs(A1,t,1), % which is the same as expm(A*funm(A,'cos'))
```

The complicated function is

$$\mathrm{e}^{\boldsymbol{A}\cos\boldsymbol{A}} = \begin{bmatrix} 4.3583 & 6.5044 & 4.3635 & -2.1326 \\ 4.3718 & 6.5076 & 4.3801 & -2.116 \\ 4.2653 & 6.4795 & 4.2518 & -2.2474 \\ -8.6205 & -12.984 & -8.6122 & 4.3832 \end{bmatrix}.$$

Exercises

1. Jordanian matrix is a very practical matrix in matrix analysis courses. The general form of the matrix is described as

$$\boldsymbol{J} = \begin{bmatrix} -\alpha & 1 & 0 & \cdots & 0 \\ 0 & -\alpha & 1 & \cdots & 0 \\ \vdots & \vdots & \vdots & \ddots & \vdots \\ 0 & 0 & 0 & \cdots & -\alpha \end{bmatrix}, \quad \text{e.g.,} \quad \boldsymbol{J}_1 = \begin{bmatrix} -5 & 1 & 0 & 0 & 0 \\ 0 & -5 & 1 & 0 & 0 \\ 0 & 0 & -5 & 1 & 0 \\ 0 & 0 & 0 & -5 & 1 \\ 0 & 0 & 0 & 0 & -5 \end{bmatrix}.$$

Construct matrix $\boldsymbol{J}_1$ with the MATLAB function `diag()`.

2. Nilpotent matrix is a special matrix defined as $\boldsymbol{H}_n = \begin{bmatrix} 0 & 1 & 0 & \cdots & 0 \\ 0 & 0 & 1 & \cdots & 0 \\ \vdots & \vdots & \vdots & \ddots & \vdots \\ 0 & 0 & 0 & \cdots & 1 \\ 0 & 0 & 0 & \cdots & 0 \end{bmatrix}$. Verify for any pre-specified n, $\boldsymbol{H}_n^i = 0$ is satisfied for all $i \geqslant n$.

3. Can you recognize from the way of display whether a matrix is a numeric matrix or a symbolic matrix? If $\boldsymbol{A}$ is a numeric matrix and $\boldsymbol{B}$ is a symbolic matrix, can you predict whether the product $\boldsymbol{C}$=$\boldsymbol{A}*\boldsymbol{B}$ is a numeric matrix or a symbolic matrix? Verify this through a simple example.

4. Compute the determinant of a Vandermonde matrix $\boldsymbol{A} = \begin{bmatrix} a^4 & a^3 & a^2 & a & 1 \\ b^4 & b^3 & b^2 & b & 1 \\ c^4 & c^3 & c^2 & c & 1 \\ d^4 & d^3 & d^2 & d & 1 \\ e^4 & e^3 & e^2 & e & 1 \end{bmatrix}$, and find the simplified results.

5. Input matrices $\boldsymbol{A}$ and $\boldsymbol{B}$ in MATLAB, and convert them into symbolic matrices.

$$\boldsymbol{A}=\begin{bmatrix}5&7&6&5&1&6&5\\2&3&1&0&0&1&4\\6&4&2&0&6&4&4\\3&9&6&3&6&6&2\\10&7&6&0&0&7&7\\7&2&4&4&0&7&7\\4&8&6&7&2&1&7\end{bmatrix},\quad \boldsymbol{B}=\begin{bmatrix}3&5&5&0&1&2&3\\3&2&5&4&6&2&5\\1&2&1&1&3&4&6\\3&5&1&5&2&1&2\\4&1&0&1&2&0&1\\-3&-4&-7&3&7&8&12\\1&-10&7&-6&8&1&5\end{bmatrix}.$$

6. Check whether the matrices given in the above exercise are singular or not. Find the rank, determinant, trace and inverse matrices for them. Check whether the inverse matrices are correct or not.

7. Find the characteristic polynomials, eigenvalues and eigenvectors for the matrices $\boldsymbol{A}$ and $\boldsymbol{B}$ in Exercise 4.5. Validate Hamilton-Caylay Theorem and explain how the error, if any, may be eliminated.

8. Perform singular value decompositions, LU decompositions and orthogonal decompositions to the matrices $\boldsymbol{A}$ and $\boldsymbol{B}$ in Exercise 4.5.

9. For arbitrary matrices

$$\boldsymbol{A}_1=\begin{bmatrix}a_{11}&a_{12}&a_{13}\\a_{21}&a_{22}&a_{23}\\a_{31}&a_{32}&a_{33}\end{bmatrix},\ \boldsymbol{A}_2=\begin{bmatrix}a_{11}&a_{12}&a_{13}&a_{14}\\a_{21}&a_{22}&a_{23}&a_{24}\\a_{31}&a_{32}&a_{33}&a_{34}\\a_{41}&a_{42}&a_{43}&a_{44}\end{bmatrix}$$

verify the Hamilton-Caylay Theorem.

10. Perform LU factorization and SVD decomposition to the following matrices

$$\boldsymbol{A}=\begin{bmatrix}8&0&1&1&6\\9&2&9&4&0\\1&5&9&9&8\\9&9&4&7&9\\6&9&8&9&6\end{bmatrix},\quad \boldsymbol{B}=\begin{bmatrix}1&2&2&2\\1&1&2&0\\1&1&1&0\\0&0&2&0\end{bmatrix}.$$

11. Compute the eigenvalues, eigenvectors and singular values of the following matrices.

$$\boldsymbol{A}=\begin{bmatrix}2&7&5&7&7\\7&4&9&3&3\\3&9&8&3&8\\5&9&6&3&6\\2&6&8&5&4\end{bmatrix},\quad \boldsymbol{B}=\begin{bmatrix}703&795&980&137&661\\547&957&271&12&284\\445&523&252&894&469\\695&880&876&199&65\\621&173&737&299&988\end{bmatrix}.$$

12. Please check whether the following matrices are positive-definite ones, if so, please find the Cholesky decomposed matrices.

$$\boldsymbol{A}=\begin{bmatrix}9&2&1&2&2\\2&4&3&3&3\\1&3&7&3&4\\2&3&3&5&4\\2&3&4&4&5\end{bmatrix},\quad \boldsymbol{B}=\begin{bmatrix}16&17&9&12&12\\17&12&12&2&18\\9&12&18&7&13\\12&2&7&18&12\\12&18&13&12&10\end{bmatrix}.$$

13. Perform the Jordanian transformation for $\boldsymbol{A}=\begin{bmatrix} -2 & 0.5 & -0.5 & 0.5 \\ 0 & -1.5 & 0.5 & -0.5 \\ 2 & 0.5 & -4.5 & 0.5 \\ 2 & 1 & -2 & -2 \end{bmatrix}$, and also find the corresponding transformation matrix.

14. Find the basic set of solutions of the homogenous equations

(a) $\begin{cases} 6x_1+x_2+4x_3-7x_4-3x_5=0 \\ -2x_1-7x_2-8x_3+6x_4 \quad =0 \\ -4x_1+5x_2+x_3-6x_4+8x_5=0 \\ -34x_1+36x_2+9x_3-21x_4+49x_5=0 \\ -26x_1-12x_2-27x_3+27x_4+17x_5=0, \end{cases}$ (b) $\boldsymbol{A}=\begin{bmatrix} -1 & 2 & -2 & 1 & 0 \\ 0 & 3 & 2 & 2 & 1 \\ 3 & 1 & 3 & 2 & -1 \end{bmatrix}$.

15. Find the numerical and analytical solutions to the following linear algebraic equations, and then validate the results.

$$\begin{bmatrix} 2 & -9 & 3 & -2 & -1 \\ 10 & -1 & 10 & 5 & 0 \\ 8 & -2 & -4 & -6 & 3 \\ -5 & -6 & -6 & -8 & -4 \end{bmatrix}\boldsymbol{X}=\begin{bmatrix} -1 & -4 & 0 \\ -3 & -8 & -4 \\ 0 & 3 & 3 \\ 9 & -5 & 3 \end{bmatrix}.$$

16. Check whether the equation $\begin{bmatrix} 16 & 2 & 3 & 13 \\ 5 & 11 & 10 & 8 \\ 9 & 7 & 6 & 12 \\ 4 & 14 & 15 & 1 \end{bmatrix}\boldsymbol{X}=\begin{bmatrix} 1 \\ 3 \\ 4 \\ 7 \end{bmatrix}$ has a solution.

17. Find the analytical solutions to the following linear algebraic equations, and then validate the results.

$$\begin{bmatrix} 2 & 9 & 4 & 12 & 5 & 8 & 6 \\ 12 & 2 & 8 & 7 & 3 & 3 & 7 \\ 3 & 0 & 3 & 5 & 7 & 5 & 10 \\ 3 & 11 & 6 & 6 & 9 & 9 & 1 \\ 11 & 2 & 1 & 4 & 6 & 8 & 7 \\ 5 & -18 & 1 & -9 & 11 & -1 & 18 \\ 26 & -27 & -1 & 0 & -15 & -13 & 18 \end{bmatrix}\boldsymbol{X}=\begin{bmatrix} 1 & 9 \\ 5 & 12 \\ 4 & 12 \\ 10 & 9 \\ 0 & 5 \\ 10 & 18 \\ -20 & 2 \end{bmatrix}.$$

18. For the matrices $\boldsymbol{A}$ and $\boldsymbol{B}$, calculate $\boldsymbol{A}\otimes\boldsymbol{B}$ and $\boldsymbol{B}\otimes\boldsymbol{A}$. Are they equal to each other?

$$\boldsymbol{A}=\begin{bmatrix} -1 & 2 & 2 & 1 \\ -1 & 2 & 1 & 0 \\ 2 & 1 & 1 & 0 \\ 1 & 0 & 2 & 0 \end{bmatrix},\quad \boldsymbol{B}=\begin{bmatrix} 3 & 0 & 3 \\ 3 & 2 & 2 \\ 3 & 1 & 1 \end{bmatrix}.$$

19. Find the analytical and numerical solutions to the following Sylvester equation, and verify the results.

$$\begin{bmatrix} 3 & -6 & -4 & 0 & 5 \\ 1 & 4 & 2 & -2 & 4 \\ -6 & 3 & -6 & 7 & 3 \\ -13 & 10 & 0 & -11 & 0 \\ 0 & 4 & 0 & 3 & 4 \end{bmatrix}\boldsymbol{X}+\boldsymbol{X}\begin{bmatrix} 3 & -2 & 1 \\ -2 & -9 & 2 \\ -2 & -1 & 9 \end{bmatrix}=\begin{bmatrix} -2 & 1 & -1 \\ 4 & 1 & 2 \\ 5 & -6 & 1 \\ 6 & -4 & -4 \\ -6 & 6 & -3 \end{bmatrix}.$$

20. Find the analytical and numerical solutions to the so-called *Stein equation* given below and verify the results.

$$\begin{bmatrix} -2 & 2 & 1 \\ -1 & 0 & -1 \\ 1 & -1 & 2 \end{bmatrix} \boldsymbol{X} \begin{bmatrix} -2 & -1 & 2 \\ 1 & 3 & 0 \\ 3 & -2 & 2 \end{bmatrix} - \boldsymbol{X} + \begin{bmatrix} 0 & -1 & 0 \\ -1 & 1 & 0 \\ 1 & -1 & -1 \end{bmatrix} = 0.$$

21. Assume that a Riccati equation is given by $\boldsymbol{PA}+\boldsymbol{A}^{\mathrm{T}}\boldsymbol{P}-\boldsymbol{PBR}^{-1}\boldsymbol{B}^{\mathrm{T}}\boldsymbol{P}+\boldsymbol{Q}=0$, where

$$\boldsymbol{A}=\begin{bmatrix} -27 & 6 & -3 & 9 \\ 2 & -6 & -2 & -6 \\ -5 & 0 & -5 & -2 \\ 10 & 3 & 4 & -11 \end{bmatrix}, \boldsymbol{B}=\begin{bmatrix} 0 & 3 \\ 16 & 4 \\ -7 & 4 \\ 9 & 6 \end{bmatrix}, \boldsymbol{Q}=\begin{bmatrix} 6 & 5 & 3 & 4 \\ 5 & 6 & 3 & 4 \\ 3 & 3 & 6 & 2 \\ 4 & 4 & 2 & 6 \end{bmatrix}, \boldsymbol{R}=\begin{bmatrix} 4 & 1 \\ 1 & 5 \end{bmatrix}.$$

Solve the equation and verify the result.

22. Certain functions can be expressed by polynomial functions, i.e., Taylor series expansions. In these functions, if x is substituted by matrix $\boldsymbol{A}$, the nonlinear function can also be expressed for matrices. Write M-functions for the matrix function evaluation problems and verify the results with `funm()`.

(i) $\cos \boldsymbol{A} = \boldsymbol{I} - \dfrac{1}{2!}\boldsymbol{A}^2 + \dfrac{1}{4!}\boldsymbol{A}^4 - \dfrac{1}{6!}\boldsymbol{A}^6 + \cdots + \dfrac{(-1)^n}{(2n)!}\boldsymbol{A}^{2n} + \cdots$

(ii) $\arcsin \boldsymbol{A} = \boldsymbol{A} + \dfrac{1}{2\cdot 3}\boldsymbol{A}^3 + \dfrac{1\cdot 3}{2\cdot 4\cdot 5}\boldsymbol{A}^5 + \dfrac{1\cdot 3\cdot 5}{2\cdot 4\cdot 6\cdot 7}\boldsymbol{A}^7 + \dfrac{1\cdot 3\cdot 5\cdot 7}{2\cdot 4\cdot 6\cdot 8\cdot 9}\boldsymbol{A}^9 + \cdots + \dfrac{(2n)!}{2^{2n}(n!)^2(2n+1)}\boldsymbol{A}^{2n+1} + \cdots$

(iii) $\ln \boldsymbol{A} = \boldsymbol{A}-\boldsymbol{I}-\dfrac{1}{2}(\boldsymbol{A}-\boldsymbol{I})^2+\dfrac{1}{3}(\boldsymbol{A}-\boldsymbol{I})^3-\dfrac{1}{4}(\boldsymbol{A}-\boldsymbol{I})^4+\cdots+\dfrac{(-1)^{n+1}}{n}(\boldsymbol{A}-\boldsymbol{I})^n+\cdots.$

23. For an autonomous linear differential equation of the form $\dot{\boldsymbol{x}}(t) = \boldsymbol{A}\boldsymbol{x}(t)$, the analytical solution can be written as $\boldsymbol{x}(t) = \mathrm{e}^{\boldsymbol{A}t}\boldsymbol{x}(0)$. Find the analytical solution to the equation

$$\dot{\boldsymbol{x}}(t) = \begin{bmatrix} -3 & 0 & 0 & 1 \\ -1 & -1 & 1 & -1 \\ 1 & 0 & -2 & 1 \\ 0 & 0 & 0 & -4 \end{bmatrix} \boldsymbol{x}(t), \quad \boldsymbol{x}(0) = \begin{bmatrix} -1 \\ 0 \\ 3 \\ 1 \end{bmatrix}.$$

24. If a block Jordanian matrix $\boldsymbol{A}$ is given by

$$\boldsymbol{A}=\begin{bmatrix} \boldsymbol{A}_1 & & \\ & \boldsymbol{A}_2 & \\ & & \boldsymbol{A}_3 \end{bmatrix}, \quad \text{where } \boldsymbol{A}_1=\begin{bmatrix} -3 & 1 & 0 \\ 0 & -3 & 1 \\ 0 & 0 & -3 \end{bmatrix},$$

$$\boldsymbol{A}_2=\begin{bmatrix} -5 & 1 \\ 0 & -5 \end{bmatrix}, \quad \boldsymbol{A}_3=\begin{bmatrix} -1 & 1 & 0 & 0 \\ 0 & -1 & 1 & 0 \\ 0 & 0 & -1 & 1 \\ 0 & 0 & 0 & -1 \end{bmatrix}$$

find the solutions to $\mathrm{e}^{\boldsymbol{A}t}$, $\sin\left(2\boldsymbol{A}t + \dfrac{\pi}{3}\right)$, $\mathrm{e}^{\boldsymbol{A}^2 t}\boldsymbol{A}^2 + \sin(\boldsymbol{A}^3 t)\boldsymbol{A}t + \mathrm{e}^{\sin \boldsymbol{A}t}$.

25. For a given matrix $\boldsymbol{A}$, find matrix functions $\mathrm{e}^{\boldsymbol{A}t}, \sin \boldsymbol{A}t, \mathrm{e}^{\boldsymbol{A}t}\sin\left(\boldsymbol{A}^2\mathrm{e}^{\boldsymbol{A}t}t\right)$.

$$\boldsymbol{A} = \begin{bmatrix} -4.5 & 0 & 0.5 & -1.5 \\ -0.5 & -4 & 0.5 & -0.5 \\ 1.5 & 1 & -2.5 & 1.5 \\ 0 & -1 & -1 & -3 \end{bmatrix}.$$

Chapter 5

Integral Transforms and Complex Variable Functions

Integral transform technique can usually be used to map a problem from one domain into another, such that the new problem may easily be solved in the transformed domain. For instance, the Laplace transform can be used to map a linear time-invariant (LTI) ordinary differential equation (ODE) into an algebraic equation. Thus the properties of the original problem, such as stability, can easily be determined, which lays a foundation of the classical control theory. In many real applications, Fourier transforms as well as Mellin transforms and Hankel transforms are all very useful. Thus computer-aided solutions to integral transform problems deserve special attention and are the main topics of this chapter. In Section 5.1, definition and properties of Laplace transform and the inverse Laplace transform are summarized. Our focus is on the MATLAB-based solutions to Laplace transform problems. Section 5.2 presents Fourier transform and its inverse transform, again with a focus on the MATLAB-based solutions. Moreover, sine and cosine Fourier transforms and discrete Fourier transforms are briefly introduced. In Section 5.3, Mellin and Hankel transforms are introduced. Z transform and inverse Z transform are introduced in Section 5.4 with illustrative examples taken from discrete-time signals and systems. Finally, problems from complex variable functions such as poles and residues are demonstrated together with Partial fraction expansions for rational functions in Section 5.5, where an evaluation method of closed-path integrals is introduced based on the concepts of residues.

For readers who wish to check the detailed explanations of functions of complex variable, Laplace and Fourier transforms, we recommend the open source textbook [14] (Chapters 6-13, 31, 32).

5.1 Laplace Transforms and Their Inverses

Integral transform introduced by the French mathematician Pierre-Simon Laplace (1749–1827) can be used to map the ordinary differential equations into algebraic equations. Thus it established the foundation for many research areas. For instance, the Laplace transform established the basis for the modeling, analysis and synthesis of control systems. In this section, the definition and basic properties of Laplace transform and inverse Laplace transform are summarized first. Then we focus on the solutions to Laplace transform problems and their applications using MATLAB.

5.1.1 Definitions and properties

The one-sided Laplace transform of a time-domain function $f(t)$ is defined as

$$\mathscr{L}[f(t)] = \int_0^\infty f(t)\mathrm{e}^{-st}\mathrm{d}t = F(s) \tag{5.1}$$

where $\mathscr{L}[f(t)]$ is the notation of Laplace transform.

The properties of Laplace transform is summarized below without proofs.

(i) **Linear property** $\mathscr{L}[af(t) \pm bg(t)] = a\mathscr{L}[f(t)] \pm b\mathscr{L}[g(t)]$ for scalars a and b.

(ii) **Time-domain shift** $\mathscr{L}[f(t-a)] = \mathrm{e}^{-as}F(s)$.

(iii) **s-domain property** $\mathscr{L}[\mathrm{e}^{-at}f(t)] = F(s+a)$.

(iv) **Differentiation property** $\mathscr{L}[\mathrm{d}f(t)/\mathrm{d}t] = sF(s) - f(0^+)$. Generally, the nth order derivative can be obtained from

$$\mathscr{L}\left[\frac{\mathrm{d}^n}{\mathrm{d}t^n}f(t)\right] = s^nF(s) - s^{n-1}f(0^+) - s^{n-2}\frac{\mathrm{d}f(0^+)}{\mathrm{d}t} - \cdots - \frac{\mathrm{d}^{n-1}f(0^+)}{\mathrm{d}t^{n-1}}. \tag{5.2}$$

If the initial values of $f(t)$ and the other derivatives are all zero, (5.2) can be simplified as

$$\mathscr{L}[\mathrm{d}^n f(t)/\mathrm{d}t^n] = s^nF(s) \tag{5.3}$$

and this property is the crucial formula to map differential equations into algebraic equations.

(v) **Integration property** If zero conditions are assumed, $\mathscr{L}[\int_0^t f(\tau)\,\mathrm{d}\tau] = F(s)/s$. Generally, the Laplace transform of the multiple integral of $f(t)$ can be obtained from

$$\mathscr{L}\left[\int_0^t \cdots \int_0^t f(\tau)\,\mathrm{d}\tau^n\right] = \frac{F(s)}{s^n}. \tag{5.4}$$

(vi) **Initial value property** $\lim\limits_{t\to 0} f(t) = \lim\limits_{s\to\infty} sF(s)$.

(vii) **Final value property** If $F(s)$ has no pole with non-negative real part, i.e., $\mathrm{Re}(s) \geqslant 0$, then
$$\lim_{t\to\infty} f(t) = \lim_{s\to 0} sF(s).$$

(viii) **Convolution property** $\mathscr{L}[f(t) * g(t)] = \mathscr{L}[f(t)]\mathscr{L}[g(t)]$, where the convolution operator $*$ is defined as

$$f(t) * g(t) = \int_0^t f(\tau)g(t-\tau)\,\mathrm{d}\tau = \int_0^t f(t-\tau)g(\tau)\,\mathrm{d}\tau. \tag{5.5}$$

(ix) **Other properties**

$$\mathscr{L}[t^n f(t)] = (-1)^n \frac{\mathrm{d}^n F(s)}{\mathrm{d}s^n}, \quad \mathscr{L}\left[\frac{f(t)}{t^n}\right] = \int_s^\infty \cdots \int_s^\infty F(s)\,\mathrm{d}s^n. \tag{5.6}$$

If the Laplace transform of a signal $f(t)$ is $F(s)$, the inverse Laplace transform of $F(s)$ is defined as

$$f(t) = \mathscr{L}^{-1}[F(s)] = \frac{1}{\mathrm{j}2\pi} \int_{\sigma-\mathrm{j}\infty}^{\sigma+\mathrm{j}\infty} F(s)\mathrm{e}^{st}\mathrm{d}s \tag{5.7}$$

where σ is greater than the real part of the poles of function $F(s)$. The definitions of poles will be given later.

5.1.2 Computer solution to Laplace transform problems

It is hard if not impossible to write numerical programs to solve Laplace transform problems. Computer algebra systems should be used instead. For instance, the Symbolic Math Toolbox of MATLAB can be used to solve the problems easily and analytically. The procedures for solving such problems are summarized as follows:

(i) The symbolic variables such as t should be declared using the `syms` command. Then the time-domain function $f(t)$ should be defined in variable *fun*.

(ii) Call the `laplace()` function to solve the problem. Thus the Laplace transform can be obtained with the following function call

```
F=laplace(fun)        % the time variable is given in t
F=laplace(fun,v,u)    % with domain variables v, u specified
```

and the function `simple()` can be used to simply the obtained symbolic result.

(iii) For complicated problems, the returned results are difficult to read. The function `pretty()` can be used to better display the results. Also the `latex()` function can be used to convert the results into LaTeX string.

(iv) If the Laplace transform function $F(s)$ is known, it can also be described in the variable *fun*. Then the MATLAB function `ilaplace()` can be used to calculate the inverse Laplace transform of the given function. The syntaxes of the function are

```
f=ilaplace(fun)       % default variable is s
f=ilaplace(fun,u,v)   % specify the domain variables v and u
```

Example 5.1 For a given time domain function $f(t) = t^2\mathrm{e}^{-2t}\sin(t+\pi)$, compute its Laplace transform function $F(s)$.

Solution From the original problem, it can be seen that the time domain variable t should be declared first. With the MATLAB statements, the function $f(t)$ can be specified. Then the `laplace()` function can be used to derive the Laplace transform of the original function

```
>> syms t; f=t^2*exp(-2*t)*sin(t+pi); F=laplace(f)
```

and the result is as follows:

$$F(s) = -8\frac{(s+2)^2}{\left[(s+2)^2+1\right]^3} + 2\frac{1}{\left[(s+2)^2+1\right]^2}.$$

The result can also be simplified with the statement `simple(F)` and the simplified result is $F(s) = -2\dfrac{3s^2+12s+11}{(s^2+4s+5)^3}$.

Example 5.2 Assume that the original function is given by $f(x) = x^2 \mathrm{e}^{-2x}\sin(x+\pi)$, compute the Laplace transform and then take inverse Laplace transform and see whether the original function can be recovered.

Solution Similarly, the `laplace()` function can still be used

```
>> syms x w; f=x^2*exp(-2*x)*sin(x+pi); F=laplace(f,x,w)
```

and the result is the same as the one obtained in the previous example, albeit the name of the variables used are different.

If the command f_1=`ilaplace(F)` is used, the result $f_1(t) = -t^2\mathrm{e}^{-2t}\sin t$ can be obtained. Since the equation $\sin(t+\pi) = -\sin t$ holds, the original function is actually restored.

Example 5.3 Compute the inverse Laplace transform for the following complex variable function

$$G(x) = \frac{-17x^5 - 7x^4 + 2x^3 + x^2 - x + 1}{x^6 + 11x^5 + 48x^4 + 106x^3 + 125x^2 + 75x + 17}.$$

Solution The following statements can be used

```
>> syms x t;  G=(-17*x^5-7*x^4+2*x^3+x^2-x+1)...
   /(x^6+11*x^5+48*x^4+106*x^3+125*x^2+75*x+17); f=ilaplace(G,x,t)
```

However, the result is displayed as

```
-1/31709*sum((39275165+45806941*_alpha^4+156459285*_alpha+5086418*
      _alpha^5+149142273*_alpha^3+221566216*_alpha^2)*exp(_alpha*t),
      _alpha = RootOf(_Z^6+11*_Z^5+48*_Z^4+106*_Z^3+125*_Z^2+75*_Z+17))
```

and it can be seen that the `RootOf` function is returned, since the denominator polynomial function does not have analytical solutions. Thus the original problem has no symbolic solution. With the use of `vpa(f,16)` function, high-precision numerical solutions can be obtained.

Example 5.4 For the function $f(t)$ given in Example 5.1, explore the relationship between $\mathscr{L}[\mathrm{d}^5 f(t)/\mathrm{d}t^5]$ and $s^5\mathscr{L}[f(t)]$.

Solution To solve the problem, the fifth-order derivative to the given function $f(t)$ can be obtained by function `diff()`. Then the Laplace transform can be obtained

```
>> syms t s; f=t^2*exp(-2*t)*sin(t+pi);
   F=simple(laplace(diff(f,t,5)))
```

which yields $F(s) = -2\dfrac{3000 + 6825s + 6660s^2 + 960s^4 + 3471s^3 + 110s^5}{(s^2 + 4s + 5)^3}$.

Taking Laplace transform to function $f(t)$, multiplying the result by s^5, one may then subtract the above result to find the difference

```
>> F0=laplace(f); simple(F-s^5*F0)
```

and the difference obtained is $6s - 48$.

It is obvious that the difference of the two terms is not zero, which seems not to agree with (5.3). This is because the initial conditions here are non-zero. It can easily be found that $f(0) = f'(0) = f''(0) = 0$, while $f^{(3)}(0) = -6$, and $f^{(4)}(0) = 48$. Hence the difference equals $6s - 48$.

Example 5.5 Display the differentiation property of Laplace transform $\mathscr{L}\left[\dfrac{\mathrm{d}^2 f(t)}{\mathrm{d}t^2}\right]$.

Solution Some of the properties of Laplace transform can be displayed with the use of the Symbolic Math Toolbox of MATLAB. For this problem, the function $f(t)$ should be declared first, then the second-order derivative can be obtained with the use of function `diff()`. The Laplace transform of the second-order derivative can then be obtained

```
>> syms t; y=sym('f(t)')  % declare the original function
   laplace(diff(y,t,2))
```

with the display `s*(s*laplace(f(t),t,s)-f(0))-D(f)(0)`. Also the formula for the Laplace transform of the eighth-order derivative can be derived

```
>> laplace(diff(y,t,8))
```

and the formula is

```
s*(s*(s*(s*(s*(s*(s*(s*laplace(f(t),t,s)-f(0))-D(f)(0))-@@(D,2)(f)(0))-
@@(D,3)(f)(0))-@@(D,4)(f)(0))-@@(D,5)(f)(0))-@@(D,6)(f)(0))-@@(D,7)(f)(0)
```

Example 5.6 For the function $f(t) = \mathrm{e}^{-5t}\cos(2t+1)+5$, compute $\mathscr{L}[\mathrm{d}^5 f(t)/\mathrm{d}t^5]$.

Solution For the given function $f(t)$, the Laplace and then the inverse Laplace transform can be performed with the following MATLAB statements

```
>> syms t; f=exp(-5*t)*cos(2*t+1)+5;
   F=laplace(diff(f,t,5)); F=simple(F)
```

the result obtained is $F(s) = \dfrac{1475\cos 1s - 1189\cos 1 - 24360\sin 1 - 4282\sin 1s}{s^2 + 10s + 29}$. In fact, a simplified result can further be obtained if needed. For instance, collecting the terms in the numerator by using the following statements

```
>> syms s; collect(F)
```

the result can be obtained as $\dfrac{(1475\cos 1 - 4282\sin 1)\,s - 1189\cos 1 - 24360\sin 1}{s^2 + 10s + 29}$.

5.2 Fourier Transforms and Their Inverses

5.2.1 Definitions and properties

The definition of Fourier transform is

$$\mathscr{F}[f(t)] = \frac{1}{\sqrt{2\pi}} \int_{-\infty}^{\infty} f(t)\mathrm{e}^{-\mathrm{j}\omega t}\mathrm{d}t = F(\omega). \tag{5.8}$$

If the frequency domain function $F(\omega)$ is known, the inverse Fourier transform can be obtained from

$$f(t) = \mathscr{F}^{-1}[F(\omega)] = \frac{1}{\sqrt{2\pi}} \int_{-\infty}^{\infty} F(\omega)\mathrm{e}^{\mathrm{j}\omega t}\mathrm{d}\omega. \tag{5.9}$$

The following properties of Fourier transform are summarized without proofs.

(i) **Linear property** $\mathscr{F}[af(t) \pm bg(t)] = a\mathscr{F}[f(t)] \pm b\mathscr{F}[g(t)]$ for scalars a and b.

(ii) **Shift property** $\mathscr{F}[f(t \pm a)] = \mathrm{e}^{\pm \mathrm{j}a\omega}F(\omega)$.

(iii) **Complex domain shift** $\mathscr{F}[\mathrm{e}^{\pm \mathrm{j}at}f(t)] = F(\omega \mp a)$.

(iv) **Differentiation property** $\mathscr{F}[\mathrm{d}f(t)/\mathrm{d}t] = \mathrm{j}\omega F(\omega)$. Generally, the Fourier transform of the nth derivative can be evaluated from

$$\mathscr{F}\left[\frac{\mathrm{d}^n}{\mathrm{d}t^n}f(t)\right] = (\mathrm{j}\omega)^n \mathscr{F}[f(t)]. \tag{5.10}$$

(v) **Integration property** $\mathscr{F}\left[\int_{-\infty}^{t} f(\tau)\,\mathrm{d}\tau\right] = F(\omega)/(\mathrm{j}\omega)$. Generally the Fourier transform to the nth order integral can be obtained from

$$\mathscr{F}\left[\int_{-\infty}^{t} \cdots \int_{-\infty}^{t} f(\tau)\,\mathrm{d}\tau^n\right] = \frac{\mathscr{F}[f(t)]}{(\mathrm{j}\omega)^n}. \tag{5.11}$$

(vi) **Scale property** $\mathscr{F}[f(at)] = \dfrac{1}{a}F\left(\dfrac{\omega}{a}\right)$.

(vii) **Convolution property** $\mathscr{F}[f(t) * g(t)] = \mathscr{F}[f(t)]\mathscr{F}[g(t)]$, where the definition of convolution is given in (5.5).

5.2.2 Solving Fourier transform problems

Similar to Laplace transform, the variable should be declared first, then the function to be transformed can be defined in variable *fun*. With the Fourier transform solver `fourier()`, the Fourier transform can be obtained by

```
F=fourier(fun)          % Fourier transform
F=fourier(fun,v,u)      % transform the function of v into a function of u
```

Note that the definition of Fourier transform used in MATLAB is slightly different than the one defined in (5.8). In the function `fourier()`, the Fourier transform is defined as

$$\mathscr{F}[f(x)] = \int_{-\infty}^{\infty} f(x)\mathrm{e}^{-\mathrm{j}\omega x}\mathrm{d}x = F_1(\omega) \tag{5.12}$$

and it can be seen that the difference is the factor $\sqrt{2\pi}$. Throughout this chapter, the definition given in (5.12) is used.

The inverse Fourier transform can be obtained using `ifourier()` function, with the syntax

```
f=ifourier(fun)         % inverse Fourier transform
f=ifourier(fun,u,v)     % transform the function of u into a function of v
```

Also the definition of inverse Fourier transform is different from the one defined in (5.9). The inverse Fourier transform is defined as

$$f(x) = \mathscr{F}^{-1}[F_1(\omega)] = \frac{1}{2\pi}\int_{-\infty}^{\infty} F_1(\omega)\mathrm{e}^{\mathrm{j}\omega x}\mathrm{d}\omega \tag{5.13}$$

and this definition will be used throughout this chapter.

Example 5.7 Compute the Fourier transform for $f(t) = 1/(t^2 + a^2)$ where $a > 0$.

Solution The following statements can be used to find the Fourier transform.

```
>> syms t w; syms a positive
   f=1/(t^2+a^2); F=fourier(f,t,w)
```

The results are $\pi(\mathrm{e}^{-a\omega}\mathrm{Heaviside}(\omega) + \mathrm{e}^{a\omega}\mathrm{Heaviside}(-\omega))/a$, where the function $\mathrm{Heaviside}(\omega)$ is the step function of ω, which is also referred to as the *Heaviside function*. If $\omega \geqslant 0$, $\mathrm{Heaviside}(\omega)$ is 1, otherwise it is 0. When $\omega \leqslant 0$, $\mathrm{Heaviside}(-\omega)$ returns 1, otherwise it is 0. Assuming that $\omega > 0$, then `F` can be simplified as $\pi\mathrm{e}^{-a\omega}/a$. While $\omega < 0$, `F` can be simplified as $\pi\mathrm{e}^{a\omega}/a$. The result can be summarized as $\mathscr{F}[f(t)] = \dfrac{\pi\mathrm{e}^{-a|\omega|}}{a}$.

It should be noted that the simplest result obtained above cannot be derived automatically with MATLAB. The inverse Fourier transform can be obtained using the following statement

```
>> syms t w; syms a positive
   f=pi*exp(-a*abs(w))/a; f1=ifourier(f)
```

and the result is $f_1 = \dfrac{1}{a^2 + x^2}$, showing that the original function can be restored.

Example 5.8 Compute the Fourier transform for $f(t) = \sin^2(at)/t$ with $a > 0$.

Solution The Fourier transform can be obtained with

```
>> syms t w; syms a positive; f=sin(a*t)^2/t; fourier(f,t,w)
```

and the result is $\mathrm{j}\pi(\text{Heaviside}(\omega-2a)+\text{Heaviside}(\omega+2a)-2\text{Heaviside}(\omega))/2$, and again the result is based on the `Heaviside()` function. When $\omega > 2a$, the three `Heaviside()` are all 1. Thus $F(\omega)=0$. If $\omega \leqslant -2a$, the three functions are all 0, which means $F(\omega)=0$. If $0<\omega<2a$, since the second and third `Heaviside()` functions are 1, $F(\omega)=-\mathrm{j}\pi/2$. When $0>\omega>-2a$, $F(\omega)=\mathrm{j}\pi/2$. The Fourier transform of the original function can be simplified as $\mathscr{F}[f(t)]=\begin{cases}0, & |\omega|>2a\\ -\mathrm{j}\pi\,\mathrm{sign}(\omega)/2, & |\omega|<2a.\end{cases}$

Example 5.9 Now consider a more complicated problem. Assume that the function is $f(t)=\mathrm{e}^{-a|t|}/\sqrt{|t|}$. Use the MATLAB function `fourier()` and the direct integration method to solve the Fourier transform problem.

Solution Now considering first the existing function `fourier()`, the following statements can be used to solve the Fourier transform problem

```
>> syms w t; syms a positive
   f=exp(-a*abs(t))/sqrt(abs(t)); F=fourier(f,t,w)
```

which gives the string `fourier(exp(-a*abs(t))/abs(t)^(1/2),t,w)`, which means nothing has been done and the function call failed to return a solution. Now, let us try the direct integration method based on the definition of Fourier transform (5.8). The original integration interval can be divided into two intervals such that $\int_{-\infty}^{\infty}=\int_{-\infty}^{0}+\int_{0}^{\infty}$. Then, the following statements can be used to solve the Fourier transform problem

```
>> f1=exp(a*t)/sqrt(-t); f2=exp(-a*t)/sqrt(t); j=sym(sqrt(-1));
   F=int(f1*exp(-j*w*t),-inf,0)+int(f2*exp(-j*w*t),0,inf)
```

and it is also claimed that the integral failed. This example shows that not all functions have their corresponding Fourier transforms. For some functions, there exist no Fourier transforms.

5.2.3 Fourier sine and cosine transforms

The Fourier sine transform is defined as

$$\mathscr{F}_{\mathscr{S}}[f(t)]=\int_0^{\infty} f(t)\sin\omega t\,\mathrm{d}t=F_{\mathrm{s}}(\omega). \tag{5.14}$$

Fourier cosine transform is defined as

$$\mathscr{F}_{\mathscr{C}}[f(t)]=\int_0^{\infty} f(t)\cos\omega t\,\mathrm{d}t=F_{\mathrm{c}}(\omega). \tag{5.15}$$

Similarly the inverse Fourier sine/cosine transforms are defined as

$$\mathscr{F}_{\mathscr{S}}^{-1}[F_{\mathrm{s}}(t)]=\frac{2}{\pi}\int_{-\infty}^{\infty} F_{\mathrm{s}}(\omega)\sin\omega t\,\mathrm{d}\omega \tag{5.16}$$

$$\mathscr{F}_{\mathscr{C}}^{-1}[F_{\rm c}(t)] = \frac{2}{\pi}\int_{-\infty}^{\infty} F_{\rm c}(\omega)\cos\omega t\,{\rm d}\omega. \tag{5.17}$$

Similar to the original Fourier transform, the symmetrical Fourier sine and cosine transforms can also be defined by multiplying the transform formulae by $\sqrt{2/\pi}$, and by dividing the inverse formulae by $\sqrt{2/\pi}$.

There are no functions for Fourier sine and cosine transforms provided in MATLAB. Thus direct integration can be applied using the Symbolic Math Toolbox. The following examples are given for computing the Fourier sine and cosine transforms.

Example 5.10 Compute the Fourier cosine transforms to the function $f(t) = t^n{\rm e}^{-at}, a>0$ for $n=1,2,\cdots,8$.

Solution Since different values of n are to be explored, the loop structure of MATLAB can be used to solve the problem. The following statements can be given to evaluate the Fourier cosine transforms and the results are given in Table 5.1.

```
>> syms t w, syms a positive
   for n=1:8
      f=t^n*exp(-a*t); F=int(f*cos(w*t),t,0,inf); simple(F)
   end
```

TABLE 5.1: The Fourier cosine transforms for different values of n

n	$\mathscr{F}_{\mathscr{C}}[f(t)]$
1∼4	$\dfrac{a^2-\omega^2}{(a^2+\omega^2)^2}, -2\dfrac{(-a^2+3\omega^2)a}{(a^2+\omega^2)^3}, 6\dfrac{(-a^2+2a\omega+\omega^2)(-a^2-2a\omega+\omega^2)}{(a^2+\omega^2)^4}, 24\dfrac{a(a^4-10a^2\omega^2+5\omega^4)}{(a^2+\omega^2)^5}$
5,6	$-120\dfrac{(-a+\omega)(a+\omega)(a^2-4\omega a+\omega^2)(a^2+4\omega a+\omega^2)}{(a^2+\omega^2)^6}, 720\dfrac{(a^6-21a^4\omega^2+35\omega^4a^2-7\omega^6)a}{(a^2+\omega^2)^7}$
7	$5040\dfrac{(a^4+4a^3\omega-6a^2\omega^2-4a\omega^3+\omega^4)(a^4-4a^3\omega-6a^2\omega^2+4a\omega^3+\omega^4)}{(a^2+\omega^2)^8}$
8	$40320\dfrac{a(-a^2+3\omega^2)(-a^6+33a^4\omega^2-27a^2\omega^4+3\omega^6)}{(a^2+\omega^2)^9}$

From the mathematics handbook, the result is actually as follows:

$$\mathscr{F}_{\mathscr{C}}\left[t^n{\rm e}^{-at}\right] = n!\left(\frac{a}{a^2+\omega^2}\right)^{n+1}\sum_{m=0}^{[n/2]}(-1)^m C_{n+1}^{2m+1}\left(\frac{\omega}{a}\right)^{2m+1}. \tag{5.18}$$

Although in MATLAB there are no functions provided for Fourier sine and cosine transforms, the functions provided in Maple can be used instead. These functions can be called within MATLAB if the Extended Symbolic Math Toolbox is installed. The syntaxes are

F=maple('fouriersin',f,t,ω)	% Fourier sine transform
F=maple('fouriercos',f,t,ω)	% Fourier cosine transform
f=maple('invfouriersin',F,ω,t)	% inverse Fourier sine transform
f=maple('invfouriercos',F,ω,t)	% inverse Fourier cosine transform

where the `maple()` function can call Maple functions directly from MATLAB command line.

Example 5.11 Reconsider the piecewise function defined in Example 5.10. If $n = 6$, the Maple functions can be called to solve the Fourier cosine transform problem. Then the inverse transform can be tested.

Solution With the functions in Maple, the Fourier cosine transform and its inverse transform can be obtained as

```
>> syms t w; syms a positive
   f=t^6*exp(-a*t); F=maple('fouriercos',f,t,w)
```

which follows immediately that

$$\mathscr{F}_{\mathscr{C}}[f(t)] = 720\sqrt{\frac{2}{\pi}}\frac{a^7}{(a^2+\omega^2)^7}\left(1-21\frac{\omega^2}{a^2}+35\frac{\omega^4}{a^4}-7\frac{\omega^6}{a^6}\right).$$

It should be noted that the result is different from the one given in Table 5.1, since the method used to obtain the result is from using (5.15). To have the results in Table 5.1, the result here by using Maple should be divided by $\sqrt{2/\pi}$. With the use of inverse transform function, the following results will be returned

```
>> f1=maple('invfouriercos',F,w,t)
```

which indicates that the inverse transform cannot be obtained.

```
invfouriercos(720*2^(1/2)/pi^(1/2)*a^7/(a^2+w^2)^7*
    (1-21*w^2/a^2+35*w^4/a^4-7*w^6/a^6),w,t)
```

Example 5.12 Compute the Fourier cosine transform for $f(t)=\begin{cases}\cos t, & 0<x<a\\ 0, & \text{otherwise.}\end{cases}$

Solution It might be difficult to deal with piecewise functions, thus the direct integration method can be used to solve the problem. From the piecewise function, it can be seen that the value of the integrand outside the interval $(0, a)$ is zero, thus the Fourier cosine transform can be solved with the following statements

```
>> syms t w; syms a positive; f=cos(t); F=simple(int(f*cos(w*t),t,0,a))
```

and the following result is obtained

$$\mathscr{F}_{\mathscr{C}}[f(t)] = \frac{(1+\omega)\sin(-a+a\omega)+(-1+\omega)\sin(a+a\omega)}{-2+2\omega^2}.$$

Furthermore, the above result can be simplified manually to

$$\mathscr{F}_{\mathscr{C}}[f(t)] = \frac{\sin(-a+a\omega)}{2(1+\omega)}+\frac{\sin(a+a\omega)}{2(1-\omega)}.$$

5.2.4 Discrete Fourier sine, cosine transforms

Discrete Fourier sine and cosine transforms are also referred to as the *finite Fourier sine, cosine transforms*. The integration interval is changed from $t \in (0, \infty)$ into a finite one $t \in (0, a)$. Thus the transforms are defined as

$$F_s(k) = \int_0^a f(t) \sin \frac{k\pi t}{a} \, dt, \quad F_c(k) = \int_0^a f(t) \cos \frac{k\pi t}{a} \, dt. \tag{5.19}$$

Similarly, the inverse transforms are defined as

$$f(t) = \frac{2}{a} \sum_{k=1}^{\infty} F_s(k) \sin \frac{k\pi t}{a} \tag{5.20}$$

$$f(t) = \frac{1}{a} F_c(0) + \frac{2}{a} \sum_{k=1}^{\infty} F_c(k) \cos \frac{k\pi t}{a}. \tag{5.21}$$

The inverse transforms are no longer integrals. They are the summations of infinite series. The finite transforms can also be obtained with the Symbolic Math Toolbox from definitions.

Example 5.13 Compute the transform for $f(t) = \begin{cases} t & t \leqslant a/2 \\ a-t & t > a/2 \end{cases}$, with $a > 0$.

Solution The discrete Fourier sine transforms can be obtained directly from

```
>> syms t k; syms a positive; f1=t; f2=a-t;
   Fs=int(f1*sin(k*pi*t/a),t,0,a/2)+int(f2*sin(k*pi*t/a),t,a/2,a);
   simple(Fs)
```

Since integer symbolic variables are not supported in the Symbolic Math Toolbox, the simplified result is then

$$\mathscr{F}_{\mathscr{S}}[f(t)] = \frac{a^2(2\sin k\pi/2 - \sin k\pi)}{k^2\pi^2}.$$

However, for integer k, $\sin k\pi \equiv 0$. Thus the results can be simplified manually as

$$\mathscr{F}_{\mathscr{S}}[f(t)] = \frac{2a^2}{k^2\pi^2} \sin \frac{k\pi}{2}.$$

5.3 Other Integral Transforms

Apart from the well established Laplace and Fourier transforms, there are other integral transforms, such as Mellin transform and Hankel transform. The standard Symbolic Math Toolbox does not provide the functions for these problems. Thus one can solve the problems either by direct integration, or by relevant Maple functions.

5.3.1 Mellin transform

The Mellin transform is defined as

$$\mathscr{M}[f(x)] = \int_0^\infty f(x)x^{z-1}\mathrm{d}x = M(z). \tag{5.22}$$

Similarly, the inverse Mellin transform is defined as

$$f(x) = \mathscr{M}^{-1}[M(z)] = \frac{1}{\mathrm{j}2\pi}\int_{c-\mathrm{j}\infty}^{c+\mathrm{j}\infty} M(z)x^{-z}\mathrm{d}z. \tag{5.23}$$

There is no Mellin transform function provided in the Symbolic Math Toolbox. Wee can solve the problem using the direct integration method. Examples are given below to show the solution process.

Example 5.14 Compute the Mellin transform to $f(t) = \ln t/(t+a)$ with $a > 0$.
Solution By definition, the following statements can be used to solve the problem

```
>> syms t z; syms a positive;
   f=log(t)/(t+a); M=simple(int(f*t^(z-1),t,0,inf))
```

and the result can be simplified as

$$\mathscr{M}[f(t)] = a^{z-1}\pi\left(\ln a \sin \pi z - \pi \cos \pi z\right)\csc^2 \pi z.$$

Example 5.15 Compute the Mellin transform to $f(t) = 1/(t+a)^n$, with $a > 0$. For different integer n, find the general form of the transform.

Solution The following statements can be used to get the Mellin transform problem for $n = 1, 2, \cdots, 8$,

```
>> syms t z; syms a positive
   for i=1:8, f=1/(t+a)^i; disp(int(f*t^(z-1),t,0,inf)), end
```

which gives respectively

$a^{z-1}\pi \csc \pi z$
$-a^{-2+z}\pi(z-1)\csc \pi z$
$1/2a^{-3+z}\pi(-2+z)(z-1)\csc \pi z$
$-1/6a^{z-4}\pi(z-1)(-2+z)(-3+z)\csc \pi z$
$1/24a^{-5+z}\pi(z-4)(-3+z)(-2+z)(z-1)\csc \pi z$
$-1/120a^{-6+z}\pi(-5+z)(z-4)(-3+z)(-2+z)(z-1)\csc \pi z$
$1/720a^{-7+z}\pi(-6+z)(-5+z)(z-4)(-3+z)(-2+z)(z-1)\csc \pi z$
$-1/5040a^{-8+z}\pi(-7+z)(-6+z)(-5+z)(z-4)(-3+z)(-2+z)(z-1)\csc \pi z.$

Thus, by inspection, it can be concluded that the Mellin transform is generally given by

$$\mathscr{M}\left[\frac{1}{(t+a)^n}\right] = \frac{(-1)^{k-1}\pi}{(n-1)!}a^{z-n}\prod_{i=1}^{n-1}(z-i)\csc \pi z.$$

The Mellin transform and its inverse can be obtained using the `mellin()` and `invmellin()` functions provided in Maple. The syntaxes of the functions are

```
Fun=maple('mellin',fun,t,z)        % Mellin transform
fun=maple('invmellin',Fun,z,t)     % inverse Mellin transforms
```

Example 5.16 Consider the function in Example 5.15. If $n = 8$, compute the Mellin transform with Maple. Then perform inverse Mellin transform to the results.

Solution With the Maple function, the problem can be solved with the following statements

```
>> syms t z; syms a positive
   F=maple('mellin',1/(t+a)^8,t,z), f1=maple('invmellin',F,z,t)
```

and it is found that $F(z) = \dfrac{-\pi(z-1)!}{5040(-8+z)!}a^{-8+z}\csc\pi z$. However, it is not known why the inverse transform function call fails to recover the original function for this example.

5.3.2 Hankel transform solutions

Hankel transform is another frequently used integral transform. The νth order Hankel transform is defined as

$$\mathscr{H}[f(t)] = \int_0^\infty tf(t)J_\nu(\omega t)\,\mathrm{d}t = H_\nu(\omega) \tag{5.24}$$

where $J_\nu(\cdot)$ is the Bessel function of order ν. The inverse Hankel transform of order ν can be defined as

$$\mathscr{H}^{-1}[H(\omega)] = \int_0^\infty \omega H_\nu(\omega)J_\nu(\omega t)\,\mathrm{d}\omega. \tag{5.25}$$

The functions `hankel()` and `invhankel()` in Maple can be used to evaluate the Hankel transform and the inverse Hankel transform with the following syntaxes

```
Fun=maple('hankel',fun,t,ω,ν)        % evaluate Hankel transform
fun=maple('invhankel',Fun,ω,t,ν)     % evaluate inverse Hankel transform
```

Example 5.17 Compute the zeroth-order Hankel transform to the function $f(t) = t^a$.

Solution With the existing Maple function, the following statements can be used in finding the zeroth-order Hankel transform

```
>> syms t w; syms a positive; f=t^a; F=maple('hankel',f,t,w,0)
```

and the transform result can be obtained as

$$\mathscr{H}[f(t)] = \frac{1}{\pi}\omega^{-1-a}\sin\left(\frac{1}{4}\pi\,(2a+3)\right)2^{a+1/2}\Gamma^2\left(\frac{1}{2}a+\frac{3}{4}\right).$$

Currently, the `hankel()` function in Maple can only be used to find the Hankel transform to t^a type of functions. The transform to other types of functions may not yield the simplest results. For instance, the zeroth-order Hankel transform can be applied to the function $f(t) = a/(a^2+t^2)^{3/2}$

```
>> syms t w a; F=maple('hankel',a/(a^2+t^2)^(3/2),t,w,0)
```

and the following results can be obtained

$$\frac{-9\sqrt{\omega}\mathscr{H}\left(\left[\frac{3}{4}\right],\left[\frac{1}{4},1\right],\frac{1}{4}a^2\omega^2\right)\Gamma^4\left(\frac{3}{4}\right)-4\omega^2\pi^{3/2}\mathscr{H}\left(\left[\frac{3}{2}\right],\left[\frac{7}{4},\frac{7}{4}\right],\frac{1}{4}a^2\omega^2\right)a^{3/2}}{\sqrt{a}\sqrt{\pi}\left(9\Gamma\left(\frac{3}{4}\right)\right)^2}$$

where $\mathscr{H}(\cdot)$ is the hyper geometric function. In fact, the zeroth-order Hankel transform to the given function $f(t)$ is $\mathrm{e}^{-a\omega}$. Thus, care should be taken when using `hankel()` function in Maple.

5.4 Z Transforms and Their Inverses

5.4.1 Definitions and properties of Z transforms and inverses

The Z transform of a discrete sequence $f(k)$, $k = 1, 2, \cdots$ is defined as

$$\mathscr{Z}[f(k)] = \sum_{k=0}^{\infty} f(k)z^{-k} = F(z). \tag{5.26}$$

Similar to the presentation of Laplace and Fourier transforms, the properties of Z transforms are summarized below without proofs.

(i) **Linear property** $\mathscr{Z}[af(k) \pm bg(k)] = a\mathscr{Z}[f(k)] \pm b\mathscr{Z}[g(k)]$ for any scalars a and b.

(ii) **Time domain translation property** $\mathscr{Z}[f(k-n)] = z^{-n}F(z)$.

(iii) **Z-domain proportional property** $\mathscr{Z}[r^{-k}f(k)] = F(rz)$.

(iv) **Frequency domain differentiation property** $\mathscr{Z}[kf(k)] = -z\dfrac{\mathrm{d}F(z)}{\mathrm{d}z}$.

(v) **Frequency domain integration property** $\mathscr{Z}\left[\dfrac{f(k)}{k}\right] = \displaystyle\int_z^{\infty}\frac{F(\omega)}{\omega}\,\mathrm{d}\omega$.

(vi) **Initial value property** $\displaystyle\lim_{k\to 0} f(k) = \lim_{z\to\infty} F(z)$.

(vii) **Final value property** If $F(z)$ has no poles outside of the unit circle,

$$\lim_{k\to\infty} f(k) = \lim_{z\to 1}(z-1)F(z). \tag{5.27}$$

(viii) **Convolution property** $\mathscr{Z}[f(k) * g(k)] = \mathscr{Z}[f(k)]\mathscr{Z}[g(k)]$ where the operator $*$ for discrete signals is defined as

$$f(k) * g(k) = \sum_{l=0}^{\infty} f(k)g(k-l). \tag{5.28}$$

For a Z transform function $F(z)$, the inverse Z transform is defined as

$$f(k) = \mathscr{Z}^{-1}[f(k)] = \frac{1}{\mathrm{j}2\pi}\oint F(z)z^{k-1}\,\mathrm{d}z. \tag{5.29}$$

5.4.2 Computations of Z transform

Z transform and its inverse can be obtained directly with the functions `ztrans()` and `iztrans()` provided in the Symbolic Math Toolbox. The syntaxes are respectively

```
F=ztrans(fun,k,z)     % Z transform
F=iztrans(fun,z,k)    % inverse Z transform
```

and if there is only one variable in *fun*, the arguments k and z can be omitted.

Example 5.18 Compute the Z transform for $f(kT) = akT - 2 + (akT+2)\mathrm{e}^{-akT}$.

Solution The Z transform can be obtained directly with the statements

```
>> syms a T k  % declare the symbolic variables
   f=a*k*T-2+(a*k*T+2)*exp(-a*k*T);  % define the discrete function
   F=ztrans(f);     % compute the Z transform
```

and the result obtained is

$$\mathscr{Z}[f(kT)] = \frac{aTz}{(z-1)^2} - 2\frac{z}{z-1} + \frac{aTz\mathrm{e}^{-aT}}{(z-\mathrm{e}^{-aT})^2} + 2\,z\mathrm{e}^{aT}\left(\frac{z}{\mathrm{e}^{-aT}} - 1\right)^{-1}.$$

Example 5.19 Consider the function $F(z) = \dfrac{q}{(z^{-1}-p)^m}$. Find the inverse Z transforms for different values of m, and then try to summarize the general formula for it.

Solution Let us try to solve the problem for $m = 1, 2, \cdots, 8$. The loop structure should be used and the inverse Z transforms can be obtained by

```
>> syms p q z
   for i=1:8, disp(simple(iztrans(q/(1/z-p)^i))), end
```

which gives

$-q/p(1/p)^n$

$q/p^2(1+n)(1/p)^n$

$-1/2q(1/p)^n(1+n)(2+n)/p^3$

$1/6q(1/p)^n(3+n)(2+n)(1+n)/p^4$

$-1/24q(1/p)^n(4+n)(3+n)(2+n)(1+n)/p^5$

$1/120q(1/p)^n(5+n)(4+n)(3+n)(2+n)(1+n)/p^6$
$-1/720q(1/p)^n(6+n)(5+n)(4+n)(3+n)(2+n)(1+n)/p^7$
$1/5040q(1/p)^n(7+n)(6+n)(5+n)(4+n)(3+n)(2+n)(1+n)/p^8$.

Summarizing the above results, by inspection, the general form of the Z transform is

$$\mathscr{Z}^{-1}\left[\frac{q}{(z^{-1}-p)^m}\right] = \frac{(-1)^m q}{(m-1)!\, p^{n+m}} \prod_{i=1}^{m-1}(n+i).$$

5.5 Solving Complex Variable Function Problems

By name, complex variable functions are those whose independent variables are complex numbers. Since the complex data type is the fundamental data type of MATLAB, the previous MATLAB functions in calculus and linear algebra can be used directly in complex variable function related problems. In this section, the mapping display of complex variable functions are introduced first, followed by the concept of poles and residues. Finally the closed-path integral of complex variable function is studied.

5.5.1 Complex variable functions and mapping visualization

The mapping graphics of complex variable functions are different from the 3D graphics discussed in Chapter 2. One should generate polar grid with the `cplxgrid()` function, and the mapping surface can be shown with `cplxmap()` function. The syntaxes of the functions are

```
z=cplxgrid(n)   % generates polar grids
cplxmap(z,f)    % draw the 3D complex mapping surface
```

Example 5.20 Draw the 3D mapping surface of the complex variable function $f(z) = z^3 \sin z^2$.

Solution The mapping surface of the given function can be displayed with the following statements, as shown in Figure 5.1.

```
>> z=cplxgrid(50); f=z.^3.*sin(z.^2); cplxmap(z,f)
```

5.5.2 Concept and computation of residues

Before introducing the ideas of residues, the concept of analytic complex functions is introduced. The complex function $f(z)$ is said to be analytic, if it is holomorphic, and the derivatives are finite, at all the points within the complex region. The points which make the function $f(z)$ not analytic are referred to as the *singular points*. These points are also known as the *poles*

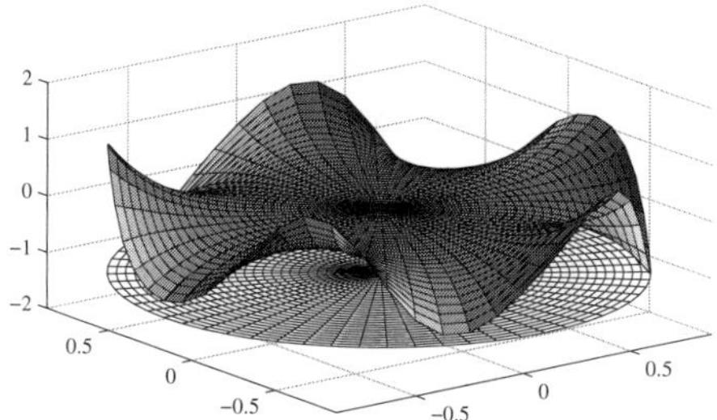

FIGURE 5.1: Mapping visualization of a complex variable function

in the complex plane. Assume that $z = a$ is a pole of the function $f(z)$, and there exists the smallest positive integer m such that the new function $(z-a)^m f(z)$ is analytic at point $z = a$, then point $z = a$ is referred to as the *pole of multiplicity m*.

If $z = a$ is a single pole, then the *residue* is defined as

$$\operatorname{Res}\Big[f(z), z = a\Big] = \lim_{z \to a} (z-a) f(z). \tag{5.30}$$

If $z = a$ is a pole of multiplicity m, the residue is defined as

$$\operatorname{Res}\Big[f(z), z = a\Big] = \lim_{z \to a} \frac{1}{(m-1)!} \frac{\mathrm{d}^{m-1}}{\mathrm{d}z^{m-1}} \Big[f(z)(z-a)^m\Big]. \tag{5.31}$$

Thus the evaluation of residues can be made very easy. The following statements can be used in evaluating the residues to different kinds of poles.

```
c=limit(F*(z-a),z,a)                                % single pole
c=limit(diff(F*(z-a)^m,z,m-1)/prod(1:m-1),z,a)      % m-multiple
```

Example 5.21 Compute the residues of the function

$$f(z) = \frac{1}{z^3(z-1)} \sin\left(z + \frac{\pi}{3}\right) \mathrm{e}^{-2z}.$$

Solution It can be seen from the original function that $z = 0$ is a pole of multiplicity 3, while $z = 1$ is a single pole. Thus the following MATLAB statements can be used to evaluate the residues at the two poles

```
>> syms z; f=sin(z+pi/3)*exp(-2*z)/(z^3*(z-1))
   F1=limit(diff(f*z^3,z,2)/prod(1:2),z,0), F2=limit(f*(z-1),z,1)
```

where the residue at $z = 0$ is $-\dfrac{\sqrt{3}}{4} + \dfrac{1}{2}$, and the residue at $z = 1$ is $\dfrac{1}{2}\mathrm{e}^{-2}\sin 1 + \dfrac{\sqrt{3}}{2}\mathrm{e}^{-2}\cos 1$.

Example 5.22 Compute the residue of the function $f(z) = \dfrac{\sin z - z}{z^6}$.

Solution It seems that the point $z = 0$ is a pole of multiplicity 6. Thus the following statements can be used to evaluate the residue of the function

```
>> syms z; f=(sin(z)-z)/z^6; R=limit(diff(f*z^6,z,5)/prod(1:5),z,0)
```

with the residue $R = 1/120$. In fact, the multiplicity of 6 selected is too conservative. According to the definition, one may try to evaluate the limit in (5.31) for each value of k such that the smallest integer k which makes the limit return a finite value can be found. For this example, it can be found that $k = 2$ makes the limits tend to infinity. And from $k = 3$ and onwards, the limits will all return the value of $1/120$. Thus the multiplicity of the pole is 3, and the residue is $1/120$. Also it is interesting to note that for this example, even if k is selected as a very large number, the value of the limit is $1/120$.

```
>> syms z; f=(sin(z)-z)/z^6; R2=limit(diff(f*z^2,z,1)/prod(1:1),z,0)
   R3=limit(diff(f*z^3,z,2)/prod(1:2),z,0),
   R20=limit(diff(f*z^20,z,19)/prod(1:19),z,0)
```

Consider the Taylor series expansion to the $\sin z$ function. It can be seen that

$$f(z) = \frac{(z - z^3/6 + z^5/120 - z^7/5040 + \cdots) - z}{z^6} = \frac{1/6 + z^3/120 - z^5/5040 + \cdots}{z^3}.$$

Thus the multiplicity of the pole is obviously 3 rather than 6. In practical applications, even if it might be difficult to find the exact multiplicity, a large value can be selected first and the correct residue can be found.

Example 5.23 Compute the residues of the function $f(z) = \dfrac{1}{z \sin z}$.

Solution The function should be analyzed first. For $k = 2$, the limit is finite, thus the multiplicity of the pole at $z = 0$ is 2. The residue can then be found from

```
>> syms z; f=1/(z*sin(z)); c0=limit(diff(f*z^2,z,1),z,0)
```

and the residue for $z = 0$ is 0.

Further it can be found that the function $f(z)$ is not analytic at the points $z = \pm k\pi$, with k of any positive integer, and these points are single poles. Since symbolic integers are not supported in MATLAB, many different values of k can be selected and the residues for these poles can be obtained such that

```
>> k=[-4 4 -3 3 -2 2 -1 1]; c=[];
   for kk=k; c=[c,limit(f*(z-kk*pi),z,kk*pi)]; end; c
```

and it can be seen that for the vector $\boldsymbol{k} = [-4, 4, -3, 3, -2, 2, -1, 1]$, the residues evaluated are $\boldsymbol{c} = \left[-\dfrac{1}{4\pi}, \dfrac{1}{4\pi}, \dfrac{1}{3\pi}, -\dfrac{1}{3\pi}, -\dfrac{1}{2\pi}, \dfrac{1}{2\pi}, \dfrac{1}{\pi}, -\dfrac{1}{\pi}\right]$. It can then be concluded that $\text{Res}[f(z), z = \pm k\pi] = \pm(-1)^k \dfrac{1}{k\pi}$.

5.5.3 Partial fraction expansion for rational functions

Consider the rational function

$$G(x) = \frac{B(x)}{A(x)} = \frac{b_1 x^m + b_2 x^{m-1} + \cdots + b_m x + b_{m+1}}{x^n + a_1 x^{n-1} + a_2 x^{n-2} + \cdots + a_{n-1} x + a_n} \tag{5.32}$$

where a_i and b_i are all constants. The concept of coprimeness of rational functions is very important. The two polynomials $A(x)$ and $B(x)$ are coprime if there does not exist any common divisor. The greatest common divisor of two polynomials can be very difficult to find manually. However, with help of the `gcd()` function provided in the Symbolic Math Toolbox of MATLAB, the greatest common divisor C can easily be found with the syntax `C=gcd(A,B)`, where A and B are the two polynomials. If C is a polynomial, then the two polynomials are not coprime. The two polynomials can then be simplified to A/C and B/C, respectively.

Example 5.24 Check whether the two polynomials $A(x), B(x)$ are coprime or not

$A(x) = x^4 + 7x^3 + 13x^2 + 19x + 20$

$B(x) = x^7 + 16x^6 + 103x^5 + 346x^4 + 655x^3 + 700x^2 + 393x + 90.$

Solution The greatest common divisor can be obtained with the `gcd()` function directly

```
>> syms x; A=x^4+7*x^3+13*x^2+19*x+20;
   B=x^7+16*x^6+103*x^5+346*x^4+655*x^3+700*x^2+393*x+90;
   d=gcd(A,B)
```

and it can be seen that the greatest common divisor is $d(x) = (x + 5)$. Thus the two polynomials are not coprime. The two polynomials can easily be reduced with

```
>> simple(A/d), simple(B/d)
```

then $A(x)/d(x) = x^3 + 2x^2 + 3x + 4$, and $B(x)/d(x) = (x + 2)(x + 3)^2(x + 1)^3$.

If the polynomials $A(x)$ and $B(x)$ are coprime, and the roots $-p_i, i = 1, 2, \cdots, n$ of the polynomial equation $A(x) = 0$ are all distinct, the original rational function $G(x)$ can be expanded into the following form

$$G(x) = \frac{r_1}{x + p_1} + \frac{r_2}{x + p_2} + \cdots + \frac{r_n}{x + p_n} \tag{5.33}$$

and the expansion is referred to as the *partial fraction expansion*. In the expression, r_i are the residues, denoted as $\mathrm{Res}[G(-p_i)]$, which can be obtained from the limit formula such that

$$r_i = \mathrm{Res}[G(-p_i)] = \lim_{x \to -p_i} G(s)(x + p_i). \tag{5.34}$$

If the term $(x+p_i)^k$ exists in the denominator, i.e., $-p_i$ is the pole of multiplicity k, the corresponding sub-expansion can be written as

$$\frac{r_i}{x+p_i}+\frac{r_{i+1}}{(x+p_i)^2}+\cdots+\frac{r_{i+k-1}}{(x+p_i)^k}. \tag{5.35}$$

Thus, the values r_{i+j-1} can be evaluated from the following formula

$$r_{i+j-1}=\frac{1}{(k-1)!}\lim_{x\to -p_i}\frac{\mathrm{d}^{j-1}}{\mathrm{d}x^{j-1}}\left[G(x)(x+p_i)^k\right],\ j=1,2,\cdots,k. \tag{5.36}$$

A numerical function `residue()` is provided in MATLAB, which can be used in the partial fraction expansion of a given rational function $G(x)$. The syntax of the function is `[r,p,K]=residue(b,a)`, where $\boldsymbol{a}=$`[`$1,a_1,a_2,\cdots,a_n$`]`, and $\boldsymbol{b}=$`[`$b_1,b_2,\cdots,b_m$`]`. The returned arguments $\boldsymbol{r}$ and $\boldsymbol{p}$ are vectors containing the r_i coefficients and $-p_i$ poles, in (5.33). If repeated poles exist, the r_i terms can be replaced with the coefficients in (5.35). The argument k is the direct term, which for the function satisfying $k<n$, K will return an empty matrix. The function can be used to automatically judge whether the pole $-p_i$ is repeated, so as to arrange the values of r_i.

Example 5.25 Compute the partial fraction expansion for the following function

$$G(s)=\frac{s^3+2s^2+3s+4}{s^6+11s^5+48s^4+106s^3+125s^2+75s+18}.$$

Solution The following statements can be used to find the partial fraction expansion

```
>> n=[1,2,3,4]; d=[1,11,48,106,125,75,18]; format long
   [r,p,k]=residue(n,d); [n,d1]=rat(r); [n,d1,p]
```

with $\boldsymbol{n}^{\mathrm{T}}=[-17,-7,2,1,-1,1]$, $\boldsymbol{d}_1^{\mathrm{T}}=[8,4,1,8,2,2]$, $\boldsymbol{p}^{\mathrm{T}}=[-3,-3,-2,-1,-1,-1]$, where $\boldsymbol{p}$ is the vector of poles, $\boldsymbol{n}$, $\boldsymbol{d}_1$ are the corresponding numerators and denominators of the coefficients $\boldsymbol{r}$. It can be seen that -3 is a pole of multiplicity 2, -2 is a single pole, while -1 is a pole of multiplicity 3. Thus the partial fraction expansion can be written as

$$G(s)=-\frac{17}{8(s+3)}-\frac{7}{4(s+3)^2}+\frac{2}{s+2}+\frac{1}{8(s+1)}-\frac{1}{2(s+1)^2}+\frac{1}{2(s+1)^3}.$$

Example 5.26 Compute the partial fraction expansion of the following function

$$G(s)=\frac{2s^7+2s^3+8}{s^8+30s^7+386s^6+2772s^5+12093s^4+32598s^3+52520s^2+45600s+16000}.$$

Solution With the `residue()` function, the numerical solutions can be obtained. And for this example, the partial fraction expansion can be obtained

```
>> n=[2,0,0,0,2,0,0,8]; d=[1,30,386,2772,12093,32598,52520,45600,16000];
   [r,p]=residue(n,d)
```

and due to the limitations in the numerical results, it might be difficult to find the multiplicity of poles. Thus the exact partial fraction expansion cannot be obtained.

From the results obtained, it can be approximately assumed that $p_1 = -5$ and $p_2 = -4$ are poles of multiplicity 3, while $p_3=-2$ and $p_4=-1$ are single poles. Thus the partial fraction expansion can be written as

$$\frac{49995.9030930686}{(s+5)}+\frac{28488.5832580441}{(s+5)^2}+\frac{13040.9999762507}{(s+5)^3}-\frac{50473.1527861460}{(s+4)}$$
$$\frac{21449.5555022347}{(s+4)^2}-\frac{5481.3333201362}{(s+4)^3}+\frac{1.2222222224}{(s+2)}+\frac{0.0023148148}{(s+1)}.$$

Clearly, a better analytical or symbolic function is expected without numerical issues that may cause the pole multiplicity issues. With the help of the Symbolic Math Toolbox, the following analytical function `residue()` can be written based on (5.34) and (5.36). The function can then be used to solve a partial fraction expansion problem. The function should be placed under the `@sym` directory.

```
function f=residue(F,s)
f=sym(0); if nargin==1, syms s; end
[num,den]=numden(F); x0=solve(den);
[x,ii]=sort(double(x0)); x0-x0(ii); x=[x0;rand(1)];
kvec=find(diff(double(x))~=0); ee=x(kvec);
kvec=[kvec(1); diff(kvec(:,1))];
a0=limit(den/s^length(x0),s,inf); % coefficient of the highest order
F1=num/(a0*prod(s-x0)); % rearrange f(x) in the denominator to write f1(x)
for i=1:length(kvec),
   for j=1:kvec(i), m=kvec(i); s0=ee(i);
      k=limit(diff(F1*(s-s0)^m,s,j-1),s,s0); % the f1(x) is used instead
      f=f+k/(s-s0)^(m-j+1)/factorial(j-1);
end, end
```

and the syntax of the function is `f=residue(fun,s)`, where *fun* is the analytical expression of the rational function, and s is the independent variable. The returned argument f is the partial fraction expansion.

It should be noted that, if the analytical solutions to the equation of denominator polynomial $D(x)=0$ cannot be obtained, high-precision numerical solutions should be used instead. Assume that `vpa()` function can be used to find all the poles $x_i, i=1,2,\cdots,n$, then the original function can be reorganized such that the term $(x-\hat{x}_0)^m$ can really cancel the poles in (5.37), and the correct residue can finally be obtained

$$\operatorname{Res}[f(\hat{x}_0)] = \lim_{x\to\hat{x}_0}\frac{1}{(m-1)!}\frac{\mathrm{d}^{m-1}}{\mathrm{d}x^{m-1}}\left[f(\hat{x}_0)(x-\hat{x}_0)^m\right]. \tag{5.37}$$

Example 5.27 Compute the partial fraction expansion to $f(s)$ in Example 5.25.

Solution The following statements can be used to solve the partial expansion problem

```
>> syms s;
```

```
G=(s^3+2*s^2+3*s+4)/(s^6+11*s^5+48*s^4+106*s^3+125*s^2+75*s+18);
G1=residue(G,s)
```

and the result below is exactly the same as the one obtained earlier

$$G_1(s) = -\frac{17}{8(s+3)} - \frac{7}{4(s+3)^2} + \frac{2}{s+2} + \frac{1}{8(s+1)} - \frac{1}{2(s+1)^2} + \frac{1}{2(s+1)^3}.$$

Example 5.28 Now consider again the rational function $G(x)$ defined in Example 5.26. Write the partial fraction expansion using analytical methods.

Solution In Example 5.26, the numerical approach was used and the results may not be accurate. Thus the problem can be explored with the symbolic `residue()` function

```
>> syms s
   G=(2*s^7+2*s^3+8)/(s^8+30*s^7+386*s^6+2772*s^5+12093*s^4+...
      32598*s^3+52520*s^2+45600*s+16000);
   f=residue(G);
```

and then the expected partial fraction expansion is

$$\frac{13041}{(s+5)^3} + \frac{341863}{12(s+5)^2} + \frac{7198933}{144(s+5)} - \frac{16444}{3(s+4)^3} + \frac{193046}{9(s+4)^2}$$
$$- \frac{1349779}{27(s+4)} + \frac{11}{9(s+2)} + \frac{1}{432(s+1)}.$$

Compared with the results obtained in Example 5.26, the new result is more convincing. The difference between the expansion and the original function can be found with `simple(`f`-`G`)` which is 0.

Example 5.29 For the non-coprime rational function in Example 5.24, the overloaded `residue()` function can be used to write out the partial fraction expansion to the rational function $G(x) = A(x)/B(x)$.

Solution The following statements can be specified to solve the problem

```
>> syms x; A=x^4+7*x^3+13*x^2+19*x+20;
   B=x^7+16*x^6+103*x^5+346*x^4+655*x^3+700*x^2+393*x+90;
   residue(A/B,x)
```

and it can be seen that

$$\frac{A(x)}{B(x)} = -\frac{7}{4\,(x+3)^2} - \frac{17}{8\,(x+3)} + \frac{2}{(x+2)} + \frac{1}{2\,(x+1)^3} - \frac{1}{2\,(x+1)^2} + \frac{1}{8\,(x+1)}$$

where in the results, the term regarding $x+5$ does not exist at all, since simplification was performed within the `residue()` function already.

Example 5.30 Compute the partial fraction expansion to the following function

$$G(x) = \frac{-17x^5 - 7x^4 + 2x^3 + x^2 - x + 1}{x^6 + 11x^5 + 48x^4 + 106x^3 + 125x^2 + 75x + 17}.$$

Solution The numerical `residue()` can be used first

```
>> num=[-17 -7 2 1 -1 1]; den=[1 11 48 106 125 75 17];
   [r,p,k]=residue(num,den); [r,p,k]
```

and the partial fraction expansion can be written as

$$\frac{-556.256530687201}{x+3.261731010738}+\frac{0.212556796963}{x+0.520859605293}$$
$$\frac{0.879464926195-\mathrm{j}5.497076257858}{x+2.53094582005-\mathrm{j}0.39976310545}+\frac{0.879464926195+\mathrm{j}5.497076257858}{x+2.53094582005+\mathrm{j}0.39976310545}$$
$$+\frac{268.64252201892+\mathrm{j}349.12310949979}{x+2.53094582005-\mathrm{j}0.39976310545}+\frac{268.64252201892-\mathrm{j}349.12310949979}{x+2.53094582005+\mathrm{j}0.39976310545}.$$

If the symbolic function residue() is used instead

```
>> syms x;  G=(-17*x^5-7*x^4+2*x^3+x^2-x+1)...
      /(x^6+11*x^5+48*x^4+106*x^3+125*x^2+75*x+17);
   G1=residue(G,x)
```

more accurate partial fraction expansion can be obtained

$$\frac{0.21255679696263229441850086860134}{x+0.520859605293200521173180894658}-\frac{556.256530686961041694760596634}{x+3.26173101073851870211029033154}$$
$$+\frac{0.879464926194620659866107689053+\mathrm{j}5.49707625785732772459122493102}{x+1.0777588719352924223318783254+\mathrm{j}0.602106591060761485593453095064}$$
$$+\frac{0.879464926194620659866107689053-\mathrm{j}5.49707625785732772459122493102}{x+1.0777588719352924223318783254-\mathrm{j}0.602106591060761485593453095064}$$
$$+\frac{268.642522018804584040304940194-\mathrm{j}349.12310949952681814422714893}{x+2.5309458200488479660263860615+\mathrm{j}0.39976310544985541814054726452}$$
$$+\frac{268.642522018804584040304940194+\mathrm{j}349.12310949952681814422714893}{x+2.5309458200488479660263860615-\mathrm{j}0.39976310544985541814054726452}.$$

5.5.4 Inverse Laplace transform using partial fraction expansions

It has been shown that the Symbolic Math Toolbox function ilaplace() can tackle the inverse Laplace transform to the rational function problems very well. However, for a class of problems where complex roots exist, it has been shown in Example 5.3 that the results obtained directly will have very poor readability.

It is found by observation of the results in Example 5.30 that some terms can be expressed by $(a+\mathrm{j}b)/(s+c+\mathrm{j}d)$, and there is also a complex conjugate term $(a-\mathrm{j}b)/(s+c-\mathrm{j}d)$. Thus the two terms can be simplified such that

$$(a+\mathrm{j}b)\mathrm{e}^{(c+\mathrm{j}d)t}+(a-\mathrm{j}b)\mathrm{e}^{(c-\mathrm{j}d)t}=\alpha\mathrm{e}^{ct}\sin(dt+\phi) \tag{5.38}$$

where $\alpha=-2\sqrt{a^2+b^2}$, and $\phi=-\tan^{-1}(b/a)$.

Based on such an algorithm, a numerical function pfrac() can be written to enhance the facilities provided in the original function residue(). The listings of the new function are

```
function [R,P,K]=pfrac(num,den)
[R,P,K]=residue(num,den);
```

```
for i=1:length(R),
   if imag(P(i))>eps, a=real(R(i)); b=imag(R(i));
      R(i)=-2*sqrt(a^2+b^2); R(i+1)=-atan2(a,b);
   elseif abs(imag(P(i)))<eps, R(i)=real(R(i));
end, end
```

with the syntax `[r,p,K]=pfrac(num,den)`, where the definitions of $\boldsymbol{p}$ and K are exactly the same as the `residue()` function, and $\boldsymbol{r}$ is defined slightly differently. If p_i is real, then r_i is the same with the definition in `residue()` function. However, if p_i is complex, then r_i and r_{i+1} return respectively the values of α and ϕ.

Example 5.31 Reconsider the problem in Example 5.30. Alternatively we use

```
>> num=[-17,-7,2,1,-1,1]; den=[1,11,48,106,125,75,17];
   [r,p,k]=pfrac(num,den); format long e; [r,p]
```

and the result is shown below

$$\begin{aligned}\mathscr{L}^{-1}[F(s)] = & -556.25653068675\mathrm{e}^{-3.26173101073386t} + 2.1255679696\mathrm{e}^{-0.5208596052932t} \\ & -881.03518709\mathrm{e}^{-2.530945820048808t}\sin(0.399763105449995t - 0.6558508770733) \\ & -11.13396711709\mathrm{e}^{-1.07775887193353t}\sin(0.6021065910608t - 2.9829493242804)\end{aligned}$$

which is in a much more readable form.

5.5.5 Computing closed-path integrals

Now consider the closed-path integral

$$\oint_\Gamma f(z)\,\mathrm{d}z \tag{5.39}$$

where Γ is a closed-path in a counterclockwise direction. Suppose that the closed-path encircles m poles, p_i, $(i = 1, 2, \cdots, m)$. The residues $\mathrm{Res}[f(p_i)]$ of the poles can be obtained using the MATLAB statements given earlier. The closed-path integral of the $f(z)$ function can be calculated from

$$\oint_\Gamma f(z)\,\mathrm{d}z = \mathrm{j}2\pi\sum_{i=1}^m \mathrm{Res}[f(p_i)]. \tag{5.40}$$

If Γ is in a clockwise direction, $f(z)$ should be multiplied by -1.

Example 5.32 Compute the closed-path integral on $|z| = 6$, and $f(z)$ is defined as

$$f(z) = \frac{2z^7 + 2z^3 + 8}{z^8+30z^7+386z^6+2772z^5+12093z^4+32598z^3+52520z^2+45600z+16000}.$$

Solution It can be seen from Example 5.28 that the partial fraction expansion of the original function is

$$f(z) = \frac{13041}{(z+5)^3} + \frac{341863}{12(z+5)^2} + \frac{7198933}{144(z+5)} - \frac{16444}{3(z+4)^3} + \frac{193046}{9(z+4)^2} - \frac{1349779}{27(z+4)} + \frac{11}{9(z+2)} + \frac{1}{432(z+1)}.$$

Thus the poles $p_1 = -1$ and $p_2 = -2$ are single poles, and $p_3 = -4$, $p_4 = -5$ are poles of multiplicity 3. Thus the residues of the poles are the coefficients of the first degree terms in the expression. Also it is known that the poles are all encircled by the $|z| = 6$ path. The closed-path integral solution can be found from

$$\oint_{|z|=6} f(z)\,\mathrm{d}z = \mathrm{j}2\pi \left[\frac{7198933}{144} - \frac{1349779}{27} + \frac{11}{9} + \frac{1}{432}\right] = \mathrm{j}4\pi.$$

With the path integral method discussed in Chapter 3, the integral can be evaluated directly. The path Γ of the circle $|z| = 6$ can be expressed as $z = 6\cos t + \mathrm{j}6\sin t$, $t \in [0, 2\pi]$. Thus the following statements can be given to find $I = \mathrm{j}4\pi$.

```
>> syms z t
   G=(2*z^7+2*z^3+8)/(z^8+30*z^7+386*z^6+2772*z^5+12093*z^4+...
       32598*z^3+52520*z^2+45600*z+16000);
   F=subs(G,z,6*cos(t)+6*sin(t)*sqrt(-1));
   I=int(F*diff(6*cos(t)+6*sin(t)*sqrt(-1),t),t,0,2*pi)
```

Example 5.33 Compute the closed-path integral $\oint_\Gamma \frac{1}{(z+\mathrm{j}1)^{10}(z-1)(z-3)}\,\mathrm{d}z$, where the path Γ is the counterclockwise path $|z| = 2$.

Solution It can be seen that the original function has single poles at $z = 1$ and $z = 3$. Also the pole at $z = -\mathrm{j}1$ is a pole of multiplicity 10. The poles $z = 1$ and $z = -\mathrm{j}1$ are encircled by Γ, and $z = 3$ is not. Thus the closed-path integral can be evaluated with the following statements

```
>> i=sym(sqrt(-1)); syms z t; f=1/((z+i)^10*(z-1)*(z-3));
   r1=limit(diff(f*(z+i)^10,z,9)/prod(1:9),z,-i);
   r2=limit(f*(z-1),z,1); a=2*pi*i*(r1+r2)
```

and the integral can be evaluated as $a = (237/312500000+\mathrm{j}779/78125000)\pi$, and the results obtained seem not to agree with the results $-\mathrm{j}\pi/(3+\mathrm{j}1)^{10}$ in other textbooks. However, it can be shown from the following statements that the difference is zero.

```
>> a+pi*i/(3+i)^10
```

With direct path integral method, the same result can be obtained.

```
>> F=subs(f,z,2*cos(t)+2*sin(t)*sqrt(-1))
   I=int(F*diff(2*cos(t)+2*sin(t)*sqrt(-1),t),t,0,2*pi)
```

If the path Γ is changed to $|z| = 4$, all three poles are encircled by the path. Thus the closed-path integral can be obtained with

```
>> r3=limit(f*(z-3),z,3); b=2*pi*i*(r1+r2+r3)
```

and the integral is 0, which can also be confirmed by direct integration method.

```
>> F=subs(f,z,4*cos(t)+4*sin(t)*sqrt(-1))
   I=int(F*diff(4*cos(t)+4*sin(t)*sqrt(-1),t),t,0,2*pi)
```

In traditional complex function courses, the single variable integrals can also be converted to the closed-path integrals of complex functions by variable substitutions. For instance, the integral $\displaystyle\int_0^{\infty}\frac{\sin x}{x}\,\mathrm{d}x$ is often solved in this way. With the use of MATLAB function `int()`, the above problem can be solved directly. Thus it is not recommended to convert it to complex functions.

Exercises

1. Perform Laplace transforms for the following functions

$$\text{(i) } f(t)=\frac{\sin\alpha t}{t},\ \text{(ii) } f(t)=t^5\sin\alpha t,\ \text{(iii) } f(t)=t^8\cos\alpha t,$$

$$\text{(iv) } f(t)=t^6\mathrm{e}^{\alpha t},\quad \text{(v) } f(t)=5\mathrm{e}^{-at}+t^4\mathrm{e}^{-at}+8\mathrm{e}^{-2t},$$

$$\text{(vi) } f(t)=\mathrm{e}^{\beta t}\sin(\alpha t+\theta),\ \text{(vii) } f(t)=\mathrm{e}^{-12t}+6\mathrm{e}^{9t}.$$

2. Take inverse transforms for the problems solved above and see whether the corresponding original function can be restored.
3. The following properties are also given for Laplace transforms. Verify for different values of n, that the following formula are satisfied.

$$\text{(i) } \mathscr{L}[t^n f(t)]=(-1)^n\frac{\mathrm{d}^n\mathscr{L}[f(t)]}{\mathrm{d}s^n},\quad \text{(ii) } \mathscr{L}[t^{n-1/2}]=\frac{\sqrt{\pi}(2n-1)!}{2^n}s^{-n-1/2}$$

4. Perform inverse Laplace transforms to the following $F(s)$.

$$\text{(i) } F(s)=\frac{1}{\sqrt{s^2(s^2-a^2)(s+b)}},\quad \text{(ii) } F(s)=\sqrt{s-a}-\sqrt{s-b},$$

$$\text{(iii) } F(s)=\ln\frac{s-a}{s-b},\quad \text{(iv) } F(s)=\frac{1}{\sqrt{s}(s+a)},\quad \text{(v) } F(s)=\frac{3a^2}{s^3+a^3},$$

$$\text{(vi) } F(s)=\frac{(s-1)^8}{s^7},\quad \text{(vii) } F(s)=\ln\frac{s^2+a^2}{s^2+b^2}$$

$$\text{(viii) } F(s)=\frac{s^2+3s+8}{\prod_{i=1}^{8}(s+i)},\quad \text{(ix) } F(s)=\frac{1}{2}\frac{s+\alpha}{s-\alpha}$$

5. Show the Laplace transforms where the non-integer power of s is introduced, which is the fundamental of fractional-order calculus.

$$\text{(i) } \mathscr{L}[t^\gamma]=\frac{\Gamma(\gamma+1)}{s^{\gamma+1}},\ \text{one should check different values of } \gamma$$

$$\text{(ii) } \mathscr{L}\left[\frac{1}{\sqrt{t}\,(1+at)}\right]=\frac{\pi}{a}\,\mathrm{e}^{s/a}\mathrm{erfc}\left(\sqrt{s/a}\right)\ \text{for } a>0.$$

6. One of the applications of Laplace transform is that it can be used in solving

linear constant differential equations with zero initial conditions, using the property $\mathscr{L}[\mathrm{d}^n f(t)/\mathrm{d}t^n] = s^n \mathscr{L}[f(t)]$. Solve the differential equation $y''(t) + 3y'(t) + 2y(t) = \mathrm{e}^{-t}$, $y(0) = y'(0) = 0$ using Laplace transforms.

7. Perform Fourier transforms to the following functions, and then perform inverse Fourier transforms to see whether the original functions can be restored.
   (i) $f(x) = x^2(3\pi - 2|x|)$, (ii) $f(t) = t^2(t - 2\pi)^2$,
   (iii) $f(t) = \mathrm{e}^{-t^2}$, (iv) $f(t) = t\mathrm{e}^{-|t|}$
8. Perform Fourier sine and cosine transforms for the following functions and then perform inverse transformation and see whether the original functions can be restored.
   (i) $f(t) = \mathrm{e}^{-t}\ln t$, (ii) $f(x) = \dfrac{\cos x^2}{x}$, (iii) $f(x) = \ln\dfrac{1}{\sqrt{1+x^2}}$
   (iv) $f(x) = x(a^2 - x^2)$, $a > 0$, (v) $f(x) = \cos kx$.
9. Compute the discrete Fourier sine and cosine transforms for the functions (i) $f(x) = \mathrm{e}^{kx}$, and (ii) $f(x) = x^3$.
10. Write the Mellin transform for the function $f(x) = \begin{cases} \sin(a\ln x), & x \leqslant 1 \\ 0, & \text{otherwise.} \end{cases}$
11. Perform Z transforms to the time sequences $f(kT)$, and verify the results.
    (i) $f(kT) = \cos(kaT)$, (ii) $f(kT) = (kT)^2\mathrm{e}^{-akT}$, (iii) $f(kT) = \dfrac{1}{a}(akT - 1 + \mathrm{e}^{-akT})$
    (iv) $f(kT) = \mathrm{e}^{-akT} - \mathrm{e}^{-bkT}$, (v) $f(kT) = 1 - \mathrm{e}^{-akT}(1 + akT)$.
12. Perform inverse Z transforms to the following functions.
    (i) $F(z) = \dfrac{10z}{(z-1)(z-2)}$, (ii) $F(z) = \dfrac{z^{-1}(1 - \mathrm{e}^{-aT})}{(1 - z^{-1})(1 - z^{-1}\mathrm{e}^{-aT})}$
    (iii) $F(z) = \dfrac{z}{(z-a)(z-1)^2}$, (iv) $F(z) = \dfrac{Az[z\cos\beta - \cos(\alpha T - \beta)]}{z^2 - 2z\cos(\alpha T) + 1}$.
13. Take inverse Laplace transform to the following functions, then take Z transform and verify the results.
    (i) $G(s) = \dfrac{b}{s^2(s+a)}$, (ii) $G(s) = \dfrac{b}{s^2(s+a)^2}\dfrac{1 - \mathrm{e}^{-Ts}}{s}$.
14. For $G(s) = \dfrac{1}{(s+1)^3}$, if one substitutes $s = \dfrac{2(z-1)}{T(z+1)}$ into $G(s)$, the function $H(z)$ can be obtained. This kind of transform is referred to as *bilinear transform*. For $T = 1/2$, find $H(z)$. One may also assume that $z = \dfrac{1 + Ts/2}{1 - Ts/2}$, inverse bilinear transform can be performed. Check whether the original function can be restored.
15. Show that

$$\mathscr{Z}\left\{1 - \mathrm{e}^{-akT}\left[\cos(bkT) + \frac{a}{b}\sin(bkT)\right]\right\} = \frac{z(Az + B)}{(z-1)(z^2 - 2\mathrm{e}^{-aT}\cos(bT)z + \mathrm{e}^{-2aT})}$$

where

$$A = 1 - \mathrm{e}^{-aT}\cos(bT) - \tfrac{a}{b}\mathrm{e}^{-aT}\sin(bT)$$

$$B = \mathrm{e}^{-2aT} + \frac{a}{b}\mathrm{e}^{-aT}\sin(bT) - \mathrm{e}^{-aT}\cos(bT).$$

16. For the function

$$f(x) = \frac{x^2+4x+3}{x^5+4x^4+3x^3+2x^2+5x+2}\mathrm{e}^{-5x}$$

find the poles and their multiplicities and compute the residues for each pole.

17. Judge whether the following pairs of polynomials are coprime or not. If not, find the terms which can simplify $B(s)/A(s)$.

(i) $B(x) - 3x^4 + x^5 - 11x^3 + 51x^2 - 62x + 24$

$A(x) = x^7 - 12x^6 + 26x^5 + 140x^4 - 471x^3 - 248x^2 + 1284x - 720$

(ii) $B(x) = 3x^6 - 36x^5 + 120x^4 + 90x^3 - 1203x^2 + 2106x - 1080$

$A(x) = x^9+15x^8+79x^7+127x^6-359x^5-1955x^4-3699x^3-3587x^2-1782x-360.$

18. Perform partial fraction expansions for the following functions

(i) $f(x) = \dfrac{3x^4-21x^3+45x^2-39x+12}{x^7+15x^6+96x^5+340x^4+720x^3+912x^2+640x+192}$

(ii) $f(s) = \dfrac{s+5}{s^8+21s^7+181s^6+839s^5+2330s^4+4108s^3+4620s^2+3100s+1000}$

(iii) $f(x) = \dfrac{3x^6-36x^5+120x^4+90x^3-1203x^2+2106x-1080}{x^7+13x^6+52x^5+10x^4-431x^3-1103x^2-1062x-360}.$

19. Find the residues of the following functions at poles

(i) $f(z) = \dfrac{1-\sin z\mathrm{e}^{-2z}}{z^7\sin(z-\pi/3)}(z^4+10z^3+35z^2+50z+24)$

(ii) $f(z) = \dfrac{(z-3)^4}{z^4+5z^3+9z^2+7z+2}(\sin z - \mathrm{e}^{-3z})$

(iii) $f(z) = \dfrac{(1-\cos 2z)(1-\mathrm{e}^{-z^2})}{z^3\sin z}.$

20. Evaluate the closed-path integrals

(i) $\displaystyle\oint_\Gamma \frac{z^{15}}{(z^2+1)^2(z^4+2)^3}\,\mathrm{d}z$, where Γ is the positive circle $|z|=3$;

(ii) $\displaystyle\oint_\Gamma \frac{z^3}{1+z}\mathrm{e}^{1/z}\,\mathrm{d}z$, where Γ is the positive circle $|z|=2$.

(iii) $\displaystyle\oint_\Gamma \frac{\cos z(1-\mathrm{e}^{-z^2})\sin(3z+2)}{z\sin z}\,\mathrm{d}z$, where Γ is the positive circle $|z|=1$.

Chapter 6

Nonlinear Equations and Numerical Optimization Problems

The solutions to linear algebraic equations have been discussed extensively in Chapter 4. However, most frequently encountered are nonlinear equations in sciences and engineering problems. Solving nonlinear equations could be computationally expensive; therefore, the solving of approximate linear equations was indispensable especially in the early times when the computers were not powerful enough. Today, with the rapid development of computing technology, solving directly nonlinear equations is becoming increasingly important. In this chapter, we will focus on MATLAB solutions to nonlinear equations and optimization problems. In Section 6.1, solutions to nonlinear algebraic equations will be presented. The graphical method for nonlinear equations with one and two unknown variables will be given first, and quasi-analytical solutions will be studied for polynomial equations and equations convertible to polynomial equations. Numerical solutions to nonlinear equations and nonlinear matrix equations will also be discussed in this section.

The so-called *optimization* is to find the values of certain variables such that the preselected objective function takes maximum or minimum. Optimization technique is very useful in scientific research and engineering practice. Optimization problems can be classified as unconstrained optimization problems and constrained optimization problems. In Section 6.2, MATLAB-based solutions to unconstrained optimization problems will be given. Graphical and numerical methods will be presented and the concept of global optimum solution and local optimum solutions will be illustrated. In Section 6.3, constrained optimization problems will be studied and MATLAB-based solution methods will be presented. The concept of feasible regions will be introduced. In this section, the linear programming, quadratic programming and general nonlinear programming will be studied and MATLAB-based solutions will be illustrated through examples. In Section 6.4, the idea of programming problems will be further extended to integer programming and mixed integer programming problems. Mixed linear programming solutions will be studied and also a MATLAB solver based on the *branch-and-bound* algorithm is used for solving nonlinear mixed integer programming problems. Binary programming problems are also studied. In Section 6.5, a special type of optimization problems — the linear matrix inequality problem, is fully discussed and the solution methods are presented. Note that in Chapter

10, the global optimization methods will be presented based on evolution methods.

For readers who wish to check the detailed explanations of various solution techniques for nonlinear equations, we recommend the free textbook [4] (Chapter 9). For optimization theory and numerical optimization methods, the free textbook [15] is highly recommended. For LMIs, we suggest the free textbook [16]. We also found the online resource for deciding the right software for optimization problems [17] useful and interesting.

6.1 Nonlinear Algebraic Equations

6.1.1 Graphical method for solving nonlinear equations

The implicit functions with one and two variables can be drawn easily using the MATLAB function `ezplot()`. With `ezplot()`, the nonlinear equations can be shown graphically and its real solutions can be obtained by extracting the coordinates of the intersections of the curves. The graphical methods are restricted only to nonlinear equations with one or two variables. Nonlinear equations with more than two variables have to be solved numerically or for some special cases, symbolically.

Graphically solving nonlinear equations of single variable

The function `ezplot()` can be used to draw the curve from the implicit function $f(x) = 0$. The real solutions can be identified from the intersections of the curves with the line $y = 0$.

Example 6.1 Solve the equation $\mathrm{e}^{-3t}\sin(4t+2)+4\mathrm{e}^{-0.5t}\cos 2t = 0.5$ using graphical method and examine the accuracy of the solutions.

Solution The function `ezplot()` can be used to draw the curve of the function as shown in Figure 6.1 (a). The intersections with the horizontal axis are the solutions to the original nonlinear equation.

```
>> ezplot('exp(-3*t)*sin(4*t+2)+4*exp(-0.5*t)*cos(2*t)-0.5',[0 5])
   line([0,5],[0,0]) % draw the horizontal axis as well
```

From the curve it can be observed that there are three real solutions, p_1, p_2, p_3, over the interval $t \in (0, 5)$. One may zoom the area around a particular solution until the horizontal axis reads the same scale. The horizontal scale is then regarded as a solution. An example of the zoomed curve is shown in Figure 6.1 (b) where a solution $t = 0.6738$ can be obtained. Substituting this reading back to the equation

```
>> t=0.6738; exp(-3*t)*sin(4*t+2)+4*exp(-0.5*t)*cos(2*t)-0.5
```

the error can be found as -2.9852×10^{-4}. So, for this example, the achieved accuracy of the solution is not quite as high. Similar methods can be used to find and validate other solutions.

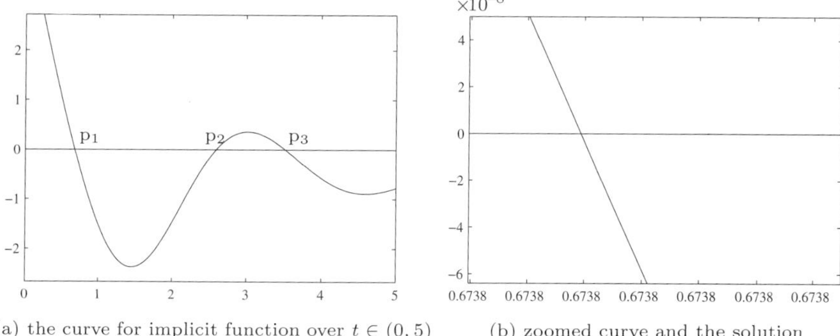

(a) the curve for implicit function over $t \in (0,5)$ (b) zoomed curve and the solution

FIGURE 6.1: Graphical solutions to an equation with single variable

Graphically solving nonlinear equations of two variables

Nonlinear equations with two variables can also be solved easily using the graphical method. Use `ezplot()` function to draw the solutions to the first equation. Then use `hold on` command to hold the graphics window such that the plot from `ezplot()` for the second nonlinear equation is superimposed to the first one. The intersections of the two sets of curves are then the solutions to the original nonlinear equations. The solutions can be read out graphically using the zooming method illustrated earlier.

Example 6.2 Solve graphically the equations $\begin{cases} x^2 e^{-xy^2/2} + e^{-x/2}\sin(xy) = 0 \\ x^2\cos(x+y^2) + y^2 e^{x+y} = 0. \end{cases}$

Solution The graphical method can be used to solve the above nonlinear simultaneous equations. The first equation can be displayed with the direct use of the implicit function drawing command `ezplot()`, as shown in Figure 6.2 (a)

```
>> ezplot('x^2*exp(-x*y^2/2)+exp(-x/2)*sin(x*y)') % the first equation
```

Use the command `hold on` to ensure the curves will not be removed. Then, `ezplot()` draws the solutions to the second equation. The curves will then be superimposed on the curves obtained earlier, as shown in Figure 6.2 (b).

```
>> hold on; ezplot('y^2 *cos(y+x^2) +x^2*exp(x+y)')
```

The intersections are then the solutions to the nonlinear equation sets. In this way, all the real solutions to the given simultaneous equations can be displayed. To get the coordinates of a certain point, for instance point B in Figure 6.2 (b), one may zoom the interested region again and again until all the scales on the x- and y-axes read the same, as shown in Figure 6.3 (a). Thus the solution at point B is $x = -0.7327, y = 1.5619$.

From Figure 6.2 (b) it is also found that most of the solutions are located in the fourth quadrant. Thus a large area can be chosen. For instance, the rectangular region $(0,0), (8,-10)$ can be selected and the solutions in the new area can be displayed as shown in Figure 6.3 (b).

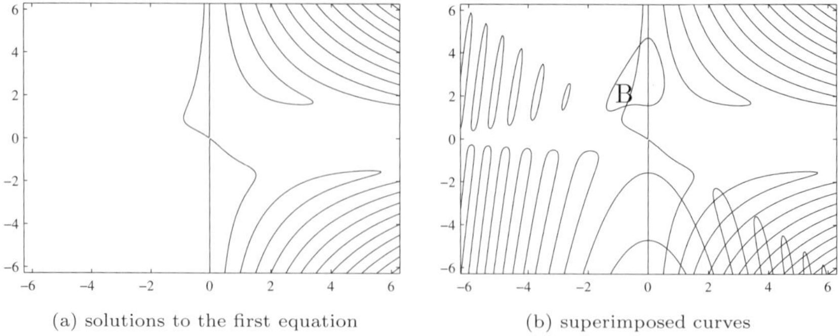

(a) solutions to the first equation (b) superimposed curves

FIGURE 6.2: Graphical solutions to the nonlinear equations

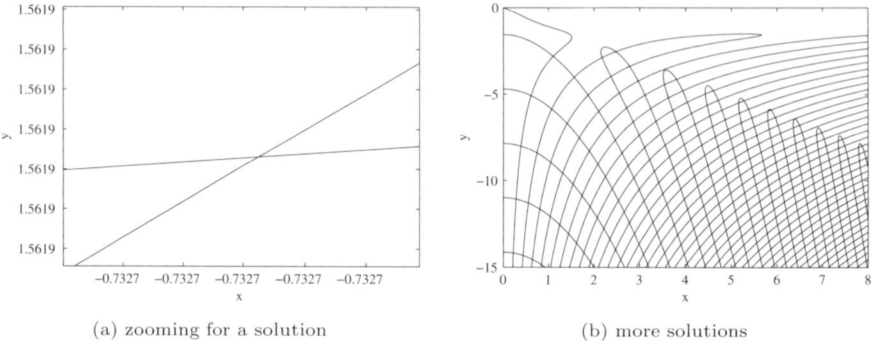

(a) zooming for a solution (b) more solutions

FIGURE 6.3: Graphical solutions to the nonlinear equations

```
>> ezplot('x^2*exp(-x*y^2/2)+exp(-x/2)*sin(x*y)',[0,8,-15,0])
   hold on, ezplot('y^2 *cos(y+x^2) +x^2*exp(x+y)',[0,8,-15,0])
```

6.1.2 Quasi-analytical solutions to polynomial-type equations

Before illustrating solutions to polynomial equations, let us consider an example.

Example 6.3 Solve the equations $\begin{cases} x^2+y^2-1=0 \\ 0.75x^3-y+0.9=0 \end{cases}$ using the graphical method.

Solution Using the graphical method, the two curves for the two equations can be displayed easily using the following statements, as shown in Figure 6.4. The intersections are the solutions to the original equations.

```
>> ezplot('x^2+y^2-1'); hold on % solutions to the first equation
   ezplot('0.75*x^3-y+0.9')     % the second equation
```

It can be seen from the curves in Figure 6.4 that there are two intersections. However, it cannot be simply concluded that the original equations have only two

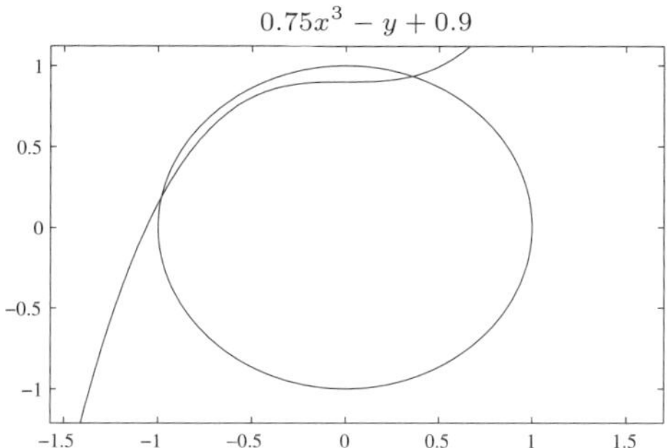

FIGURE 6.4: Solutions using graphical method

solutions. One may solve y from the second equation and find that y is a function of x^3. Substituting the equation into the first one, it can be concluded that the equation can be converted into a polynomial equation of x, with the highest degree of 6. Thus the polynomial equation must have 6 roots. The graphical method can only be used to find the real solutions. No information on complex roots is available from the graphical method. Thus the graphical methods sometimes are not adequate in solving polynomial equations.

The function `solve()` provided in the Symbolic Math Toolbox of MATLAB is quite effective in finding the solutions to polynomial-type equations. The function can be used in finding all the solutions to the simultaneous equations which can be converted to polynomial equations. The syntaxes of the function are

```
S=solve(eqn1,eqn2,···,eqnn)                   % the simplest syntax
[x,y,···]=solve(eqn1,eqn2,···,eqnn)           % direct solutions
[x,y,···]=solve(eqn1,eqn2,···,eqnn,'x,y,···') % variables specified
```

where eqn_i is the symbolic representation of the ith equation to be solved. In this way, simultaneous equations can easily be represented. In the first statement, a structure variable `S` is returned, and the solutions are members of `S`. For instance, `S.x` and `S.y`.

Example 6.4 Solve again the equations in Example 6.3.

Solution The `solve()` function can be used in solving the equations such that

```
>> syms x y; [x,y]=solve('x^2+y^2-1=0','75*x^3/100-y+9/10=0')
```

and the solutions are found as

$$x = \begin{bmatrix} .35696997189122287798839037801365 \\ .86631809883611811016789809418 65 + \mathrm{j}1.2153712664671427801318378 5444 \\ -.55395176056834560077984413882 7 + \mathrm{j}.35471976465080793456863789934 9 \\ -.98170264842676789676449828873194 \\ -.55395176056834560077984413882 7 - \mathrm{j}.35471976465080793456863789934 9 \\ .86631809883611811016789809418 65 - \mathrm{j}1.2153712664671427801318378 5444 \end{bmatrix}$$

$$y = \begin{bmatrix} .93411585960628007548796029415446 \\ -1.4916064075658223174787216959 + j.70588200721402267753918827138 8 \\ .92933830226674362852985276677 2 + j.21143822185895923615623381762 21 \\ .19042035099187730240977756415289 \\ .92933830226674362852985276677 2 - j.21143822185895923615623381762 21 \\ -1.4916064075658223174787216959 - j.70588200721402267753918827138 8 \end{bmatrix}.$$

For this high-degree polynomial-type equation, according to the well-known Abel-Ruffini Theorem, there exist no analytical solutions. The Symbolic Math Toolbox of MATLAB can be used to obtain high-precision solutions. These types of solutions are referred to as the *quasi-analytical solutions*. It can be seen that apart from the two sets of real solutions, there are yet other sets of complex conjugate solutions to the original nonlinear equations. These solutions cannot be obtained using graphical or other search algorithms.

The following statements can be given to verify the accuracies of the solutions

```
>> [eval('x.^2+y.^2-1') eval('75*x.^3/100-y+9/10')]'
```

and the error to each equation can be obtained as

$$\begin{bmatrix} -.1\times10^{-31} & .5\times10^{-30}+j.1\times10^{-30} & 0.-j0. & 0. & 0.-j0. & .5\times10^{-30}-j.1\times10^{-30} \\ 0. & 0.-j0 & 0.-j0. & 0. & 0.-j0. & 0.-j0. \end{bmatrix}$$

where each column corresponds to a pair of (x_i, y_i) solutions. It can be seen that the results thus obtained are extremely accurate, with an accuracy impossible to achieve by double-precision arithmetics.

Example 6.5 The polynomial-type equations with more variables can also be obtained using the `solve()` function. Find the solutions to the following equations

$$\begin{cases} x + 3y^3 + 2z^2 = 1/2 \\ x^2 + 3y + z^3 = 2 \\ x^3 + 2z + 2y^2 = 2/4. \end{cases}$$

Solution The equations given are with three variables x, y, z. It can be seen that there are only polynomial terms, thus it can theoretically be converted into polynomial equations of a single variable. The following statements can be used to find the quasi-analytical solutions to the given equations.

```
>> [x,y,z]=solve('x+3*y^3+2*z^2=1/2',...
                 'x^2+3*y+z^3=2','x^3+2*z+2*y^2=2/4')
```

In fact, the original equations can be converted into a single polynomial equation with a single variable with a degree of 27. Thus the quasi-analytical solutions can be obtained using the above statements. Substituting the solutions back to the original equations, it can be found that the error could be as small as 6.9146×10^{-26}. So, the equations can perfectly be solved using the quasi-analytical approach.

```
>> err=[x+3*y.^3+2*z.^2-1/2, x.^2+3*y+z.^3-2, x.^3+2*z+2*y.^2-2/4];
   norm(double(eval(err)))
```

In fact, the terms given in the equations can also be written as a product of polynomials. For instance, if the last equation is given by $x^3 + 2zy^2 = 2/4$, with the product of polynomials such as zy^2, the solutions to the original equations can also be found by the direct use of the `solve()` function. The statements can be changed to

```
>> [x,y,z]=solve('x+3*y^3+2*z^2=1/2','x^2+3*y+z^3=2','x^3+2*z*y^2=2/4')
   err=[x+3*y.^3+2*z.^2-1/2, x.^2+3*y+z.^3-2, x.^3+2*z.*y.^2-2/4];
   norm(double(eval(err)))    % norm of the error
```

and quasi-analytical solutions can be found and the norm of the error for the new equations can be as small as 6.4156×10^{-26}.

Example 6.6 Solve the following equations where the reciprocals to the variables are involved

$$\begin{cases} \dfrac{1}{2}x^2+x+\dfrac{3}{2}+2\dfrac{1}{y}+\dfrac{5}{2y^2}+3\dfrac{1}{x^3}=0 \\ \dfrac{y}{2}+\dfrac{3}{2x}+\dfrac{1}{x^4}+5y^4=0. \end{cases}$$

Solution It is not likely possible to solve this kind of complicated equation without the help of powerful computer mathematics languages. However, with the following statements the quasi-analytical solutions can be obtained easily.

```
>> syms x y;
   f1=x^2/2+x+3/2+2/y+5/(2*y^2)+3/x^3; f2=y/2+3/(2*x)+1/x^4+5*y^4;
   [x0,y0]=solve(f1,f2)
```

and it can be seen that there are 26 pairs of solutions. Substituting all the solutions back to the original equations, one can immediately find that the norm of the error is 6.3172×10^{-30}, which means that the solutions are very accurate.

```
>> err=[subs(f1,{x,y},{x0,y0}) subs(f2,{x,y},{x0,y0})];
   norm(double(err))
```

Example 6.7 Solve the equations with constants $\begin{cases} x^2+ax^2+6b+3y^2=0 \\ y=a+x+3. \end{cases}$

Solution The `solve()` function can be used directly to solve the equations, even if it contains extra variables. The solutions to the problem can be obtained by the direct use of the function calls such that

```
>> syms a b x y; [x,y]=solve('x^2+a*x^2+6*b+3*y^2=0','y=a+(x+3)','x,y')
```

and the solutions can be written as

$$\begin{cases} x=\dfrac{-6a-18\pm2\sqrt{-21a^2-45a-27-24b-6ab-3a^3}}{2(4+a)} \\ y=a+\dfrac{-6a-18\pm2\sqrt{-21a^2-45a-27-24b-6ab-3a^3}}{2(4+a)}+3. \end{cases}$$

In fact, the method may apply to third- or fourth-degree equations as well. However, the solutions are usually too complicated to display.

It should be noted that the analytical or quasi-analytical solution methods introduced in the previous subsections are not general-purpose. They can only be used in dealing with problems convertible to high-degree polynomial equations with a single variable. Furthermore, for most nonlinear equations, we cannot expect to find all the possible solutions.

6.1.3 Numerical solutions to general nonlinear equations

A numerical solution function `fsolve()` provided in MATLAB can be used to search for a real solution to given nonlinear equations. The syntax of the function is

```
x=fsolve(fun,x0)                                  % simple syntax
[x,f,flag,out]=fsolve(fun,x0,opt,p1,p2,···)       % formal full syntax
```

where *fun* can either be an M-function, an anonymous function or an inline function describing the equations to be solved. The variable $\boldsymbol{x}_0$ is the initial search point for the solution. A real solution to the equations can be obtained by searching method from the initial point $\boldsymbol{x}_0$ using numerical algorithms. If a solution is successfully found, the returned `flag` is greater than 0, otherwise the search is not successful.

For more complicated problems, the solution control option `opt` can be used to select methods and control accuracies in searching the solution. The `opt` variable is defined as a structured variable, with the commonly used members explained in Table 6.1. The following syntaxes can be used in modifying the contents in the control options

```
opt=optimset;                                 % get default control template
opt.TolX=1e-10; or set(opt,'TolX',1e-10)      % set control parameters
```

where some of the members such as `MaxFunEvals` are problem dependent, which is usually set to 100 to 200 times the number of variables. The user may change the options using the above mentioned function calls.

TABLE 6.1: Control options for equation solutions and optimizations

member name	explanation to the options
`Display`	To control whether the intermediate results are displayed, with the values `'off'` for no display, `'iter'` for display in each iteration, `'notify'` for alert at none convergence, and `'final'` for final results display only
`GradObj`	To indicate whether the gradient information is used in optimization. The options are `'off'` and `'on'`, with `'off'` the default
`LargeScale`	To indicate whether large-scale algorithms are used, with options `'on'` and `'off'`. For problems with only a few variables, it should be set to `'off'`
`MaxIter`	The maximum allowed iterations for equation solution and optimization. This value can be increased for problems failed to converge within the current control options
`MaxFunEvals`	The maximum allowed times of objective function calls
`TolFun`	The error tolerance of objective functions
`TolX`	The error tolerance of the solutions

Example 6.8 Solve the equations in Example 6.3 using numerical algorithms.

Solution Before solving such equations, the variables should be selected such that the unknown variables to be solved are assigned as a vector. Selecting the variables $p_1 = x$, $p_2 = y$, the original equations can be represented by an anonymous function, and then one may select the initial values at $\boldsymbol{p}_0 = [1, 2]^{\mathrm{T}}$. The function `fsolve()` can be used directly to solve the original equations and find a solution.

```
>> f=@(p)[p(1)*p(1)+p(2)*p(2)-1; 0.75*p(1)^3-p(2)+0.9];
   OPT=optimset; OPT.LargeScale='off';
   [x,Y,c,d]=fsolve(f,[1; 2],OPT),
```

The solution found is $\boldsymbol{x} = [0.35696997, 0.93411586]^{\mathrm{T}}$, with the error $\boldsymbol{Y} = [0.1215 \times 10^{-9}, 0.0964 \times 10^{-9}]$. It can also be found by examining the `d` argument that 21 function calls are made. Thus the algorithm is quite effective.

Similarly, the original equations can also be described by the inline function or by M-file. With the inline functions and anonymous functions, there is no need to create a separate M-file for each problem, which makes the file management more tidy and convenient.

If the initial values are changed to $\boldsymbol{p}_0 = [-1, 0]^{\mathrm{T}}$, then by using

```
>> [x,Y,c,d]=fsolve(f,[-1,0]',OPT); x, Y, kk=d.funcCount
```

another solution is found at $\boldsymbol{x} = [-0.981703, 0.1904204]^{\mathrm{T}}$, and this time 15 function calls are made and the norm of the error vector is 0.5618×10^{-10}. In this example, it can be seen that the selection of initial values may lead to other solutions.

Example 6.9 The Lambert function is a special function defined as $w = \mathrm{lam}(x)$, where w is the solution to the equation $we^w = x$ for a given variable x. For different values of x, solve the Lambert equation and then draw the relationship between w and x.

Solution The solution to this problem can be obtained by taking loops for various values of x. The following statements can be given and the curves shown in Figure 6.5 can be obtained such that the Lambert function curve can then be obtained.

```
>> y=[]; xx=0:.05:10; x0=0; h=optimset; h.Display='off';
   for x=xx
      f=@(w)w.*exp(w)-x; y1=fsolve(f,x0,h); x0=y1; y=[y,y1];
   end
   plot(xx,y)
```

A MATLAB function `lambertw()` is provided in the Symbolic Math Toolbox which can be used to evaluate the Lambert equations directly. The following statements can be used instead to draw the Lambert curves and the results are exactly the same as the ones obtained in Figure 6.5.

```
>> y0=lambertw(xx); plot(xx,y0)
```

Example 6.10 Consider again the equation $\mathrm{e}^{-3t}\sin(4t+2) + 4\mathrm{e}^{-0.5t}\cos 2t = 0.5$ defined in Example 6.1. Find the solutions using numerical methods for a better accuracy.

Solution Using the `solve()` function

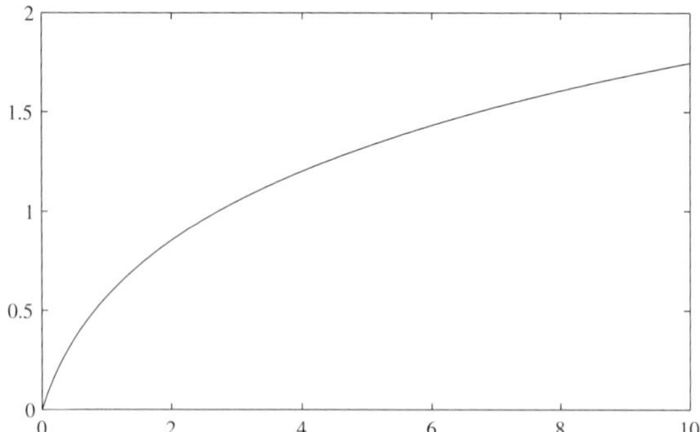

FIGURE 6.5: Solution of the Lambert function

```
>> syms t x; solve(exp(-3*t)*sin(4*t+2)+4*exp(-0.5*t)*cos(2*t)-0.5)
```

it can be found that the solution is $t = .6737457050013475670296022042 7474$. It is obvious that the nonlinear equation has no analytical solutions. Graphical methods shown in Example 6.1 can be used to find the numerical solutions. However, the accuracy achieved by graphical method may not be very high. From the approximate solution by graphical approach $t = 3.5203$, better results can be obtained by directly using the `fsolve()` function.

By combining the graphical and numerical methods, it can be seen that a better solution can be found

```
>> y=@(t)exp(-3*t).*sin(4*t+2)+4*exp(-0.5*t).*cos(2*t)-0.5;
   ff=optimset; [t,f]=fsolve(y,3.5203,ff)
```

such that $t = 3.52026389294877$ and $f = -6.06378\times 10^{-10}$. The solution found is much more accurate than the graphical method. To get even better approximations, one can further modify the control options with the following statements

```
>> ff=optimset; ff.TolX=1e-16; ff.TolFun=1e-30;
   [t,f]=fsolve(y,3.5203,ff)
```

and the new solution is $t = 3.52026389244155$ with $f = 0$.

6.1.4 Nonlinear matrix equations

In Section 4.4.4, a special form of nonlinear matrix equation, algebraic Riccati equation, is discussed. However, the solution is based on a very specialized algorithm, which cannot be extended to other forms of nonlinear matrix equations. For instance, if the equation is changed to

$$\boldsymbol{AX} + \boldsymbol{XD} - \boldsymbol{XBX} + \boldsymbol{C} = 0 \tag{6.1}$$

or even a tricky form

$$\boldsymbol{AX} + \boldsymbol{XD} - \boldsymbol{XBX}^{\mathrm{T}} + \boldsymbol{C} = 0 \tag{6.2}$$

the `are()` function is no longer applicable. So here, a nonlinear matrix equation solution method is given for solving general nonlinear matrix equations.

Example 6.11 Consider again the Riccati equation in Example 4.45, where the matrices are given below

$$\boldsymbol{A}=\begin{bmatrix}-2 & 1 & -3\\ -1 & 0 & -2\\ 0 & -1 & -2\end{bmatrix},\ \boldsymbol{B}=\begin{bmatrix}2 & 2 & -2\\ -1 & 5 & -2\\ -1 & 1 & 2\end{bmatrix},\ \boldsymbol{C}=\begin{bmatrix}5 & -4 & 4\\ 1 & 0 & 4\\ 1 & -1 & 5\end{bmatrix}.$$

Solution The unknown variable $\boldsymbol{X}$ in Riccati equation is a matrix, however, the `fsolve()` function can only deal with unknown vectors. Thus the first thing to do is to transform the matrix $\boldsymbol{X}$ into a vector $\boldsymbol{x}$. The simplest way is by $\boldsymbol{x}=\boldsymbol{X}$(:), i.e., by rearranging the matrix into a vector. O the other hand, when describing the Riccati equations, the best way is to keep using matrix $\boldsymbol{X}$. Thus within the function, one should restore the matrix $\boldsymbol{X}$ using the `reshape()` function. The function describing the errors in the Riccati equation can be written as

```
function y=new_are(x,A,B,C)
X=reshape(x,size(A)); y1=A'*X+X*A-X*B*X+C; y=y1(:);
```

where $\boldsymbol{A}$, $\boldsymbol{B}$, and $\boldsymbol{C}$ can be regarded as additional arguments. From the equation function, the Riccati equation can be solved with another function written as

```
function X=solve_are(A,B,C,x0)
if nargin==3, x0=rand(size(A)); end
x=fsolve(@new_are,x0(:),[],A,B,C); X=reshape(x,size(A));
```

It can be seen from the new solver that a random initial vector $\boldsymbol{x}_0$ is assigned. The following statements can be used to solve the Riccati equation

```
>> A=[-2,1,-3; -1,0,-2; 0,-1,-2]; B=[2,2,-2; -1 5 -2; -1 1 2];
   C=[5 -4 4; 1 0 4; 1 -1 5]; X=solve_are(A,B,C),
   norm(A'*X+X*A-X*B*X+C)
```

It is surprising that, apart from the solution in Example 4.45, another solution may also be found and it is verified that the error norm is 1.0406×10^{-15}

$$\boldsymbol{X}=\begin{bmatrix}-0.1538 & 0.10866 & 0.46226\\ 2.0277 & -1.7437 & 1.3475\\ 1.9003 & -1.7513 & 0.50571\end{bmatrix}.$$

Since Riccati equation is a quadratic equation, which is similar to the case of quadratic equation, more than one solutions exist. However, the new solution cannot be found using the `are()` function.

Example 6.12 Now consider the new Riccati-like equation given in (6.2), where

$$\boldsymbol{A}=\begin{bmatrix}2 & 1 & 9\\ 9 & 7 & 9\\ 6 & 5 & 3\end{bmatrix},\ \boldsymbol{B}=\begin{bmatrix}0 & 3 & 6\\ 8 & 2 & 0\\ 8 & 2 & 8\end{bmatrix},\ \boldsymbol{C}=\begin{bmatrix}7 & 0 & 3\\ 5 & 6 & 4\\ 1 & 4 & 4\end{bmatrix},\ \boldsymbol{D}=\begin{bmatrix}3 & 9 & 5\\ 1 & 2 & 9\\ 3 & 3 & 0\end{bmatrix}.$$

Find and verify all the possible solutions.

Solution To date, there are no existing algorithms for solving such types of equations. Similar to the above example, an M-function can be written to describe the equation

```
function y=new_are1(x,A,B,C,D)
X=reshape(x,size(A)); y1=A*X+X*D-X*B*X.'+C; y=y1(:);
```

where $\boldsymbol{A}$, $\boldsymbol{B}$, $\boldsymbol{C}$ and $\boldsymbol{D}$ are additional arguments. From the equation function, the Riccati-like equation can be solved with the following function:

```
function X=solve_are1(A,B,C,D,x0)
if nargin==4, x0=rand(size(A)); end
x=fsolve(@new_are1,x0(:),[],A,B,C,D); X=reshape(x,size(A));
```

Through repeated runs of the following statements

```
>> A=[2,1,9; 9,7,9; 6,5,3]; B=[0,3,6; 8,2,0; 8,2,8];
   C=[7,0,3; 5,6,4; 1,4,4]; D=[3,9,5; 1,2,9; 3,3,0];
   X=solve_are1(A,B,C,D), norm(A*X+X*D-X*B*X.'+C)
```

three solutions can be found. Note that $\boldsymbol{X}_2$ is more difficult to find

$$\boldsymbol{X}_1=\begin{bmatrix}1.7539 & 1.2408 & -0.00023348\\ 2.2114 & 3.3662 & -0.72222\\ 0.86565 & 1.8109 & -0.26194\end{bmatrix},\quad \boldsymbol{X}_2=\begin{bmatrix}6.74 & -0.36997 & 0.8394\\ -1.7679 & -0.25863 & 1.4835\\ 1.7761 & -0.39974 & 1.0043\end{bmatrix},$$

$$\boldsymbol{X}_3=\begin{bmatrix}-0.43386 & 0.40803 & 0.14075\\ 1.3621 & -2.5373 & 1.4561\\ -1.0243 & 0.97048 & -1.0438\end{bmatrix}.$$

6.2 Unconstrained Optimization Problems

Unconstrained optimization problems are considered as simpler than constrained optimization problems. The mathematical description of unconstrained problems is that

$$\min_{\boldsymbol{x}} f(\boldsymbol{x}) \tag{6.3}$$

where $\boldsymbol{x}=[x_1,x_2,\cdots,x_n]^{\mathrm{T}}$ are referred to as *decision variables*, or *optimization variables* and $f(\cdot)$ is referred to as the *objective function*. The task is to find a vector $\boldsymbol{x}$ such that the value of the objective function $f(\boldsymbol{x})$ is minimized. Thus the optimization problem is also referred to as the *minimization problem*. In fact the definition of minimization problem may not lose the generality. For instance, the maximization problem can be converted to minimization problem simply by multiplying -1 to its objective function. Thus the optimization problems studied in this book are for minimization problems.

6.2.1 Analytical solutions and graphical solution methods

It is known from advanced mathematics courses that the necessary conditions for an unconstrained optimization problem are that at the optimum point $\boldsymbol{x}^*$, the first-order derivatives of the objective function are all 0's. Thus the following simultaneous equations can be established

$$\left.\frac{\partial f}{\partial x_1}\right|_{\boldsymbol{x}=\boldsymbol{x}^*}=0,\ \left.\frac{\partial f}{\partial x_2}\right|_{\boldsymbol{x}=\boldsymbol{x}^*}=0,\ \cdots,\ \left.\frac{\partial f}{\partial x_n}\right|_{\boldsymbol{x}=\boldsymbol{x}^*}=0. \tag{6.4}$$

Solving the above equations, the extremum points can be obtained. In fact, the extremum points obtained may not be all minimum points, some of the points may actually be maximum points. Minimum points can be judged by taking the second-order derivatives, where positive second-order derivative means that minimum points are obtained. For single variable functions, the analytical method can be considered. However, for multivariable problems, solving the equations derived may be even more difficult than solving the optimization problem itself.

The graphical solutions to optimization problems with one variable are quite straightforward. The derivative function can be drawn first and the minimum points can be read from the curves. Functions with two variables may also be solved using graphical methods. However, for problems with three or more variables, graphical methods may not be applicable.

Example 6.13 From the equation $f(t)=\mathrm{e}^{-3t}\sin(4t+2)+4\mathrm{e}^{-0.5t}\cos(2t)-0.5$ studied in Example 6.1, use graphical and analytical methods to study the optimality of the function.

Solution The first-order derivative of the objective function can be derived first and with the function `ezplot()`, the first-order derivative can be drawn over the interval $t\in[0,4]$ as shown in Figure 6.6 (a).

```
>> syms t; y=exp(-3*t)*sin(4*t+2)+4*exp(-0.5*t)*cos(2*t)-0.5;
   y1=diff(y,t); ezplot(y1,[0,4])
```

In fact, finding the solution points of $y_1(t)=0$ is not easier than the direct finding of optimum points. With the graphical method, two points A_1 and A_2 can be found. From the first-order derivative, it can be seen that the point A_1 has a positive second-order derivative, thus it corresponds to the minimum point, while for point A_2, a negative second-order derivative corresponds to the maximum point. The value of A_1 can be obtained using the following statements

```
>> t0=solve(y1), ezplot(y,[0,4]) % draw the first-order derivative
   y2=diff(y1); y0=subs(y2,t,t0) % verify positive 2nd-order derivative
```

where $t_0=1.4528424981725411893375778$, and $y_0=7.8553420253333360137946 44$, which means that the point has a positive second-order derivative. Therefore, the minimum point has been obtained. It can further be confirmed from Figure 6.6 (b) that A_1 is the minimum point, and A_2 is the maximum point.

Since the original problem is a nonlinear function, analytical solutions may not be obtained. Note that the solution to the first-order derivative equation is not easier than the direct solution to the unconstrained optimization problem. For practical applications rather than demonstrations, such a graphical method is not recommended. Direct solutions to the optimization problems are recommended instead.

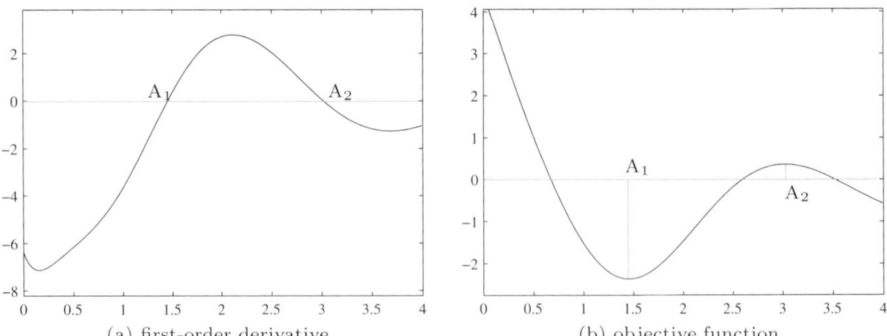

FIGURE 6.6: Minimum solution using graphical method

6.2.2 Numerical solution of unconstrained optimization using MATLAB

An unconstrained optimization problem solver, `fminsearch()`, is provided in MATLAB. Moreover, a similar function `fminunc()` is also provided in the Optimization Toolbox. Both functions have the same syntaxes. For instance

```
x=fminunc(fun,x0)                                % simplest statement
[x,f,flag,out]=fminunc(fun,x0,opt,p1,p2,···)     % more general form
```

where the input and output arguments are very similar to the `fsolve()` function described earlier. The control options are also the same. The improved simplex algorithm in Reference [18] is used to solve the optimization problem. This method is an effective method in solving unconstrained optimization problems. The following examples will be given for illustrations.

Example 6.14 For a function with two variables given by $z = (x^2-2x)\mathrm{e}^{-x^2-y^2-xy}$, find the minimum with MATLAB functions, and interpret the solutions graphically.

Solution The variables in the objective function are x, y, not the same as in the unconstrained optimization definition. A vector $\boldsymbol{x}$ should be defined by the variable substitutions such that $x_1 = x$, and $x_2 = y$. Thus the objective function can be declared with the anonymous function. The following statements can be used to find the optimal solution

```
>> f=@(x)(x(1)^2-2*x(1))*exp(-x(1)^2-x(2)^2-x(1)*x(2));
   x0=[2; 1]; x=fminsearch(f,x0)
```

and the solution is $\boldsymbol{x} = [0.6110, -0.3056]^{\mathrm{T}}$.

Similarly, the same problem can be solved with the `fminunc()` function, where

```
>> x=fminunc(f,x0)
```

and the solution obtained is $\boldsymbol{x} = [0.6110, -0.3055]^{\mathrm{T}}$.

Normally, the number of the function calls in the `fminunc()` function is much less than the `fminsearch()` function, since a more effective algorithm is used in `fminunc()`. Thus if one installs the Optimization Toolbox, it is suggested that the function `fminunc()` be used for unconstrained optimization problems.

To illustrate the solution process graphically, one may even modify the existing MATLAB function temporarily. For instance, use `edit private/fminusub` command to open the `fminsub` file in the private sub-directory and add `x'` command in the `callOutputFcn` function such that one can monitor the search process. If $\boldsymbol{x}_0 = [2, 1]^{\mathrm{T}}$ is used as the initial search point, the following intermediate points can be monitored

x_i	2	0.2401	−0.1398	0.2168	0.3355	0.5514	0.6129	0.6111
y_i	1	1.0502	0.5752	1.0210	−0.5508	−0.1775	−0.3053	−0.3058

and the search trajectory can be superimposed on the contours of the given function using the following statements, as shown in Figure 6.7.

```
>> xx=[2  0.2401 -0.1398 0.2168  0.3355  0.5514  0.6129  0.6111
       1  1.0502  0.5752 1.0210 -0.5508 -0.1775 -0.3053 -0.3058];
   [x,y]=meshgrid(-3:.1:3, -2:.1:2);
   z=(x.^2-2*x).*exp(-x.^2-y.^2-x.*y);
   contour(x,y,z,30); line(xx(1,:),xx(2,:))
   h=line(xx(1,:),xx(2,:)); set(h,'Marker','o')
```

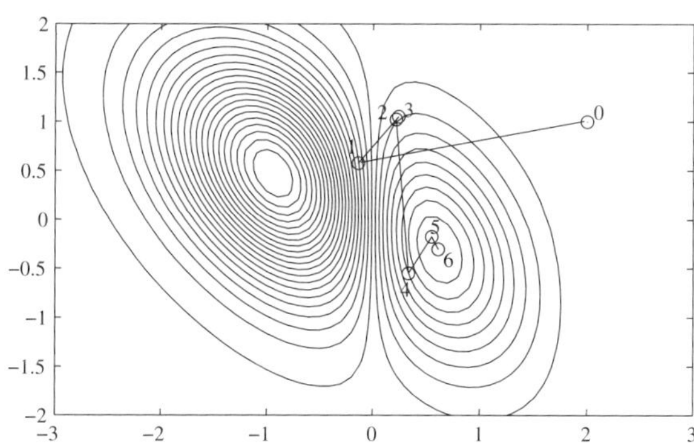

FIGURE 6.7: Search trajectory

6.2.3 Global minimum and local minima

Now consider a function with a single variable x. The necessary condition for a minimum point to exist is that $\mathrm{d}f(x)/\mathrm{d}x = 0$. However, the points satisfying such a condition may not be unique. However, if a search method is used, only one such point may be found from a given initial point. The global minimum and local minima are illustrated through the following example.

Example 6.15 Consider a function defined as $y(t) = \mathrm{e}^{-2t}\cos 10t + \mathrm{e}^{-3t-6}\sin 2t$, with $t \geqslant 0$. Compute the minimum point from different initial values. Then establish the concept of global minimum and local minima.

Solution For the given objective function, an anonymous function can be defined

```
>> f=@(t)exp(-2*t).*cos(10*t)+exp(-3*(t+2)).*sin(2*t);
```

If the initial search point is selected as $t_0 = 1$, the following statements can be used for finding the minimum point for the given objective function

```
>> t0=1; [t1,f1]=fminsearch(f,t0)
```

and it can be found that $t_1 = 0.9228$, $f_1 = -0.1547$. If another initial point $t_0 = 0.1$ is chosen, the following commands can be used

```
>> t0=0.1; [t2,f2]=fminsearch(f,t0)
```

which returns $t_2 = 0.2945$, $f_2 = -0.5436$.

It can be seen that for different initial search points, different optimum points are obtained. However, it is found by comparing the values of the objective functions at these optimum points, the point t_1 corresponds to a larger function value than the point t_2. Thus it is found that the point t_1 is not really the optimum point. These optimum points are usually referred to as the *local minima*. Selecting other initial points, such as $t_0 = 1.5, 2, 2.5, \cdots$ may lead to other local minima. With the function `ezplot()`, the objective function can be drawn within the interval $t \in (0, 2.5)$, as shown in Figure 6.8 (a). The global minimum and local minima can be indicated on the curve.

```
>> syms t; y=exp(-2*t)*cos(10*t)+exp(-3*(t+2))*sin(2*t);
   ezplot(y,[0,2.5]); ylim([-0.6,1])
```

It can be seen from Figure 6.8 (a) that the point t_2 is the global minimum point within the interval $t \geqslant 0$. The global minimum cannot be guaranteed by the optimization functions such as `fminsearch()`.

For larger intervals, however, for instance, for $t \geqslant -0.5$, the objective function can be redrawn as shown in Figure 6.8 (b). It can be seen that the point t_2 is no longer the global minimum. In fact, there is to date no algorithm that can guarantee a global minimum solution to an arbitrarily given objective function.

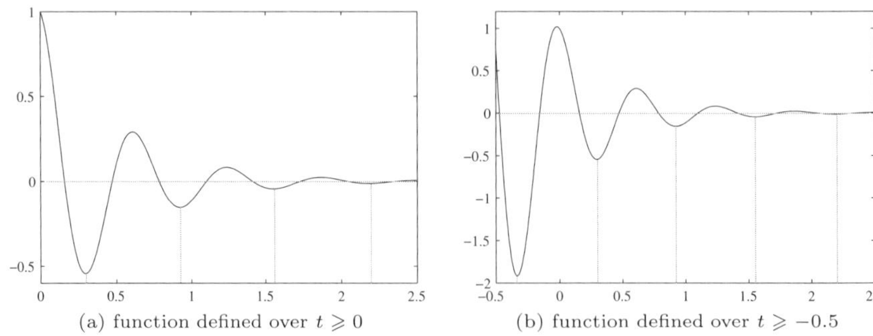

(a) function defined over $t \geqslant 0$ (b) function defined over $t \geqslant -0.5$

FIGURE 6.8: Illustration to global minimum and local minima

```
>> ezplot(y,[-0.5,2.5]); set(gca,'Ylim',[-2,1.2])
   t0=-0.2; [t3,f3]=fminsearch(f,t0); [t3 f3]
```

In this new range of t, the optimal solution becomes $t_3 = -0.3340$, $f_3 = -1.9163$.

It can be seen from the above example that sometimes the solutions obtained directly from the function calls may be local minimum, rather than global minimum. Thus different initial conditions should be tested to possibly improve the solutions. Genetic algorithm-based optimization technique can significantly improve the globalness of the minima, since many initial conditions are tested simultaneously. However, even genetic algorithm-based algorithms cannot guarantee the global solution. In Section 10.3, introductions and application illustrations will be given on genetic algorithms and other evolution-based optimization approaches.

6.2.4 Solving optimization problems using gradient information

Sometimes, the convergence speed for solving optimization problems may be very low. The exact optimum may not even be obtained using the information provided in the objective function alone. Thus the gradient information can be used to improve the optimization process.

In some functions of the Optimization Toolbox, the gradient information can also be provided describing the objective function. In this case, two arguments are returned with the first one still describing the objective function, and the second one for the gradients. Furthermore, in this case, the `GradObj`, a member of the control options should be set to `'on'`. The optimization solver can then be used to solve the optimization problems with the gradient information.

Example 6.16 Consider the Rosenbrock function

$$f(x_1, x_2) = 100(x_2 - x_1^2)^2 + (1 - x_1)^2.$$

Solve the unconstrained optimization problem for the function.

Solution It can be seen that the objective function is made of two squared terms. Thus when $x_2 = x_1 = 1$, the objective function takes its global minimum. Three-dimensional contours can be drawn as shown in Figure 6.9.

```
>> [x,y]=meshgrid(0.5:0.01:1.5); z=100*(y.^2-x).^2+(1-x).^2;
   contour3(x,y,z,100), zlim([0,310])
```

From the contours, the minimum is located in a very narrow valley. Thus Rosenbrock function is also known as the *banana function*. In the valley, the change of the objective function is extremely slow, thus it is very challenging to optimization algorithms. This function is often used as a *benchmark problem* to test whether the optimization algorithm is good or not. The following statements can be used to solve the optimization problem without the gradient information.

```
>> f=@(x)100*(x(2)-x(1)^2)^2+(1-x(1))^2;
   ff=optimset; ff.TolX=1e-10; ff.TolFun=1e-20; x=fminunc(f,[0;0],ff)
```

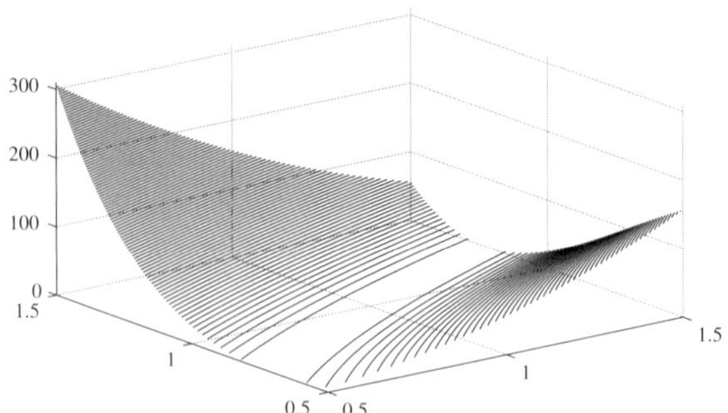

FIGURE 6.9: Three-dimensional contours for the Rosenbrock function

The best solution obtained is $x_1 = 0.999995588$, and $x_2 = 0.999991167$. It can be seen that although a very tough error tolerance is used, the exact minimum still cannot be obtained. Thus the information of gradients should be used to solve the problem.

For the given Rosenbrock function, the gradient matrix can easily be obtained by

```
>> syms x1 x2; f=100*(x2-x1^2)^2+(1-x1)^2;
   J=jacobian(f,[x1,x2])
```

and the Jacobian matrix is $\boldsymbol{J} = [-400(x_2 - x_1^2)x_1 - 2 + 2x_1, 200x_2 - 200x_1^2]$. Thus the gradient matrix $\boldsymbol{J}$ can be described in the objective function, which can be rewritten as

```
function [y,Gy]=c6fun3(x)
y=100*(x(2)-x(1)^2)^2+(1-x(1))^2;
Gy=[-400*(x(2)-x(1)^2)*x(1)-2+2*x(1); 200*x(2)-200*x(1)^2];
```

It should be noted that since two returned arguments are expected in the function, neither inline function nor anonymous function can be used to describe the new objective function. M-function is the only way of describing the gradients. The following statements can be used to solve the optimization problem:

```
>> ff.GradObj='on'; x=fminunc(@c6fun3,[0;0],ff)
```

and the solution found is $\boldsymbol{x} = [0.99999999999999, 0.99999999999997]^{\mathrm{T}}$. It can be seen that with the gradient information, the optimization is significantly speeded up, and the accuracy is also significantly improved, which approaches to the analytical values. Such an accuracy cannot be obtained without specifying the gradient information. However, in some other applications, the derivation and programming of gradient information could be very difficult, if not impossible, and one has to directly solve the optimization problem without the gradient information.

6.2.5 Optimization problems with bounded constraints

The unconstrained optimization problems discussed earlier are purely unconstrained problems. In actual applications, it is more practical to formulate unconstrained optimization problems having decision variables with known bounds. Of course, with bound constrains on the decision variables, the optimization problems are no longer unconstrained. The mathematical description to this special type of optimization problems is

$$\min_{\boldsymbol{x} \text{ s.t. } \boldsymbol{x}_{\rm m}\leqslant\boldsymbol{x}\leqslant\boldsymbol{x}_{\rm M}} f(\boldsymbol{x}) \tag{6.5}$$

where the notation s.t. means *subject to*, indicating the constraints that should be satisfied. The MATLAB function `fminsearchbnd()`, developed by John D'Errico, can be used to solve the problems. The function can be freely downloaded from the MATLAB File-exchange website, also provided on the companion CD. The syntax of the function is

```
x=fminsearchbnd(fun,x0,xm,xM)
[x,f,flag,out]=fminsearchbnd(fun,x0,xm,xM,opt,p1,p2,···)
```

If no upper- or lower-bounds are specified, they can be set to an empty matrix `[]`.

Example 6.17 Consider again the Rosenbrock function in Example 6.16. Apparently the global solution is $x_1 = x_2 = 1$. If one further assumes that the allowed intervals for x_1 and x_2 are respectively $x_1 \in (2,4)$ and $x_2 \in (3,6)$, solve the optimization problem.

Solution From the allowed ranges of x_1 and x_2, one may immediately find that $\boldsymbol{x}_{\rm m} = [2,3]^{\rm T}$ and $\boldsymbol{x}_{\rm M} = [4,6]^{\rm T}$. Thus the following statements can be used to solve the problem and it is found that $\boldsymbol{x} = [2, 3.9999996]^{\rm T}$.

```
>> f=@(x)[100*(x(2)-x(1)^2)^2+(1-x(1))^2]; xm=[2,3]; xM=[4,6];
   x=fminsearchbnd(f,[0,0],xm,xM)
```

6.3 Constrained Optimization Problems

A general description to constrained optimization problems is

$$\min_{\boldsymbol{x} \text{ s.t. } \boldsymbol{G}(\boldsymbol{x})\leqslant\boldsymbol{0}} f(\boldsymbol{x}) \tag{6.6}$$

where $\boldsymbol{x} = [x_1, x_2, \cdots, x_n]^{\rm T}$. The interpretation of such a description is that, under the constraints $\boldsymbol{G}(\boldsymbol{x}) \leqslant 0$, a decision vector $\boldsymbol{x}$ is expected which minimizes the objective function $f(\boldsymbol{x})$. In practical optimization problems, the constraints could be very complicated. For instance, it can be equalities or inequalities, and it can also be linear or nonlinear. Sometimes, these functions may not easily be described by mathematical functions.

6.3.1 Constraints and feasibility regions

The solution area $\boldsymbol{x}$ satisfying the constraints $\boldsymbol{G}(\boldsymbol{x}) \leqslant \boldsymbol{0}$ is referred to as the *feasible region*. A function of two variables will be given below and the feasible regions will be illustrated graphically.

Example 6.18 Consider the optimization problem with two variables given below. Study the optimization problem using the graphical method.

$$\max_{\boldsymbol{x} \text{ s.t. } \begin{cases} 9 \geqslant x_1^2+x_2^2 \\ x_1+x_2 \leqslant 1 \end{cases}} (-x_1^2 - x_2)$$

Solution From the given constraints, the initial square region $[-3, 3]$ can be selected and grids can be made. Then the objective function for unconstraint problems can be obtained using the following statements.

```
>> [x1,x2]=meshgrid(-3:.1:3);  % grid data generation
   z=-x1.^2-x2;                % calculate objective function over the grids
```

When the constraints are introduced, one may remove the point from the three-dimensional surface by setting the values to NaN. Thus the following statements can be given to remove the function points out of the constraint regions. The solutions can then be found graphically using the following statements.

```
>> i=find(x1.^2+x2.^2>9); z(i)=NaN; %find x1^2+x2^2>9 points and assign NaN
   i=find(x1+x2>1); z(i)=NaN; %find x1+x2>1 points and assign values to NaN
   surf(x1,x2,z); shading interp; % draw the allowed surface
```

The three-dimensional surface plot can be drawn as shown in Figure 6.10 (a). If one wants to observe the three-dimensional surface from the top, then the command view(0,90) can be used and the two-dimensional projection can be obtained as shown in Figure 6.10 (b). The region shown on the graph is the *feasible region*. The maximum value in this feasible region is the solution to the constrained optimization problem. Using the graphical method, it can be found that the solution is $x_1 = 0$, $x_2 = -3$. The maximum value of the function is 3.

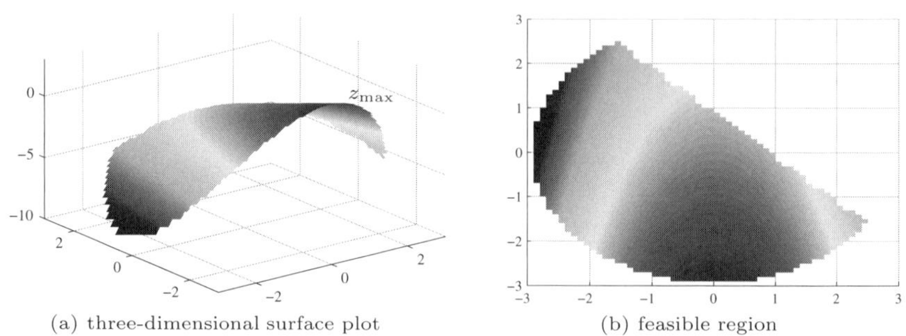

(a) three-dimensional surface plot (b) feasible region

FIGURE 6.10: Graphical solution to a 2D optimization problem

For problems with one or two variables, the graphical solution is of course the most effective and straightforward method. However, for problems with more variables, the graphical solution method cannot be used. Numerical searching methods should be used instead and, again, there is to date no method to judge whether the solutions obtained is global or not.

6.3.2 Solving linear programming problems

Linear programming problems are special constrained programming problems whose objective function and constraints are all linear with respect to $\boldsymbol{x}$. The general mathematical description is

$$\min_{\boldsymbol{x} \text{ s.t. } \begin{cases} \boldsymbol{A}\boldsymbol{x} \leqslant \boldsymbol{B} \\ \boldsymbol{A}_{\mathrm{eq}}\boldsymbol{x} = \boldsymbol{B}_{\mathrm{eq}} \\ \boldsymbol{x}_{\mathrm{m}} \leqslant \boldsymbol{x} \leqslant \boldsymbol{x}_{\mathrm{M}} \end{cases}} \boldsymbol{f}^{\mathrm{T}}\boldsymbol{x}. \tag{6.7}$$

In order to solve the problem effectively, the constraints can further be classified to linear equality constraints $\boldsymbol{A}_{\mathrm{eq}}\boldsymbol{x} = \boldsymbol{B}_{\mathrm{eq}}$ and linear inequality constraints $\boldsymbol{A}\boldsymbol{x} \leqslant \boldsymbol{B}$. Moreover, the upper- and lower-bounds $\boldsymbol{x}_{\mathrm{M}}$ and $\boldsymbol{x}_{\mathrm{m}}$ can be introduced, such that $\boldsymbol{x}_{\mathrm{m}} \leqslant \boldsymbol{x} \leqslant \boldsymbol{x}_{\mathrm{M}}$.

For inequality constraints, the standard description under MATLAB is the "$\leqslant$" relationship. If certain constraints are given by "$\geqslant$" relationship, then -1 can be multiplied to both sides of the inequality such that the inequalities can be converted to the "$\leqslant$" relationship.

Linear programming problems are the simplest and most widely used constrained optimization problems. There are many algorithms suggested for this kind of optimization problem. The simplex algorithm is proven to be the most effective algorithm and with the well established function `linprog()`, the linear programming problems can easily be solved. The syntax of the function is

```
[x,f_opt,flag,c]=linprog(f,A,B,A_eq,B_eq,x_m,x_M,x_0,OPT,p_1,p_2,···)
```

where $\boldsymbol{f}$, $\boldsymbol{A}$, $\boldsymbol{B}$, $\boldsymbol{A}_{\mathrm{eq}}$, $\boldsymbol{B}_{\mathrm{eq}}$, $\boldsymbol{x}_{\mathrm{m}}$, $\boldsymbol{x}_{\mathrm{M}}$ are the same as in the above general formulation. The argument $\boldsymbol{x}_0$ is the user selected initial search point, which can be omitted in linear programming problems. If certain constraints do not exist, replace the matrices with empty ones. The argument `OPT` is the control option to further specify the function call. In this function, additional arguments $p_1, p_2, \cdots$ can also be used to pass more information to the objective function and constraints. After the function call, the optimized results are returned in variable $\boldsymbol{x}$, and the optimum objective function is returned in f_{opt}. A positive `flag` indicates the successful function call.

Example 6.19 Solve the following linear programming problem:

$$\min_{\boldsymbol{x} \text{ s.t. } \begin{cases} 2x_2+x_3+4x_4+2x_5 \leqslant 54 \\ 3x_1+4x_2+5x_3-x_4-x_5 \leqslant 62 \\ x_1,x_2 \geqslant 0,,x_3 \geqslant 3.32,x_4 \geqslant 0.678,x_5 \geqslant 2.57 \end{cases}} (-2x_1 - x_2 - 4x_3 - 3x_4 - x_5).$$

Solution For this linear programming problem, the objective function can be defined by the vector $\boldsymbol{f}=[-2,-1,-4,-3,-1]^{\mathrm{T}}$. There are two inequality constraints, thus

$$\boldsymbol{A}=\begin{bmatrix}0 & 2 & 1 & 4 & 2\\ 3 & 4 & 5 & -1 & -1\end{bmatrix}, \text{ and } \boldsymbol{B}=\begin{bmatrix}54\\ 62\end{bmatrix}.$$

Since there are no equality constraints, the matrices $\boldsymbol{A}_{\mathrm{eq}}$ and $\boldsymbol{B}_{\mathrm{eq}}$ can be declared as empty matrices. From the original problem, it is also seen that the lower-bound of $\boldsymbol{x}$ can be defined as $\boldsymbol{x}_{\mathrm{m}}=[0,0,3.32,0.678,2.57]^{\mathrm{T}}$, and since there is no upper-bound, $\boldsymbol{x}_{\mathrm{M}}$ can be assigned to an empty vector. Thus the problem can be solved directly using the following MATLAB statements and results can be obtained immediately

```
>> f=-[2 1 4 3 1]'; A=[0 2 1 4 2; 3 4 5 -1 -1];
   B=[54; 62]; Ae=[]; Be=[]; xm=[0,0,3.32,0.678,2.57];
   ff=optimset; ff.LargeScale='off'; % turn off the large-scale algorithm
   ff.TolX=1e-15; ff.TolFun=1e-20; ff.TolCon=1e-20;
   [x,f_opt,key,c]=linprog(f,A,B,Ae,Be,xm,[],[],ff)
```

where $\boldsymbol{x}=[19.785,0,3.32,11.385,2.57]^{\mathrm{T}}$, $f_{\mathrm{opt}}=-89.5750$, and `key`=1, meaning that the solution is successful. In fact only 5 steps are used in finding the optimal solution for this problem.

Example 6.20 Consider the linear programming problem with four variables defined as follows. Solve it using the Optimization Toolbox of MATLAB.

$$\max_{\boldsymbol{x}\ \text{s.t.}\begin{cases}x_1/4-60x_2-x_3/50+9x_4\leqslant 0\\ -x_1/2+90x_2+x_5/50-3x_4\geqslant 0\\ x_3\leqslant 1,x_1\geqslant -5,x_2\geqslant -5,x_3\geqslant -5,x_4\geqslant -5\end{cases}} \left[\frac{3}{4}x_1-150x_2+\frac{1}{50}x_3-6x_4\right]$$

Solution In the original problem, the maximization is required. It should be converted first to the minimization problem by multiplying -1 to the objective function. The new objective function can be rewritten as $-3x_1/4+150x_2-x_3/50+6x_4$. From the linear programming problem formulation, it can be seen that $\boldsymbol{f}=[-3/4,150,-1/50,6]^{\mathrm{T}}$.

For the given constraints, $x_i\geqslant -5$ will lead to a lower-bound vector $\boldsymbol{x}_{\mathrm{m}}$=`[-5;-5;-5;-5]`. Similarly, the upper-bound vector $\boldsymbol{x}_{\mathrm{M}}$=`[Inf;Inf;1;Inf]`, with `Inf` for $+\infty$. Also in the two inequality constraints, the second one is given by the $\geqslant$ relationship. We should multiply both sides by -1 to convert it into the $\leqslant$ inequality. Thus the matrices for the inequalities can be written as

$$\boldsymbol{A}=\begin{bmatrix}1/4 & -60 & -1/50 & 9\\ 1/2 & -90 & -1/50 & 3\end{bmatrix}, \ \boldsymbol{B}=\begin{bmatrix}0\\ 0\end{bmatrix}.$$

Since there are no equality constraints, one should assume that $\boldsymbol{A}_{\mathrm{eq}}$=`[]`, $\boldsymbol{B}_{\mathrm{eq}}$=`[]`. Thus the original linear programming problem can be solved using the following statements

```
>> f=[-3/4,150,-1/50,6]; Aeq=[]; Beq=[];
   A=[1/4,-60,-1/50,9; 1/2,-90,-1/50,3]; B=[0;0];
   xm=[-5;-5;-5;-5]; xM=[Inf;Inf;1;Inf]; ff=optimset;
   ff.TolX=1e-15; ff.TolFun=1e-20; TolCon=1e-20;
   [x,f_opt,key,c]=linprog(f,A,B,Aeq,Beq,xm,xM,[0;0;0;0],ff)
```

and with 10 iterations, the solution $\boldsymbol{x} = [-5, -0.194667, 1, -5]^{\mathrm{T}}$ can be found. The objective function reaches its minimum at $f_{\mathrm{opt}} = -55.47$.

Example 6.21 Consider a tricky double subscripted linear programming problem given below. Find the optimal solution to the problem.

$$\min_{\boldsymbol{x}\ \text{s.t.}\begin{cases} x_{11}+x_{12}+x_{13}+x_{14}\geqslant 15 \\ x_{12}+x_{13}+x_{14}+x_{21}+x_{22}+x_{23}\geqslant 10 \\ x_{13}+x_{14}+x_{22}+x_{23}+x_{31}+x_{32}\geqslant 20 \\ x_{14}+x_{23}+x_{32}+x_{41}\geqslant 12 \\ x_{ij}\geqslant 0,\ (i=1,2,3,4, j=1,2,3,4)\end{cases}} 2800(x_{11}+x_{21}+x_{31}+x_{41})+4500(x_{12}+x_{22}+x_{32})+6000(x_{13}+x_{23})+7300x_{14}$$

Solution Since the `linprog()` function can only be used to solve single subscripted problems, the original problem should be converted by rearranging the variables such that $x_1 = x_{11}, x_2 = x_{12}, x_3 = x_{13}, x_4 = x_{14}, x_5 = x_{21}, x_6 = x_{22}, x_7 = x_{23}, x_8 = x_{31}, x_9 = x_{32}, x_{10} = x_{41}$. The original problem can then be rewritten as

$$\min_{\boldsymbol{x}\ \text{s.t.}\begin{cases} -(x_1+x_2+x_3+x_4)\leqslant -15 \\ -(x_2+x_3+x_4+x_5+x_6+x_7)\leqslant -10 \\ -(x_3+x_4+x_6+x_7+x_8+x_9)\leqslant -20 \\ -(x_4+x_7+x_9+x_{10})\leqslant -12 \\ x_i\geqslant 0,\ i=1,2,\cdots,12\end{cases}} 2800(x_1 + x_5 + x_8 + x_{10}) + 4500(x_2 + x_6 + x_9) + 6000(x_3 + x_7) + 7300x_4.$$

The following MATLAB statements can then be used in finding the solutions to the optimization problem

```
>> f=2800*[1 0 0 0 1 0 0 1 0 1]+4500*[0 1 0 0 0 1 0 0 1 0]+...
     6000*[0 0 1 0 0 0 1 0 0 0]+7300*[0 0 0 1 0 0 0 0 0 0];
   A=-[1 1 1 1 0 0 0 0 0 0; 0 1 1 1 1 1 1 0 0 0;
       0 0 1 1 0 1 1 1 1 0; 0 0 0 1 0 0 1 0 1 1];
   B=-[15; 10; 20; 12]; xm=[0 0 0 0 0 0 0 0 0 0]; Aeq=[]; Beq=[];
   x=linprog(f,A,B,Aeq,Beq,xm)
```

and it is found that $\boldsymbol{x} = [4.2069, 0, 0, 10.7931, 0, 0, 0, 8, 1.2069, 0.0000]$. Converting the solutions back to the double subscripted format, one has $x_{11} = 4.2069$, $x_{14} = 10.7931$, $x_{31} = 8$, $x_{32} = 1.2069$, and the rest of the variables are all zeros.

6.3.3 Solving quadratic programming problems

Quadratic programming problems are another category of simple constrained optimization problems. In quadratic programming problems, the objective function contains the quadratic form of vector $\boldsymbol{x}$. The constraints are still linear. The general form of quadratic programming is

$$\min_{\boldsymbol{x}\ \text{s.t.}\begin{cases}\boldsymbol{A}\boldsymbol{x}\leqslant\boldsymbol{B}\\ \boldsymbol{A}_{\mathrm{eq}}\boldsymbol{x}=\boldsymbol{B}_{\mathrm{eq}}\\ \boldsymbol{x}_{\mathrm{m}}\leqslant\boldsymbol{x}\leqslant\boldsymbol{x}_{\mathrm{M}}\end{cases}} \left(\frac{1}{2}\boldsymbol{x}^{\mathrm{T}}\boldsymbol{H}\boldsymbol{x} + \boldsymbol{f}^{\mathrm{T}}\boldsymbol{x}\right). \tag{6.8}$$

Comparing the problem in linear programming, it can be seen that a quadratic form $\boldsymbol{x}^{\mathrm{T}}\boldsymbol{H}\boldsymbol{x}$ is introduced to describe x_i^2 and $x_i x_j$ terms. The `quadprog()` function can be used to solve the quadratic programming problems, with the following syntax

$$\texttt{[}\boldsymbol{x}\texttt{,}f_{\mathrm{opt}}\texttt{,flag,c]=quadprog(}\boldsymbol{H}\texttt{,}\boldsymbol{f}\texttt{,}\boldsymbol{A}\texttt{,}\boldsymbol{B}\texttt{,}\boldsymbol{A}_{\mathrm{eq}}\texttt{,}\boldsymbol{B}_{\mathrm{eq}}\texttt{,}\boldsymbol{x}_{\mathrm{m}}\texttt{,}\boldsymbol{x}_{\mathrm{M}}\texttt{,}\boldsymbol{x}_0\texttt{,OPT,}p_1\texttt{,}p_2\texttt{,}\cdots\texttt{)}$$

where in the function call, $\boldsymbol{H}$ matrix should be declared, while the other arguments are exactly the same as those in linear programming problems.

Example 6.22 Solve the following quadratic programming problems with four variables:

$$\min_{\boldsymbol{x} \text{ s.t.} \begin{cases} x_1+x_2+x_3+x_4 \leqslant 5 \\ 3x_1+3x_2+2x_3+x_4 \leqslant 10 \\ x_1,x_2,x_3,x_4 \geqslant 0 \end{cases}} \left[(x_1-1)^2+(x_2-2)^2+(x_3-3)^2+(x_4-4)^2\right].$$

Solution In order to solve the original problem, the objective function should be rewritten such that

$$\begin{aligned} f(x) &= x_1^2-2x_1+1+x_2^2-4x_2+4+x_3^2-6x_3+9+x_4^2-8x_4+16 \\ &= x_1^2+x_2^2+x_3^2+x_4^2-2x_1-4x_2-6x_3-8x_4+30. \end{aligned}$$

Since the constant term does not affect the optimization problem solving, it can be omitted safely. Thus the matrices for quadratic programming can then be defined as $\boldsymbol{H}=\mathrm{diag}([2,2,2,2])$, and $\boldsymbol{f}^{\mathrm{T}}=[-2,-4,-6,-8]$. Thus the following statements can be used to solve this quadratic programming problem

```
>> f=[-2,-4,-6,-8]; H=diag([2,2,2,2]);
   OPT=optimset; OPT.LargeScale='off'; % turn off the large-scale algorithm
   A=[1,1,1,1; 3,3,2,1]; B=[5;10]; Aeq=[]; Beq=[]; LB=zeros(4,1);
   [x,f_opt]=quadprog(H,f,A,B,Aeq,Beq,LB,[],[],OPT)
```

with the solutions $\boldsymbol{x}=[0,0.6667,1.6667,2.6667]^{\mathrm{T}}$, $f_{\mathrm{opt}}=-23.6667$.

It should be noted that since there is a factor 1/2 in the quadratic term, the generation of matrix $\boldsymbol{H}$ should be prepared with care. Thus the diagonal terms in $\boldsymbol{H}$ should be 2's instead of 1's in this example. It should also be noted that the constant term was removed in performing the optimization computation. The constant should be added back to the optimum value such that the objective function is 6.3333.

6.3.4 Solving general nonlinear programming problems

The general nonlinear programming problems are formulated as follows:

$$\min_{\boldsymbol{x} \text{ s.t. } \boldsymbol{G}(\boldsymbol{x})\leqslant 0} f(\boldsymbol{x}) \tag{6.9}$$

where $\boldsymbol{x}=[x_1,x_2,\cdots,x_n]^{\mathrm{T}}$. For simplicity, the constraints can further be classified into linear inequalities and equalities, upper- and lower-bounds, and nonlinear equalities and inequalities. The original constraints can thus be rewritten as

$$\min_{\boldsymbol{x} \text{ s.t. } \begin{cases} \boldsymbol{A}\boldsymbol{x} \leqslant \boldsymbol{B} \\ \boldsymbol{A}_{\rm eq}\boldsymbol{x} = \boldsymbol{B}_{\rm eq} \\ \boldsymbol{x}_{\rm m} \leqslant \boldsymbol{x} \leqslant \boldsymbol{x}_{\rm M} \\ \boldsymbol{C}(\boldsymbol{x}) \leqslant 0 \\ \boldsymbol{C}_{\rm eq}(\boldsymbol{x}) = 0 \end{cases}} f(\boldsymbol{x}). \tag{6.10}$$

A MATLAB function `fmincon()` can be used to solve general nonlinear programming problems. The syntax of the function is

```
[x,f_opt,flag,c]=fmincon(fun,x0,A,B,Aeq,Beq,xm,xM,...
                         CFun,OPT,p1,p2,···)
```

where *fun* is the M-function or anonymous function to describe the objective function. The argument $\boldsymbol{x}_0$ is the initial search point. The definitions of $\boldsymbol{A}$, $\boldsymbol{B}$, $\boldsymbol{A}_{\rm eq}$, $\boldsymbol{B}_{\rm eq}$, $\boldsymbol{x}_{\rm m}$, $\boldsymbol{x}_{\rm M}$ are the same as in (6.10). The argument *CFun* is the M-function to describe the nonlinear constraints, with two returned arguments, indicating the inequality and equality constraints respectively. The argument `OPT` is the control options. The returned arguments are exactly the same as in other optimization functions in MATLAB.

Example 6.23 Solve the following nonlinear programming problem:

$$\min_{\boldsymbol{x} \text{ s.t. } \begin{cases} x_1^2+x_2^2+x_3^2-25=0 \\ 8x_1+14x_2+7x_3-56=0 \\ x_1,x_2,x_3 \geqslant 0 \end{cases}} \left[1000 - x_1^2 - 2x_2^2 - x_3^2 - x_1x_2 - x_1x_3\right].$$

Solution In this problem, there are nonlinear constraints, thus the original problem is not a quadratic programming problem. Nonlinear programming solvers should be used to solve this problem. The objective function can be expressed by an anonymous function. Moreover, since the two constraints are all equalities, thus the constraint functions can be expressed as

```
function [c,ceq]=opt_con1(x)
ceq=[x(1)*x(1)+x(2)*x(2)+x(3)*x(3)-25; 8*x(1)+14*x(2)+7*x(3)-56];
c=[];
```

where the nonlinear inequalities and nonlinear equalities are returned respectively in arguments `c` and `ceq`. Since there is no inequality constraint, the argument `c` is then assigned to an empty matrix.

Having declared the constraints, the matrices $\boldsymbol{A}$, $\boldsymbol{B}$, $\boldsymbol{A}_{\rm eq}$, $\boldsymbol{B}_{\rm eq}$ are now all empty matrices. The lower-bound vector can be written as $\boldsymbol{x}_{\rm m} = [0,0,0]^{\rm T}$. If the initial point is selected as $\boldsymbol{x}_0 = [1,1,1]^{\rm T}$, the problem can then be solved using the following statements

```
>> f=@(x)1000-x(1)*x(1)-2*x(2)*x(2)-x(3)*x(3)-x(1)*x(2)-x(1)*x(3);
   ff=optimset; ff.LargeScale='off';
   ff.TolFun=1e-30; ff.TolX=1e-15; ff.TolCon=1e-20;
   x0=[1;1;1]; xm=[0;0;0]; xM=[]; A=[]; B=[]; Aeq=[]; Beq=[];
   [x,f_opt,c,d]=fmincon(f,x0,A,B,Aeq,Beq,xm,xM,@opt_con1,ff)
```

with $\boldsymbol{x} = [3.5121, 0.2170, 3.5522]^{\mathrm{T}}$, and $f_{\mathrm{opt}} = 961.7151$. Totally 113 calls to the objective functions are made during the solution process.

Since the second constraint is in fact linear, it can be removed from the nonlinear constraint function. Thus the constraint function can be simplified as

```
function [c,ceq]=opt_con2(x)
ceq=x(1)*x(1)+x(2)*x(2)+x(3)*x(3)-25; c=[];
```

and the linear equality constraint can be declared in the following statements and the solution obtained using the following is exactly the same as the one obtained previously.

```
>> x0=[1;1;1]; Aeq=[8,14,7]; Beq=56;
   [x,f_opt,c,d]=fmincon(f,x0,A,B,Aeq,Beq,xm,xM,@opt_con2,ff)
```

Example 6.24 Consider again the optimization problem in Example 6.23. Solve the problem using gradients and compare the results with those in the original example.

Solution For the given objective function $f(x)$, the gradient vector, or the Jacobian matrix, can be derived

```
>> syms x1 x2 x3; f=1000-x1*x1-2*x2*x2-x3*x3-x1*x2-x1*x3;
   J=jacobian(f,[x1,x2,x3])
```

and the gradient matrix can be written as

$$\boldsymbol{J} = \left[\frac{\partial f}{\partial x_1}, \frac{\partial f}{\partial x_2}, \frac{\partial f}{\partial x_3}\right]^{\mathrm{T}} = \begin{bmatrix} -2x_1 - x_2 - x_3 \\ -4x_2 - x_1 \\ -2x_3 - x_1 \end{bmatrix}.$$

With the gradients, the objective function can then be rewritten as

```
function [y,Gy]=opt_fun2(x)
y=1000-x(1)*x(1)-2*x(2)*x(2)-x(3)*x(3)-x(1)*x(2)-x(1)*x(3);
Gy=[-2*x(1)-x(2)-x(3); -4*x(2)-x(1); -2*x(3)-x(1)];
```

where Gy returns the gradient vector of the objective function. The following statements

```
>> x0=[1;1;1]; xm=[0;0;0]; xM=[]; A=[]; B=[]; Aeq=[]; Beq=[];
   ff=optimset; ff.GradObj='on'; ff.LargeScale='off';
   ff.TolFun=1e-30; ff.TolX=1e-15; ff.TolCon=1e-20;
   [x,f_opt,c,d]=fmincon(@opt_fun2,x0,A,B,Aeq,Beq,xm,xM,@opt_con1,ff)
```

find the solution $\boldsymbol{x} = [3.5121, 0.2170, 3.5522]^{\mathrm{T}}$, and $f_{\mathrm{opt}} = 961.7151$. In total, 86 calls to the objective functions are made during the solution process.

For this example, if the gradients are used, 86 iterations are needed which are fewer than the one without gradients, where 113 iterations are needed. However, taking into account the time needed in deriving and coding the gradients, the actual time required may be much greater. Thus sometimes it may not be worthwhile to use gradient information for certain applications.

6.4 Mixed Integer Programming Problems

In many applications, all or part of the decision variables may be required to be integers for the optimization problems, which are referred to as the *integer programming problems*. If only part of the decision variables are required to be integers, they are referred to as *mixed integer programming problems*. In some applications, the allowable values for the decision variables are either 0's or 1's, and are referred to as *binary programming problems*, also known as the *0-1 programming problems*.

6.4.1 Solving mixed integer programming problems

One of the most frequently used algorithms for mixed integer programming problems is the *branch-and-bound algorithm*. Details of the algorithm is omitted and an existing MATLAB function `bnb20()`, developed by Koert Kuipers of Groningen University in The Netherlands, can be used directly. The function can be downloaded freely from The MathWorks file-exchange website. The toolbox is also provided with the companion CD. The syntax of the function is

`[err,`f`,`$\boldsymbol{x}$`]=bnb20(`*fun*`,`$\boldsymbol{x}_0$`,intlist,`$\boldsymbol{x}_{\rm m}$`,`$\boldsymbol{x}_{\rm M}$`,`$\boldsymbol{A}$`,`$\boldsymbol{B}$`,`$\boldsymbol{A}_{\rm eq}$`,`$\boldsymbol{B}_{\rm eq}$`,CFun)`

where most of the arguments in the function call are the same as in other optimization functions. The function `fmincon()` is called by the `bnb20()` function. The variables $\boldsymbol{x}$ and f are returned respectively and if `err` is empty, the results obtained are correct. The argument `intlist` is crucial in addressing the mixed integer programming problems, where `intlist(`i`)=1` is used to indicate that the variable x_i should be an integer, and 0 for a real number.

There are some limitations in the function, for example, anonymous functions cannot be used, and those terms required as integers should also be specified as integers in the corresponding $\boldsymbol{x}_{\rm m}$ and $\boldsymbol{x}_{\rm M}$ terms. The returned variables $\boldsymbol{x}$ may not be exactly integers. The following statements should be used before and after the function call to truncate the unwanted digits to find the integer solutions

Before `ix=(intlist==1); xm(ix)=ceil(xm(ix)); xM(ix)=floor(xM(ix));`

After `if length(err)==0, x(ix)=round(x(ix)), end`

Example 6.25 Consider again the linear programming problem in Example 6.19. If additionally x_i are required to be integers, then the original problem becomes an integer linear programming problem. Solve such an integer programming problem.

Solution An M-function can be written to describe the objective function

```
function f=c6miopt(x)
f=-[2 1 4 3 1]*x;
```

Since all the x_i's are expected to be integers, all the elements in the `intlist` vector should be set to 1's. Also the upper-bound can no longer be set to `Inf`, and finite values should be assigned instead. For instance, the upper-bounds can be assigned to 20000. The following statements can then be used

```
>> A=[0 2 1 4 2; 3 4 5 -1 -1]; intlist=ones(5,1); Aeq=[]; Beq=[];
   B=[54; 62]; xm=[0,0,3.32,0.678,2.57]'; % original lower-bounds
   ix=(intlist==1); xm(ix)=ceil(xm(ix)); xM=20000*ones(5,1); x0=xm;
   [errmsg,f,X]=bnb20('c6miopt',x0,intlist,xm,xM,A,B,Aeq,Beq);
   if length(errmsg)==0, X(ix)=round(X(ix)), end
```

and the solution obtained is $\boldsymbol{x} = [19, 0, 4, 10, 5]^{\mathrm{T}}$.

For small-scale problems, the enumeration method can be used. One may assume that the maximum values of $\boldsymbol{x}_{\mathrm{M}}$ are 20's. The following statements can be used to enumerate all the possible combinations of the x_i's and for all the feasible combinations, the objective functions are sorted to find the minimum solutions

```
>> [x1,x2,x3,x4,x5]=ndgrid(1:20,0:20,4:20,1:20,3:20);
   i=find((2*x2+x3+4*x4+2*x5<=54) & (3*x1+4*x2+5*x3-x4-x5<=62));
   f=-2*x1(i)-x2(i)-4*x3(i)-3*x4(i)-x5(i); [fmin,ii]=sort(f);
   ind=i(ii(1)); x=[x1(ind),x2(ind),x3(ind),x4(ind),x5(ind)]
```

and the solution is $\boldsymbol{x} = [19, 0, 4, 10, 5]^{\mathrm{T}}$, which is the same as the one obtained above. We make the following remarks on this example:

(i) The global optimum is found within the intervals $x_i \in [0, 20]$ and cannot be extended to larger intervals. If one wants to modify the value of 20 to 30, there will be significant increase in both computational effort and memory allocations. For instance, the memory required for the five variables x_i will be $31^5 \times 5 \times 8/2^{20} = 1092.1$MB. Thus enumerate methods are not suitable to this kind of problem.

(ii) Apart from the optimum solutions, there are other combinations of $\boldsymbol{x}$ which may have slightly larger objective functions than the optimum one. These combinations can be referred to as the *sub-optimum solutions.*

For instance, the sub-optimum solutions to the example are found as

```
>> fmin(1:10)'
   in=i(ii(1:15)); x=[x1(in),x2(in),x3(in),x4(in),x5(in),fmin(1:15)]
```

and the top 10 objective functions are $-89, -88, -88, -88, -88, -88, -88, -88, -87, -87$. Thus apart from the optimum solution at -89, there are several others that can achieve the objective function -88, and some others with values of -87. The sub-optimum solutions are only solvable by enumerate methods and not applicable to search methods. The sub-optimum results can be obtained in Table 6.2.

If x_1, x_4, x_5 are constrained to be integers while the other two variables can be arbitrarily chosen, then the original problem can be converted into a mixed integer programming problem. The argument `intlist` should be modified accordingly and the following statements can then be used to solve this new example problem

```
>> intlist=[1,0,0,1,1]'; ix=(intlist==1);
   [errmsg,f,X]=bnb20('c6miopt',x0,intlist,xm,xM,A,B,Aeq,Beq);
   if length(errmsg)==0, X(ix)=round(X(ix)), end;
```

TABLE 6.2: Optimum and sub-optimum solutions

x_1	x_2	x_3	x_4	x_5	f	x_1	x_2	x_3	x_4	x_5	f	x_1	x_2	x_3	x_4	x_5	f
19	0	4	10	5	−89	19	0	4	9	7	−88	11	0	8	10	3	−87
18	0	4	11	3	−88	16	0	6	8	8	−88	10	0	9	9	4	−87
17	0	5	10	4	−88	20	0	4	7	11	−88	8	0	10	9	4	−87
15	0	6	10	4	−88	15	0	6	10	3	−87	5	0	12	8	5	−87
12	0	8	9	5	−88	13	0	7	10	3	−87	18	0	4	10	5	−87

where $\boldsymbol{x} = [19, 0, 3.8, 11, 3]^{\mathrm{T}}$.

Example 6.26 The well established Rosenbrock function can be modified slightly such that $f(\boldsymbol{x}) = 100(x_2 - x_1^2)^2 - (4.5543 - x_2)^2$. Solve for the integers x_1 and x_2, such that

$$\min_{\boldsymbol{x} \text{ s.t. } \begin{cases} -100 \leqslant x_1 \leqslant 100 \\ -100 \leqslant x_2 \leqslant 100 \end{cases}} f(\boldsymbol{x}).$$

Solution The following M-function can be written to describe the objective function

```
function y=c6fun1(x)
y=100*(x(2)-x(1)^2)^2+(4.5543-x(1))^2;
```

The function `bnb20()` can then be called to solve the problem. It should be noted that the anonymous functions are not supported in `bnb20()` function. The original problem can then be solved using the following MATLAB statements

```
>> x0=[1;1]; xm=-100*[1;1]; xM=100*[1;1];
   A=[]; B=[]; Aeq=[]; Beq=[]; intlist=[1,1]';
   [errmsg,f,x]=bnb20('c6fun1',x0,intlist,xm,xM,A,B,Aeq,Beq);
   if length(errmsg)==0, x=round(x), end
```

with the solutions $\boldsymbol{x} = [5, 25]^{\mathrm{T}}$.

It can be seen that during the computation, there exists a small error in x_2. The converting statements should be used to remove the minor errors. It should also be noted that the search interval is often very important. If it is not correctly chosen, for instance if $x_{1,2} \in [-20, 20]$ is chosen, the following statement may fail to find the global solution

```
>> xm=-20*[1;1]; xM=20*[1;1];
   [errmsg,f,x]=bnb20('c6fun1',x0,intlist,xm,xM,A,B,Aeq,Beq);
   if length(errmsg)==0, x=round(x), end
```

which is $\boldsymbol{x} = [4, 16]^{\mathrm{T}}$.

In fact, for such a small-scale problem, the enumeration method can be used instead to find the global optimum. For instance, if larger intervals are searched such that $x_{1,2} \in (-200, 200)$, the enumeration method can be used to find the solution

```
>> [x1,x2]=meshgrid(-200:200); f=100*(x2-x1.^2).^2+(4.5543-x1).^2;
   [fmin,i]=sort(f(:));   x=[x1(i(1)),x2(i(1))]
```

which is $\boldsymbol{x} = [5, 25]^{\mathrm{T}}$. Clearly, the result is exactly the same as those by using other methods.

6.4.2 Solving binary programming problems

The so-called binary programming is the optimization problem, where the decision variables x_i to be optimized are 0's and 1's. It seems quite easy to solve the binary programming problems, since one may try to substitute all the possible combinations of each variable and the optimum point can be found by comparing the calculated objective functions. In fact, this brutal force method is only applicable to small-scale problems. For large-scale problems, for instance if there are n variables to be optimized, the size of the possible combinations will be $2^n n$ which might be prohibitive to run on an average computer. Thus specially design search methods should be used instead.

A new function `bintprog()` provided in MATLAB 7.0 can be used to solve binary linear programming problems. However, this function is not applicable for binary nonlinear programming problems. The syntax of the function is $\boldsymbol{x}$=`bintprog(`$\boldsymbol{f}$,$\boldsymbol{A}$,$\boldsymbol{B}$,$\boldsymbol{A}_{\mathrm{eq}}$,$\boldsymbol{B}_{\mathrm{eq}}$`)`.

Example 6.27 Solve the following binary linear programming problem.

$$\min_{\boldsymbol{x} \text{ s.t.} \begin{cases} x_1+2x_2-x_3\leqslant 2 \\ x_1+4x_2+x_3\leqslant 4 \\ x_1+x_2\leqslant 3 \\ 4x_2+x_3\leqslant 6 \end{cases}} (-3x_1 + 2x_2 + 5x_3).$$

Solution From the linear programming model, the required vectors and matrices $\boldsymbol{f}$, $\boldsymbol{A}$ and $\boldsymbol{B}$ can easily be constructed and the binary linear programming problem can be solved from

```
>> f=[-3,2,-5]; A=[1 2 -1; 1 4 1; 1 1 0; 0 4 1]; B=[2;4;5;6];
   x=bintprog(f,A,B,[],[])
```

and the solution is $\boldsymbol{x} = [1, 0, 1]^{\mathrm{T}}$.

For such a small-scale problem, the enumeration method can be used to test for all possible combinations of the variables whether the constraints are satisfied. Then by simple sorting to the feasible combinations, according to the values of the objective function, the optimum solution can be found. In this case, the global optimal solution can be obtained

```
>> [x1,x2,x3]=meshgrid([0,1]);
   i=find((x1+2*x2-x3<=2)&(x1+4*x2+x3<=4)&(x1+x2<=3)&(4*x1+x3<=6));
   f=-3*x1(i)+2*x2(i)-5*x3(i); [fmin,ii]=sort(f);
   index=i(ii(1)); x=[x1(index),x2(index),x3(index)]
```

with $\boldsymbol{x} = [1\ 0\ 1]^{\mathrm{T}}$. Moreover, all feasible solutions to the problem can be found using the following statements, which were not possible to be found using other methods

```
>> x=[x1(i(ii)),x2(i(ii)),x3(i(ii))]; [x fmin]
```

where $\boldsymbol{x}_1 = [1,0,1]^{\mathrm{T}}, f(\boldsymbol{x}_1) = -8, \boldsymbol{x}_2 = [0,0,1]^{\mathrm{T}}, f(\boldsymbol{x}_2) = -5, \boldsymbol{x}_3 = [1,0,0]^{\mathrm{T}}, f(\boldsymbol{x}_3) = -3, \boldsymbol{x}_4 = [0,0,0]^{\mathrm{T}}, f(\boldsymbol{x}_4) = 0, \boldsymbol{x}_5 = [0,1,0]^{\mathrm{T}}, f(\boldsymbol{x}_5) = 2$.

Direct calls to `bnb20()` functions, with the lower- and upper-bounds $\boldsymbol{x}_{\mathrm{m}}$, $\boldsymbol{x}_{\mathrm{M}}$ set to zeros and ones vectors respectively, can also solve the binary programming problems.

Example 6.28 Solve the binary linear programming problem in Example 6.27 with the `bnb20()` function.

Solution A MATLAB function can be written for the objective function

```
function y=c6milp(x)
y=[-3,2,-5]*x;
```

When the lower- and upper-bounds $\boldsymbol{x}_{\mathrm{m}}$, $\boldsymbol{x}_{\mathrm{M}}$ set to zeros and ones vectors respectively, the binary programming problem can also be solved for the example using the following statements

```
>> xm=[0;0;0]; xM=[1;1;1]; intlist=[1;1;1]; x0=xm;
   A=[1 2 -1; 1 4 1; 1 1 0; 0 4 1]; B=[2;4;5;6];
   [err,f,x]=bnb20('c6milp',x0,intlist,xm,xM,A,B)
```

where $\boldsymbol{x} = [1,0,1]^{\mathrm{T}}$ and $f = -8$. The results are exactly the same as the one in the previous example. In fact, the last two constraints are redundant and can be removed.

Since the Optimization Toolbox of MATLAB is usually restricted to the ordinary optimization problems and not applicable to integer programming problems and the free toolboxes sometimes may fail to solve certain problems, a third party product, the TOMLAB Toolbox[1] of the TOMLAB Optimization Inc., can be used to better solve mixed integer programming problems.

6.5 Linear Matrix Inequalities

The theory of linear matrix inequalities (LMI) has been attracting the attention of research communities for a decade especially from researchers in the control systems community[16]. The concept of LMI and its applications are based on the fact that LMIs can be reduced to linear programming problems which can easily be solved by computers[19].

In this section, the concepts of LMI will be presented first, followed by the MATLAB solution examples using the Robust Control Toolbox, as well as a free YALMIP Toolbox.

[1]A trial version can be downloaded from `http://tomopt.com/`. It can be evaluated by the users free for 21 days.

6.5.1 A general introduction to LMIs

Linear matrix inequalities can be generally described as

$$\boldsymbol{F}(\boldsymbol{x}) = \boldsymbol{F}_0 + x_1\boldsymbol{F}_1 + \cdots + x_m\boldsymbol{F}_m < 0 \tag{6.11}$$

where $\boldsymbol{x} = [x_1, \cdots, x_m]^{\mathrm{T}}$ is the coefficient vector of a polynomial. It is also referred to as a *decision vector*. The matrices $\boldsymbol{F}_i$ are Hermitian matrices. If the LMI matrix $\boldsymbol{F}(\boldsymbol{x})$ is a negative-definite matrix, the solution set is convex, i.e.,

$$\boldsymbol{F}[\alpha\boldsymbol{x}_1 + (1-\alpha)\boldsymbol{x}_2] = \alpha\boldsymbol{F}(\boldsymbol{x}_1) + (1-\alpha)\boldsymbol{F}(\boldsymbol{x}_2) < 0 \tag{6.12}$$

where $\alpha > 0, 1-\alpha > 0$. The solution is also referred to as the *feasible solution*. From two such LMIs $\boldsymbol{F}_1(\boldsymbol{x}) < 0$ and $\boldsymbol{F}_2(\boldsymbol{x}) < 0$, a single LMI can be constructed such that

$$\begin{bmatrix} \boldsymbol{F}_1(\boldsymbol{x}) & 0 \\ 0 & \boldsymbol{F}_2(\boldsymbol{x}) \end{bmatrix} < 0. \tag{6.13}$$

It can be seen that several LMIs $\boldsymbol{F}_i(\boldsymbol{x}) < 0, i = 1, 2, \cdots, k$ can be combined into a single LMI such that $\boldsymbol{F}(\boldsymbol{x}) < 0$, where

$$\boldsymbol{F}(\boldsymbol{x}) = \begin{bmatrix} \boldsymbol{F}_1(\boldsymbol{x}) & & & \\ & \boldsymbol{F}_2(\boldsymbol{x}) & & \\ & & \ddots & \\ & & & \boldsymbol{F}_k(\boldsymbol{x}) \end{bmatrix} < 0. \tag{6.14}$$

6.5.2 Lyapunov inequalities

Consider first the Lyapunov stability problem. Lyapunov theory states that for a given positive-definite matrix $\boldsymbol{Q}$, if the Lyapunov equation

$$\boldsymbol{A}^{\mathrm{T}}\boldsymbol{X} + \boldsymbol{X}\boldsymbol{A} = -\boldsymbol{Q} \tag{6.15}$$

has positive-definite solution $\boldsymbol{X}$, the matrix $\boldsymbol{A}$ is stable, i.e., all the eigenvalues of the matrix are located in the left-hand-side of the complex plane. The previous equation can also be converted into a Lyapunov inequality.

$$\boldsymbol{A}^{\mathrm{T}}\boldsymbol{X} + \boldsymbol{X}\boldsymbol{A} < 0 \tag{6.16}$$

Since $\boldsymbol{X}$ is a symmetrical matrix, a vector $\boldsymbol{x}$ containing the $n(n+1)/2$ elements can be used to describe the original matrix such that

$$x_i = X_{i,1},\ \ i = 1, \cdots, n,\ x_{n+i} = X_{i,2},\ i = 2, \cdots, n, \cdots \tag{6.17}$$

and it follows that

$$x_{(2n-j+2)(j-1)/2+i} = X_{i,j},\ \ j = 1, 2, \cdots, n,\ \ i = j, j+1, \cdots, n. \tag{6.18}$$

A MATLAB function can be created for the above conversion as follows:

```
function F=lyap2lmi(A0)
if prod(size(A0))==1, n=A0;
   for i=1:n, for j=1:n,
      i1=int2str(i);j1=int2str(j); eval(['syms a' i1 j1]),
      eval(['A(' i1 ',' j1 ')=a' i1 j1,';'])
   end, end
else, n=size(A0,1); A=A0; end
vec=0; for i=1:n, vec(i+1)=vec(i)+n-i+1; end
for k=1:n*(n+1)/2,  X=zeros(n);
   i=find(vec>=k); i=i(1)-1; j=i+k-vec(i)-1;
   X(i,j)=1; X(j,i)=1; F(:,:,k)=A.'*X+X*A;
end
```

The function can be called by `F=lyap2lmi(A)`, where $\boldsymbol{A}$ can be a double-precision matrix. Note that if $\boldsymbol{A}$ is simply an integer, it indicates a square symbolic matrix to be established. The returned argument $\boldsymbol{F}$ is a three-dimensional array and $\boldsymbol{F}(:,:,i)$ is the $\boldsymbol{F}_i$ matrix.

Example 6.29 If in the Lyapunov inequality $\boldsymbol{A}=\begin{bmatrix}1&2&3\\4&5&6\\7&8&0\end{bmatrix}$, find its LMI representation. For a 3×3 matrix $\boldsymbol{A}$, display its LMI form.

Solution The matrix $\boldsymbol{A}$ is entered by the statements

```
>> A=[1,2,3; 4,5,6; 7,8,0]; F=lyap2lmi(A)
```

and it can be found that the $\boldsymbol{F}_i$ matrices are

$$x_1\begin{bmatrix}2&2&3\\2&0&0\\3&0&0\end{bmatrix}+x_2\begin{bmatrix}8&6&6\\6&4&3\\6&3&0\end{bmatrix}+x_3\begin{bmatrix}14&8&1\\8&0&2\\1&2&6\end{bmatrix}$$
$$+x_4\begin{bmatrix}0&4&0\\4&10&6\\0&6&0\end{bmatrix}+x_5\begin{bmatrix}0&7&4\\7&16&5\\4&5&12\end{bmatrix}+x_6\begin{bmatrix}0&0&7\\0&0&8\\7&8&0\end{bmatrix}<0.$$

For a symbolic 3×3 matrix, the following statement can be given instead

```
>> F=lyap2lmi(3)
```

such that the LMI can be written as

$$x_1\begin{bmatrix}2a_{11}&a_{12}&a_{13}\\a_{12}&0&0\\a_{13}&0&0\end{bmatrix}+x_2\begin{bmatrix}2a_{21}&a_{22}+a_{11}&a_{23}\\a_{22}+a_{11}&2a_{12}&a_{13}\\a_{23}&a_{13}&0\end{bmatrix}+x_3\begin{bmatrix}2a_{31}&a_{32}&a_{33}+a_{11}\\a_{32}&0&a_{12}\\a_{33}+a_{11}&a_{12}&2a_{13}\end{bmatrix}$$
$$+x_4\begin{bmatrix}0&a_{21}&0\\a_{21}&2a_{22}&a_{23}\\0&a_{23}&0\end{bmatrix}+x_5\begin{bmatrix}0&a_{31}&a_{21}\\a_{31}&2a_{32}&a_{33}+a_{22}\\a_{21}&a_{33}+a_{22}&2a_{23}\end{bmatrix}+x_6\begin{bmatrix}0&0&a_{31}\\0&0&a_{32}\\a_{31}&a_{32}&2a_{33}\end{bmatrix}<0.$$

Some nonlinear inequalities can be converted into LMIs, too. For instance, for a partitioned matrix $\boldsymbol{F}(\boldsymbol{x})=\left[\begin{array}{c:c}\boldsymbol{F}_{11}(\boldsymbol{x})&\boldsymbol{F}_{12}(\boldsymbol{x})\\\hdashline\boldsymbol{F}_{21}(\boldsymbol{x})&\boldsymbol{F}_{22}(\boldsymbol{x})\end{array}\right]$, if $\boldsymbol{F}_{11}(\boldsymbol{x})$ is a square matrix, the following three cases are equivalent:

$$F(x) < 0 \tag{6.19}$$

$$F_{11}(x) < 0, \quad F_{22}(x) - F_{21}(x)F_{11}^{-1}(x)F_{12}(x) < 0 \tag{6.20}$$

$$F_{22}(x) < 0, \quad F_{11}(x) - F_{12}(x)F_{22}^{-1}(x)F_{21}(x) < 0. \tag{6.21}$$

The above property is known as *Schur complement*.

Consider an algebraic Riccati inequality

$$A^{\mathrm{T}}X + XA + (XB - C)R^{-1}(XB - C^{\mathrm{T}})^{\mathrm{T}} < 0 \tag{6.22}$$

where $R = R^{\mathrm{T}} > 0$. Due to its quadratic term, it is not an LMI. However, with Schur complement, the original nonlinear inequality can be converted equivalently into the following LMIs

$$X > 0, \quad \left[\begin{array}{c|c} A^{\mathrm{T}}X + XA & XB - C^{\mathrm{T}} \\ \hline B^{\mathrm{T}}X - C & -R \end{array}\right] < 0. \tag{6.23}$$

6.5.3 Classification of LMI problems

LMI problems can be classified into three typical problems, i.e., the feasible solution problems, linear objective function minimization problems and the generalized eigenvalue problems.

(i) **Feasible solution problem** The so-called *feasible solution problem* is in fact the feasible region problem in optimization, i.e., for the inequality

$$F(x) < 0 \tag{6.24}$$

find a feasible solution. The feasible solution problem is to find the solution $F(x) < t_{\min}I$, where the minimum $t_{\min}$ is to be found. If $t_{\min} < 0$ can be found, there exist solutions to the original problem, otherwise there is no feasible solution.

(ii) **Linear objective function minimization problems** Consider the problem

$$\min_{x \text{ s.t. } F(x)<0} c^{\mathrm{T}}x. \tag{6.25}$$

Since the constraints are given as LMIs and the objective function is also linear, this problem can also be solved using ordinary linear programming methods.

(iii) **The generalized eigenvalue problems** The generalized eigenvalue problem is most commonly seen in LMI optimizations. Recall the generalized eigenvalue problem in Chapter 4, which is expressed as $Ax = \lambda Bx$. Such a problem can be expressed by general matrix functions as

$A(x) < \lambda B(x)$, and λ can be regarded as the generalized eigenvalue. Thus the optimization problem becomes

$$\min_{\lambda, x \text{ s.t. } \begin{cases} A(x)<\lambda B(x) \\ B(x)>0 \\ C(x)<0 \end{cases}} \lambda \tag{6.26}$$

Other constraints can be written as $C(x) < 0$. The generalized eigenvalue problem can be expressed as a special LMI problem.

6.5.4 LMI problem solutions with MATLAB

The LMI solver in MATLAB is currently provided in the Robust Control Toolbox. However, the way of describing the LMIs are quite complicated. An example is presented to show in detail the uses of the LMI solver.

The following procedures are used to describe LMIs in MATLAB:

(i) **Create an LMI model** An LMI framework can be established with `setlmis([])` function. So, a framework can be established in MATLAB workspace.

(ii) **Define the decision variables** The decision variables can be declared by `lmivar()` function, with `P=lmivar(key,[n1,n2])`, where `key` specifies the type of the decision matrix, with `key`=2 for an ordinary $n_1 \times n_2$ matrix P, while `key`=1 for an $n_1 \times n_1$ symmetrical matrix. If `key`=1, n_1 and n_2 are both vectors, P is a block diagonal symmetrical matrix. If `key`=3, P is a special matrix, which is not discussed in this book. The interested readers may refer to the Robust Control Toolbox manual[20].

(iii) **Describe LMIs in partitioned form** The LMIs can be described by the `lmiterm()` function and its syntax is quite complicated

```
lmiterm([k,i,j,P],A,B,flag)
```

where k is the number of the LMIs. Since an LMI problem may be described by several LMIs, one should number each of them. If an LMI is given $G_k(x) > 0$, then k should be described by $-k$. A term in a block in the partitioned matrix can be described by `lmiterm()` function, with i, j representing respectively the row and column numbers of the block. P is the declared decision variables, and the matrices A, B indicate the matrices in the term APB. If `flag` is assigned as `'s'`, the symmetrical term $APB + (APB)^{\mathrm{T}}$ is specified. If the whole term is a constant matrix, P is set to 0, and matrix B is omitted.

(iv) **Confirm the LMI model** After all the LMIs are declared by the `lmiterm()` function, `G=getlmis` can be used to confirm the LMI model G.

(v) **Solve the LMI problem** For the declared G model, the LMI optimization problems can be solved in one of the following three forms

```
[t_min,x]=feasp(G,options,target)              % feasible solution
[c_opt,x]=mincx(G,c,options,x_0,target)         % linear objective function
[λ,x]=gevp(G,nlfc,options,λ_0,x_0,target)       % generalized eigenvalues
```

The solution $\boldsymbol{x}$ thus obtained is a vector, and the `dec2mat()` function can be used to extract the matrix. The control variable `options` is defined as a 5-element vector, whose first element declares the precision requirement, with its default value of 10^{-5}.

Example 6.30 For the Riccati inequality $\boldsymbol{A}^{\mathrm{T}}\boldsymbol{X}+\boldsymbol{X}\boldsymbol{A}+\boldsymbol{X}\boldsymbol{B}\boldsymbol{R}^{-1}\boldsymbol{B}^{\mathrm{T}}\boldsymbol{X}+\boldsymbol{Q}<0$, where

$$\boldsymbol{A}=\begin{bmatrix}-2 & -2 & -1\\ -3 & -1 & -1\\ 1 & 0 & -4\end{bmatrix},\ \boldsymbol{B}=\begin{bmatrix}-1 & 0\\ 0 & -1\\ -1 & -1\end{bmatrix},\ \boldsymbol{Q}=\begin{bmatrix}-2 & 1 & -2\\ 1 & -2 & -4\\ -2 & -4 & -2\end{bmatrix},\ \boldsymbol{R}=\boldsymbol{I}_2$$

find a feasible positive-definite solution $\boldsymbol{X}$.

Solution The original nonlinear matrix inequality is obvious not an LMI. Using the Schur complement, this Riccati inequality can be expressed by a partitioned LMI as follows. Also since a positive-definite solution is expected, the second LMI can be established

$$\left[\begin{array}{c:c}\boldsymbol{A}^{\mathrm{T}}\boldsymbol{X}+\boldsymbol{X}\boldsymbol{A}+\boldsymbol{Q} & \boldsymbol{X}\boldsymbol{B}\\ \hdashline \boldsymbol{B}^{\mathrm{T}}\boldsymbol{X} & -\boldsymbol{R}\end{array}\right]<0,\quad \text{and}\quad \boldsymbol{X}>0.$$

One may number the Riccati inequality as no. 1, and the positive-definite inequality as no. 2. Thus one can set k to 1 and 2 respectively in using the `lmiterm()` function. It should also be noted that the matrix $\boldsymbol{X}$ is a 3×3 symmetrical matrix. Thus the feasible solution to the original problem can be obtained with the following statements. It should also be noted that since the second inequality is $\boldsymbol{X}>0$, its number should be -2 instead of 2.

```
>> A=[-2,-2,-1; -3,-1,-1; 1,0,-4]; B=[-1,0; 0,-1; -1,-1];
   Q=[-2,1,-2; 1,-2,-4; -2,-4,-2]; R=eye(2);
   setlmis([]);                    % create a blank LTI framework
   X=lmivar(1,[3 1]);              % declare X as a 3 × 3 symmetrical matrix
   lmiterm([1 1 1 X],A',1,'s')     % (1,1)th block, 's' means A^T X + XA
   lmiterm([1 1 1 0],Q)            % (1,1)th, appended by constant matrix Q
   lmiterm([1 1 2 X],1,B)          % (1,2)th, meaning XB
   lmiterm([1 2 2 0],-1)           % (2,2)th, meaning -R
   lmiterm([-2,1,1,X],1,1)         % the second inequality meaning X > 0
   G=getlmis;                      % complete the LTI framwork setting
   [tmin b]=feasp(G);              % solve the feasible problem
   X=dec2mat(G,b,X)                % extract the solution matrix X
```

It is found that $t_{\min}=-0.2427$, and the feasible solution to the original problem is

$$\boldsymbol{X}=\begin{bmatrix}1.0329 & 0.4647 & -0.23583\\ 0.4647 & 0.77896 & -0.050684\\ -0.23583 & -0.050684 & 1.4336\end{bmatrix}.$$

It is worth mentioning that, due to possible problems in the new Robust Control Toolbox, if the command `lmiterm([1 2 1 X],B',1)` is used to describe the symmetrical term in the first inequality, wrong results were found. Thus the symmetrical terms should not be described again in solving LMI problems.

6.5.5 Optimization of LMI problems by YALMIP Toolbox

The YALMIP Toolbox released by Dr. Johan Löfberg is a more flexible general purpose optimization language in MATLAB, with support also for LMI problems[21]. The description of LMI problems is much simpler and more straightforward than the ones in the Robust Control Toolbox. The YALMIP Toolbox can be downloaded for free from The MathWorks Inc.'s file-exchange site. The toolbox is also provided with the companion CD.

Decision variables can be declared in YALMIP Toolbox with `sdpvar()` function, which can be called in the following ways

```
X=sdpvar(n)              % symmetrical matrix description
X=sdpvar(n,m)            % rectangular matrix declaration
X=sdpvar(n,n,'full')     % declaration of a square asymmetrical matrix
```

The decision variables declared previously can further be treated, for instance, `hankel()` function can be applied on a decision vector to form the decision matrix in Hankel form. Similarly the functions `intvar()` and `binvar()` can be used to declare integer and binary variables respectively, thus integer programming and binary programming problems can be handled.

For `sdpvar` decision variables, the symbols `[` and `]` can be used to describe LMIs. If there are many LMIs, they can be joined together with the `,` sign, to form a single LMI representation.

An objective function, when necessary, can also be described, and the LMI optimization programs can be solved with the following syntaxes

```
s=solvesdp(F)              % find a feasible solution
s=solvesdp(F,f)            % optimization problem with objective function f
s=solvesdp(F,f,options)    % options are allowed such as algorithm selection
```

where F is the collection of constraints. After the solution, the command `X=double(X)` can be used to extract the solution matrix $\boldsymbol{X}$.

Example 6.31 With the use of the YALMIP Toolbox, the problem in Example 6.30 can be solved using simpler commands such that

```
>> A=[-2,-2,-1; -3,-1,-1; 1,0,-4]; B=[-1,0; 0,-1; -1,-1];
   Q=[-2,1,-2; 1,-2,-4; -2,-4,-2]; R=eye(2); X=sdpvar(3);
   F=[[A'*X+X*A+Q, X*B; B'*X, -R]<0, X>0];
   sol=solvesdp(F); X=double(X)
```

and it can be seen that the results are exactly the same.

Example 6.32 Solve the linear programming problem in Example 6.19 again.

Solution For simplicity, the original problem can be rewritten as

$$\min_{\boldsymbol{x} \text{ s.t.} \begin{cases} 2x_2+x_3+4x_4+2x_5 \leqslant 54 \\ 3x_1+4x_2+5x_3-x_4-x_5 \leqslant 62 \\ x_1,x_2 \geqslant 0,,x_3 \geqslant 3.32,x_4 \geqslant 0.678,x_5 \geqslant 2.57 \end{cases}} (-2x_1 - x_2 - 4x_3 - 3x_4 - x_5).$$

It is obvious that $\boldsymbol{x}$ is a 5×1 column vector, and the original problem can be solved with the following statements

```
>> x=sdpvar(5,1);
   F=[2*x(2)+x(3)+4*x(4)+2*x(5)<=54,
      3*x(1)+4*x(2)+5*x(3)-x(4)-x(5)<=62,
      x>=[0;0;3.32;0.678;2.57]];
   sol=solvesdp(F,-[2 1 4 3 1]*x); x=double(x)
```

and the solution is $\boldsymbol{x} = [19.785, 0, 3.32, 11.385, 2.57]^{\mathrm{T}}$, which is exactly the same as the one obtained in the original example. Now assuming that the decision variables are integers with the `intvar()` function, the integer linear programming problem can be solved by the following scripts:

```
>> x=intvar(5,1);
   F=[2*x(2)+x(3)+4*x(4)+2*x(5)<=54,
      3*x(1)+4*x(2)+5*x(3)-x(4)-x(5)<=62,
      x>=[0;0;3.32;0.678;2.57]];
   sol=solvesdp(F,-[2 1 4 3 1]*x); x=double(x)
```

which gives the solution $\boldsymbol{x} = [19, 0, 4, 10, 5]^{\mathrm{T}}$, the same as the one in Example 6.25.

Example 6.33 For a linear system $(\boldsymbol{A}, \boldsymbol{B}, \boldsymbol{C}, \boldsymbol{D})$, its $\mathcal{H}_\infty$ norm can directly be evaluated by `norm()` function. The norm evaluation problem can also be posed into the following LMI framework

$$\min_{\gamma,\ \boldsymbol{P} \text{ s.t.} \begin{cases} \begin{bmatrix} \boldsymbol{A}^{\mathrm{T}}\boldsymbol{P} + \boldsymbol{P}\boldsymbol{A} & \boldsymbol{P}\boldsymbol{B} & \boldsymbol{C}^{\mathrm{T}} \\ \boldsymbol{B}^{\mathrm{T}}\boldsymbol{P} & -\gamma\boldsymbol{I} & \boldsymbol{D}^{\mathrm{T}} \\ \boldsymbol{C} & \boldsymbol{D} & -\gamma\boldsymbol{I} \end{bmatrix} < 0 \\ \boldsymbol{P} > 0 \end{cases}} \gamma. \tag{6.27}$$

Find the $\mathcal{H}_\infty$ norm for the system

$$\boldsymbol{A} = \begin{bmatrix} -4 & -3 & 0 & -1 \\ -3 & -7 & 0 & -3 \\ 0 & 0 & -13 & -1 \\ -1 & -3 & -1 & -10 \end{bmatrix}, \ \boldsymbol{B} = \begin{bmatrix} 0 \\ -4 \\ 2 \\ 5 \end{bmatrix}, \ \boldsymbol{C} = [\,0,\ 0,\ 4,\ 0\,],\ D = 0.$$

Solution With the YALMIP Toolbox, the $\mathcal{H}_\infty$ norm is computed to 0.4640, which is quite close to the value obtained by the `norm()` function.

```
>> A=[-4,-3,0,-1; -3,-7,0,-3; 0,0,-13,-1; -1,-3,-1,-10];
   B=[0; -4; 2; 5]; C=[0,0,4,0]; D=0; gam=sdpvar(1); P=sdpvar(4);
   F=[[A*P+P*A',P*B,C'; B'*P,-gam,D'; C,D,-gam]<0, P>0];
   sol=solvesdp(F,gam); double(gam), norm(ss(A,B,C,D),'inf')
```

Exercises

1. Find the solutions to the following equations, and verify the accuracy of the solutions.

(i) $\begin{cases} x_1^2 - x_2 - 1 = 0 \\ (x_1 - 2)^2 + (x_2 - 0.5)^2 - 1 = 0 \end{cases}$ (ii) $\begin{cases} x^2y^2 - zxy - 4x^2yz^2 = xz^2 \\ xy^3 - 2yz^2 = 3x^3z^2 + 4xzy^2 \\ y^2x - 7xy^2 + 3xz^2 = x^4zy \end{cases}$

2. Solve graphically the following equations, and verify the results.

(i) $\mathrm{e}^{-(x+1)^2+\pi/2}\sin(5x+2) = 0$ (ii) $(x^2 + y^2 + xy)\mathrm{e}^{-x^2-y^2-xy} = 0$.

3. Find by numerical methods the solutions to the above problems, and verify the results.

4. Find c such that the integral $\int_0^1 (\mathrm{e}^x - cx)^2 \,\mathrm{d}x$ is minimized.

5. Find all the solutions to the following modified Riccati equation, and verify the results:

$$\boldsymbol{AX} + \boldsymbol{XD} - \boldsymbol{XBX} + \boldsymbol{C} = 0$$

where

$$\boldsymbol{A} = \begin{bmatrix} 2 & 1 & 9 \\ 9 & 7 & 9 \\ 6 & 5 & 3 \end{bmatrix}, \ \boldsymbol{B} = \begin{bmatrix} 0 & 3 & 6 \\ 8 & 2 & 0 \\ 8 & 2 & 8 \end{bmatrix}, \ \boldsymbol{C} = \begin{bmatrix} 7 & 0 & 3 \\ 5 & 6 & 4 \\ 1 & 4 & 4 \end{bmatrix}, \ \boldsymbol{D} = \begin{bmatrix} 3 & 9 & 5 \\ 1 & 2 & 9 \\ 3 & 3 & 0 \end{bmatrix}.$$

6. Solve the unconstrained optimization problems $\min_{\boldsymbol{x}} f(\boldsymbol{x})$, where

$$f(\boldsymbol{x}) = 100(x_2 - x_1^2)^2 + (1 - x_1)^2 + 90(x_4 - x_3^2) + (1 - x_3^2)^2 + \\ 10.1\left[(x_2 - 1)^2 + (x_4 - 1)^2\right] + 19.8(x_2 - 1)(x_4 - 1).$$

7. A set of challenging benchmark problems for evaluating optimization algorithms can be solved using MATLAB. Solve the following unconstrained optimization problems with MATLAB:

(i) De Jong's problems[22]

$$J = \min_{\boldsymbol{x}} \boldsymbol{x}^{\mathrm{T}}\boldsymbol{x} = \min_{\boldsymbol{x}}(x_1^2 + x_2^2 + \cdots + x_p^2), \quad \text{where } x_i \in [-512, 512]$$

where $i = 1, \cdots, p$, with theoretic solution $x_1 = \cdots = x_p = 0$.

(ii) Griewangk's benchmark problem

$$J = \min_{\boldsymbol{x}} \left(1 + \sum_{i=1}^{p} \frac{x_i^2}{4000} - \prod_{i=1}^{p} \cos\frac{x_i}{\sqrt{i}}\right), \quad \text{where } x_i \in [-600, 600].$$

(iii) Ackley's benchmark problem[23]

$$J = \min_{\boldsymbol{x}} \left[20 + \mathrm{e} - 20\exp\left(-0.2\sqrt{\frac{1}{p}\sum_{i=1}^{p} x_i^2}\right) - \exp\left(\frac{1}{p}\sum_{i=1}^{p}\cos 2\pi x_i\right)\right].$$

8. Consider the Rastrigin function[24]

$$f(x_1, x_2) = 20 + x_1^2 + x_2^2 - 10(\cos\pi x_1 + \cos\pi x_2).$$

3D surface plot for the objective function can be shown. An initial point can be selected from the plot, such that a good minimization to the problem can be found. Understand the dependency of the optimum points with respect to initial values.

9. Solve the nonlinear programming problem with graphical methods and verify the results using numerical methods.

$$\min_{\boldsymbol{x} \text{ s.t.} \begin{cases} x_1-x_2+2\geqslant 0 \\ -x_1^2+x_2-1\geqslant 0 \\ x_1\geqslant 0, x_2\geqslant 0 \end{cases}} (x_1^3+x_2^2-4x_1+4).$$

10. Try to solve the following linear programming problems:

(i) $$\min_{\boldsymbol{x} \text{ s.t.} \begin{cases} 4x_1-x_2+2x_3-x_4=-2 \\ x_1+x_2-x_3+2x_4\leqslant 14 \\ 2x_1-3x_2-x_3-x_4\geqslant -2 \\ x_{1,2,3}\geqslant -1,\ x_4 \text{ unconstrained} \end{cases}} -3x_1+4x_2-2x_3+5x_4$$

(ii) $$\min_{\boldsymbol{x} \text{ s.t.} \begin{cases} x_1+x_2+x_3+x_4=4 \\ -2x_1+x_2-x_3-x_6+x_7=1 \\ 3x_2+x_3+x_5+x_7=9 \\ x_{1,2,\cdots,7}\geqslant 0 \end{cases}} x_6+x_7.$$

11. Solve the following quadratic programming problems and also illustrate the solutions using graphical methods:

(i) $$\min_{\boldsymbol{x} \text{ s.t.} \begin{cases} x_1+x_2\leqslant 3 \\ 4x_1+x_2\leqslant 9 \\ x_{1,2}\geqslant 0 \end{cases}} 2x_1^2-4x_1x_2+4x_2^2-6x_1-3x_2$$

(ii) $$\min_{\boldsymbol{x} \text{ s.t.} \begin{cases} -x_1+x_2=1 \\ x_1+x_2\leqslant 2 \\ x_{1,2}\geqslant 0 \end{cases}} (x_1-1)^2+(x_2-2)^2.$$

12. Solve numerically the following nonlinear programming problems:

(i) $$\min_{\boldsymbol{x} \text{ s.t.} \begin{cases} x_1+x_2\leqslant 0 \\ -x_1x_2+x_1+x_2\geqslant 1.5 \\ x_1x_2\geqslant -10 \\ -10\leqslant x_1,x_2\leqslant 10 \end{cases}} \mathrm{e}^{x_1}(4x_1^2+2x_2^2+4x_1x_2+2x_2+1)$$

(ii) $$\min_{\boldsymbol{x} \text{ s.t.} \begin{cases} 0.003079x_1^3x_2^3x_5-\cos^3 x_6\geqslant 0 \\ 0.1017x_3^3x_4^3-x_5^2\cos^3 x_6\geqslant 0 \\ 0.09939(1+x_5)x_1^3x_2^2-\cos^2 x_6\geqslant 0 \\ 0.1076(31.5+x_5)x_3^3x_4^2-x_5^2\cos^2 x_6\geqslant 0 \\ x_3x_4(x_5+31.5)-x_5[2(x_1+5)\cos x_6+x_1x_2x_5]\geqslant 0 \\ 0.2\leqslant x_1\leqslant 0.5, 14<\leqslant x_2\leqslant 22, 0.35\leqslant x_3\leqslant 0.6, \\ 16\leqslant x_4\leqslant 22, 5.8\leqslant x_5\leqslant 6.5, 0.14\leqslant x_6\leqslant 0.2618 \end{cases}} \frac{1}{2\cos x_6}\left[x_1x_2(1+x_5)+x_3x_4\left(1+\frac{31.5}{x_5}\right)\right].$$

13. Solve the constrained optimization problem $\boldsymbol{q}$ and k[25].

(i) $\min k$

$$\underset{\boldsymbol{q},w,k}{\text{s.t.}}\begin{cases} q_3+9.625q_1w+16q_2w+16w^2+12-4q_1-q_2-78w=0 \\ 16q_1w+44-19q_1-8q_2-q_3-24w=0 \\ 2.25-0.25k\leqslant q_1\leqslant 2.25+0.25k \\ 1.5-0.5k\leqslant q_2\leqslant 1.5+0.5k \\ 1.5-1.5k\leqslant q_3\leqslant 1.5+1.5k \end{cases}$$

(ii) $\min k$

$$\underset{\boldsymbol{q},k}{\text{s.t.}}\begin{cases} g(\boldsymbol{q})\leqslant 0 \\ 800-800k\leqslant q_1\leqslant 800+800k \\ 4-2k\leqslant q_2\leqslant 4+2k \\ 6-3k\leqslant q_3\leqslant 6+3k \end{cases}$$

where

$$g(\boldsymbol{q}) = 10q_2^2q_3^3 + 10q_2^3q_3^2 + 200q_2^2q_3^2 + 100q_2^3q_3 + q_1q_2q_3^2 + q_1q_2^2q_3 + 1000q_2q_3^3 \\ + 8q_1q_3^2 + 1000q_2^2q_3 + 8q_1q_2^2 + 6q_1q_2q_3 - q_1^2 + 60q_1q_3 + 60q_1q_2 - 200q_1.$$

14. Solve the following integer linear programming problems:

(i) $\max \quad (592x_1 + 381x_2 + 273x_3 + 55x_4 + 48x_5 + 37x_6 + 23x_7)$

$$\underset{\boldsymbol{x}}{\text{s.t.}}\begin{cases} \boldsymbol{x}\geqslant 0 \\ 3534x_1+2356x_2+1767x_3+589x_4+528x_5+451x_6+304x_7\leqslant 119567 \end{cases}$$

(ii) $\max \quad (120x_1 + 66x_2 + 72x_3 + 58x_4 + 132x_5 + 104x_6).$

$$\underset{\boldsymbol{x}}{\text{s.t.}}\begin{cases} x_1+x_2+x_3=30 \\ x_4+x_5+x_6=18 \\ x_1+x_4=10 \\ x_2+x_5\leqslant 18 \\ x_3+x_6\geqslant 30 \\ x_{1,\cdots,6}\geqslant 0 \end{cases}$$

15. Solve the following binary linear programming problems and verify the results in problems (i) and (ii) using the enumeration methods.

(i) $\min \quad (5x_1 + 7x_2 + 10x_3 + 3x_4 + x_5)$

$$\underset{\boldsymbol{x}}{\text{s.t.}}\begin{cases} x_1-x_2+5x_3+x_4-4x_5\geqslant 2 \\ -2x_1+6x_2-3x_3-2x_4+2x_5\geqslant 0 \\ -2x_2+2x_3-x_4-x_5\leqslant 1 \\ 0\leqslant x_i\leqslant 1 \end{cases}$$

(ii) $\min \quad (-3x_1 - 4x_2 - 5x_3 + 4x_4 + 4x_5 + 2x_6).$

$$\underset{\boldsymbol{x}}{\text{s.t.}}\begin{cases} x_1-x_6\leqslant 0 \\ x_1-x_5\leqslant 0 \\ x_2-x_4\leqslant 0 \\ x_2-x_5\leqslant 0 \\ x_3-x_4\leqslant 0 \\ x_1+x_2+x_3\leqslant 2 \\ 0\leqslant x_i\leqslant 1 \end{cases}$$

16. Solve the following binary linear programming problem:

$$\max \quad -\boldsymbol{f}\boldsymbol{x}$$

$$\underset{\boldsymbol{x}}{\text{s.t.}}\begin{cases} [\boldsymbol{A}_1, \boldsymbol{A}_2]\boldsymbol{x} \leqslant \begin{bmatrix} 600 \\ 600 \end{bmatrix} \\ 0\leqslant x_i\leqslant 1 \end{cases}$$

where

$$\boldsymbol{A}_1 = \begin{bmatrix} 45 & 0 & 85 & 150 & 65 & 95 & 30 & 0 & 170 & 0 & 40 & 25 & 20 & 0 \\ 30 & 20 & 125 & 5 & 80 & 25 & 35 & 73 & 12 & 15 & 15 & 40 & 5 & 10 \end{bmatrix}$$

$$\boldsymbol{A}_2 = \begin{bmatrix} 0 & 25 & 0 & 0 & 25 & 0 & 165 & 0 & 85 & 0 & 0 & 0 & 0 & 100 \\ 10 & 12 & 10 & 9 & 0 & 20 & 60 & 40 & 50 & 36 & 49 & 40 & 19 & 150 \end{bmatrix}$$

$\boldsymbol{f} = [1898, 440, 22507, 270, 14148, 3100, 4650, 30800, 615, 4975, 1160, 4225, 510, 11880, 479, 440, 490, 330, 110, 560, 24355, 2885, 11748, 4550, 750, 3720, 1950, 10500]$.

17. Solve the following optimization problem using Robust Control Toolbox and YALIMP:

$$\min_{\boldsymbol{X} \text{ s.t. } \begin{cases} \begin{bmatrix} \boldsymbol{A}^{\mathrm{T}}\boldsymbol{X}+\boldsymbol{X}\boldsymbol{A}+\boldsymbol{Q} & \boldsymbol{X}\boldsymbol{B} \\ \boldsymbol{B}^{\mathrm{T}}\boldsymbol{X} & -\boldsymbol{I} \end{bmatrix} < 0 \\ \boldsymbol{X} < 0 \end{cases}} \operatorname{tr}(\boldsymbol{X})$$

where $\boldsymbol{A} = \begin{bmatrix} -1 & -2 & 1 \\ 3 & 2 & 1 \\ 1 & -2 & -1 \end{bmatrix}$, $\boldsymbol{B} = \begin{bmatrix} 1 \\ 0 \\ 1 \end{bmatrix}$, $\boldsymbol{Q} = \begin{bmatrix} 1 & -1 & 0 \\ -1 & -3 & -12 \\ 0 & -12 & -36 \end{bmatrix}$.

18. Solve the following linear matrix inequalities

$$\begin{cases} \boldsymbol{P}^{-1} > 0, \quad \text{or equavelently} \quad \boldsymbol{P} > 0 \\ \boldsymbol{A}_1\boldsymbol{P} + \boldsymbol{P}\boldsymbol{A}_1^{\mathrm{T}} + \boldsymbol{B}_1\boldsymbol{Y} + \boldsymbol{Y}^{\mathrm{T}}\boldsymbol{B}_1^{\mathrm{T}} < 0 \\ \boldsymbol{A}_2\boldsymbol{P} + \boldsymbol{P}\boldsymbol{A}_2^{\mathrm{T}} + \boldsymbol{B}_2\boldsymbol{Y} + \boldsymbol{Y}^{\mathrm{T}}\boldsymbol{B}_2^{\mathrm{T}} < 0 \end{cases}$$

where

$$\boldsymbol{A}_1 = \begin{bmatrix} -1 & 2 & -2 \\ -1 & -2 & 1 \\ -1 & -1 & 0 \end{bmatrix}, \ \boldsymbol{B}_1 = \begin{bmatrix} -2 \\ 1 \\ -1 \end{bmatrix}, \ \boldsymbol{A}_2 = \begin{bmatrix} 0 & 2 & 2 \\ 2 & 0 & 2 \\ 2 & 0 & 1 \end{bmatrix}, \ \boldsymbol{B}_2 = \begin{bmatrix} -1 \\ -2 \\ -1 \end{bmatrix}.$$

Chapter 7

Differential Equation Problems

Differential equations include ordinary differential equations (ODEs) and partial differential equations (PDEs). It is interesting to note that the phrase "extraordinary differential equations" is sometimes used to indicate differential equations of non-integer orders based on fractional calculus, which is covered in Section 10.6 of the last chapter. In this chapter, we focus on integer order differential equations.

Before Isaac Newton and Gottfried Wilhelm Leibniz invented calculus, people characterized the nature simply by geometry and algebra. With calculus, a dynamic system view of the nature around us becomes possible. Today, differential equations are the most widely used mathematical tools for describing dynamic systems evolving either along time or along spatial variables. Differential equations also provide a solid foundation for mathematical modeling in many scientific and engineering disciplines. Linear differential equations and a few special low-order nonlinear differential equations may have analytical solutions. However, generally speaking, most nonlinear equations have no analytical or close form solutions. Thus numerical techniques should be adopted for solving these equations. In Section 7.1, analytical solutions to a special class of ordinary differential equations (ODEs) will be explored. It will be explained how linear time-invariant (LTI) ODEs can be solved in MATLAB. Moreover, analytical solutions to a very special first-order nonlinear differential equation is presented. For linear state-space equations the analytical solutions can be obtained by vector integration methods. In Section 7.2, numerical algorithms for first-order explicit ODEs will be presented, with an illustrative MATLAB implementation. How to convert other types of ODEs into first-order explicit equations will be discussed. Most interestingly, solutions to several different types of special ODEs will be discussed in Section 7.3, which include stiff ODEs, differential algebraic equations (DAEs), implicit ODEs as well as delay differential equations (DDEs). In Sections 7.4 and 7.5, boundary value problems (BVPs) of ODEs and PDEs will be discussed respectively with extensive demonstrative examples. Section 7.6 briefly discusses the modeling and solution of ODEs using Simulink environment with different kinds of example ODEs. Solutions to the Simulink models are obtained by automatic simulation methods based on Simulink's extensive libraries of building blocks. We demonstrate that, theoretically speaking, ODEs of almost any complexity can be solved numerically in Simulink.

For readers who wish to check the detailed explanations of various aspects of differential equations, we recommend the free textbooks [26] (Chapters 17, 18 and 20) and [27]. The open source free textbook [14] is also an excellent resource for learning differential equations (Chapters 14-34 for ODE and Chapters 35-47 for PDE.)

7.1 Analytical Solution Methods for Special Classes of ODEs

7.1.1 Mathematical descriptions

The general description of the linear time-invariant ordinary differential equations is given by

$$\begin{aligned}\frac{\mathrm{d}^n y(t)}{\mathrm{d}t^n}+a_1\frac{\mathrm{d}^{n-1}y(t)}{\mathrm{d}t^{n-1}}+a_2\frac{\mathrm{d}^{n-2}y(t)}{\mathrm{d}t^{n-2}}+\cdots+a_{n-1}\frac{\mathrm{d}y(t)}{\mathrm{d}t}+a_n y(t)=\\ b_n\frac{\mathrm{d}^n u(t)}{\mathrm{d}t^n}+b_{n-1}\frac{\mathrm{d}^{n-1}u(t)}{\mathrm{d}t^{n-1}}+\cdots+b_1\frac{\mathrm{d}u(t)}{\mathrm{d}t}+b_0 u(t)\end{aligned} \tag{7.1}$$

where a_i, b_i are constants. For practical reasons, the derivative order in the right-hand-side should be less than n for the system to be strictly proper. In this case, b_n, b_{n-1} and so on can be set to zeros. Note that, if we use $D=\dfrac{\mathrm{d}}{\mathrm{d}t}$ and $D^k=\left(\dfrac{\mathrm{d}}{\mathrm{d}t}\right)^k$, the above ODE can be written in a compact form using the symbolic differential operator D as follows

$$Q(D)y(t)=P(D)u(t) \tag{7.2}$$

where

$$\begin{aligned}Q(D)&=D^n+a_1D^{n-1}+a_2D^{n-2}+\cdots+a_{n-1}D+a_n\\ P(D)&=b_nD^n+b_{n-1}D^{n-1}+\cdots+b_1D+b_0.\end{aligned} \tag{7.3}$$

From the properties introduced in Section 5.1, for zero initial condition problems, one has $\mathscr{L}[\mathrm{d}^m y(t)/\mathrm{d}t^m]=s^m\mathscr{L}[y(t)]=s^mY(s)$ and $\mathscr{L}[\mathrm{d}^m u(t)/\mathrm{d}t^m]=s^m\mathscr{L}[u(t)]=s^mU(s)$. Thus the above LTI ODE can be mapped into the following algebraic polynomial equation:

$$Y(s)Q(s)=U(s)P(s) \tag{7.4}$$

where $Q(s)=s^n+a_1s^{n-1}+a_2s^{n-2}+\cdots+a_{n-1}s+a_n$ and $P(s)=b_ns^n+b_{n-1}s^{n-1}+\cdots+b_1s+b_0$. Note that, if initial conditions are not all zeros, extra attention should be paid during the conversion. The neat form of (7.4) with the same coefficients as (7.1) is still possible. However, by re-arranging the terms, we can get

$$Y(s)Q(s)=Y_0(s)+U_0(s)+U(s)P(s) \tag{7.5}$$

where $Y_0(s)$ and $U_0(s)$ are due to non-zero initial conditions in signals $y(t)$ and $u(t)$, respectively.

The following polynomial equation

$$Q(s) = s^n + a_1 s^{n-1} + a_2 s^{n-2} + \cdots + a_{n-1}s + a_n = 0 \tag{7.6}$$

is referred to as the *characteristic equation.* If all the roots r_i in the characteristic equation can be obtained and they are all distinct, the general form of the analytical solution to the corresponding ordinary differential equation can be written as

$$y(t) = y_{\rm ZI}(t) + y_{\rm ZS}(t) \tag{7.7}$$

where $y_{\rm ZI}(t)$ is due to initial conditions when zero input (ZI) is considered which can be expressed as

$$y_{\rm ZI}(t) = C_1 \mathrm{e}^{r_1 t} + C_2 \mathrm{e}^{r_2 t} + \cdots + C_n \mathrm{e}^{r_n t} \tag{7.8}$$

where C_i's are the undetermined coefficients related to initial conditions. When r_j is a repeated root with multiplicity m_j, the term $C_j \mathrm{e}^{r_j t}$ should be replaced by

$$\sum_{k=0}^{m_j-1} C_{jk} t^k \mathrm{e}^{r_j t}. \tag{7.9}$$

So, in general, let us assume that there are N distinct roots r_j, $j = 1, 2, \cdots, N$, and each distinct root r_j has its multiplicity m_j with $m_j \geqslant 1$ and $\sum_{j=1}^{N} m_j = n$. Then,

$$y_{\rm ZI}(t) = \sum_{j=1}^{N} \sum_{k=0}^{m_j-1} C_{jk} t^k \mathrm{e}^{r_j t}. \tag{7.10}$$

$y_{\rm ZS}(t)$ is called the zero-state (ZS) response when all initial conditions are assumed zeros. $y_{\rm ZS}(t)$ can be obtained by convolution of input signal $u(t)$ and $h(t)$, the impulse response of the system (7.1) as follows:

$$y_{\rm ZS}(t) = u(t) * h(t) = \int_0^{\infty} u(\tau) h(t - \tau) \mathrm{d}\tau. \tag{7.11}$$

The original LTI ODE can be solved in purely time domain when $h(t)$ is obtained. The procedure for obtaining $h(t)$ is simply

$$h(t) = b_n \delta(t) + P(D) y_0(t) \tag{7.12}$$

where $\delta(t)$ is the impulse function also known as Dirac's delta function. Note that b_n is usually zero. $y_0(t)$ is obtained from $y_{\rm ZI}(t)$ by applying the following special set of initial conditions dedicated for the purpose of getting $h(t)$:

$$y_0^{(n-1)}(0) = 1, \quad y_0^{(n-2)}(0) = y_0^{(n-3)}(0) = \cdots = y_0(0) = 0. \tag{7.13}$$

So, it is quite tedious to get the solution of the original ODE in time domain. When in s-domain, the solution process is much easier as shown in the next subsections. In particular, it is also obvious that using computer mathematics languages is indispensable.

Let us investigate (7.4). From the well-known Abel-Ruffini Theorem, it is known that the polynomial equation with degree less than or equal to 4 has analytical solutions. Thus it can be concluded that low-order LTI ODEs have analytical solutions. For higher-order equations, one may combine the numerical and analytical approaches to find quasi-analytical solutions with high accuracy. In this section, symbolic math-based analytical solution approaches will be presented first.

7.1.2 Analytical solution methods

The function `dsolve()` provided in the Symbolic Math Toolbox in MATLAB can be used to symbolically solve a class of ODEs with mixed initial and boundary conditions. The syntax of the function is

```
y=dsolve(fun_1, fun_2, ..., fun_m)
y=dsolve(fun_1, fun_2, ..., fun_m,'x')          % variables are assigned
```

where the string variables fun_i can be used to describe not only ODEs, but also initial and boundary conditions. When describing ODEs in `dsolve()`, the symbol `D4y` is used to denote $y^{(4)}(t)$. One can also use `D2y(2)=3` to denote given condition as $\ddot{y}(2) = 3$. With `dsolve()`, the analytical solutions to a class of ODEs can easily be found. If the variable in the ODE is x rather than t, this should be declared in the function call statement.

Example 7.1 Let the input signal be defined by $u(t) = \mathrm{e}^{-5t}\cos(2t+1)+5$. Find the general solution to the following ODE

$$y^{(4)}(t) + 10y^{(3)}(t) + 35\ddot{y}(t) + 50\dot{y}(t) + 24y(t) = 5\ddot{u}(t) + 4\dot{u}(t) + 2u(t).$$

Solution To solve the original ODE using `dsolve()`, the right-hand-side of the ODE should be evaluated first given $u(t)$. Then the original ODE can be solved using the following statements

```
>> syms t y; u=exp(-5*t)*cos(2*t+1)+5;
   uu=5*diff(u,t,2)+4*diff(u,t)+2*u;
   y=dsolve(['D4y+10*D3y+35*D2y+50*Dy+24*y=',char(uu)])
```

where in the above statements, the right-hand side of the ODE has been evaluated in a string variable `uu`. The final solution to this ODE is found as

$$y(t) = \frac{5}{12} - \frac{343}{520}\mathrm{e}^{-5t}\cos(2t+1) - \frac{547}{520}\mathrm{e}^{-5t}\sin(2t+1)$$
$$+C_1\mathrm{e}^{-4t} + C_2\mathrm{e}^{-3t} + C_3\mathrm{e}^{-2t} + C_4\mathrm{e}^{-t}$$

where C_i's are undetermined constants. Given initial and boundary conditions, the constants C_i can then be solved uniquely.

Now let us assume that the following conditions are given $y(0) = 3, \dot{y}(0) = 2, \ddot{y}(0) = y^{(3)}(0) = 0$. Use the following statements to find the solution to the original ODE.

```
>> y=dsolve(['D4y+10*D3y+35*D2y+50*Dy+24*y=',char(uu)],'y(0)=3',...
      'Dy(0)=2','D2y(0)=0','D3y(0)=0')
```

which is

$$y(t) = \frac{5}{12} - \frac{343}{520}\mathrm{e}^{-5t}\cos(2t+1) - \frac{547}{520}\mathrm{e}^{-5t}\sin(2t+1) + \left(-\frac{445}{26}\cos 1 - \frac{51}{13}\sin 1 - \frac{69}{2}\right)\mathrm{e}^{-2t} + \left(-\frac{271}{30}\cos 1 + \frac{41}{15}\sin 1 - \frac{25}{4}\right)\mathrm{e}^{-4t} + \left(\frac{179}{8}\cos 1 + \frac{5}{8}\sin 1 + \frac{73}{3}\right)\mathrm{e}^{-3t} + \left(\frac{133}{30}\cos 1 + \frac{97}{60}\sin 1 + 19\right)\mathrm{e}^{-t}.$$

With the powerful Symbolic Math Toolbox, solutions to some seemingly impossible to solve ODEs can be obtained easily. For instance, let $y(0) = 1/2, \dot{y}(\pi) = 1, \ddot{y}(2\pi) = 0, \dot{y}(2\pi) = 1/5$, the analytical solution to this ODE can be obtained

```
>> y=dsolve(['D4y+10*D3y+35*D2y+50*Dy+24*y=',char(uu)],'y(0)=1/2',...
      'Dy(pi)=1','D2y(2*pi)=0','Dy(2*pi)=1/5')
```

It is possible to display the analytical form of the undetermined coefficients C_i but each coefficient will take about 10 lines at least. In this situation, a fairly accurate approximate representation to these terms can be used such that the quasi-analytical solution to the equation can be found with `vpa(ans)`, where the solution obtained above can be expressed by

$$y(t) = \frac{5}{12} - \frac{343}{520}\mathrm{e}^{-5t}\cos(2t+1) - \frac{547}{520}\mathrm{e}^{-5t}\sin(2t+1) - 219.1291604\mathrm{e}^{-t} + 442590.9052\mathrm{e}^{-4t} + 31319.63786\mathrm{e}^{-2t} - 473690.0889\mathrm{e}^{-3t}.$$

Example 7.2 The above ODE contains only real poles. In fact, in the `dsolve()` function, ODEs containing complex poles can also be solved. Now consider the following ODE

$$y^{(5)}(t) + 5y^{(4)}(t) + 12y^{(3)}(t) + 16\ddot{y}(t) + 12\dot{y}(t) + 4y(t) = \dot{u}(t) + 3u(t)$$

where the input signal $u(t) = \sin t$, and $y(0) = \dot{y}(0) = \ddot{y}(0) = y^{(3)}(0) = y^{(4)}(0) = 0$. Try to use the analytical solution approach for this ODE.

Solution One can solve the differential equation using the following statements

```
>> syms t y; u=sin(t); uu=3*diff(u)+3*u;
   y=dsolve(['D5y+5*D4y+12*D3y+16*D2y+12*Dy+4*y=' char(uu)],...
      'y(0)=0','Dy(0)=0','D2y(0)=0','D3y(0)=0','D4y(0)=0')
```

The analytical solution can then be found

$$y(t) = -\frac{12}{25}\cos t - \frac{9}{25}\sin t + \frac{57}{50}\mathrm{e}^{-t}\sin t + \frac{12}{25}\mathrm{e}^{-t}\cos t + \frac{3}{5}t\mathrm{e}^{-t}\sin t - \frac{3}{10}t\mathrm{e}^{-t}\cos t$$

or even more simply, the solution can be manually simplified as

$$y(t) = -\frac{12}{25}\cos t - \frac{9}{25}\sin t + \left(\frac{57}{50} + \frac{3}{5}t\right)\mathrm{e}^{-t}\sin t + \left(\frac{12}{25} - \frac{3}{10}t\right)\mathrm{e}^{-t}\cos t.$$

Example 7.3 Find the analytical solution to the following simultaneous differential equations:

$$\begin{cases} \ddot{x}(t) + 2\dot{x}(t) = x(t) + 2y(t) - \mathrm{e}^{-t} \\ \dot{y}(t) = 4x(t) + 3y(t) + 4\mathrm{e}^{-t}. \end{cases}$$

Solution The linear differential equation set can also be solved directly using `dsolve()` function. For instance, the above equation can be solved with

```
>> [x,y]=dsolve('D2x+2*Dx=x+2*y-exp(-t)','Dy=4*x+3*y+4*exp(-t)')
```

and it can be found that the solutions are

$$\begin{cases} x(t) = -6t\mathrm{e}^{-t} + C_1\mathrm{e}^{-t} + C_2\mathrm{e}^{(1+\sqrt{6})t} + C_3\mathrm{e}^{-(-1+\sqrt{6})t} \\ y(t) = 6t\mathrm{e}^{-t} - C_1\mathrm{e}^{-t} + 2\left(2+\sqrt{6}\right)C_2\mathrm{e}^{(1+\sqrt{6})t} + 2\left(2-\sqrt{6}\right)C_3\mathrm{e}^{-(-1+\sqrt{6})t} + \dfrac{1}{2}\mathrm{e}^{-t}. \end{cases}$$

7.1.3 Applications of Laplace transforms

Reconsider the linear differential equation with constant coefficients given in (7.1). Using Laplace transform, we can arrive at (7.5). Note that, usually, the input signal $u(t) = 0$ when $t \leqslant 0^-$. So, $U_0(s) = 0$. However, due to initial internal states, $Y_0(s)$ is in general not zero. But any way, the Laplace transform of the output signal $y(t)$ can be written as

$$Y(s) = \frac{Y_0(s) + U(s)P(s)}{Q(s)}. \tag{7.14}$$

Thus the analytical solution to the original ODE can be obtained by taking the inverse Laplace transform of $Y(s)$, i.e., $y(t) = \mathscr{L}^{-1}[Y(s)]$. Note that $Y(s)$ is a rational function and the `residue()` function can be used to perform partial fraction expansion. Then, the inverse Laplace transform function `ilaplace()` can be used to find $y(t)$. If the denominator polynomial of $Y(s)$ contains complex roots, in order to increase the readability of the result obtained, the function `pfrac()` can be used to further process the results.

Example 7.4 Find the analytical solution to the differential equation in Example 7.1 where the output $y(t)$ and its derivatives are all zero at time $t = 0$. For simplicity, the original ODE is rewritten as

$$y^{(4)}(t) + 10y^{(3)}(t) + 35\ddot{y}(t) + 50\dot{y}(t) + 24y(t) = 5\ddot{u}(t) + 4\dot{u}(t) + 2u(t)$$

and assume that $u(t) = \mathrm{e}^{-5t}\cos(2t+1) + 5$.

Solution Taking Laplace transform to both sides of the differential equation, the following formula can be obtained easily.

$$s^4Y(s) + 10s^3Y(s) + 35s^2Y(s) + 50sY(s) + 24Y(s) = 5s^2U(s) + 4sU(s) + 2U(s).$$

Substituting $U(s)$ into $Y(s)$, then taking the inverse Laplace transform, it seems that the analytical solution to the original equation can be found.

However, the above seemingly straightforward procedure should be applied carefully. It may happen that $U_0(s)$ in (7.5) is not zero. Let us take a closer look at this example.

Since $u(t)$ function is given, before Laplace transform, derivatives to $u(t)$ should be evaluated first, then take the Laplace transform to the result as follows:

```
>> syms t; u=exp(-5*t)*cos(2*t+1)+5;          % define the input
   uu=laplace(5*diff(u,2)+4*diff(u)+2*u); % Laplace transform
   uu=collect(simple(uu));                % simplification
```

and it can be seen the correct Laplace transform is

$$\frac{290 + (87\cos 1 + 92\sin 1 + 10)\, s^2 + (619\cos 1 + 100 + 286\sin 1)\, s}{(s^2 + 10s + 29)\, s}.$$

The above method takes care of $U_0(s)$ automatically which is suggested. However, if we use

```
>> u1=laplace(u);  % takes the input Laplace transform first
   u2=(5*s^2+4*s+2)*u1;   % the do P(s)U(s)
   simple(uu-u2) % show the difference
```

we get the difference as follows:

```
>> -25*s-5*cos(1)*s-20+21*cos(1)+10*sin(1)
```

which alerts us to proceed carefully with the right-hand side when $P(s)$ is not a constant.

Having obtained the formula on the right-hand side, the rational representation of $Y(s)$ can be obtained. Then perform partial fraction expansion and take inverse Laplace transform

```
>> syms s; G=uu/(s^4+10*s^3+35*s^2+50*s+24);
   Y=residue(G,s); y=ilaplace(Y); y=simple(y)
```

and finally the correct analytical solution can be found as

$$\begin{aligned} y(t) = {} & \left(\frac{97}{60}\sin 1 + \frac{133}{30}\cos 1 - \frac{5}{3}\right)\mathrm{e}^{-t} + \left(\frac{343}{520}\sin 1 - \frac{547}{520}\cos 1\right)\mathrm{e}^{-5t}\sin(2t) + \frac{5}{12} \\ & + \left(-\frac{547}{520}\sin 1 - \frac{343}{520}\cos 1\right)\cos(2t)\,\mathrm{e}^{-5t} + \left(\frac{41}{15}\sin 1 - \frac{271}{30}\cos 1 + \frac{5}{12}\right)\mathrm{e}^{-4t} \\ & + \left(-\frac{51}{13}\sin 1 - \frac{445}{26}\cos 1 + \frac{5}{2}\right)\mathrm{e}^{-2t} + \left(\frac{5}{8}\sin 1 + \frac{179}{8}\cos 1 - \frac{5}{3}\right)\mathrm{e}^{-3t}. \end{aligned}$$

If partial fraction expansion is not used, the following statements can be used alternatively by using the `dsolve()` function. Thus the same solution can be found.

```
>> syms t; u=exp(-5*t)*cos(2*t+1)+5;   % define the input signal
   uu=5*diff(u,2)+4*diff(u)+2*u;        % evaluate the right-hand side
   y1=dsolve(['D4y+10*D3y+35*D2y+50*Dy+24*y=',char(uu)],...
          'y(0)=0','Dy(0)=0','D2y(0)=0','D3y(0)=0');
```

From the analytical solution obtained above, the function `ezplot()` can then be used directly to draw the solution curve, as shown in Figure 7.1.

```
>> ezplot(y,[0,10]);  axis([0,10,0,0.6])
```

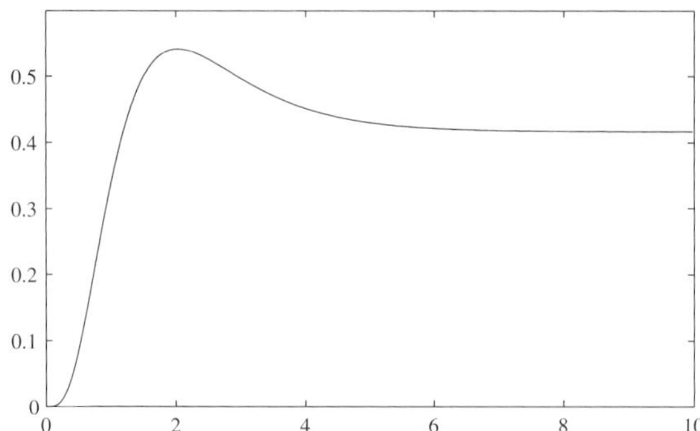

FIGURE 7.1: Solution curve of the ODE

7.1.4 Analytical solutions to LTI state-space equations

The linear time-invariant (LTI) state-space model of a given dynamic system can be described by

$$\begin{cases} \dot{\boldsymbol{x}}(t) = \boldsymbol{A}\boldsymbol{x}(t) + \boldsymbol{B}\boldsymbol{u}(t) \\ \boldsymbol{y}(t) = \boldsymbol{C}\boldsymbol{x}(t) + \boldsymbol{D}\boldsymbol{u}(t) \end{cases} \quad \text{for given } \boldsymbol{x}_0 \tag{7.15}$$

where $\boldsymbol{A}, \boldsymbol{B}, \boldsymbol{C}, \boldsymbol{D}$ are constant matrices. The analytical solution is

$$\boldsymbol{x}(t) = \mathrm{e}^{\boldsymbol{A}(t-t_0)}\boldsymbol{x}(t_0) + \int_{t_0}^{t} \mathrm{e}^{\boldsymbol{A}(t-\tau)}\boldsymbol{B}\boldsymbol{u}(\tau)\,\mathrm{d}\tau. \tag{7.16}$$

Given an input function $\boldsymbol{u}(t)$, the analytical solution to the above vector form first-order ODE can be obtained by matrix integrals.

Example 7.5 Given the input signal $u(t) = 2 + 2\mathrm{e}^{-3t}\sin 2t$, find the analytical solution to the system given by

$$\boldsymbol{A} = \begin{bmatrix} -19 & -16 & -16 & -19 \\ 21 & 16 & 17 & 19 \\ 20 & 17 & 16 & 20 \\ -20 & -16 & -16 & -19 \end{bmatrix}, \quad \boldsymbol{B} = \begin{bmatrix} 1 \\ 0 \\ 1 \\ 2 \end{bmatrix}, \quad \boldsymbol{C}^{\mathrm{T}} = \begin{bmatrix} 2 \\ 1 \\ 0 \\ 0 \end{bmatrix}, \quad \boldsymbol{D} = 0, \quad \boldsymbol{x}_0 = \begin{bmatrix} 0 \\ 1 \\ 1 \\ 2 \end{bmatrix}.$$

Solution From (7.16), the analytical solution can be found by using the following scripts

```
>> syms t tau; u=2+2*exp(-3*tau)*sin(2*tau);
   A=[-19,-16,-16,-19; 21,16,17,19; 20,17,16,20; -20,-16,-16,-19];
   B=[1; 0; 1; 2]; C=[2 1 0 0]; x0=[0; 1; 1; 2];
   y=C*(expm(A*t)*x0+int(expm(A*(t-tau))*B*u,tau,0,t)); simple(y)
```

and the analytical solution is

$$y(t) = -\frac{135}{8}\mathrm{e}^{-3t}\cos 2t + \frac{77}{4}\mathrm{e}^{-3t}\sin 2t + \frac{127}{4}t\mathrm{e}^{-t} + 4t^2\mathrm{e}^{-t} + \frac{119}{8}\mathrm{e}^{-t} + 57\mathrm{e}^{-3t} - 54.$$

7.1.5 Analytical solutions to special nonlinear differential equations

Very few nonlinear differential equations can be analytically solved by the dsolve() function. An example for the analytical solution to a special nonlinear differential equation is demonstrated. Another example is also presented where the analytical solution is not possible.

Example 7.6 Find the analytical solution to the first-order nonlinear differential equation $\dot{x}(t) = x(t)(1 - x^2(t))$.

Solution Such a simple nonlinear equation can be solved using the function dsolve().

```
>> syms t x;  x=dsolve('Dx=x*(1-x^2)')
```

Thus the analytical solutions are $x(t) = \pm 1/\sqrt{1 + C_1 \mathrm{e}^{-2t}}$.

This is a lucky case. Now, let us slightly change the original ODE, for instance, by adding 1 to the right-hand side of the original ODE. The following statements can be used to solve the modified ODE. With no surprise, no solution can be found by using dsolve().

```
>> syms x;  x=dsolve('Dx=x*(1-x^2)+1','x')
```

Example 7.7 Try to find the analytical solution to the well-known *Van der Pol equation* $\dfrac{\mathrm{d}^2 y(t)}{\mathrm{d}t^2} + \mu(y^2(t) - 1)\dfrac{\mathrm{d}y(t)}{\mathrm{d}t} + y(t) = 0$.

Solution From the previous examples, it seems that the function dsolve() is quite powerful in finding analytical solutions to many ODEs. Here we are trying to solve the Van der Pol nonlinear ODE. The following statements

```
>> syms y mu; y=dsolve(['D2y+mu*(y^2-1)*Dy+y=','y'])
```

can be executed but from the message, the effort failed, since there is no analytical solution at all to the given Van der Pol equation.

It can be seen that the function dsolve() cannot be used to solve general nonlinear ODEs. Thus to solve nonlinear ODEs, numerical methods have to be used. In the following sections of this chapter, we shall concentrate on numerical solutions of various ODEs.

7.2 Numerical Solutions to ODEs

In the previous section, analytical solutions for a limited class of ODEs were discussed and demonstrated. It is also noted that there is no analytical solutions to most nonlinear differential equations. Thus numerical algorithms should be used.

7.2.1 Overview of numerical solution algorithms

A large category of numerical solution algorithms for ODEs is for the so-called initial value problems (IVPs). The typical IVPs are *vector form first-order explicit ODEs* given by

$$\dot{\boldsymbol{x}}(t) = \boldsymbol{f}(t, \boldsymbol{x}(t)), \quad \boldsymbol{x}(t_0) = \boldsymbol{x}_0 \tag{7.17}$$

where $\boldsymbol{x}^{\mathrm{T}}(t) = [x_1(t), x_2(t), \cdots, x_n(t)]$ is referred to as the *state vector*, $\boldsymbol{x}(t_0) = [x_1(t_0), \cdots, x_n(t_0)]^{\mathrm{T}}$ is the given initial value vector, and $\boldsymbol{f}^{\mathrm{T}}(\cdot) = [f_1(\cdot), f_2(\cdot), \cdots, f_n(\cdot)]$ can be any nonlinear functions. In this section, we will explore numerical algorithms for solving $\boldsymbol{x}(t)$, $t \in [t_0, t_f]$, where t_f is also referred to as the *terminal time*.

Numerical error and step-size issues

For the above IVPs, the Euler's algorithm is obviously the most straightforward algorithm. Although the algorithm is very simple, understanding such an algorithm will help us to better understand other complicated algorithms.

Assume that at time t_0, the state vector is written as $\boldsymbol{x}(t_0)$. Given a small calculation step h, the left-hand-side of the ODE can be approximately written as $(\boldsymbol{x}(t_0+h) - \boldsymbol{x}(t_0))/(t_0+h-t_0)$. Substituting it back to the original ODE, the approximate solution at time $t_0 + h$ can be written as

$$\hat{\boldsymbol{x}}(t_0 + h) = \boldsymbol{x}(t_0) + h\boldsymbol{f}(t_0, \boldsymbol{x}(t_0)). \tag{7.18}$$

The above approximate solution certainly contains errors. Thus the state vector at time $t_0 + h$ should be written as

$$\boldsymbol{x}(t_0 + h) = \hat{\boldsymbol{x}}(t_0 + h) + \boldsymbol{R}_0 = \boldsymbol{x}_0 + h\boldsymbol{f}(t, \boldsymbol{x}_0) + \boldsymbol{R}_0 \tag{7.19}$$

where the vector $\boldsymbol{R}_0$ is the approximation error. Denoting $\boldsymbol{x}_1 = \boldsymbol{x}(t_0 + h)$, $\hat{\boldsymbol{x}}_1 = \hat{\boldsymbol{x}}(t_0+h)$ approximates the state vector at time t_0+h. In this numerical procedure, one can simply denote the numerical solution by $\boldsymbol{x}_1$.

The state vector at time t_k is denoted by $\boldsymbol{x}_k$. At time $t_k + h$, the Euler's algorithm generates a new state vector

$$\boldsymbol{x}_{k+1} = \boldsymbol{x}_k + h\boldsymbol{f}(t, \boldsymbol{x}_k). \tag{7.20}$$

Thus a recursive algorithm can be used to evaluate the solutions in each time instant over the given time interval $t \in [0, t_f]$. In this way, the numerical solutions at time instances $t_0 + h, t_0 + 2h, \cdots$ can be obtained.

To increase the accuracy of the solutions, one may reduce the step size h. However, one cannot expect to reduce the step-size unlimitedly. The following two reasons should be considered:

(i) **Slowing down in computation** If an extremely small step-size is selected, the number of computation points over the simulation interval will significantly increase.

(ii) **Increasing the cumulative numerical errors** No matter how small the step-size is, the round-off error in numerical solution is unavoidable. As pointed out earlier, decreasing the step-size means the increase in computation points, which also means that the cumulative error and error propagation may increase. The relationship among the step-size, round-off error, cumulative error and the overall error are sketched in Figure 7.2 (a) for illustration.

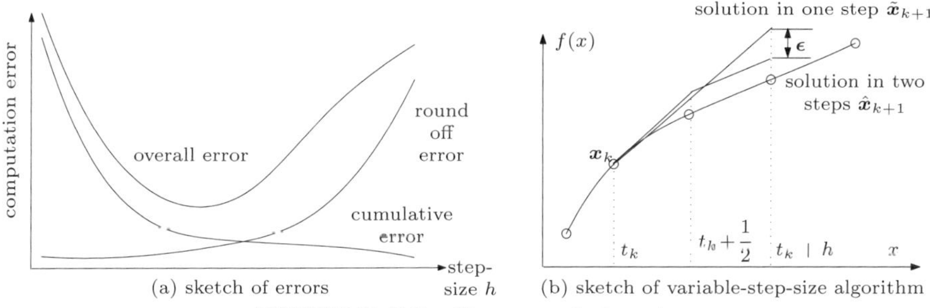

(a) sketch of errors (b) sketch of variable-step-size algorithm

FIGURE 7.2: Errors and step-sizes

Thus in order to effectively solve the ODEs, the following considerations should be in place:

(i) **Suitable step-size selection** If simple algorithms such as the Euler's algorithm are used, one should choose suitable step-sizes, neither too large, nor too small.

(ii) **Improved algorithms** Since the above simple Euler's algorithm is a rectangular approximation of the original integration problem, thus the accuracy is rather low when the step-size is not small enough. It usually cannot effectively solve the original ODE problem. Other advanced algorithms should be adopted to replace the Euler's algorithm. Successful algorithms include the Runge-Kutta's algorithm and the Adams' algorithm, etc.

(iii) **Variable-step-size algorithms** It has been suggested that one should "suitably" choose the step-size. The concept of "suitable step-size" is very vague. The selection of step-size sometimes depends on the experience and the ODE problem at hand. In fact, many ODE numerical algorithms allow the use of variable-step-size computation. When the error detected is small, a relatively large step-size can be chosen. However, when the detected error is large, a smaller step-size will be used instead. Using this variable step-size mechanism, fast and accurate algorithms can be devised for ordinary differential equations.

The basic idea of variable-step-size algorithm is illustrated in Figure 7.2 (b).

If at the current time instant t_k, the state vector is $\boldsymbol{x}_k$, then the state vector $\tilde{\boldsymbol{x}}_{k+1}$ at time instant $t_k + h$ can be obtained. On the other hand, if one divides the step-size by two halves, the state vector at time $t_k + h$ obtained by integrating two steps using half step-size $h/2$ is denoted by $\hat{\boldsymbol{x}}_{k+1}$. The error of the state vectors using the two methods is $\boldsymbol{\epsilon} = |\hat{\boldsymbol{x}}_{k+1} - \tilde{\boldsymbol{x}}_{k+1}|$, and if it is smaller than the preassigned error tolerance, then the original step-size can be used and accordingly, the step-size can be increased. If the error is too large, the step-size should be decreased to ensure the computational accuracy. Thus the above method considers both the computation speed and the computation accuracy. Such algorithms are referred to as *adaptive or variable step-size algorithm.*

7.2.2 Fixed-step Runge-Kutta algorithm and its MATLAB implementation

Fourth-order fixed-step Runge-Kutta algorithm is the classical ODE solving algorithm, often taught in numerical analysis or system simulation courses. It has been considered an effective and easy to implement numerical algorithm. Four additional intermediate variables are introduced such that

$$\begin{cases} \boldsymbol{k}_1 = h\boldsymbol{f}(t_k, \boldsymbol{x}_k) \\ \boldsymbol{k}_2 = h\boldsymbol{f}\left(t_k + \dfrac{h}{2}, \boldsymbol{x}_k + \dfrac{\boldsymbol{k}_1}{2}\right) \\ \boldsymbol{k}_3 = h\boldsymbol{f}\left(t_k + \dfrac{h}{2}, \boldsymbol{x}_k + \dfrac{\boldsymbol{k}_2}{2}\right) \\ \boldsymbol{k}_4 = h\boldsymbol{f}(t_k + h, \boldsymbol{x}_k + \boldsymbol{k}_3) \end{cases} \tag{7.21}$$

where h is the fixed step-size. The above fourth-order Runge-Kutta algorithm calculates the state vector at the next step using

$$\boldsymbol{x}_{k+1} = \boldsymbol{x}_k + \frac{1}{6}(\boldsymbol{k}_1 + 2\boldsymbol{k}_2 + 2\boldsymbol{k}_3 + \boldsymbol{k}_4). \tag{7.22}$$

Thus, with the use of the recursive algorithm, the initial value problem at time instants $t_0 + h, t_0 + 2h, \cdots$, within the time interval $t \in [t_0, t_h]$ can be solved numerically step by step.

Based on the above algorithm, its MATLAB implementation is listed as follows:

```
function [tout,yout]=rk_4(odefile,tspan,y0)
ts=tspan; t0=ts(1); th=ts(2); yout=[]; tout=[]; y0=y0(:);
if length(ts)==3, h=ts(3); else, h=(ts(2)-ts(1))/100; th=ts(2); end
for t=[t0:h:th]
   k1=h*feval(odefile,t,y0); k2=h*feval(odefile,t+h/2,y0+0.5*k1);
   k3=h*feval(odefile,t+h/2,y0+0.5*k2); k4=h*feval(odefile,t+h,y0+k3);
   y0=y0+(k1+2*k2+2*k3+k4)/6; yout=[yout; y0']; tout=[tout; t];
end
```

where tspan can be given in two ways. One way is to specify an evenly distributed time vector and the other is to use tspan=[t_0,t_h,h], where t_0 and t_h are the initial and final time, h the step-size. The argument *odefile* provides the file name for describing the ODE's right-hand-side (RHS) function. The argument $\boldsymbol{y}_0$ provides the initial state vector. The time vector and the matrix containing states at each time instant are returned in arguments tout and yout, respectively.

The above algorithm seems to be rather simple. However, from a numerical accuracy point of view, it may not be a good algorithm, since numerical accuracy during the computation is not monitored. Thus the overall numerical accuracy of the algorithm cannot be guaranteed. In later examples, this algorithm will be compared with variable-step algorithm.

7.2.3 Numerical solution to first-order vector ODEs

Runge-Kutta-Felhberg algorithm

German mathematician Erwin Felhberg proposed an improved version of the algorithm based on the traditional Runge-Kutta algorithm[28]. Within each computation step, six evaluations of the $f_i(\cdot)$ function are performed to ensure high-precision and numerical stability. The algorithm is also known as the *4/5 Runge-Kutta-Felhberg algorithm.* Assuming that the current step-size is h_k, the six intermediate variables $\boldsymbol{k}_i$ are evaluated by the following formula:

$$\boldsymbol{k}_i = h_k \boldsymbol{f}\left(t_k + \alpha_i h_k, \boldsymbol{x}_k + \sum_{j=1}^{i-1} \beta_{ij} \boldsymbol{k}_j\right), \quad i = 1, 2, \cdots, 6 \tag{7.23}$$

where t_k is the current time instant, and intermediate parameters α_i, β_{ij} and other parameters are given in Table 7.1. The parameter pairs α_i, β_{ij} are also referred to as the *Dormand-Prince pairs.* The state vector at the next time instant is obtained from

TABLE 7.1: Coefficients in 4/5 Runge-Kutta-Felhberg algorithm

α_i	β_{ij}					γ_i	γ_i^*
0						16/135	25/216
1/4	1/4					0	0
3/8	3/32	9/32				6656/12825	1408/2565
12/13	1932/2197	−7200/2197	7296/2197			28561/56430	2197/4104
1	439/216	−8	3680/513	−845/4104		−9/50	−1/5
1/2	−8/27	2	−3544/2565	1859/4104	−11/40	2/55	0

$$\boldsymbol{x}_{k+1} = \boldsymbol{x}_k + \sum_{i=1}^{6} \gamma_i \boldsymbol{k}_i. \tag{7.24}$$

Of course, this algorithm seems to be a fixed-step type like the RK4 algorithm in the previous subsection. However, in practical applications, an error vector $\boldsymbol{\epsilon}_k = \sum_{i=1}^{6} (\gamma_i - \gamma_i^*)\boldsymbol{k}_i$ can be obtained, and the step-size can be adjusted according to this error vector. This algorithm is often referred to as the *adaptive step-size algorithm*. It can not only ensure the accuracy of numerical solutions but also provide faster speed.

In contrast with the concepts in feedback control, the fixed-step algorithm is somewhat like an open-loop control structure. It does not care whether the error of solution is acceptable or not using the identical step-size throughout the computation period. However, the variable-step algorithm is similar to the closed-loop control concept. It monitors the errors in the numerical solutions and whenever necessary, the step-size can be adjusted according to the estimated computation error.

MATLAB functions for solving ordinary differential equations

The MATLAB function `ode45()` is the most widely used initial value problem solver. In the algorithm, the variable-step 4/5 Runge-Kutta-Felhberg algorithm is implemented. The syntaxes of the function are

```
[t,x]=ode45(fun,[t0,tf],x0)                        % direct solutions
[t,x]=ode45(fun,[t0,tf],x0,options)                % control options
[t,x]=ode45(fun,[t0,tf],x0,options,p1,p2,···)      % additional arguments
```

where the ODE should be expressed using an M-function *fun*, an anonymous function or an inline function. The structures of the functions will be explained through several examples later. The time span $[t_0, t_f]$ describes the time interval for numerical solutions. If only one value is provided, it means that the final value t_f with the default setting of initial time at $t_0 = 0$. For initial value problems, one should also specify the initial state vector $\boldsymbol{x}_0$. Note that this function allows the selection of $t_f < t_0$, where t_0 can be regarded as the final time, with t_f the initial time. Also $\boldsymbol{x}_0$ can be regarded as the final value of the equations. Thus this function can also be used for final value problems.

One of the most important procedures in solving numerically the initial value problem is that the first-order explicit ODE should be described correctly in MATLAB. The structure of the function is as follows

```
function xd=fun(t,x)                     % without additional arguments
function xd=fun(t,x,flag,p1,p2,···)      % with additional arguments
```

where t is the time variable. Note that even the original ODE is time-independent, and the argument t should also be used. Otherwise mismatch in variables will occur. The variable $\boldsymbol{x}_d$ is the derivative of the state vector. The variable `flag` can be used to control the solution process. It should be noted that if anonymous function is used, the argument `flag` should be omitted.

In actual ODE solving process, sometimes one may further assign some control options. This can be done with the use of the variable `options`. The initial `options` template can be obtained with the function `odeset()`. This template is a structured variable, with many member components. Some of the frequently used member components are given in Table 7.2.

TABLE 7.2: Control parameters in differential equation solutions

Parameters	Descriptions to the parameters
`RelTol`	Upper-bound of the relative error tolerance. The default value is 0.001, i.e., % relative error. In most applications, this value should be reduced to ensure that the results are accurate
`AbsTol`	A vector controlling the permissible absolute error in states. The default value is 10^{-6}. Of course, this value can be changed to improve the accuracy of solution
`MaxStep`	Maximum allowed step-size
`Mass`	Mass matrix, which is used in describing differential algebraic equations
`Jacobian`	The function describing the Jacobian matrix $\partial \boldsymbol{f}/\partial \boldsymbol{x}$. If the Jacobian matrix is known, the simulation process can speed up

Two methods can be used in modifying the `options` template. One is by the use of the function `odeset()`, and the other is by the modification of the member component directly. For instance, if one wants to assign the relative error tolerance to 10^{-7}, either of the following statements can be used

```
options=odeset('RelTol',1e-7);          % assign it by odeset() function
options=odeset; options.RelTol=1e-7;   % by direct assignment
```

In many ODE solution applications, sometimes additional arguments may be used for flexible testing of various parameter combinations. These additional arguments, denoted by $p_1, p_2, \cdots, p_m$, should be properly declared and passed to the MATLAB description for the RHS of the ODE. When calling the ODE solver, the arguments should also be matched in the same order.

Example 7.8 Consider the well-known *Lorenz equation* given by

$$\begin{cases} \dot{x}_1(t) = -\beta x_1(t) + x_2(t)x_3(t) \\ \dot{x}_2(t) = -\rho x_2(t) + \rho x_3(t) \\ \dot{x}_3(t) = -x_1(t)x_2(t) + \sigma x_2(t) - x_3(t) \end{cases}$$

where $\beta = 8/3, \rho = 10, \sigma = 28$. The initial values are given by $x_1(0) = x_2(0) = 0$, $x_3(0) = \epsilon$, and ϵ is a very small positive number, i.e., $\epsilon = 10^{-10}$. Find the numerical solutions to the IVP of the given ODEs.

Solution It is obvious that these ODEs are nonlinear time-invariant with no analytical solutions. Numerical solutions should be pursued. For the ODEs, an anonymous function can be prepared as follows for use with the function `ode45()` to directly solve the ODEs.

```
>> f=@(t,x)[-8/3*x(1)+x(2)*x(3); -10*x(2)+10*x(3); ...
         -x(1)*x(2)+28*x(2)-x(3)];
   t_final=100; x0=[0;0;1e-10];
   [t,x]=ode45(f,[0,t_final],x0); plot(t,x),
   figure; plot3(x(:,1),x(:,2),x(:,3)); grid
   axis([10 42 -20 20 -20 25]);  % setting the axis ranges
```

In the above MATLAB statements, t_final denotes the terminating time, x0 is the initial state vector. The first drawing command shows the dynamical curves of the states versus time, as in Figure 7.3 (a). In the second drawing command, the three-dimensional phase space trajectory is visualized as shown in Figure 7.3 (b). So, the seemingly difficult nonlinear ODE IVP can be solved using only a few MATLAB statements.

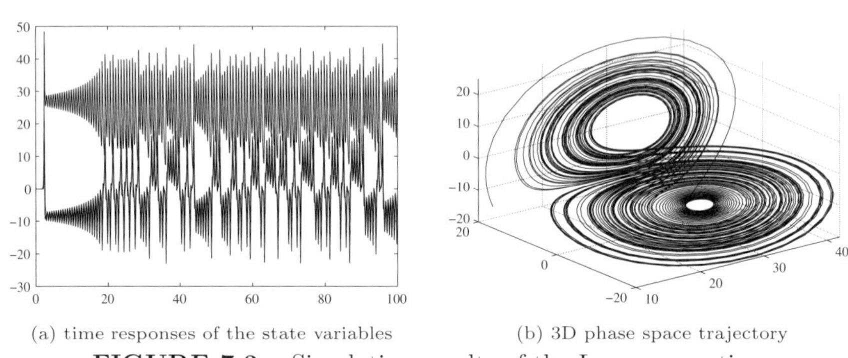

(a) time responses of the state variables (b) 3D phase space trajectory

FIGURE 7.3: Simulation results of the Lorenz equations

Furthermore, the best command to visualize the three-dimensional trajectory is by the use of the function `comet3(x(:,1),x(:,2),x(:,3))`, where the animated display shows the trends of the trajectory.

From the above example, we can observe that if the ODEs to be solved can be expressed by first-order explicit ODEs, using `ode45()` can immediately find the numerical solutions. Thus preparing a MATLAB function describing the ODEs is a crucial step in solving the initial value problems.

Solving ODEs in MATLAB with additional arguments

In the ODE solving process, frequently one may introduce additional arguments so that when the parameters in the ODEs are changed, they can be modified through the additional arguments rather than having to modify the MATLAB function itself. For instance, the Lorenz equations in Example 7.8 contain parameters such as β, ρ, and σ and they can all be considered as additional arguments. Thus when their values change, one does not have to modify the MATLAB function describing the Lorenz equations. The following example illustrates the method and benefits of using additional arguments in ODE solver.

Example 7.9 Write a MATLAB function to describe the Lorenz equations in Example 7.8 with additional arguments. Then use the new function to find the numerical solutions for another set of parameters $\beta = 2, \rho = 5$, and $\sigma = 20$.

Solution Select the variables β, ρ, and σ as the additional arguments. The new anonymous function for the ODEs can then be written as

```
>> f=@(t,x,beta,rho,sigma)[-beta*x(1)+x(2)*x(3);...
    -rho*x(2)+rho*x(3); -x(1)*x(2)+sigma*x(2)-x(3)];
```

From the above new function, the following statements can be used instead for the numerical solutions

```
>> t_final=100; x0=[0;0;1e-10];
   b1=8/3; r1=10; s1=28;  % one may also use other variable names
   [t,x]=ode45(f,[0,t_final],x0,[],b1,r1,s1); plot(t,x),
   figure; plot3(x(:,1),x(:,2),x(:,3)); axis([10 42 -20 20 -20 25]);
```

From the above statements, it can be seen that the additional arguments can easily be passed to the RHS function. Moreover, the `options` variable is now replaced by an empty matrix, which means that the control parameters need not be changed.

With the use of MATLAB functions with additional arguments, the Lorenz equations with other values of β, ρ, and σ can be solved directly, without the need of modifying the existing anonymous function. For instance, for $\beta = 1, \rho = 5$, and $\sigma = 20$, the following statements can be used to find the numerical solutions. The time responses and phase space trajectory obtained are shown in Figures 7.4 (a) and (b), respectively.

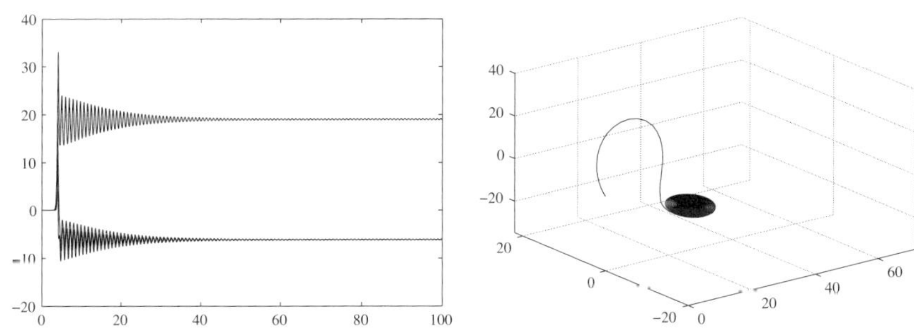

(a) time responses of state variables (b) three-dimensional phase trajectory

FIGURE 7.4: Results of Lorenz equations under a new set of parameters

```
>> t_final=100; x0=[0;0;1e-10];
   b2=2; r2=5; s2=20;
   [t2,x2]=ode45(f,[0,t_final],x0,[],b2,r2,s2); plot(t2,x2),
   figure; plot3(x2(:,1),x2(:,2),x2(:,3)); axis([0 72 -20 22 -35 40]);
```

7.2.4 Transforms to standard ODEs

From the above description and introduction, it can be seen that only the first-order vector form explicit ODEs can be solved using the relevant ODE solvers in MATLAB. If the ODEs are described by high-order differential equations, one has to convert the ODEs first into the first-order vector form explicit ODEs. In this subsection, two different cases are explored.

Manipulating a single high-order ODE

Assume that a high-order ordinary differential equation can be expressed as

$$y^{(n)} = f(t, y, \dot{y}, \cdots, y^{(n-1)}) \tag{7.25}$$

and the initial conditions of the output signal and its derivatives are $y(0), \dot{y}(0), \cdots, y^{(n-1)}(0)$. One may select a set of state variables $x_1 = y, x_2 = \dot{y}, \cdots, x_n = y^{(n-1)}$, such that the high-order ODE can be converted into the following first-order vector form explicit ODEs

$$\begin{cases} \dot{x}_1 = x_2 \\ \dot{x}_2 = x_3 \\ \qquad \vdots \\ \dot{x}_n = f(t, x_1, x_2, \cdots, x_n) \end{cases} \tag{7.26}$$

with initial states $x_1(0) = y(0), x_2(0) = \dot{y}(0), \cdots, x_n(0) = y^{(n-1)}(0)$. Thus the converted ODEs can directly be solved using the numerical methods introduced earlier.

Example 7.10 Consider again the Van der Pol equation $\ddot{y} + \mu(y^2 - 1)\dot{y} + y = 0$. If the initial conditions $y(0) = -0.2, \dot{y}(0) = -0.7$ are given, solve numerically the Van der Pol equation for different μ's.

Solution Since the MATLAB functions illustrated earlier can only deal with first-order explicit differential equations, thus conversion should be made before the problem can be solved numerically using the MATLAB ODE solvers. For this example, by choosing the state variables $x_1 = y, x_2 = \dot{y}$, the original ODE can be converted to

$$\begin{cases} \dot{x}_1 = x_2 \\ \dot{x}_2 = -\mu(x_1^2 - 1)x_2 - x_1. \end{cases}$$

It is not wise to write one function for each interested value of μ. The concept of additional parameters should be introduced so as to pass the value of μ to the ODE right-hand-side function. Thus an anonymous function can be written. The calling syntax of `ode45()` is by additional argument. The initial state vector is given by $\boldsymbol{x} = [-0.2, -0.7]^{\mathrm{T}}$ and the final solution can be obtained from

```
>> f=@(t,x,mu)[x(2); -mu*(x(1)^2-1)*x(2)-x(1)];
   x0=[-0.2;-0.7]; t_final=20;
```

```
mu=1; [t1,y1]=ode45(f,[0,t_final],x0,[],mu);
mu=2; [t2,y2]=ode45(f,[0,t_final],x0,[],mu);
plot(t1,y1,t2,y2,':')
figure; plot(y1(:,1),y1(:,2),y2(:,1),y2(:,2),':')
```

and for $\mu = 1, 2$, the time responses and the phase plane trajectories are shown respectively in Figures 7.5 (a) and (b). It can be seen that both the phase plane trajectories settle down on a common closed-path. The closed-path is referred to as the *limit cycle*.

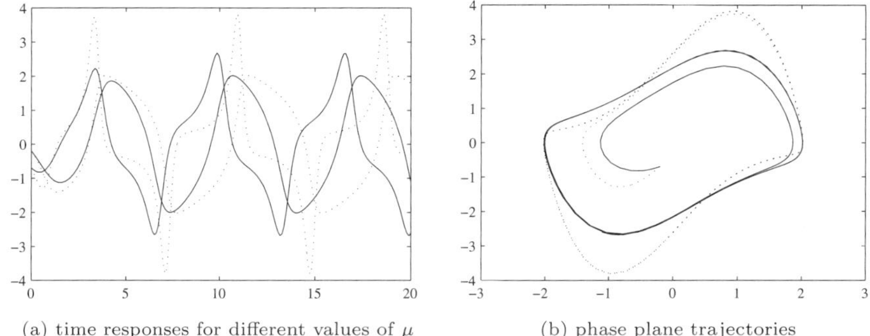

(a) time responses for different values of μ (b) phase plane trajectories

FIGURE 7.5: Van der Pol equation solutions for different values of μ's

If one changes the value of μ, say, $\mu = 1000$, and sets the terminate time $t_f = 3000$, the following statements can be used to find the numerical solution to the corresponding Van der Pol equation.

```
>> x0=[2;0]; t_final=3000;
   mu=1000; [t,y]=ode45(f,[0,t_final],x0,[],mu);
```

However, even after a long wait, the solutions cannot be found, since the step-size used might be too small and too many computation points may be involved. Thus this kind of ODE IVPs may not be suitably solvable by `ode45()` functions. In the next section, stiff equations-based algorithms will be presented to solve the problem.

Manipulating multiple high-order ODEs

Now consider the differential equation sets composed of several high-order differential equations. For example,

$$\begin{cases} x^{(m)} = f(t, x, \dot{x}, \cdots, x^{(m-1)}, y, \cdots, y^{(n-1)}) \\ y^{(n)} = g(t, x, \dot{x}, \cdots, x^{(m-1)}, y, \cdots, y^{(n-1)}). \end{cases} \tag{7.27}$$

Let us still select the state variables $x_1 = x, x_2 = \dot{x}, \cdots, x_m = x^{(m-1)}, x_{m+1} = y, x_{m+2} = \dot{y}, \cdots, x_{m+n} = y^{(n-1)}$. Thus the original high-order ODEs can be converted to

$$\begin{cases} \dot{x}_1 = x_2 \\ \quad\vdots \\ \dot{x}_m = f(t, x_1, x_2, \cdots, x_{m+n}) \\ \dot{x}_{m+1} = x_{m+2} \\ \quad\vdots \\ \dot{x}_{m+n} = g(t, x_1, x_2, \cdots, x_{m+n}). \end{cases} \tag{7.28}$$

Thus the desirable first-order vector form explicit ODEs can be obtained. The following example illustrates the conversion and solution process.

Example 7.11 The trajectory (x, y) of the Apollo satellite satisfies the following equation sets[29]

$$\ddot{x} = 2\dot{y} + x - \frac{\mu^*(x+\mu)}{r_1^3} - \frac{\mu(x-\mu^*)}{r_2^3}, \quad \ddot{y} = -2\dot{x} + y - \frac{\mu^* y}{r_1^3} - \frac{\mu y}{r_2^3}$$

where $\mu = 1/82.45$, $\mu^* = 1 - \mu$, $r_1 = \sqrt{(x+\mu)^2 + y^2}$, $r_2 = \sqrt{(x-\mu^*)^2 + y^2}$. It is also known that the initial values are given by $x(0) = 1.2$, $\dot{x}(0) = 0$, $y(0) = 0$, $\dot{y}(0) = -1.04935751$. Solve the differential equations and draw the trajectory of (x, y).

Solution The state variables are chosen as $x_1 = x, x_2 = \dot{x}, x_3 = y, x_4 = \dot{y}$. Thus the first-order explicit ODEs can be found as follows

$$\begin{cases} \dot{x}_1 = x_2 \\ \dot{x}_2 = 2x_4 + x_1 - \mu^*(x_1+\mu)/r_1^3 - \mu(x_1-\mu^*)/r_2^3 \\ \dot{x}_3 = x_4 \\ \dot{x}_4 = -2x_2 + x_3 - \mu^* x_3/r_1^3 - \mu x_3/r_2^3 \end{cases}$$

where $r_1 = \sqrt{(x_1+\mu)^2 + x_3^2}$, $r_2 = \sqrt{(x_1-\mu^*)^2 + x_3^2}$, and $\mu = 1/82.45, \mu^* = 1 - \mu$.

From the above mathematical equations obtained, the MATLAB function describing the original ODEs is prepared as follows:

```
function dx=apolloeq(t,x)
mu=1/82.45; mu1=1-mu;
r1=sqrt((x(1)+mu)^2+x(3)^2); r2=sqrt((x(1)-mu1)^2+x(3)^2);
dx=[x(2);
    2*x(4)+x(1)-mu1*(x(1)+mu)/r1^3-mu*(x(1)-mu1)/r2^3;
    x(4);
   -2*x(2)+x(3)-mu1*x(3)/r1^3-mu*x(3)/r2^3];
```

Since there are some intermediate computation steps involved, the anonymous or inline functions are not suitable. Using `ode45()` function, the numerical solutions of the equations are as follows, with the trajectory shown in Figure 7.6 (a).

```
>> x0=[1.2; 0; 0; -1.04935751];
   tic, [t,y]=ode45(@apolloeq,[0,20],x0); toc
   length(t), plot(y(:,1),y(:,3))
```

For this example, the elapsed time is 0.33 seconds. Also the number of points calculated is returned by the use of `length()` function. For this example, the number of points is 689.

In fact, the obtained trajectory is not correct, since in the simulation control parameters, the relative error tolerance `RelTol` is too large. To have more accurate results, one usually should reduce that value, for instance, to 1×10^{-6}. Let us use the following statements to solve the ODEs again:

```
>> options=odeset; options.RelTol=1e-6;
   tic, [t1,y1]=ode45(@apolloeq,[0,20],x0,options); toc
   length(t1), plot(y1(:,1),y1(:,3)),
```

The elapsed time is 0.701 seconds, with 1873 computation points. The new trajectory obtained is shown in Figure 7.6 (b). It can be seen that the different results are obtained and further reducing the error tolerance will yield the same numerical results. It can be concluded that it is always necessary to validate the simulation results after simulation by reducing the error tolerance `RelTol`.

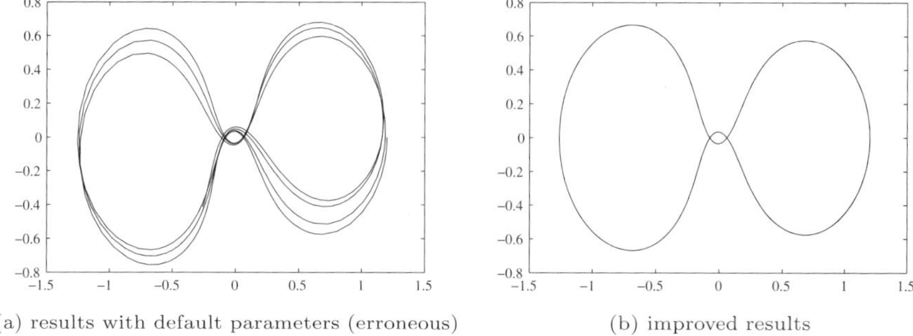

(a) results with default parameters (erroneous) (b) improved results

FIGURE 7.6: The trajectories of Apollo with different numerical accuracy specifications

The following statements can be used to find the step-sizes used in the simulation process, as shown in Figure 7.7. The minimum step-size is 1.8927×10^{-4}.

```
>> plot(t1(1:end-1),diff(t1)), min(diff(t1))
```

From the above step-size curve, it can be seen that when variable-step-size algorithms are used, the step-size can be adapted according to the error accuracy requirements. In part of the simulation period, step-sizes larger than 0.03 are used. However, in order to keep precision under control, at some points, very small step-size of 2×10^{-4} is used. If fixed-step algorithms are used, in order to make sure that the simulation results reliable, a small step-size of 2×10^{-4} should be used in the whole simulation process. The computation required becomes a total of 10^5 steps, which is 56 times more than the number of points used in the variable-step algorithm. Therefore, variable-step-size approaches are more effective.

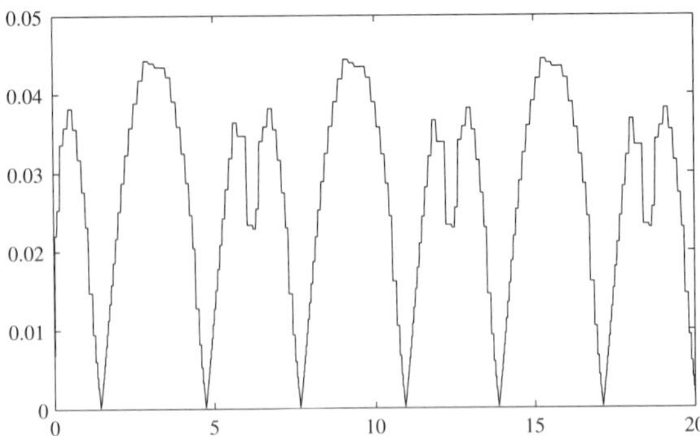

FIGURE 7.7: The step-sizes over time in the simulation process

Example 7.12 Solve the Apollo problem using fixed-step Runge-Kutta algorithm.

Solution To use fixed-step algorithms, one should consider two problems first: (i) how to select the step-size, (ii) how to ensure the accuracy in computation. The first problem can only be solved by trial-and-error approach. When selecting step-sizes, small step-size can be tested. However, the computational cost could be very high. For instance, a step-size of 0.01 will produce the trajectory shown in Figure 7.8 (a). The computation time is 2.654 second.

```
>> x0=[1.2; 0; 0; -1.04935751];
   tic, [t1,y1]=rk_4(@apolloeq,[0,20,0.01],x0); toc
   plot(y1(:,1),y1(:,3))      % draw trajectory
```

It is obvious that the results thus obtained are wrong, thus a smaller step-size should be used instead. If one selects a smaller step 0.001, more accurate results can be obtained and the trajectory is shown in Figure 7.8 (b). However, the time required is 97.079 seconds, which is about 138 times the time required for the variable-step computation.

```
>> tic, [t2,y2]=rk_4(@apolloeq,[0,20,0.001],x0); toc
   plot(y2(:,1),y2(:,3))      % draw trajectory
```

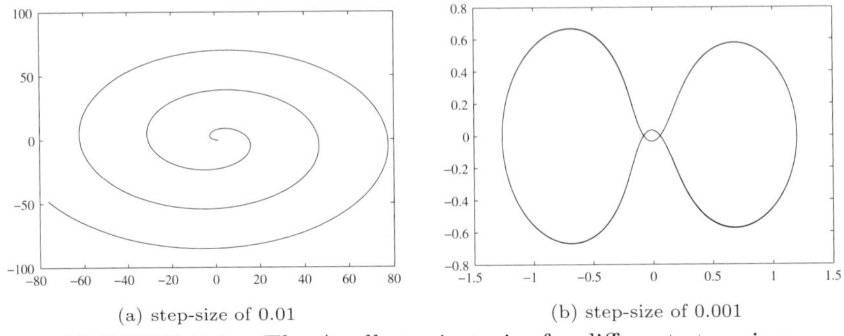

(a) step-size of 0.01 (b) step-size of 0.001

FIGURE 7.8: The Apollo trajectories for different step-sizes

In fact, strictly speaking, the results obtained using the fixed-step algorithm cannot satisfy the requirement of 10^{-6} relative error tolerance, although the shape of the responses is similar to the one obtained from variable-step ODE solvers.

The following example illustrates how to transform two simultaneous high-order ODEs into the standard first-order vector form ODEs involving symbolic computation.

Example 7.13 Consider the following high-order ODEs

$$\begin{cases} \ddot{x} + 2\dot{y}x = 2\ddot{y} \\ \ddot{x}\dot{y} + 3\dot{x}\ddot{y} + x\dot{y} - y = 5. \end{cases}$$

Convert it into the first-order explicit ODEs.

Solution In the above two equations, both $\ddot{x}$ and $\ddot{y}$ are cross-related. Let us still select the state variables as $x_1 = x, x_2 = \dot{x}, x_3 = y, x_4 = \dot{y}$. Our purpose is to eliminate one of the high-order terms and solve for the other. Thus it can be found that the analytical expression of $\ddot{y}$ is

$$\ddot{y} = \dot{y}x + \frac{\ddot{x}}{2}.$$

Substituting it into the section equation, $\ddot{x}$ can be found that

$$\ddot{x} = \frac{2y + 10 - 2x\dot{y} - 6x\dot{x}\dot{y}}{2\dot{y} + 3\dot{x}}.$$

Thus the state equation can be written as

$$\dot{x}_2 = \frac{2x_3 + 10 - 2x_1x_4 - 6x_1x_2x_4}{2x_4 + 3x_2}, \quad \dot{x}_4 = \frac{x_3 + 5 - x_1x_4 + 2x_1x_4^2}{2x_4 + 3x_2}.$$

Summarizing, the converted first-order explicit differential equation can be written as

$$\begin{cases} \dot{x}_1 = x_2 \\ \dot{x}_2 = \dfrac{2x_3 + 10 - 2x_1x_4 - 6x_1x_2x_4}{2x_4 + 3x_2} \\ \dot{x}_3 = x_4 \\ \dot{x}_4 = \dfrac{x_3 + 5 - x_1x_4 + 2x_1x_4^2}{2x_4 + 3x_2}. \end{cases}$$

In fact, some of the above equations may not be easily solved manually. Sometimes one may solve the conversion problem using Symbolic Math Toolbox. For simplicity, denote `dx`$=\ddot{x}$ and `dy`$=\ddot{y}$, thus `dx` and `dy` are $\dot{x}_2$ and $\dot{x}_4$. The following statements can be used to solve the equations

```
>> syms x1 x2 x3 x4
   [dx,dy]=solve('dx+2*x4*x1=2*dy','dx*x4+3*x2*dy+x1*x4-x3=5','dx,dy')
```

The derivatives of the variables can be written as

$$\dot{x}_2 = -2\frac{3x_4x_1x_2 - 5 + x_4x_1 - x_3}{3x_2 + 2x_4}, \quad \dot{x}_4 = \frac{2x_4^2x_1 + 5 - x_4x_1 + x_3}{3x_2 + 2x_4}.$$

It can be seen that the results are exactly the same as the previous results.

For more complicated problems, solving the corresponding algebraic equations manually might be extremely difficult, if not impossible. Thus, algebraic equations may be used in the MATLAB function describing the first-order explicit ODEs. These mixed differential algebraic equations (DAEs) and their numerical solution methods will be discussed in the following section.

Transformation of differential matrix equations

In some real applications, matrix-type ODEs are preferred. For instance, the Lagrange equation can be expressed as

$$\boldsymbol{M}\ddot{\boldsymbol{X}}+\boldsymbol{C}\dot{\boldsymbol{X}}+\boldsymbol{K}\boldsymbol{X}=\boldsymbol{F}u(t) \tag{7.29}$$

where $\boldsymbol{M},\boldsymbol{C},\boldsymbol{K}$ are $n\times n$ matrices, and $\boldsymbol{X},\boldsymbol{F}$ are $n\times 1$ column vectors. Introducing the vectors $\boldsymbol{x}_1=\boldsymbol{X}$ and $\boldsymbol{x}_2=\dot{\boldsymbol{X}}$, it is found that $\dot{\boldsymbol{x}}_1=\boldsymbol{x}_2$, and $\dot{\boldsymbol{x}}_2=\ddot{\boldsymbol{X}}$. From (7.29), it is found that $\ddot{\boldsymbol{X}}=\boldsymbol{M}^{-1}\Big[\boldsymbol{F}u(t)-\boldsymbol{C}\dot{\boldsymbol{X}}-\boldsymbol{K}\boldsymbol{X}\Big]$. The new state vector can be expressed as $\boldsymbol{x}=[\boldsymbol{x}_1^{\mathrm{T}},\boldsymbol{x}_2^{\mathrm{T}}]^{\mathrm{T}}$, the state-space equation can be rewritten as

$$\dot{\boldsymbol{x}}(t)=\begin{bmatrix}\boldsymbol{x}_2(t)\\ \boldsymbol{M}^{-1}\Big[\boldsymbol{F}u(t)-\boldsymbol{C}\boldsymbol{x}_2(t)-\boldsymbol{K}\boldsymbol{x}_1(t)\Big]\end{bmatrix} \tag{7.30}$$

which is exactly an explicit vector form first-order ODE directly solvable with the corresponding MATLAB ODE solver functions.

Example 7.14 Consider the inverted pendulum model expressed as[30]

$$\boldsymbol{M}(\boldsymbol{\theta})\ddot{\boldsymbol{\theta}}+\boldsymbol{C}(\boldsymbol{\theta},\dot{\boldsymbol{\theta}})\dot{\boldsymbol{\theta}}=\boldsymbol{F}(\boldsymbol{\theta})$$

where $\boldsymbol{\theta}=[a,\theta_1,\theta_2]^{\mathrm{T}}$, and a is the position of the cart, and θ_1,θ_2 are the angles of the two bars. The matrices are given by

$$\boldsymbol{M}(\boldsymbol{\theta})=\begin{bmatrix} m_c+m_1+m_2 & (0.5m_1+m_2)L_1\cos\theta_1 & 0.5m_2L_2\cos\theta_2\\ (0.5m_1+m_2)L_1\cos\theta_1 & (m_1/3+m_2)L_1^2 & 0.5m_2L_1L_2\cos\theta_1\\ 0.5m_2L_2\cos\theta_2 & 0.5m_2L_1L_2\cos\theta_1 & m_2L_2^2/3\end{bmatrix}$$

$$\boldsymbol{C}(\boldsymbol{\theta},\dot{\boldsymbol{\theta}})=\begin{bmatrix}0 & -(0.5m_1+m_2)L_1\dot{\theta}_1\sin\theta_1 & -0.5m_2L_2\dot{\theta}_2\sin\theta_2\\ 0 & 0 & 0.5m_2L_1L_2\dot{\theta}_2\sin(\theta_1-\theta_2)\\ 0 & -0.5m_2L_1L_2\dot{\theta}_1\sin(\theta_1-\theta_2) & 0\end{bmatrix}$$

$$\boldsymbol{F}(\boldsymbol{\theta})=\begin{bmatrix}u(t)\\ (0.5m_1+m_2)L_1g\sin\theta_1\\ 0.5m_2L_2g\sin\theta_2\end{bmatrix}.$$

The parameters for a particular experimental system are $m_c=0.85$kg, $m_1=0.04$kg, $m_2=0.14$kg, $L_1=0.1524$m, and $L_2=0.4318$m. Find the step response of the system.

Solution The coefficient matrices $\boldsymbol{M}(\theta_1,\theta_2),\boldsymbol{C}(\theta_1,\theta_2)$ and $\boldsymbol{F}(\theta_1,\theta_2)$ contain the nonlinear terms of the state vector $\boldsymbol{x}$, for instance the cosine terms of θ_1. Introducing additional variables $\boldsymbol{x}_1=\boldsymbol{\theta}$ and $\boldsymbol{x}_2=\dot{\boldsymbol{\theta}}$, the new state vector $\boldsymbol{x}=[\boldsymbol{x}_1^{\mathrm{T}},\boldsymbol{x}_2^{\mathrm{T}}]^{\mathrm{T}}$ can be constructed and the explicit first-order differential equation can be established

```
function dx=inv_pendulum(t,x,u,mc,m1,m2,L1,L2,g)
M=[mc+m1+m2, (0.5*m1+m2)*L1*cos(x(2)), 0.5*m2*L2*cos(x(3))
```

```
    (0.5*m1+m2)*L1*cos(x(2)),(m1/3+m2)*L1^2,0.5*m2*L1*L2*cos(x(2))
    0.5*m2*L2*cos(x(3)),0.5*m2*L1*L2*cos(x(2)),m2*L2^2/3];
C=[0,-(0.5*m1+m2)*L1*cos(x(5))*sin(x(2)),-0.5*m2*L2*x(6)*sin(x(3))
    0, 0, 0.5*m2*L1*L2*x(6)*sin(x(2)-x(3))
    0, -0.5*m2*L1*L2*x(5)*sin(x(2)-x(3)), 0];
F=[u; (0.5*m1+m2)*L1*g*sin(x(2)); 0.5*m2*L2*g*sin(x(3))];
dx=[x(4:6); inv(M)*(F-C*x(4:6))];
```

The input signal $u(t)$ is a step signal, and the following statements can be used to solve numerically the ODE. The results are shown in Figures 7.9 (a) and (b).

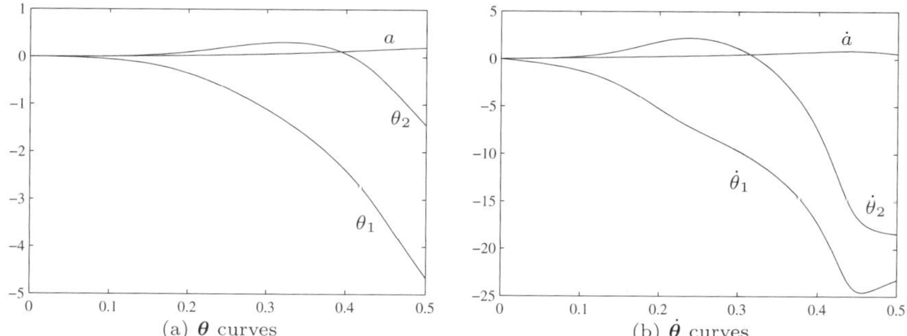

(a) $\boldsymbol{\theta}$ curves (b) $\dot{\boldsymbol{\theta}}$ curves

FIGURE 7.9: Step responses of the inverted pendulum

```
>> opt=odeset; opt.RelTol=1e-8; u=1; mc=0.85;
   m1=0.04; m2=0.14; L1=0.1524; L2=0.4318; g=9.81;
   [t,x]=ode45(@inv_pendulum,[0,10],zeros(6,1),opt,...
             u,mc,m1,m2,L1,L2,g);
   plot(t,x(:,1:3)), figure; plot(t,x(:,4:6))
```

It should be noted that the inverted pendulum system is naturally unstable. A properly designed input signal constructed from the state vector information should be applied to stabilize the system.

Furthermore, if the matrices $\boldsymbol{M}$, $\boldsymbol{C}$, $\boldsymbol{K}$ and $\boldsymbol{F}$ are independent of $\boldsymbol{X}$, the original matrix type ODE becomes a linear ODE. The linear state-space equation can easily be derived through the following simple transformation

$$\begin{bmatrix} \dot{\boldsymbol{x}}_1(t) \\ \dot{\boldsymbol{x}}_2(t) \end{bmatrix} = \begin{bmatrix} \boldsymbol{0} & \boldsymbol{I} \\ -\boldsymbol{M}^{-1}\boldsymbol{K} & -\boldsymbol{M}^{-1}\boldsymbol{C} \end{bmatrix} \begin{bmatrix} \boldsymbol{x}_1(t) \\ \boldsymbol{x}_2(t) \end{bmatrix} + \begin{bmatrix} \boldsymbol{0} \\ \boldsymbol{M}^{-1}\boldsymbol{F} \end{bmatrix} u(t). \tag{7.31}$$

7.2.5 Validation of numerical solutions to ODEs

It has been shown through examples that if the control parameters are not properly chosen, the solutions may not even be correct. Thus the numerical solutions should be validated. However, since the equations have no analytic solutions, an alternative method to validate the solution is by setting the control parameters to different values and checking whether they yield the same results. The most effective control parameter is the `RelTol` property. The default value for it is 10^{-3}, which is usually too large for many applications. It can be set to 10^{-6} or even 10^{-8}. This will not add too much computation effort. Finally, selecting different ODE solvers may cross-validate the results.

7.3 Numerical Solutions to Special Ordinary Differential Equations

From the introduction and examples in Section 7.2, one can easily convert a given ODE into first-order vector form explicit ODEs. The function `ode45()` can then be used to solve the ODEs. However, it is also shown from some examples, for instance the Van der Pol equation with $\mu = 1000$ cannot be solved by using the `ode45()` function. Thus other types of ODEs should be introduced, for instance, the stiff equations. Special MATLAB functions can be used to solve the stiff equation problems. Moreover, some special types of differential equations such as differential algebraic equations, implicit differential equations and delay differential equations will be discussed in this section.

7.3.1 Solutions of stiff ODEs

In many differential equations, some states change very rapidly while others may change very slowly. This type of differential equations is usually referred to as the *stiff equations*. The function `ode45()` is not suitable for stiff equations. An alternative function, `ode15s()`, can be used instead and this function has exactly the same syntax as that of the `ode45()` function.

Example 7.15 Find the numerical solutions to the Van der Pol equation, when $\mu = 1000$.

Solution Similar to the statements given earlier, if the function `ode15s()` is used, the states can be obtained directly in 1.943 seconds and the time responses of the states are shown in Figure 7.10.

```
>> h_opt=odeset; h_opt.RelTol=1e-6; x0=[2;0]; t_final=3000;
   f=@(t,x,mu)[x(2); -mu*(x(1)^2-1)*x(2)-x(1)];
   tic, mu=1000; [t,y]=ode15s(f,[0,t_final],x0,h_opt,mu); toc
   plot(t,y(:,1)); figure; plot(t,y(:,2))
```

Example 7.16 In classical textbooks regarding numerical solutions to ordinary differential equations, the following ODE is regarded as stiff.

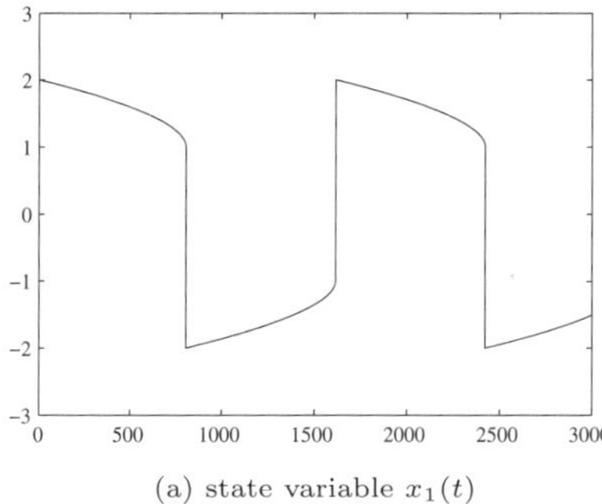

(a) state variable $x_1(t)$

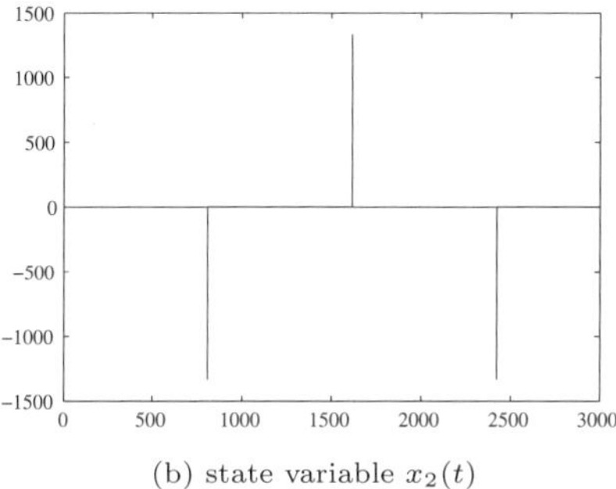

(b) state variable $x_2(t)$

FIGURE 7.10: Solutions of Van der Pol equation with $\mu = 1000$

$$\dot{\boldsymbol{y}} = \begin{bmatrix} -21 & 19 & -20 \\ 19 & -21 & 20 \\ 40 & -40 & -40 \end{bmatrix} \boldsymbol{y}, \quad \boldsymbol{y}_0 = \begin{bmatrix} 1 \\ 0 \\ -1 \end{bmatrix}.$$

Find the numerical solutions with MATLAB.

Solution The analytical solutions of the equations can be found symbolically by the following statements

```
>> syms t; A=sym([-21,19,-20; 19,-21,20; 40,-40,-40]);
   y0=[1; 0; -1]; y=expm(A*t)*y0
```

which yields

$$\boldsymbol{y}(t) = \begin{bmatrix} 0.5\mathrm{e}^{-2t} + 0.5\mathrm{e}^{-40t}(\cos 40t + \sin 40t) \\ 0.5\mathrm{e}^{-2t} - 0.5\mathrm{e}^{-40t}(\cos 40t + \sin 40t) \\ \mathrm{e}^{-40t}(\sin 40t - \cos 40t) \end{bmatrix}.$$

Now let us consider the numerical solutions to the same problem. Prepare an anonymous function, and find the numerical solutions by the following statements

```
>> opt=odeset; opt.RelTol=1e-6;
   f=@(t,x)[-21,19,-20; 19,-21,20; 40,-40,-40]*x;
   tic,[t,y]=ode45(f,[0,1],[1;0;-1],opt); toc
   x1=exp(-2*t); x2=exp(-40*t).*cos(40*t); x3=exp(-40*t).*sin(40*t);
   y1=[0.5*x1+0.5*x2+0.5*x3, 0.5*x1-0.5*x2-0.5*x3, -x2+x3];
   plot(t,y,t,y1,':')
```

and it can be seen that only 0.16 seconds are used to find the solutions using the `ode45()` function. The analytical and numerical solutions can be shown in Figure 7.11 (a). It can be seen that the accuracy of the results are rather high, and computation speed is also very high. The stiffness of the problem seems to be not very obvious. This is because the variable-step algorithm is used and the step-size can adaptively be modified so the stiffness of the problem may not cause serious issues. However, if a fixed-step algorithm is used, for instance, the fourth-order Runge-Kutta algorithm is applied, the following statements

```
>> tic,[t2,y2]=rk_4(f,[0,1,0.01],[1;0;-1]); toc
   plot(t,y1,t2,y2,':')
```

produce numerical results shown in Figure 7.11 (b) together with the analytical curves. The elapsed time is 0.21 seconds, which is slightly longer than the direct use of `ode45()` function. From the results it can be seen that the fixed-step approach gives erroneous results if the step-size if large. Further reducing the step-size until $h = 0.0001$ seconds, one can still see the difference between the analytical and numerical results. However, this time, the time required is about 26 seconds, which is 162 times longer than the time required for the `ode45()` function. Thus in practical applications, the variable-step algorithm should be used whenever possible.

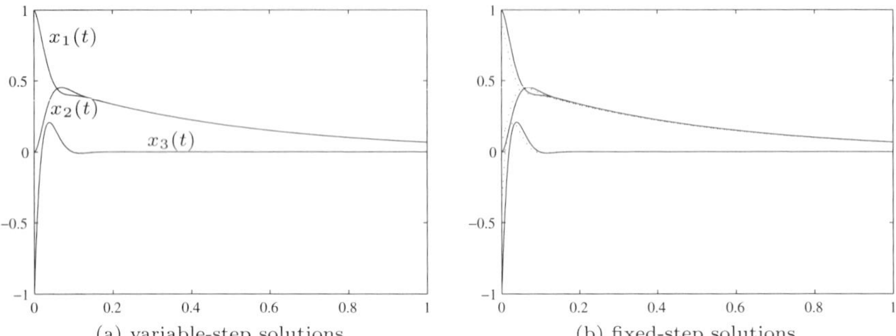

FIGURE 7.11: Comparisons of solutions of a traditional stiff equation

It can further be tested that in the fixed-step algorithm, if the step-size is 0.00001, the time required may as long as 8124 seconds, more than two hours.

It can be concluded that for many conventional stiff equations, one may still try to use the ordinary ODE solvers such as `ode45()` rather than the stiff equation solver. If for some examples, the time consumed using `ode45()` is too much, the stiff equation solver should be used instead. This will be illustrated in the following Example 7.17.

Example 7.17 Consider the following ODEs

$$\begin{cases} \dot{y}_1 = 0.04(1-y_1) - (1-y_2)y_1 + 0.0001(1-y_2)^2 \\ \dot{y}_2 = -10^4 y_1 + 3000(1-y_2)^2 \end{cases}$$

where the initial values are $y_1(0) = 0, y_2(0) = 1$. Within the time interval $t \in (0, 100)$, find a suitable algorithm for the IVP.

Solution For the given differential equations, an anonymous function can be written to describe the right-hand-side functions, then the following statements can be issued in the MATLAB command window to find the numerical solutions.

```
>> f=@(t,y)[0.04*(1-y(1))-(1-y(2))*y(1)+0.0001*(1-y(2))^2;
            -10^4*y(1)+3000*(1-y(2))^2];
   tic, [t2,y2]=ode45(f,[0,100],[0;1]); toc
   length(t2), plot(t2,y2)
```

After a long wait for about 105.14 seconds, the numerical solutions can be obtained. The number of points computed is 356,941. The numerical solutions are shown in Figure 7.12 (a). It is found that the function `ode45()` takes too much time. Also the step-sizes can be obtained from the following statements

```
>> plot(t2(1:end-1),diff(t2))
```

and the step-sizes are shown in Figure 7.12 (b). It can be seen that during the whole simulation period the step-sizes are very small. In most of the time interval, the step-sizes are about 0.004, which increases significantly the computation time. The adaptation in the step-size also consumes a tremendous amount of time.

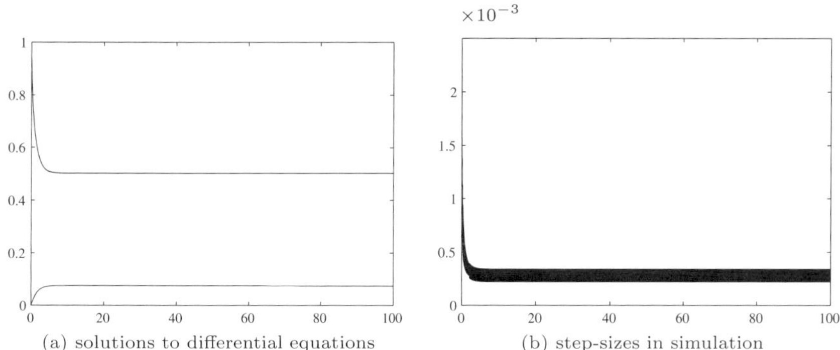

FIGURE 7.12: Simulation using the 4/5 Runge-Kutta-Felhberg algorithm

Now consider the use of `ode15s()` instead, the following statements can be used

```
>> opt=odeset; opt.RelTol=1e-6;
   tic,[t1,y1]=ode15s(f,[0,100],[0;1],opt); toc
   length(t1), plot(t1,y1)
```

and it can be seen that the time required is reduced significantly to 0.281 seconds, which is 1/375 of the time by `ode45()`. The number of points computed is 169. The curves obtained are almost identical to the previous results.

7.3.2 Solutions of implicit differential equations

The so-called *implicit differential equations* are those not convertible into the first-order vector form explicit ODEs as in (7.17). In earlier versions of MATLAB, solvers for these implicit ODEs were not provided. The following two examples demonstrate the implicit ODE solution procedures.

Example 7.18 Consider the following implicit ODE

$$\begin{cases} \sin x_1\dot{x}_1 + \cos x_2\dot{x}_2 + x_1 = 1 \\ -\cos x_2\dot{x}_1 + \sin x_1\dot{x}_2 + x_2 = 0 \end{cases}$$

where $x_1(0) = x_2(0) = 0$. Find the numerical solutions to the IVP.

Solution Let $\boldsymbol{x} = [x_1, x_2]^{\mathrm{T}}$. The matrix form of the ODE can be written as

$$\boldsymbol{A}(\boldsymbol{x})\dot{\boldsymbol{x}} = \boldsymbol{B}(\boldsymbol{x}), \text{ where } \boldsymbol{A}(\boldsymbol{x}) = \begin{bmatrix} \sin x_1 & \cos x_2 \\ -\cos x_2 & \sin x_1 \end{bmatrix}, \quad \boldsymbol{B}(\boldsymbol{x}) = \begin{bmatrix} 1 - x_1 \\ -x_2 \end{bmatrix}.$$

If $\boldsymbol{A}(\boldsymbol{x})$ is a non-singular matrix for all $\boldsymbol{x}$, the first-order vector form explicit ODE can be expressed by $\dot{\boldsymbol{x}} = \boldsymbol{A}^{-1}(\boldsymbol{x})\boldsymbol{B}(\boldsymbol{x})$. Using the methods stated earlier, the numerical solutions can be obtained. However, it is not possible to prove theoretically that the matrix $\boldsymbol{A}(\boldsymbol{x})$ is a non-singular matrix. Thus one may initially assume that matrix $\boldsymbol{A}(\boldsymbol{x})$ is non-singular. During the ODE solving process, if there is no error message displayed, it will indicate that for the solutions the matrix $\boldsymbol{A}(\boldsymbol{x})$ is non-singular. Thus the solutions obtained are valid. If however error messages appear, then the solutions obtained are useless.

For the ODE to be solved, an anonymous function can be written and the following statements can be used

```
>> f=@(t,x)inv([sin(x(1)) cos(x(2)); -cos(x(2)) sin(x(1))])...
          *[1-x(1); -x(2)]; opt=odeset; opt.RelTol=1e-6;
   [t,x]=ode45(f,[0,10],[0; 0],opt); plot(t,x)
```

The obtained state variables are shown in Figure 7.13. Since in the solution process, no error messages are given, this indicates that for the solution points, the matrix $\boldsymbol{A}(t)$ is not singular. Thus the solutions obtained may be valid.

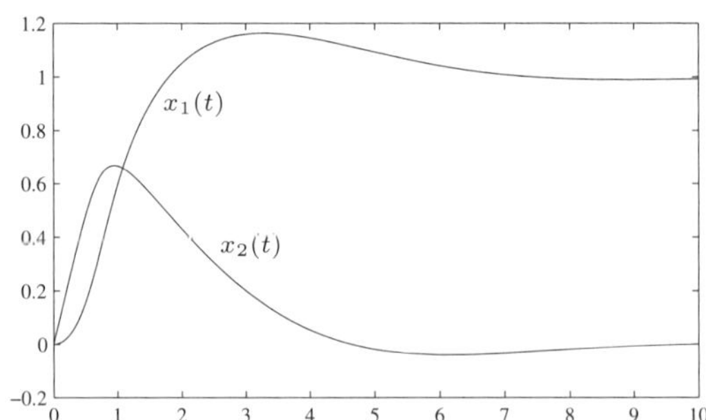

FIGURE 7.13: Time responses of the state variables

Example 7.19 The previous example is very simple and can be converted into explicit ODEs immediately. Now consider a set of more complicated implicit ODEs given by

$$\begin{cases} \ddot{x}\sin\dot{y} + \ddot{y}^2 = -2xy\mathrm{e}^{-\dot{x}} + x\ddot{x}\dot{y} \\ x\ddot{x}\ddot{y} + \cos\ddot{y} = 3y\dot{x}\mathrm{e}^{-x}. \end{cases}$$

The state variables can be selected as $x_1 = x, x_2 = \dot{x}, x_3 = y, x_4 = \dot{y}$. The initial states of the equation are $\boldsymbol{x} = [1, 0, 0, 1]^{\mathrm{T}}$. Find the numerical solutions of the IVP.

Solution Obviously the analytical solutions for state derivatives $\dot{x}_2$ and $\dot{x}_4$ cannot be written out as in Example 7.13. Hence a numerical-based converting algorithm is introduced such that the numerical solutions for each state variable $\dot{\boldsymbol{x}}$ can be obtained.

From the original equations, assume that $p_1 = \ddot{x}$, $p_2 = \ddot{y}$, then the following equation can be written as

$$\begin{cases} p_1 \sin x_4 + p_2^2 + 2x_1 x_3 \mathrm{e}^{-x_2} - x_1 p_1 x_4 = 0 \\ x_1 p_1 p_2 + \cos p_2 - 3x_3 x_2 \mathrm{e}^{-x_1} = 0 \end{cases}$$

and from the following MATLAB statements, the algebraic equation solution statements can be embedded into the MATLAB function describing the differential equations. The RHS function can then be written as

```
function dy=c7impode(t,x)
dx=@(p,x)[p(1)*sin(x(4))+p(2)^2+2*x(1)*x(3)*exp(-x(2))-x(1)*...
    p(1)*x(4); x(1)*p(1)*p(2)+cos(p(2))-3*x(3)*x(2)*exp(-x(1))];
ff=optimset; ff.Display='off'; dx1=fsolve(dx,x([1,3]),ff,x);
dy=[x(2); dx1(1); x(4); dx1(2)];
```

Once calling the function, from the input arguments $\boldsymbol{x}$, the newly defined variables p_1, p_2 can be used in the anonymous function, and with the use of the `fsolve()` function, the variables p_1 and p_2 can be numerically solved. In the above MATLAB statements, $\boldsymbol{x}$ is used as the additional argument. Thus from this function, the variables p_i can be obtained. Since the obtained p_1, p_2 are in fact the derivatives of the state variables such that $p_1 = \dot{x}_2$, $p_2 = \dot{x}_4$, the explicit differential equations can then be obtained and the corresponding MATLAB scripts are given in the function as well.

Once the explicit differential equations are obtained, the solutions to the implicit differential equations can be solved numerically using the following statements and the time responses of the states are shown in Figure 7.14.

```
>> [t,x]=ode15s(@c7impode,[0,2],[1,0,0,1]); plot(t,x)
```

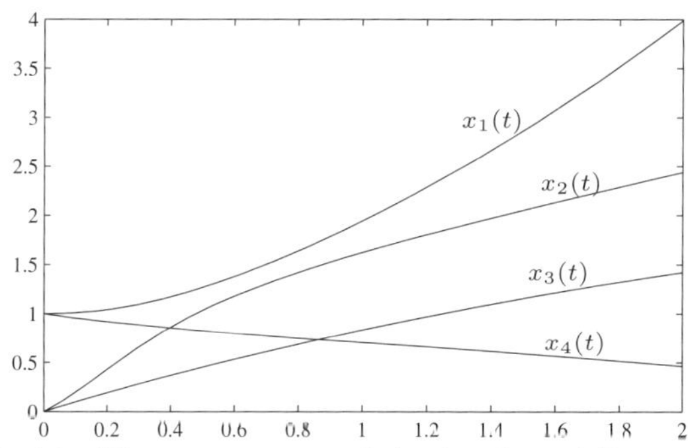

FIGURE 7.14: Time responses of the implicit differential equations

Starting from MATLAB 7.0, a new function `ode15i()` is provided for the solutions of implicit differential equations. If the mathematical descriptions to the implicit differential equations can be written as

$$\boldsymbol{F}[t, \boldsymbol{x}(t), \dot{\boldsymbol{x}}(t)] = 0, \text{ and } \boldsymbol{x}(0) = \boldsymbol{x}_0, \ \dot{\boldsymbol{x}}(0) = \dot{\boldsymbol{x}}_0 \tag{7.32}$$

then the function *fun* can be used to describe the implicit ODEs. The new MATLAB function `decic()` can be used to solve the undefined initial conditions. Then, the solver function `ode15i()` can be used to solve the implicit differential equations.

```
[x0*,xd0*]=decic(fun,t0,x0,x0F,xd0,xd0F)  % find consistent initial values
res=ode15i(fun,tspan,x0*,xd0*)            % solve implicit equations
```

The solution process of the implicit ODEs is different from the explicit ODEs. In numerically solving implicit ODEs, the initial state variables and their derivatives should both be declared and they cannot be assigned arbitrarily, otherwise there might be conflicting initial conditions. Before the solution, the $2n$ initial values ($\boldsymbol{x}_0$, $\dot{\boldsymbol{x}}_0$) can only have n independent ones. The rest of the values should be solved from the corresponding implicit algebraic equations. Thus in the actual solution process, if one cannot determine the values $\dot{\boldsymbol{x}}_0^*$, the function `decic()` can be used to solve for compatible initial values. In the function call, ($\boldsymbol{x}_0$, $\dot{\boldsymbol{x}}_0$) can be any initial values, while $\boldsymbol{x}_0^F$ and $\dot{\boldsymbol{x}}_0^F$ are both n-dimensional column vectors and when the vector element is 1, it means that the corresponding initial value is to be maintained, otherwise it indicates that the initial value should be resolved. From the corresponding algebraic equation solver, the compatible initial values $\boldsymbol{x}_0^*$ and $\dot{\boldsymbol{x}}_0^*$ can be obtained. The implicit differential equations can then be solved with the function `ode15i()`. The returned variables `res.x` and `res.y` are respectively $\boldsymbol{t}$ and $\boldsymbol{x}$ as in other ODE solvers. The following example demonstrates the numerical solution process of an implicit ODE.

Example 7.20 Solve the implicit ODEs given in Example 7.19 using the implicit ODE solver `ode15i()`.

Solution Still select the state variables $x_1 = x, x_2 = \dot{x}, x_3 = y, x_4 = \dot{y}$ and the original ODEs can be converted into the following:

$$\begin{cases} \dot{x}_1 - x_2 = 0 \\ \dot{x}_2 \sin x_4 + \dot{x}_4^2 + 2\mathrm{e}^{-x_2} x_1 x_3 - x_1 \dot{x}_2 x_4 = 0 \\ \dot{x}_3 - x_4 = 0 \\ x_1 \dot{x}_2 \dot{x}_4 + \cos \dot{x}_4 - 3\mathrm{e}^{-x_1} x_3 x_2 = 0. \end{cases}$$

Thus, the implicit ODEs can be entered in MATLAB using anonymous function. The definitions of the initial values $\boldsymbol{x}_0$ are exactly the same as before. The function `decic()` can be used to determine the initial derivatives. Thus $\boldsymbol{x}_0^F$ should be assigned to a vector of ones. Since a consistent $\dot{\boldsymbol{x}}_0$ is expected, the indicator $\dot{\boldsymbol{x}}_0^F$ should be set to an all zero vector. The following statements can be used to solve the implicit equations:

```
>> f=@(t,x,xd)[xd(1)-x(2);
        xd(2)*sin(x(4))+xd(4)^2+2*exp(-x(2))*x(1)*x(3)-x(1)*xd(2)*x(4);
        xd(3)-x(4);
        x(1)*xd(2)*xd(4)+cos(xd(4))-3*exp(-x(1))*x(3)*x(2)];
   x0=[1,0,0,1]; xd0=[0;1;1;-1]; x0F=[1 1 1 1]; xd0F=[];  % retain x0
   [x0,xd0]=decic(f,0,x0,x0F,xd0,xd0F)       % determine xd0 from x0
   r=ode15i(f,[0,2],x0,xd0); plot(r.x,r.y) % draw the time responses
```

With the `decic()` function, the consistent initial derivatives $\dot{\boldsymbol{x}}_0 = [0, 1.6833, 1, -0.5166]^{\mathrm{T}}$ can be obtained. Then the implicit equations can be solved and the time responses of can be drawn. It can be seen that the results are exactly the same with the results shown in Figure 7.14.

7.3.3 Solutions to differential algebraic equations

The so-called *differential algebraic equation* (DAE) means that some of the differential equations are degenerated to algebraic equations. Thus these algebraic equations appear as the constraints in the differential equations. Differential algebraic equations cannot be solved directly using the methods presented earlier.

The general form of the differential equations is given by

$$\boldsymbol{M}(t,\boldsymbol{x})\dot{\boldsymbol{x}} = \boldsymbol{f}(t,\boldsymbol{x}) \tag{7.33}$$

where the right-hand-side (RHS) function description $\boldsymbol{f}(t,\boldsymbol{x})$ is exactly the same as in the previous sections. For differential algebraic equations, the matrix $\boldsymbol{M}(t,\boldsymbol{x})$ is singular. Thus in the solutions options for MATLAB functions, the `Mass` property can be used to describe the matrix $\boldsymbol{M}(t,\boldsymbol{x})$ in MATLAB. Then the differential algebraic equations can be solved.

Example 7.21 Find the solutions to the following differential algebraic equation:

$$\begin{cases} \dot{x}_1 = -0.2x_1 + x_2x_3 + 0.3x_1x_2 \\ \dot{x}_2 = 2x_1x_2 - 5x_2x_3 - 2x_2^2 \\ 0 = x_1 + x_2 + x_3 - 1 \end{cases}$$

with initial conditions $x_1(0) = 0.8$, $x_2(0) = x_3(0) = 0.1$.

Solution The last equation is an algebraic equation. It can also be regarded as a constraint among the three state variables. The matrix form of the differential algebraic equations can be written as

$$\begin{bmatrix} 1 & 0 & 0 \\ 0 & 1 & 0 \\ 0 & 0 & 0 \end{bmatrix} \begin{bmatrix} \dot{x}_1 \\ \dot{x}_2 \\ \dot{x}_3 \end{bmatrix} = \begin{bmatrix} -0.2x_1 + x_2x_3 + 0.3x_1x_2 \\ 2x_1x_2 - 5x_2x_3 - 2x_2^2 \\ x_1 + x_2 + x_3 - 1 \end{bmatrix}$$

Clearly in MATLAB the RHS function can be expressed by an anonymous function. The matrix $\boldsymbol{M}$ can also be entered into MATLAB workspace, and the following statements can be entered as well into MATLAB command window [1]

```
>> f=@(t,x)[-0.2*x(1)+x(2)*x(3)+0.3*x(1)*x(2);
      2*x(1)*x(2)-5*x(2)*x(3)-2*x(2)*x(2); x(1)+x(2)+x(3)-1];
   M=[1,0,0; 0,1,0; 0,0,0]; options=odeset; options.Mass=M;
   x0=[0.8; 0.1; 0.1]; [t,x]=ode15s(f,[0,20],x0,options);
   plot(t,x)
```

From the above statements, the differential algebraic equations can be directly solved and the time responses of the states are shown in Figure 7.15.

[1] In this problem, the solver `ode45()` generates wrong results. Thus the stiff equation based algorithm should be used instead.

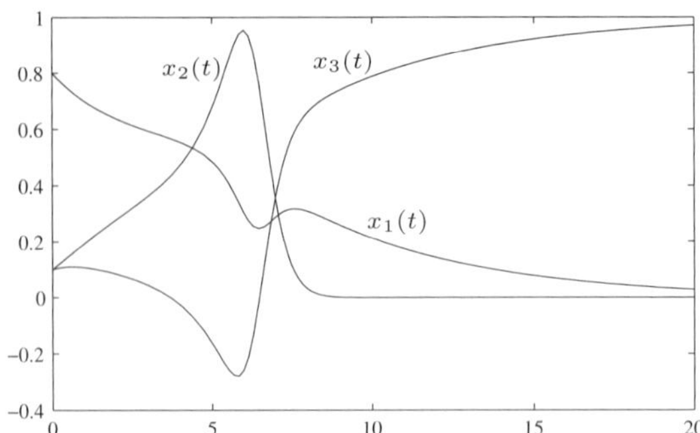

FIGURE 7.15: Numerical solutions to differential algebraic equations

In fact, some of the differential algebraic equations can be converted into lower-order explicit ODEs. For instance, in the example, from the constraint, one may immediately find that $x_3(t) = 1 - x_1(t) - x_2(t)$. Substituting it into the other two equations yields

$$\begin{cases} \dot{x}_1 = -0.2x_1 + x_2[1 - x_1(t) - x_2(t)] + 0.3x_1x_2 \\ \dot{x}_2 = 2x_1x_2 - 5x_2[1 - x_1(t) - x_2(t)] - 2x_2^2 \end{cases}$$

and the original three-state DAE can be converted into a second-order ODE. An anonymous function can be written to describe the ODE solvable by the following statements. The results using the two methods are exactly the same.

```
>> x0=[0.8; 0.1];
   fDae=@(t,x)[-0.2*x(1)+x(2)*(1-x(1)-x(2))+0.3*x(1)*x(2);...
                    2*x(1)*x(2)-5*x(2)*(1-x(1)-x(2))-2*x(2)*x(2)];
   [t1,x1]=ode45(fDae,[0,20],x0); plot(t1,x1,t1,1-sum(x1'))
```

Note that, in this converted case, even with the use of function `ode45()`, the results are still valid.

The implicit differential equation solver `ode15i()` can also be used to solve this problem under MATLAB 7.*. An anonymous function can be written to describe the implicit equation. Let $\boldsymbol{x}_0 = [1, 1, *]^{\mathrm{T}}$, where * is used to denote free values. The following statements can be used to solve compatible initial conditions. The original differential algebraic equation can be solved in this way, and the same results can be obtained. It can be seen that with the use of the method, a more straightforward solution process can be enjoyed.

```
>> f=@(t,x,xd)[xd(1)+0.2*x(1)-x(2)*x(3)-0.3*x(1)*x(2);
      xd(2)-2*x(1)*x(2)+5*x(2)*x(3)+2*x(2)^2; x(1)+x(2)+x(3)-1];
   x0=[0.8;0.1;2]; x0F=[1;1;0]; xd0=[1;1;1]; xd0F=[];
   [x0,xd0]=decic(f,0,x0,x0F,xd0,xd0F) % Compatible initial conditions
   res=ode15i(f,[0,20],x0,xd0); plot(res.x,res.y)
```

With the above solution commands, the compatible initial values are $\boldsymbol{x}(0) = [0.8, 0.1, 0.1]^{\mathrm{T}}$, and $\dot{\boldsymbol{x}}(0) = [-0.126, 0.09, 1]^{\mathrm{T}}$.

Example 7.22 Solve the implicit differential equations given in Example 7.18 using the differential algebraic equation solver.

Solution In Example 7.18, one converts the original equations to first-order explicit differential equations by inverting matrix $\boldsymbol{A}(\boldsymbol{x})$. In fact, an assumption has already been made that the matrix $\boldsymbol{A}(t)$ is non-singular. Although it happens that the assumption is correct for this example, the numerical algorithm used is not quite reliable and convincing, strictly speaking. For this kind of problem, the differential algebraic equation solvers can also be used.

For the original equations, an anonymous function can be written for the differential equation, and another anonymous function for the mass matrix. The differential algebraic equation can then be solved using the following statements:

```
>> f=@(t,x)[1-x(1); -x(2)];
   fM=@(t,x)[sin(x(1)),cos(x(2)); -cos(x(2)),sin(x(1))];
   options=odeset; options.Mass=fM; options.RelTol=1e-6;
   [t,x]=ode45(f,[0,10],[0;0],options); plot(t,x)
```

and the results obtained are exactly the same as the ones shown in Figure 7.13.

7.3.4 Solutions to delay differential equations

The general form of the *delay differential equations* is given by

$$\dot{\boldsymbol{x}}(t) = \boldsymbol{f}(t, \boldsymbol{x}(t-\tau_1), \boldsymbol{x}(t-\tau_2), \cdots, \boldsymbol{x}(t-\tau_n)) \tag{7.34}$$

where $\tau_i \geqslant 0$ are the delay constants for state variables $\boldsymbol{x}(t)$.

A MATLAB function `dde23()` [31] is provided to solve numerically delay differential equations using implicit Runge-Kutta algorithms, with the syntax

`sol=dde23(`fun_1`,`τ`,`fun_2`,[`t_0, t_f`])`

where $\boldsymbol{\tau} = [\tau_1, \tau_2, \cdots, \tau_n]$, fun_1 is the function describing the delay differential equations, and fun_2 is used to describe the state-space vector for $t \leqslant t_0$, which can either be a MATLAB function or a constant. The returned variable `sol` is a structure, with its `sol.x` and `sol.y` member components describing the time vector $\boldsymbol{t}$ and states matrix $\boldsymbol{x}$ respectively. The returned variable $\boldsymbol{x}$ is different from the $\boldsymbol{x}$ matrix from the `ode45()` function. It is arranged on a row basis instead of a column.

Example 7.23 The delay differential equations are given by

$$\begin{cases} \dot{x}(t) = 1 - 3x(t) - y(t-1) - 0.2x^3(t-0.5) - x(t-0.5) \\ \ddot{y}(t) + 3\dot{y}(t) + 2y(t) = 4x(t) \end{cases}$$

where when $t \leqslant 0$, $x(t) = y(t) = \dot{y}(t) = 0$. Solve numerically the DDEs.

Solution The values of $x(t)$, $y(t)$ at time instants $t, t-1$ and $t-0.5$ are involved in the ODEs, thus special functions are required for the DDEs. One of the straightforward way is to introduce a set of state variables, such that $x_1(t) = x(t), x_2(t) = y(t), x_3(t) = \dot{y}(t)$, then the original equation can be transformed into the following first-order vector form explicit DDEs such that

$$\begin{cases} \dot{x}_1(t) = 1 - 3x_1(t) - x_2(t-1) - 0.2x_1^3(t-0.5) - x_1(t-0.5) \\ \dot{x}_2(t) = x_3(t) \\ \dot{x}_3(t) = 4x_1(t) - 2x_2(t) - 3x_3(t). \end{cases}$$

Two delay constants $\tau_1 = 1$ and $\tau_2 = 0.5$ can be defined. Thus from the first state equation, the delay constant to the first state $x_1(t)$, both the delay constants τ_1 and τ_2 are involved. However, for the second state $x_2(t)$, only the delay constant τ_2 is used. So the following M-function can be written to express the delay differential equation

```
function dx=c7exdde(t,x,z)
xlag1=z(:,1);  % the 1st column extracts x(t-τ1)
xlag2=z(:,2);  % the 2nd column extracts x(t-τ2)
dx=[1-3*x(1)-xlag1(2)-0.2*xlag2(1)^3-xlag2(1); % xlag1(2) is x2(t-τ1)
    x(3);
    4*x(1)-2*x(2)-3*x(3)];
```

Then, the following statements can be used to find the numerical solutions to the delay differential equations

```
>> lags=[1 0.5]; tx=dde23(@c7exdde,lags,zeros(3,1),[0,10]);
   plot(tx.x,tx.y(2,:))
```

and from the commands, the time response of $y(t)$ can be obtained as shown in Figure 7.16.

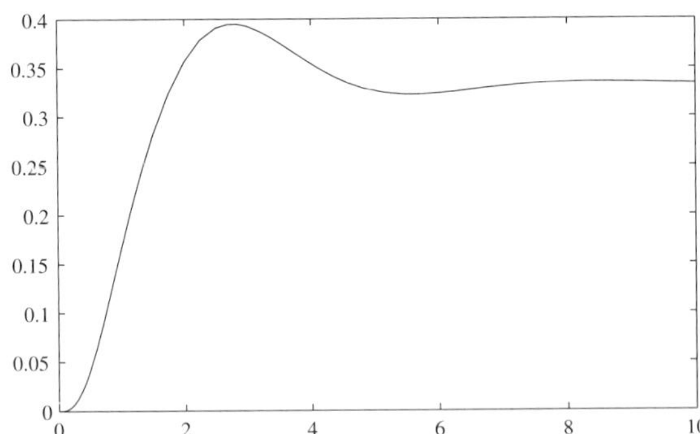

FIGURE 7.16: Numerical solutions to the delay differential equations

Example 7.24 The function `dde23()` described earlier can be used to solve most delay differential equations. However, consider the following ODE:

$$\dot{\boldsymbol{x}}(t) = \boldsymbol{A}_1\boldsymbol{x}(t-\tau_1) + \boldsymbol{A}_2\dot{\boldsymbol{x}}(t-\tau_2) + \boldsymbol{B}u(t)$$

where $\tau_1 = 0.15, \tau_2 = 0.5$. Since in the equations, the terms $\dot{\boldsymbol{x}}(t)$ and $\dot{\boldsymbol{x}}(t-\tau_2)$ are involved simultaneously, this type of differential equation is also referred to as the *neutral-type delay differential equation.* This type of equation cannot be solved using the `dde23()` function. Better tools are expected and in Section 7.6, block diagram-based solution technique using Simulink environment can be used to effectively solve this type of problems.

7.4 Solving Boundary Value Problems

In the previous sections, only initial value problems were considered, i.e., from the given $\boldsymbol{x}_0$, the state vector $\boldsymbol{x}$ at other time instants is to be evaluated. In practical situations, some states at time $t = 0$ are given while other states at $t = t_f$ are also given. This kind of ODE problem involving mixed initial and final conditions is referred to as the *boundary value problems* (BVPs). Boundary value problems cannot be solved directly from `ode45()` type functions. In this section, we focus on how to solve boundary value problems of ODEs.

7.4.1 Solutions to two-point boundary value problems

The two-point boundary value problem (TPBVP) of a given ODE involves mixed initial and final conditions. In this section, we illustrate the TPBVP solution procedures using a second order ODE as an example. General BVPs are considered in the next subsection.

The TPBVP for second-order ODE can be mathematically described as follows:

$$\ddot{y}(x) = F(x, y, \dot{y}) \tag{7.35}$$

given the solution interval $[a, b]$ and the following two boundary conditions:

$$\alpha_a y(a) + \beta_a \dot{y}(a) = \eta_a, \quad \alpha_b y(b) + \beta_b \dot{y}(b) = \eta_b. \tag{7.36}$$

In some simple cases, the boundary conditions are specified as

$$y(a) = \gamma_a, \ y(b) = \gamma_b. \tag{7.37}$$

Assume that the original TPBVP problem can be converted to an initial value problem as follows:

$$\ddot{y} = F(x, y, \dot{y}), \quad y(a) = \gamma_a, \ \dot{y}(a) = m. \tag{7.38}$$

That is, the TPBVP is converted to solving $y(b|m) = \gamma_b$, which means to evaluate the value of $y(b)$ based on the information of m. Using the following Newton's iterative algorithm the value of m can be obtained:

$$m_{i+1} = m_i - \frac{y(b|m_i) - \gamma_b}{(\partial y/\partial m)(b|m_i)} = m_i - \frac{v_1(b) - \gamma_b}{v_3(b)} \tag{7.39}$$

where $v_1 = y(x|m_i), v_2 = \dot{y}(x|m_i), v_3 = (\partial y/\partial m)(x|m_i), v_4 = (\partial \dot{y}/\partial m)(x|m_i)$, and clearly, the original TPBVP can be converted to a series of IVPs that can be solved using the numerical algorithms given in the previous sections.

$$\begin{cases} \dot{v}_1 = v_2, & v_1(a) = \gamma_a \\ \dot{v}_2 = F(x, v_1, v_2), & v_2(a) = m \\ \dot{v}_3 = v_4, & v_3(a) = 0 \\ \dot{v}_4 = \dfrac{\partial F}{\partial y}(x, v_1, v_2)v_3 + \dfrac{\partial F}{\partial \dot{y}}(x, v_1, v_2)v_4, & v_4(a) = 1 \end{cases} \tag{7.40}$$

where in order to solve explicitly $\partial F/\partial y$, $\partial F/\partial \dot{y}$, an auxiliary value of m can be introduced such that the initial value problems in (7.40) can be solved. The results can be substituted into (7.39) for one step. Then, in turn the results can be substituted into (7.40). When the values of m computed in the two algorithms meet the pre-specified error tolerance, the iteration stops and the required m is found. In this way, from (7.38), the original TPBVP can be considered as solved. The above algorithm can be implemented in the following MATLAB function

```
function [t,y]=nlbound(funcn,funcv,tspan,x0f,tol,varargin)
t0=tspan(1);tfinal=tspan(2);   ga=x0f(1); gb=x0f(2); m=1; m0=0;
while (norm(m-m0)>tol), m0=m;
   [t,v]=ode45(funcv,tspan,[ga;m;0;1],varargin);
   m=m0-(v(end,1)-gb)/(v(end,3));
end
[t,y]=ode45(funcn,tspan,[ga;m],varargin);
```

where a MATLAB user defined function `funcv()` must be prepared to describe the IVPs defined in (7.40). An example is given below to illustrate the coded algorithm.

Example 7.25 Solve the following TPBVP for the nonlinear ODE

$$\ddot{y} = F(x, y, \dot{y}) = 2y\dot{y}, \;\; y(0) = -1, \;\; y(\pi/2) = 1.$$

Solution The partial derivatives can easily be found as $\partial F/\partial y = 2\dot{y}$, $\partial F/\partial \dot{y} = 2y$. Substituting them back to the fourth equation in (7.40), it can immediately be found that $\dot{v}_4 = 2v_2v_3 + 2v_1v_4$. Thus the related functions can be expressed by anonymous functions. The following statements can be used to solve this TPBVP. The time histories of the states are shown in Figure 7.17. It can be seen that the solution $x_1(t)$ satisfies the two given boundary conditions.

```
>> f1=@(t,v)[v(2); 2*v(1)*v(2); v(4); 2*v(2)*v(3)+2*v(1)*v(4)];
   f2=@(t,x)[x(2); 2*x(1)*x(2)];
   [t,y]=nlbound(f2,f1,[0,pi/2],[-1,1],1e-8);
   plot(t,y); xlim([0,pi/2]);
```

It is known that the analytical solution of the TPBVP is $y(x) = \tan(x - \pi/4)$. The following statements can be used to check the accuracy of the above numerical results.

```
>> y0=tan(t-pi/4); norm(y(:,1)-y0)
```

The norm of the error vector is 1.6629×10^{-5}, which is satisfactory.

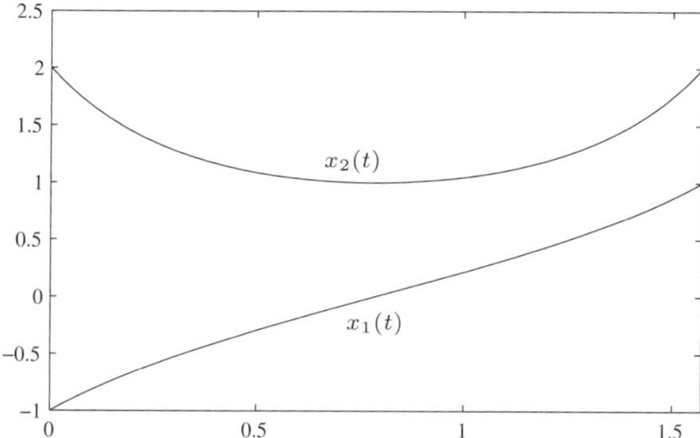

FIGURE 7.17: Solutions to a simple two-point boundary value problem

7.4.2 Solutions to general boundary value problems

The two-point boundary value problems are quite restricted, since they can only be used to deal with second-order differential equations with known parameters. Assume the ODE to be analyzed is given by

$$\dot{\boldsymbol{y}} = \boldsymbol{f}(t, \boldsymbol{y}, \boldsymbol{\theta}) \tag{7.41}$$

where the time $t \in [a, b]$; $\boldsymbol{y}$ is the state vector, and $\boldsymbol{\theta}$ is the vector of unknown parameters. The general boundary conditions are given by

$$\boldsymbol{\phi}[\boldsymbol{y}(a), \boldsymbol{y}(b), \boldsymbol{\theta}] = 0. \tag{7.42}$$

To convert the original problems into initial value problems, several algebraic equations should be solved. If the numbers of unknowns and equations are the same, numerical algebraic equation solution algorithms can be used. The boundary value problems to be solved here are more general, since the equations and undetermined parameters can be handled at the same time.

The `bvp5c()` function provided in MATLAB can be used to solve the above general boundary value problems[32]. The procedures of problem solution are summarized below:

(i) **Parameter initialization** The `bvpinit()` function can be used to initialize the BVP. The equation and the undetermined parameters can be described together in this function such that `sinit=bvpinit(`$\boldsymbol{v}$`,`$\boldsymbol{x}_0$`,`$\boldsymbol{\theta}_0$`)`, where $\boldsymbol{v}$ contains the sample times generated by $\boldsymbol{v}$`=linspace(`a`,`b`,`M`)`. M should be set to small integers for computation speed, e.g., $M = 5$. Apart from the vector $\boldsymbol{v}$, the initial search points of the state vector $\boldsymbol{x}_0$ and undetermined parameters $\boldsymbol{\theta}_0$ should also be provided.

(ii) **MATLAB descriptions to ODEs and BVPs** The description of ODEs is exactly the same as the one in the initial value problems illustrated in the previous sections. The description of boundary values in (7.42) will be demonstrated through examples in the following.

(iii) **Solving the boundary value problem** The `bvp5c()` function can be used in solving boundary value problems

```
sol=bvp5c(fun1,fun2,sinit,options,p1,p2,···)
```

where *fun1* and *fun2* are respectively the differential equations and the boundary values. The returned argument `sol` is a structured variable, whose members `sol.x` and `sol.y` store respectively the $\boldsymbol{t}$ vector and the state matrix. The member `sol.parameters` stores the undetermined parameter vector $\boldsymbol{\theta}$.

Example 7.26 Solve again the boundary value problem in Example 7.25 with the `bvp5c()` solver.

Solution Let $x_1 = y, x_2 = \dot{y}$. The first-order vector form explicit ODE can be written as $\dot{x}_1 = x_2, \dot{x}_2 = 2x_1x_2$. Anonymous functions can be used to describe the ODEs and the boundary values. The `bvp5c()` function can be used to solve directly the boundary value problem and the result is shown in Figure 7.18 (a). It can be seen that the result is exactly the same as the one in the previous example.

```
>> f1=@(t,x)[x(2); 2*x(1)*x(2)]; f2=@(xa,xb)[xa(1)+1; xb(1)-1];
   sinit=bvpinit(linspace(0,pi/2,5),rand(2,1));
   sol=bvp5c(f1,f2,sinit); plot(sol.x,sol.y)
```

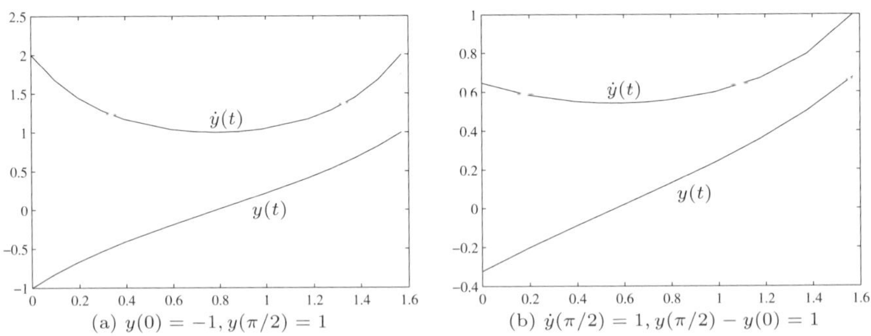

(a) $y(0) = -1, y(\pi/2) = 1$ (b) $\dot{y}(\pi/2) = 1, y(\pi/2) - y(0) = 1$

FIGURE 7.18: Solutions of boundary value problem

The `bvp5c()` function can also be used to solve more complicated boundary value problems. For instance, if the boundary value problem is changed to $\dot{y}(\pi/2) = 1$, $y(\pi/2) - y(0) = 1$, the function *f2* should be changed as follows

```
>> f2=@(xa,xb)[xb(2)-1; xb(1)-xa(1)-1];
   sol=bvp5c(f1,f2,sinit); plot(sol.x,sol.y)
```

and the results are shown in Figure 7.18 (b).

Example 7.27 Given the ODE $\begin{cases} \dot{x} = 4x - \alpha xy \\ \dot{y} = -2y + \beta xy, \end{cases}$ with initial and final conditions $x(0) = 2, y(0) = 1, x(3) = 4, y(3) = 2$, find the parameters α and β and solve the boundary value problem.

Solution Choosing the state variables $x_1 = x, x_2 = y$, the original problem can be converted into an explicit differential equation with respect to $\boldsymbol{x}$. Let $v_1 = \alpha$ and $v_2 = \beta$. The ODE and boundary value problem can then be expressed by the following anonymous functions

```
>> f=@(t,x,v)[4*x(1)+v(1)*x(1)*x(2); -2*x(2)+v(2)*x(1)*x(2)];
   g=@(ya,yb,v)[ya(1)-2; ya(2)-1; yb(1)-4; yb(2)-2];
```

and it can be seen that the description to boundary value problem is quite straightforward in MATLAB. The `bvpinit()` function should be called to initialize the time array. The initial states and parameters α and β should also be specified. Since there are two states and two undetermined parameters, thus they both can be set to `rand(2,1)`. With these initial parameters, the function `bvp5c()` can be called to solve the boundary value problem as well as the undetermined parameters α and β.

```
>> x1=[1;1]; x2=[-1;1]; sinit=bvpinit(linspace(0,3,5),x1,x2);
   options=bvpset; options.RelTol=1e-8; sol=bvp5c(f,g,sinit,options);
   sol.parameters % show undermined parameters
   plot(sol.x,sol.y); figure; plot(sol.y(1,:),sol.y(2,:));
```

The results are shown in Figure 7.19. Meanwhile, it is found that $\alpha = -2.3720, \beta = 0.8934$. From the simulation results, it is found that the boundary conditions are satisfied, which verifies the obtained results. It should be noted that if the initial vectors $\boldsymbol{x}_1$ and $\boldsymbol{x}_2$ are not properly chosen, the Jacobian matrix generated might be singular. In this case, the initial vectors should be chosen again differently so that convergent results can be obtained.

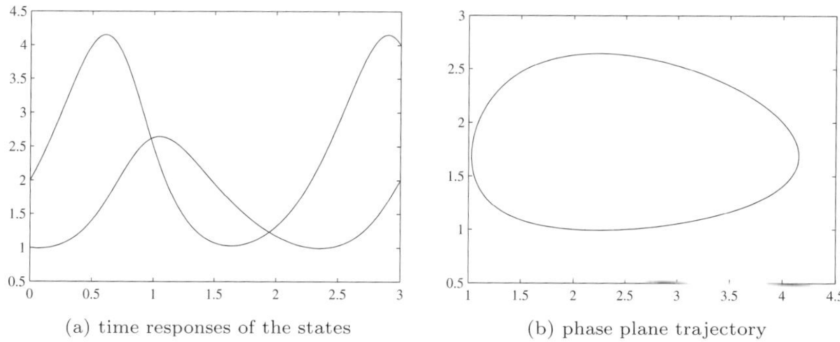

(a) time responses of the states (b) phase plane trajectory

FIGURE 7.19: Solution to the example TPBVP

7.5 Introduction to Partial Differential Equations

Apart from ordinary differential equations, partial differential equations (PDEs) are also very useful in science and engineering. In MATLAB many partial differential equations can be solved by the Partial Differential Equation Toolbox. In this section, a general introduction to partial differential equations is given and some particular types of the two-dimensional partial differential equations solvable in the easy-to-use GUI of the PDE Toolbox are illustrated.

A PDE contains 4 elements: the PDE, the initial conditions (ICs), the boundary conditions (BCs) and the domain of interest. In the following, we will introduce the solution procedures for 1D PDEs first and then 2D PDEs.

7.5.1 Solving a set of 1D PDEs

A partial differential equation solver `pdepe()` provided in MATLAB PDE Toolbox can be used to solve numerically the general 1D partial differential equations of the following form

$$\boldsymbol{c}\left(x,t,\boldsymbol{u},\frac{\partial \boldsymbol{u}}{\partial x}\right)\frac{\partial \boldsymbol{u}}{\partial t}=x^{-m}\frac{\partial}{\partial x}\left[x^{m}\boldsymbol{f}\left(x,t,\boldsymbol{u},\frac{\partial \boldsymbol{u}}{\partial x}\right)\right]+\boldsymbol{s}\left(x,t,\boldsymbol{u},\frac{\partial \boldsymbol{u}}{\partial x}\right) \tag{7.43}$$

where the PDE to be solved should be modeled in the following function `[c,f,s]=pdefun(x,t,u,ux)`, where `pdefun` is the function name, the functions $\boldsymbol{c}$, $\boldsymbol{f}$, and $\boldsymbol{s}$ can be calculated from the given arguments.

Boundary conditions can be described by the following function

$$\boldsymbol{p}(x,t,\boldsymbol{u})+\boldsymbol{q}(x,t,\boldsymbol{u}).*\boldsymbol{f}\left(x,t,\boldsymbol{u},\frac{\partial \boldsymbol{u}}{\partial x}\right)=0 \tag{7.44}$$

where `.*` operation is the MATLAB element-wise dot product. These BCs can be described by the following MATLAB function

```
[pa,qa,pb,qb]=pdebc(x,t,u,ux)
```

Apart from the above two MATLAB functions, the initial conditions should also be described. The ICs are usually defined as $\boldsymbol{u}(x,t_0)=\boldsymbol{u}_0$. Thus a simple MATLAB function can be prepared by using `u0=pdeic(x)`.

One can also select the variable vectors $\boldsymbol{x}$ and $\boldsymbol{t}$. With the use of the above functions, the `pdepe()` function can be used to solve the PDE. The syntax of the function is

```
sol=pdepe(m,@pdefun,@pdeic,@pdebc,x,t)
```

Example 7.28 Solve the following partial differential equations:

$$\begin{cases}\dfrac{\partial u_1}{\partial t}=0.024\dfrac{\partial^2 u_1}{\partial x^2}-F(u_1-u_2)\\[2mm]\dfrac{\partial u_2}{\partial t}=0.17\dfrac{\partial^2 u_2}{\partial x^2}+F(u_1-u_2)\end{cases}$$

where $F(x)=\mathrm{e}^{5.73x}-\mathrm{e}^{-11.46x}$, and the initial conditions are given by $u_1(x,0)=1$, $u_2(x,1)=0$ and the boundary conditions are

$$\frac{\partial u_1}{\partial x}(0,t)=0,\ u_2(0,t)=0,\ u_1(1,t)=1,\ \frac{\partial u_2}{\partial x}(1,t)=0.$$

Solution Comparing the given PDE with the standard form described in (7.43), we can rewrite the PDE as

$$\begin{bmatrix}1\\1\end{bmatrix}.*\frac{\partial}{\partial t}\begin{bmatrix}u_1\\u_2\end{bmatrix}=\frac{\partial}{\partial x}\begin{bmatrix}0.024\partial u_1/\partial x\\0.17\partial u_2/\partial x\end{bmatrix}+\begin{bmatrix}-F(u_1-u_2)\\F(u_1-u_2)\end{bmatrix}$$

with $m=0$, and

$$\boldsymbol{c}=\begin{bmatrix}1\\1\end{bmatrix},\quad \boldsymbol{f}=\begin{bmatrix}0.024\partial u_1/\partial x\\0.17\partial u_2/\partial x\end{bmatrix},\quad \boldsymbol{s}=\begin{bmatrix}-F(u_1-u_2)\\F(u_1-u_2)\end{bmatrix}.$$

Thus the M-function describing the PDE is written as

```
function [c,f,s]=c7mpde(x,t,u,du)
c=[1; 1]; y=u(1)-u(2); F=exp(5.73*y)-exp(-11.46*y); s=F*[-1; 1];
f=[0.024*du(1); 0.17*du(2)];
```

Referring to the boundary conditions in (7.44), the following boundary conditions can be constructed such that

$$\text{left bounds}\quad \begin{bmatrix}0\\u_2\end{bmatrix}+\begin{bmatrix}1\\0\end{bmatrix}.*\boldsymbol{f}=\begin{bmatrix}0\\0\end{bmatrix},\qquad \text{right bounds}\quad \begin{bmatrix}u_1-1\\0\end{bmatrix}+\begin{bmatrix}0\\1\end{bmatrix}.*\boldsymbol{f}=\begin{bmatrix}0\\0\end{bmatrix}$$

and the MATLAB function for the BCs can be prepared as

```
function [pa,qa,pb,qb]=c7mpbc(xa,ua,xb,ub,t)
pa=[0; ua(2)]; qa=[1;0]; pb=[ub(1)-1; 0]; qb=[0;1];
```

The initial conditions for this PDE can be given by an anonymous function.

With these three functions, if the vectors $\boldsymbol{x}$ and $\boldsymbol{t}$ are used, the following statements solve the given PDE where the solutions u_1 and u_2 are visualized in Figures 7.20 (a) and (b), respectively.

```
>> x=0:.05:1; t=0:0.05:2; m=0; u0=@(x)[1; 0];
   sol=pdepe(m,@c7mpde,u0,@c7mpbc,x,t); surf(x,t,sol(:,:,1))
   figure;  surf(x,t,sol(:,:,2))
```

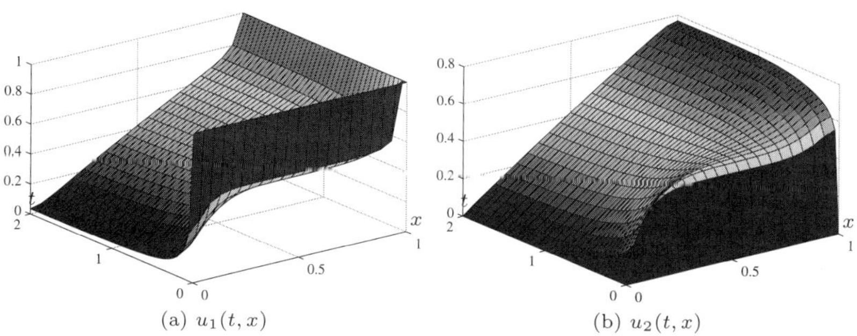

FIGURE 7.20: Solution surfaces in partial differential equations

7.5.2 Mathematical description to 2D PDEs

The PDE Toolbox can be used to solve 2D partial differential equations via a very handy GUI `pdetool`. Some of the 2D PDEs solvable via the PDE

Toolbox are introduced and examples are given to demonstrate the use of the PDE GUI.

Elliptic PDEs

The general form of an elliptic PDE is given by

$$-\mathrm{div}(c\nabla u) + au = f(\boldsymbol{x}, t) \tag{7.45}$$

where $u = u(x_1, x_2, \cdots, x_n, t) = u(\boldsymbol{x}, t)$, and ∇u is the gradient of u defined as

$$\nabla u = \left[\frac{\partial}{\partial x_1}, \frac{\partial}{\partial x_2}, \cdots, \frac{\partial}{\partial x_n}\right] u. \tag{7.46}$$

The divergence $\mathrm{div}(v)$ is defined as

$$\mathrm{div}(v) = \left(\frac{\partial}{\partial x_1} + \frac{\partial}{\partial x_2} + \cdots + \frac{\partial}{\partial x_n}\right) v. \tag{7.47}$$

Thus $\mathrm{div}(c\nabla u)$ can further be defined as

$$\mathrm{div}(c\nabla u) = \left[\frac{\partial}{\partial x_1}\left(c\frac{\partial u}{\partial x_1}\right) + \frac{\partial}{\partial x_2}\left(c\frac{\partial u}{\partial x_2}\right) + \cdots + \frac{\partial}{\partial x_n}\left(c\frac{\partial u}{\partial x_n}\right)\right]. \tag{7.48}$$

If c is a constant, the above equation can be simplified as

$$\mathrm{div}(c\nabla u) = c\left(\frac{\partial^2}{\partial x_1^2} + \frac{\partial^2}{\partial x_2^2} + \cdots + \frac{\partial^2}{\partial x_n^2}\right) u = c\Delta u \tag{7.49}$$

where Δ is also referred to as the *Laplacian operator*. Thus the elliptic PDE can further be simplified as

$$-c\left(\frac{\partial^2}{\partial x_1^2} + \frac{\partial^2}{\partial x_2^2} + \cdots + \frac{\partial^2}{\partial x_n^2}\right) u + au = f(\boldsymbol{x}, t). \tag{7.50}$$

Parabolic PDEs

The general form of a parabolic PDE is given by

$$d\frac{\partial u}{\partial t} - \mathrm{div}(c\nabla u) + au = f(\boldsymbol{x}, t). \tag{7.51}$$

If c is a constant, the above equation can be simplified as

$$d\frac{\partial u}{\partial t} - c\left(\frac{\partial^2 u}{\partial x_1^2} + \frac{\partial^2 u}{\partial x_2^2} + \cdots + \frac{\partial^2 u}{\partial x_n^2}\right) + au = f(\boldsymbol{x}, t). \tag{7.52}$$

Hyperbolic PDEs

The general form of a hyperbolic PDE is given by

$$d\frac{\partial^2 u}{\partial t^2} - \mathrm{div}(c\nabla u) + au = f(\boldsymbol{x}, t). \tag{7.53}$$

If c is a constant, the above equation can be simplified as

$$d\frac{\partial^2 u}{\partial t^2} - c\left(\frac{\partial^2 u}{\partial x_1^2} + \frac{\partial^2 u}{\partial x_2^2} + \cdots + \frac{\partial^2 u}{\partial x_n^2}\right) + au = f(\boldsymbol{x}, t). \tag{7.54}$$

It can be seen from the above three types of PDEs that the most significant difference is the order of the derivative of the function u with respect to t. If the derivative term is zero, the PDE is elliptic. The first- and second-order derivative terms correspond to parabolic and hyperbolic PDEs, respectively.

Finite element-based algorithms are implemented in the PDE Toolbox. In the elliptic PDEs, the variables c, a, d, and f can all be defined as any given functions, while in the other two types of PDEs, they must be constants.

Eigenvalue problem

An eigenvalue PDE problem is defined as

$$-\mathrm{div}(c\nabla u) + au = \lambda du. \tag{7.55}$$

If c is a constant, the above equation can be simplified as

$$-c\left(\frac{\partial^2 u}{\partial x_1^2} + \frac{\partial^2 u}{\partial x_2^2} + \cdots + \frac{\partial^2 u}{\partial x_n^2}\right) + au = \lambda du. \tag{7.56}$$

7.5.3 The GUI for the PDE Toolbox — an introduction

PDE Toolbox — an overview

A GUI for solving PDEs is provided in the PDE Toolbox. The interface can be used in solving 2D PDEs. The solution regions of interest can be drawn freely by the GUI tools using combinations of circles, ellipsis, rectangles and polygons. Moreover, the regions of interest can be organized by set operations such as union, difference and intersect, etc. 2D PDEs can easily be solved using the GUI.

Type `pdetool` in the MATLAB prompt, and a user interface shown in Figure 7.21 will be displayed. This interface can be used in solving the given PDE.

The PDE user interface has the following functions:

(i) **Menu system** A complete and comprehensive menu system is provided in the interface, which makes most of the functions callable directly from the menu items and toolbar buttons.

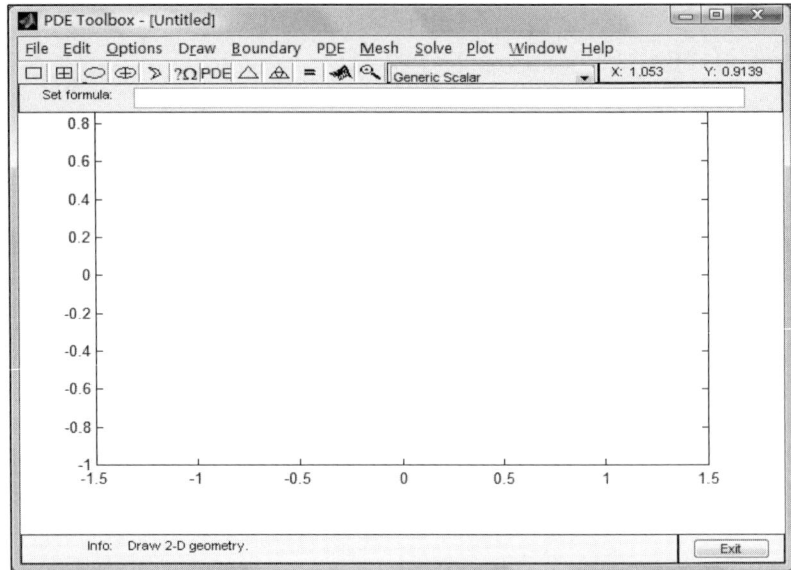

FIGURE 7.21: Graphical user interface of the PDE solver

(ii) **Toolbar** The detailed explanation to the buttons on the toolbar is shown in Figure 7.22. The toolbar can be used to define the solution regions of interest, to set parameters in the PDE, to solve the PDE and to visualize the results. The right-hand side of the interface provides a list box containing different types of solvable PDEs.

(iii) **Set formula** Set formula edit box can be used to define set operations, such as union, intersect and difference operations.

(iv) **Solution regions** The user can draw solution regions, and then solve the two-dimensional PDEs within the solution region. 3D display can also be obtained.

Drawing and defining the PDE solution region

An illustrative example is introduced to define the solution region in the GUI. One can first select the ellipses and rectangle buttons and draw the regions, defined as *sets*, as shown in Figure 7.23 (a). Then the solution region can be defined using the set operation edit box such that the contents are changed to `(R1+C1+E2)-E1`, which means removal of the set `E1` from the union of the rectangle `R1`, the ellipse `E2`, and the circle `C1`. Thus the button labeled as $\partial\Omega$ can be used to define the solution region. The menu item Boundary $\rightarrow$ Remove All Subdomain Borders can be used to remove the curves within the adjacent regions. Thus the solution region can then be obtained as shown in Figure 7.23 (b).

From the given solution region, if one clicks the Δ button, triangular mesh

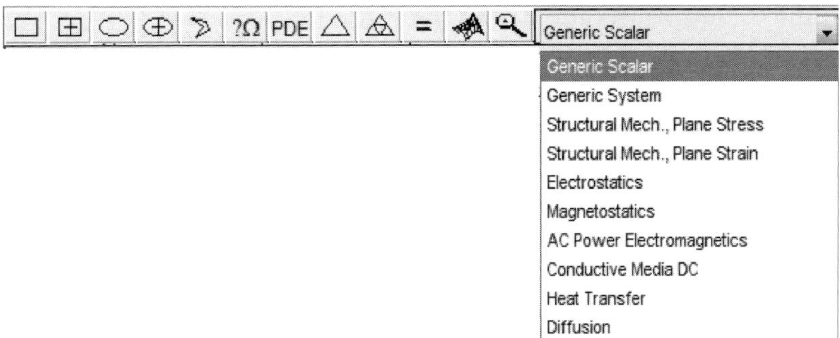

FIGURE 7.22: Toolbar in the PDE solver GUI

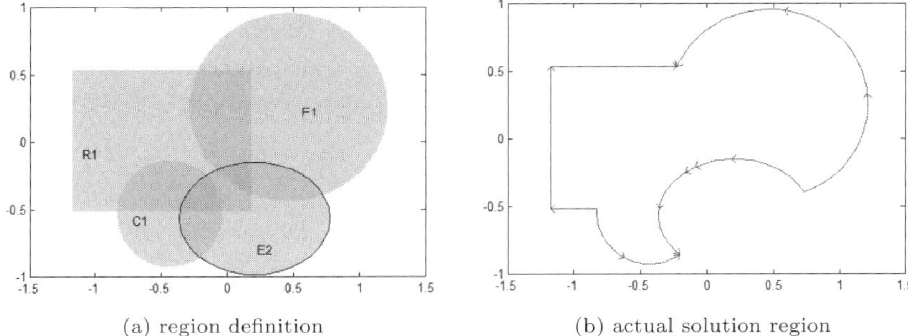

(a) region definition (b) actual solution region

FIGURE 7.23: Solution region for partial differential equations

can be generated within the solution region, as shown in Figure 7.24 (a). If one is not satisfied with the mesh, the refine mesh button can be used to add more grids to the region and the final grids can be shown in Figure 7.24 (b). It is worth mentioning that the finer the mesh, the more accurate solutions can be obtained. However, the cost is that the longer computation time is required.

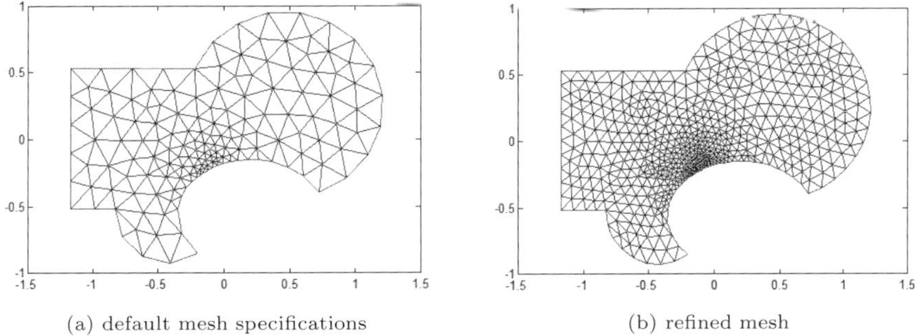

(a) default mesh specifications (b) refined mesh

FIGURE 7.24: Generation of the grids within the solution region

Boundary conditions for 2D PDEs

In the interface, the boundary conditions can be represented by the $\partial\Omega$ button. Generally speaking, the Dirichlet and Neumann types of boundary conditions are supported. These two types of boundary conditions are discussed below:

(i) **Dirichlet conditions** Dirichlet boundary conditions are described as follows:

$$h\left(\boldsymbol{x},t,u,\frac{\partial u}{\partial \boldsymbol{x}}\right)u\bigg|_{\partial\Omega}=r\left(\boldsymbol{x},t,u,\frac{\partial u}{\partial \boldsymbol{x}}\right) \tag{7.57}$$

where $\partial\Omega$ denotes the boundary of the solution region. Assume that certain conditions must be satisfied on the solution boundary, one can simply specify the functions r and h which can be either constants or functions of $\boldsymbol{x}$ and u, $\partial u/\partial \boldsymbol{x}$. For convenience, one may assume that $h=1$. In the following example, how to specify the Dirichlet boundary conditions in MATLAB will be illustrated.

(ii) **Neumann conditions** The extended form is given by

$$\left[\frac{\partial}{\partial \boldsymbol{n}}(c\nabla u)+qu\right]\bigg|_{\partial\Omega}=g \tag{7.58}$$

where $\partial u/\partial \boldsymbol{n}$ is the partial derivative of vector $\boldsymbol{x}$ in the normal direction.

If Boundary → Specify Boundary Conditions menu item is selected, a dialog box as in Figure 7.25 will show up. The boundary conditions can be specified through the dialog box. If ones wants to assign zero values on all the boundaries, fill 0 in the r edit box.

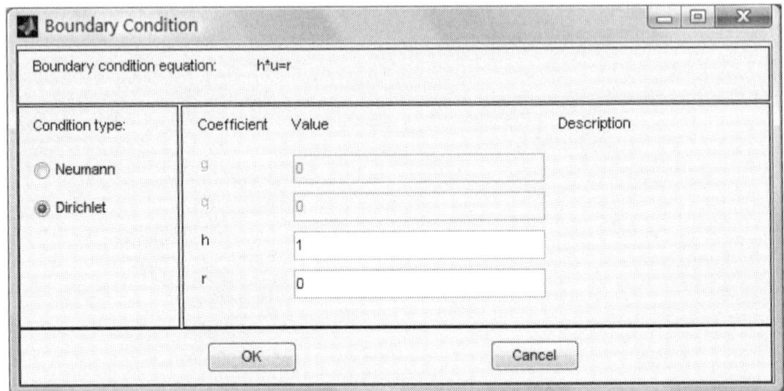

FIGURE 7.25: Boundary condition setting dialog box

PDE solution examples via PDE Toolbox

The solution region, boundary conditions can be specified using the methods discussed earlier. If the partial differential equations can be specified, the = button in the toolbar can be used to solve the partial differential equations. An example will be given below to show the solution procedures.

Example 7.29 Solve the following hyperbolic partial differential equation

$$\frac{\partial^2 u}{\partial t^2} - \frac{\partial^2 u}{\partial x^2} - \frac{\partial^2 u}{\partial y^2} + 2u = 10.$$

Solution From the given partial differential equation, one can immediately find that $c = 1, a = 2, f = 10$ and $d = 1$. Click the PDE button in the toolbar. A dialog box as in Figure 7.26 shows. From the radio button on the left, select the Hyperbolic option, and the parameters can be entered in the dialog box.

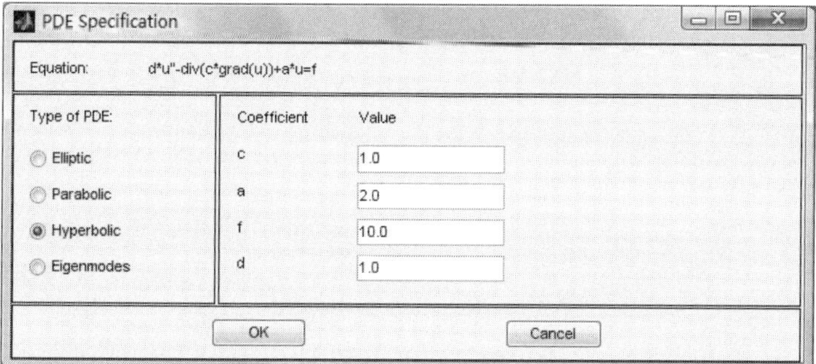

FIGURE 7.26: PDE parameters setting dialog box

If the numerical solutions are required, click the = button in the toolbar and the solutions can immediately be obtained as shown in Figure 7.27 (a), where pseudo-colors are used to denote the values of $u(x, y)$. It should be noted that only the solution $u(x, y)$ at $t = 0$ is displayed. Later on, it will be shown how the solutions at other t values can be visualized.

The boundary value conditions can be modified. For instance, the Dirichlet conditions can still be used and assume $r = 5$ on the boundaries. Use the dialog box to fill 5 into the r edit box. Solve the PDE again and the obtained results are shown in Figure 7.27 (b).

Numerical results can be visualized in many different ways. If the 3D button is clicked, a dialog box shown in Figure 7.28 appears. One can select the Contour icon to show the contours of the solutions. If the Arrows option is selected, attraction curves will be displayed. When the two options are selected simultaneously, the results are visualized in Figure 7.29 (a).

It should be noted that, in the dialog box shown in Figure 7.28, list boxes are provided for each item of the Property column. For instance, the default

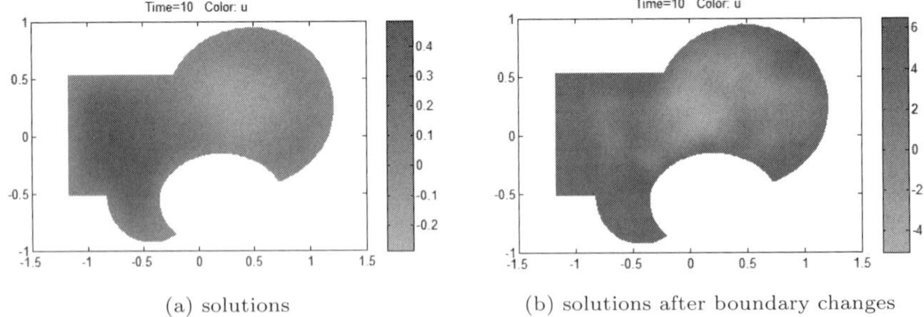

(a) solutions (b) solutions after boundary changes

FIGURE 7.27: Solutions to the partial differential equations

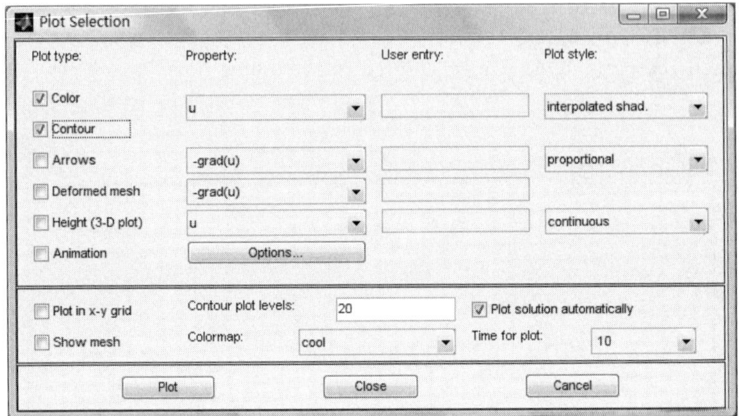

FIGURE 7.28: Dialog box for results display format settings

for the first item is u, indicating all the analyses are made for the function $u(\cdot)$. The results displayed are the function of $u(x, y)$. If other functions are to be visualized, the button ▼ to the right should be clicked. Another list box is then shown and other functions can be selected for visualization.

In the menu item Height (3d-plot), a graphics window is opened and a three-dimensional mesh grid plot is generated, as shown in Figure 7.29 (b).

Animation of the time varying solutions

The default time vector is defined as `t=0:10`. The results shown in Figure 7.27 are only at the final time $t = 10$. From the hyperbolic equations it can be seen that the solutions should also be a function of time t. Thus animation should be used to dynamically display the solutions. Let us still use the PDE in Example 7.29 to illustrate the animation process.

Use the menu item Solve → Parameters and a dialog box is displayed so as to specify the time vector. For instance, if in the edit box, one specifies the vector `0:0.1:4`, the solutions will then be made over the new time vector instead. In the dialog box shown in Figure 7.28, the Animation check box

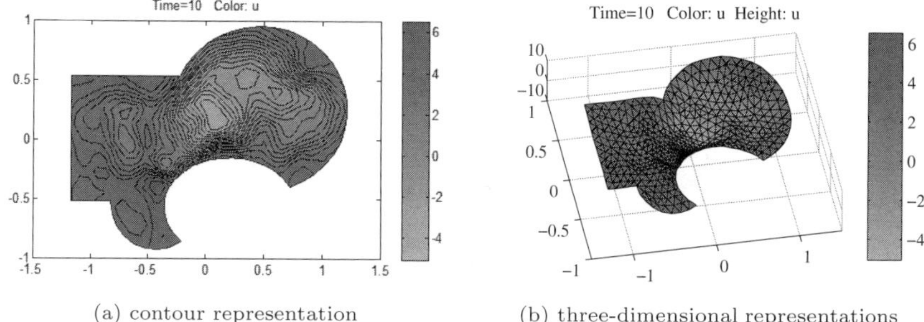

(a) contour representation (b) three-dimensional representations

FIGURE 7.29: Different representations of the solutions

can be selected. The Options button can be used to adjust the speed in animations (the default speed is 6 frames per second). The animation can then be obtained directly. The Plot → Export Movie menu item can be used to export the animation variable into the MATLAB workspace. For instance, the animation results can be saved to MATLAB variable `M`, and with the use of the `movie(M)` function, animation can be played in a MATLAB graphics window. Furthermore, using the `movie2avi(M,'myavi.avi')` command, the animation can be saved in the `myavi.avi` file for later play.

Solving PDEs when parameters are not constants

In the partial differential equations discussed earlier, c, a, d, f coefficients are all assumed constant. In practical applications, however, c, a, d, f may be functions. For elliptic PDEs, the solvers currently allow the use of function to describe the above-mentioned coefficients. The variables `x` and `y` are used to represent x_1, x_2 or x, y, while the variables `ux` and `uy` are for $\partial u/\partial x$ and $\partial u/\partial y$, respectively. They can be described by any nonlinear functions. The following example illustrates the solution process when c, a, d, f are not constants.

Example 7.30 Assume that the partial differential equations are described as

$$-\text{div}\left(\frac{1}{1+|\nabla u|^2}\nabla u\right)+(x^2+y^2)u=\mathrm{e}^{-x^2-y^2}$$

with boundaries at 0. Solve numerically the partial differential equations.

Solution It can be found that the original PDE is elliptic, with

$$c=\frac{1}{\sqrt{(1+\left(\dfrac{\partial u}{\partial x}\right)^2+\left(\dfrac{\partial u}{\partial y}\right)^2}},\quad a=x^2+y^2,\quad f=\mathrm{e}^{-x^2-y^2}$$

and the boundary conditions are 0. One may still use the `pdetool` interface shown in Figure 7.26. In the PDE type dialog box, select the Elliptic item, and in the c edit box, specify `1./sqrt(1+ux.^2+uy.^2)`. In edit boxes of a and f, specify respectively `x.^2+y.^2` and `exp(-x.^2-y.^2)`. Then open the Solve → Parameters dialog box and select the Use nonlinear solver property (it should be noted that this option is only applicable to elliptic equations), and the equal sign icon can be clicked to solve the equation. The results obtained are shown in Figures 7.30 (a) and (b).

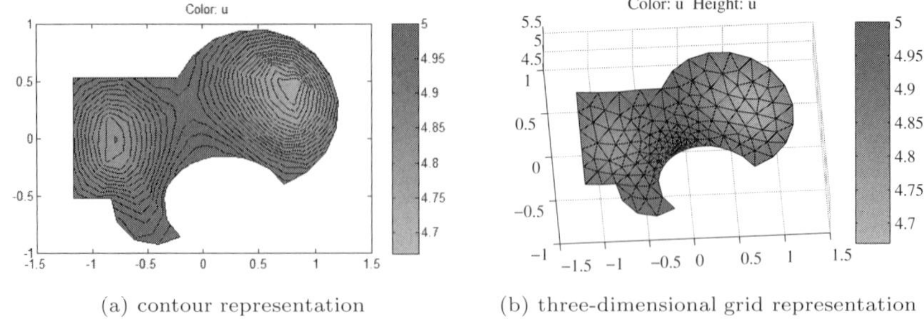

(a) contour representation (b) three-dimensional grid representation

FIGURE 7.30: Different presentations of the PDE solutions

7.6 Solving ODEs with Block Diagrams in Simulink

7.6.1 A brief introduction to Simulink

The Simulink environment was first proposed by The MathWorks Inc., in around 1990. The original name was SimuLAB[33], and it took the current name in 1992. From the name Simulink, we can see two meanings: "simu" and "link," which means to connect the blocks together, then to perform simulation for the system thus constructed. Simulink is an effective tool useful in defining different types of ODEs and other algebraic equations.

Of course, the functions provided in Simulink are not limited to ODE solvers. It can also be used to construct control systems with the existing and extended blocks. Moreover, modeling and simulation to engineering systems, such as motor and drive systems, mechanical systems and communication systems, can be carried out very easily, with the help of the Simulink blocksets and MATLAB toolboxes. Simulink, a very powerful environment, can be used to model and simulate dynamic systems of almost any complexity using Simulink block libraries and user-defined blocks[34, 35]. Only materials related to ODE solutions are discussed in this section.

The most commonly used Simulink blocks will be introduced and then examples will be given to show the modeling and simulation procedures.

7.6.2 Simulink — relevant blocks

One can issue the command `open_system('simulink')` in the MATLAB window, so that the block library window of Simulink can be opened as shown in Figure 7.31. It can be seen that block groups are also provided.

For instance, the groups such as Sources, Continuous are provided. Each of the groups further contains sub-groups or blocks. Theoretically speaking, the systems of almost any complexity can be modeled and simulated using the facilities provided in Simulink.

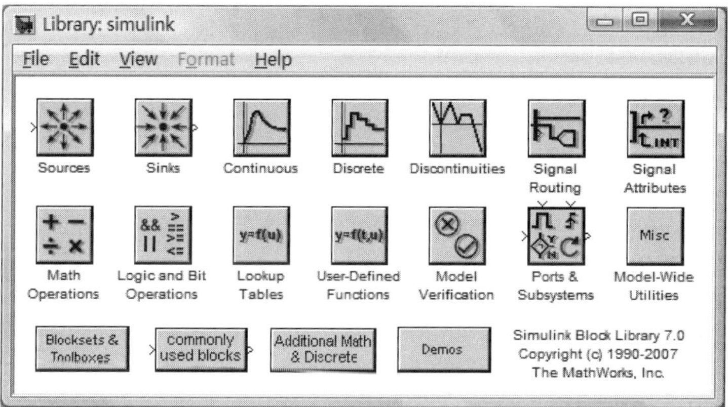

FIGURE 7.31: Main window in Simulink environment

Hundreds of Simulink blocks are provided and it is not possible to describe all of them. Here only the blocks related to differential equation modeling are summarized and some of the commonly used blocks are made into a user-group, named as `odegroup`. The command `odegroup` can be used to open the user defined blockset, as shown in Figure 7.32.

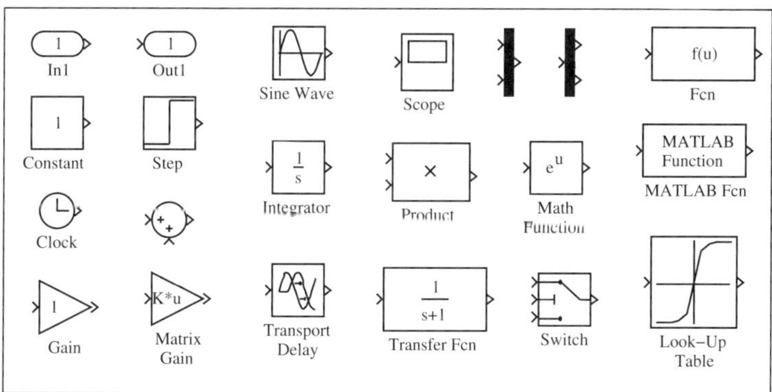

FIGURE 7.32: User-defined most commonly used blocks

The frequently used blocks are summarized below

(i) **Input and output port blocks** (In1, Out1) These blocks can generate a variable `yout` in MATLAB workspace. Any signal in the simulation model can be connected to the Scope blocks.

(ii) **Clock block** generates time t and it can be used in the modeling of time varying differential equations.

(iii) **Commonly used input blocks** The Sine block can be used in generating sinusoidal signals. The block Step can be used in generating the step signal, and the Constant block can be used to define the constant signals.

(iv) **Integrator block** (Int) The block is used to evaluate integral to the input signal. For instance, assume that the input to the ith integrator is $\dot{x}_i(t)$, the output of the integrator is then $x_i(t)$. The use of integrator is a crucial step in the modeling of differential equations. For high-order linear differential equations, it can also be modeled by Transfer Function blocks.

(v) **Transport delay blocks** (Transport Delay) The output signal is the value of the input signal at time $t - \tau$. It can be used in modeling of delay differential equations.

(vi) **Gain blocks** (Gain, Sliding Gain **and** Matrix Gain) These gain blocks are very useful in Simulink modeling. The definitions of them are different. The Gain block is used to amplify the input signal. If the input signal is u, the output of the block is Ku. The Matrix Gain block is used for vector input signal $\boldsymbol{u}$, whose output is $\boldsymbol{K}\boldsymbol{u}$. The Sliding Gain block implements a scroll bar so that the gain of the block can be changed arbitrarily using mouse dragging.

(vii) **Mathematical operation blocks** These blocks can be used to perform algebraic operations such as plus, minus, and times as well as logical operations.

(viii) **Mathematical function blocks** These blocks can be used to perform nonlinear functions such as trigonometry functions or exponential functions.

(ix) **Signal vectorization blocks** The Mux block can be used as the vector signal composed of individual input signals. The Demux block can be used to extract scalar signals from the vector signal.

7.6.3 Using Simulink for modeling and simulation of ODEs

Using suitable blocks, the differential equations can be constructed using Simulink. The solution of the ODEs can be obtained with the function `sim()`, with the following syntax

```
[t,x,y]=sim(model_name,tspan,options)
```

which is quite similar to the function `ode45()`. Examples will be given in the following to demonstrate the modeling and simulation of different types of ODEs. The Lorenz equation will be studied first, followed by the delay differential equations.

Example 7.31 Consider again the Lorenz equation studied in Example 7.8

$$\begin{cases} \dot{x}_1(t) = -\beta x_1(t) + x_2(t)x_3(t) \\ \dot{x}_2(t) = -\rho x_2(t) + \rho x_3(t) \\ \dot{x}_3(t) = -x_1(t)x_2(t) + \sigma x_2(t) - x_3(t) \end{cases}$$

where $\beta = 8/3, \rho = 10$, and $\sigma = 28$, and the initial states are $x_1(0) = x_2(0) = 0$, and $x_3(0) = 10^{-10}$. Model the ODEs with Simulink and then find the solutions using simulation technique.

Solution Since there are three first-order derivative terms, thus three integrators are used to describe respectively $\dot{x}_1(t)$, $\dot{x}_2(t)$, and $\dot{x}_3(t)$. The outputs of these blocks are $x_1(t)$, $x_2(t)$, and $x_3(t)$. The framework of the system can be established as shown in Figure 7.33 (a). Double click each integrator block and the initial state variables can be specified for each integrator.

With the framework, the states and their derivatives are defined and with the use of Mux block, the state vector can be defined such that $\boldsymbol{x}(t) = [x_1(t), x_2(t), x_3(t)]^{\mathrm{T}}$. Then the Lorenz equation can be established when the derivative terminals are assigned to suitable signals. For instance, the first equation, $\dot{x}_1(t) = -\beta x_1(t) + x_2(t)x_3(t)$, can be specified with the use of Fcn block. One can fill the edit box of Fcn block with the string `-beta*u[1]+u[2]*u[3]`. The other two equations can also be modeled in a similar way, such that the original ODEs can be constructed in Simulink as shown in Figure 7.33 (b). The numerical solutions of the equations can be obtained using simulation methods.

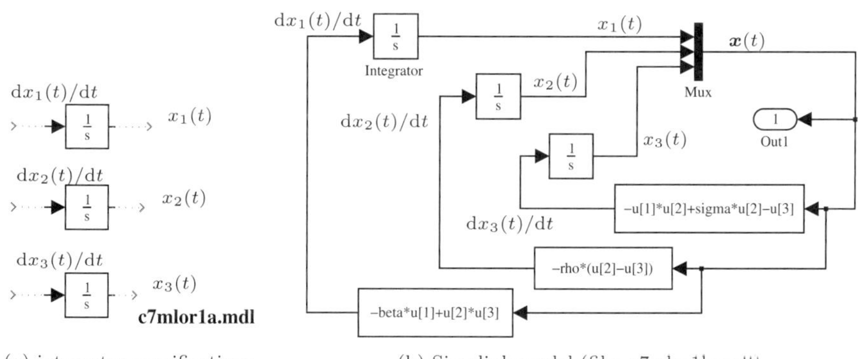

(a) integrator specifications (b) Simulink model (file: c7mlor1b.mdl)

FIGURE 7.33: Simulink model of the Lorenz equation

The following statements can be used to solve the Simulink model. The results obtained are exactly the same as the ones given in Figures 7.3 (a) and (b).

```
>> beta=8/3; rho=10; sigma=28;
   [t,x]=sim('c7mlor1b',[0,100]); plot(t,x)
   figure; plot3(x(:,1),x(:,2),x(:,3))
```

It should be noted that the variables `beta`, `rho` and `sigma` can be entered directly into MATLAB workspace, rather than specifying them as additional arguments. The initial states can be specified in relevant dialog boxes.

It is noted that for problems of small scales, the modeling of the differential equations in Simulink is more complicated than the direct use of `ode45()` function. However, for complicated ODEs, using Simulink may make the ODE solution process straightforward.

Example 7.32 Consider the delay differential equations in Example 7.23, where

$$\begin{cases} \dot{x}(t) = 1 - 3x(t) - y(t-1) - 0.2x^3(t-0.5) - x(t-0.5) \\ \ddot{y}(t) + 3\dot{y}(t) + 2y(t) = 4x(t). \end{cases}$$

Solve numerically the DDE using Simulink.

Solution The equation has already been solved in the previous example by the use of function `dde23()`, where some MATLAB functions are used. However, this method is not quite straightforward.

Now consider the first equation. One may move the $-3x(t)$ term to the left-hand side, and the original equation can be converted to

$$\dot{x}(t) + 3x(t) = 1 - y(t-1) - 0.2x^3(t-0.5) - x(t-0.5).$$

Thus the state $x(t)$ can be regarded as the output signal from the transfer function $1/(s+3)$, while the input signal to the transfer function is $1 - y(t-1) - 0.2x^3(t-0.5) - x(t-0.5)$. In the second equation, the $y(t)$ signal can be regarded as the output of block $4/(s^2+3s+2)$, while the input to the block is $x(t)$. The transport delay blocks can be connected to the signals $x(t)$ and $y(t)$ to generate the delayed signals from them. From the above analysis, the Simulink model shown in Figure 7.34 can be constructed.

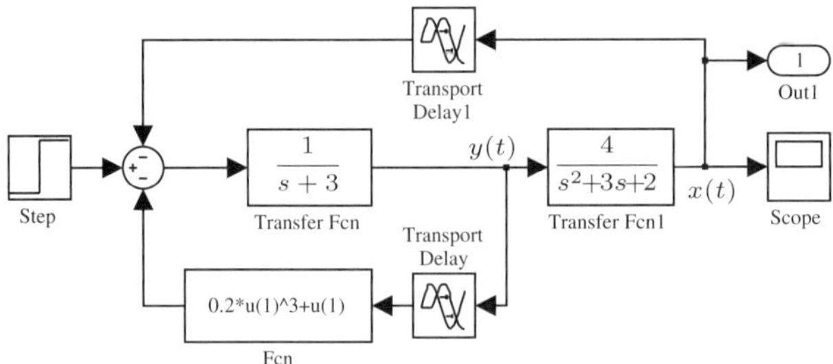

FIGURE 7.34: Simulink model (file: c7mdde2.mdl)

When the model is established, the following statements can be used to solve the differential equations. The output signal $y(t)$ obtained is exactly the same as the one obtained in Example 7.23. Also the solutions can be displayed on the Scope in the simulation model.

```
>> [t,x]=sim('c7mdde2',[0,10]); plot(t,x)
```

Of course, if one is not used to the transfer function representation, one can still assume that $x_1 = x, x_2 = y, x_3 = \dot{y}$, thus the original ordinary differential equations are converted to the first-order explicit differential equations

$$\begin{cases} \dot{x}_1(t) = 1 - x_1(t) - x_2(t-1) + 0.2x_1^3(t-0.5) - x_1(t-0.5) \\ \dot{x}_2(t) = x_3(t) \\ \dot{x}_3(t) = -4x_1(t) - 3x_3(t) - 2x_2(t). \end{cases}$$

Since there are three equations, three integrators should be used and the Simulink model can then be established using the descriptions given earlier. The same numerical results are obtained.

Example 7.33 Now let us consider the unsolved delay differential equations defined in Example 7.24, where

$$\boldsymbol{A}_1=\begin{bmatrix}-13 & 3 & -3\\ 106 & -116 & 62\\ 207 & -207 & 113\end{bmatrix},\ \boldsymbol{A}_2=\begin{bmatrix}0.02 & 0 & 0\\ 0 & 0.03 & 0\\ 0 & 0 & 0.04\end{bmatrix},\ \boldsymbol{B}=\begin{bmatrix}0\\1\\2\end{bmatrix}.$$

Model and solve it with Simulink.

Solution The equations cannot be directly solved using the `dde23()` function. Simulink-based modeling can be used to describe the equations, and it can also be simulated via Simulink. Before building the Simulink model, the following statements can be entered to define the given matrices

```
>> A1=[-13,3,-3; 106,-116,62; 207,-207,113];
   A2=diag([0.02,0.03,0.04]); B=[0; 1; 2];
```

Now consider the original equations. Define an integrator and then define the output signal to the integrator as the state vector, $\boldsymbol{x}(t)$ and the input signal is automatically the derivative of the states $\dot{\boldsymbol{x}}(t)$. Transport delay blocks can be appended to the two signals to describe the signals $\boldsymbol{x}(t-\tau_1)$ and $\dot{\boldsymbol{x}}(t-\tau_2)$. Thus the Simulink model can be constructed as shown in Figure 7.35.

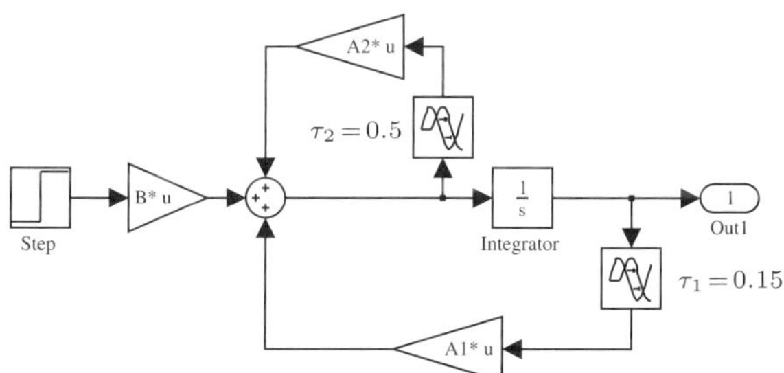

FIGURE 7.35: Neutral type DDE model (file: c7mdde3.mdl)

This neutral type DDE can easily be solved and the time responses are shown in Figure 7.36.

```
>> [t,x]=sim('c7mdde3',[0,8]); plot(t,x)
```

Exercises

1. Find the general solution to the following LTI ODE:

$$\frac{\mathrm{d}^5y(t)}{\mathrm{d}t^5}+13\frac{\mathrm{d}^4y(t)}{\mathrm{d}t^4}+64\frac{\mathrm{d}^3y(t)}{\mathrm{d}t^3}+152\frac{\mathrm{d}^2y(t)}{\mathrm{d}t^2}+176\frac{\mathrm{d}y(t)}{\mathrm{d}t}+80y(t)$$

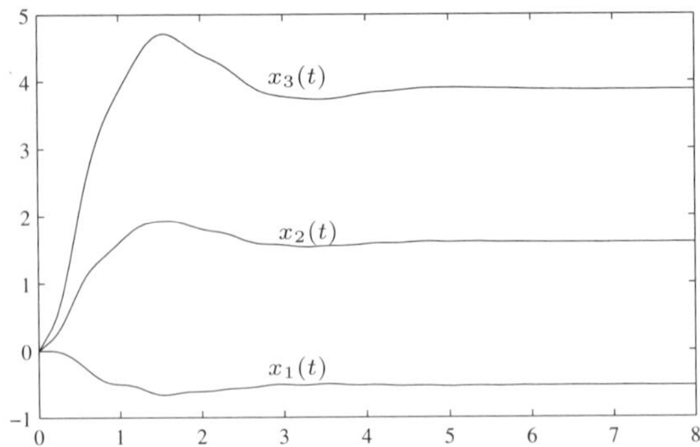

FIGURE 7.36: Solutions to delay differential equations

$$= \mathrm{e}^{-2t}\left[\sin\left(2t+\frac{\pi}{3}\right)+\cos 3t\right].$$

With the initial conditions $y(0)=1$, $y(1)=3$, $y(\pi)=2, \dot{y}(0)=1, \dot{y}(1)=2$, find the analytical solution. Verify the obtained results by back-substitution.

2. Find the general analytical solution to the following LTI simultaneous ODEs:

$$\begin{cases} \ddot{x}(t)+5\dot{x}(t)+4x(t)+3y(t)=\mathrm{e}^{-6t}\sin 4t \\ 2\dot{y}(t)+y(t)+4\dot{x}(t)+6x(t)=\mathrm{e}^{-6t}\cos 4t. \end{cases}$$

If initial and boundary conditions are $x(0)=1, x(\pi)=2, y(0)=0$, find the corresponding specific solution. Verify the results by back-substitution.

3. Find the analytical solutions to the following linear time-varying (LTV) ODEs.

(i) Legendre equation: $(1-t^2)\dfrac{\mathrm{d}^2x}{\mathrm{d}t^2}-2t\dfrac{\mathrm{d}x}{\mathrm{d}t}+n(n+1)x=0$,

(ii) Bessel equation: $t^2\dfrac{\mathrm{d}^2x}{\mathrm{d}t^2}+t\dfrac{\mathrm{d}x}{\mathrm{d}t}+(t^2-n^2)x=0$.

4. Find the analytical solution to the following nonlinear ODE:

$$\ddot{y}(x)-\left(2-\frac{1}{x}\right)\dot{y}(x)+\left(1-\frac{1}{x}\right)y(x)=x^2\mathrm{e}^{-5x}.$$

Furthermore, if the boundaries are specified by $y(1)=\pi,\ y(\pi)=1$, find the analytical solution.

5. Solve the following LTI ODEs using Laplace transform method:

$$\begin{cases} \ddot{x}(t)+\ddot{y}(t)+x(t)+y(t)=0,\ x(0)=2, y(0)=1 \\ 2\ddot{x}(t)-\ddot{y}(t)-x(t)+y(t)=\sin t,\ \dot{x}(0)=\dot{y}(0)=-1. \end{cases}$$

Compare the obtained results with those by other ODE solving methods.

6. Find the general solutions to the following ODEs.

(i) $\ddot{x}(t)+2t\dot{x}(t)+t^2x(t)=t+1$, (ii) $\dot{y}(x)+2xy(x)=x\mathrm{e}^{-x^2}$,

(iii) $y^{(3)}+3\ddot{y}+3\dot{y}+y=\mathrm{e}^{-t}\sin t$.

7. Limit cycle is a common phenomenon in nonlinear differential equations. For dynamic systems governed by nonlinear differential equations, no matter what initial values are selected, the phase trajectory will settle down on the same

closed path, which is referred to as the *limit cycle*. Solve the following differential equations and draw the limit cycle in the x-y plane:

$$\begin{cases} \dot{x} = y + x(1 - x^2 - y^2) \\ \dot{y} = -x + y(1 - x^2 - y^2). \end{cases}$$

Try different initial values, and check whether the phase plane plot converges to the same limit cycle.

8. Consider the well-known Rössler equation described as $\begin{cases} \dot{x} = -y - z \\ \dot{y} = x + ay \\ \dot{z} = b + (x - c)z, \end{cases}$

   where $a = b = 0.2$, $c = 5.7$, and $x_1(0) = x_2(0) = x_3(0)$. Draw the 3D phase trajectory and also its projection on the x-y plane. In preparing the MATLAB scripts, the parameters a, b, c are suggested to be used as additional parameters. If the parameters are changed to $a = 0.2, b = 0.5, c = 10$, solve the problem again and observe the change of phase behavior.

9. Consider the following Chua's circuit equation well-known in chaos studies:

$$\begin{cases} \dot{x} = \alpha[y - x - f(x)] \\ \dot{y} = x - y + z \\ \dot{z} = -\beta y - \gamma z \end{cases}$$

   where the nonlinear function $f(x)$ is described by

$$f(x) = bx + \frac{1}{2}(a - b)(|x + 1| - |x - 1|), \text{ and } a < b < 0.$$

   Prepare an M-function to describe the above ODEs, and draw the phase trajectory for the parameters $\alpha = 9, \beta 100/7, \gamma = 0, a = -8/7, b = -5/7$, and initial conditions $x(0) = -2.121304, y(0) = -0.066170, z(0) = 2.881090$.

10. For the Lotka-Volterra's predator-prey equations $\begin{cases} \dot{x}(t) = 4x(t) - 2x(t)y(t) \\ \dot{y}(t) = x(t)y(t) - 3y(t) \end{cases}$

    with initial conditions $x(0) = 2, y(0) = 3$, solve the time responses of $x(t)$ and $y(t)$, and also plot the phase plane trajectory.

11. Select appropriate state variables to convert the following high-order ODEs into first-order vector-form explicit ODEs. Solve the ODEs and plot the phase trajectories.

$$\text{(i)} \begin{cases} \ddot{x} = -x - y - (3\dot{x})^2 + \dot{y}^3 + 6\ddot{y} + 2t \\ y^{(3)} = -\ddot{y} - \dot{x} - \mathrm{e}^{-x} - t \end{cases}, \text{ with } \begin{cases} x(1) = 2, \dot{x}(1) = -4 \\ y(1) = -2, \dot{y}(1) = 7 \\ \ddot{y}(1) = 6, \end{cases}$$

$$\text{(ii)} \begin{cases} \ddot{x} - 2xz\dot{x} = 3x^2yt^2 \\ \ddot{y} - \mathrm{e}^y\dot{y} = 4xt^2z \\ \ddot{z} - 2t\dot{z} = 2t\mathrm{e}^{-xy} \end{cases}, \text{ with } \begin{cases} \dot{z}(1) = \dot{x}(1) = \dot{y}(1) = 2 \\ \dot{z}(1) = x(1) = y(1) = 3. \end{cases}$$

12. Solve the following nonlinear time-varying ODE

$$y^{(3)} + ty\ddot{y} + t^2\dot{y}y^2 = \mathrm{e}^{-ty}, \; y(0) = 2, \; \dot{y}(0) = \ddot{y}(0) = 0$$

    and plot $y(t)$. Select the fixed-step Runge-Kutta algorithm for solving the same problem as the baseline and compare in speed and accuracy with other MATLAB ODE solvers for this problem.

13. Find the analytical and numerical solutions of the following high-order simultaneous linear time-invariant ordinary differential equations and compare the results:
$$\begin{cases} \ddot{x}(t) = -2x(t) - 3\dot{x}(t) + \mathrm{e}^{-5t}, & x(0) = 1, \dot{x}(0) = 2 \\ \ddot{y}(t) = 2x(t) - 3y(t) - 4\dot{x}(t) - 4\dot{y}(t) - \sin t, & y(0) = 3, \dot{y}(0) = 4. \end{cases}$$

14. Consider the following nonlinear ODEs:
$$\begin{cases} \ddot{u}(t) = -u(t)/r^3(t) \\ \ddot{v}(t) = -v(t)/r^3(t) \end{cases}$$
where $r(t) = \sqrt{u^2(t) + v^2(t)}$, and $u(0) = 1, \dot{u}(0) = 2, \dot{v}(0) = 2, v(0) = 1$. Select a set of state variables and convert the original ODEs to the form solvable by MATLAB. Plot the curves of $u(t), v(t)$, and the phase plane trajectory.

15. Consider the following implicit ODEs[36]
$$\begin{cases} \dot{u}_1 = u_3 \\ \dot{u}_2 = u_4 \\ 2\dot{u}_3 + \cos(u_1 - u_2)\dot{u}_4 = -\mathrm{g}\sin u_1 - \sin(u_1 - u_2)u_4^2 \\ \cos(u_1 - u_2)\dot{u}_3 + \dot{u}_4 = -\mathrm{g}\sin u_2 + \sin(u_1 - u_2)u_3^2 \end{cases}$$
where $u_1(0) = 45, u_2(0) = 30, u_3(0) = u_4(0) = 0$, and g$= 9.81$. Solve the above initial value problem and plot the time responses to the given initial states.

16. Numerically solve the following implicit ODEs
$$\begin{cases} \dot{x}_1\ddot{x}_2\sin(x_1x_2) + 5\ddot{x}_1\dot{x}_2\cos(x_1^2) + t^2x_1x_2^2 = \mathrm{e}^{-x_2^2} \\ \ddot{x}_1x_2 + \ddot{x}_2\dot{x}_1\sin(x_1^2) + \cos(\ddot{x}_2x_2) = \sin t \end{cases}$$
where $x_1(0) = 1, \dot{x}_1(0) = 1, x_2(0) = 2, \dot{x}_2(0) = 2$. Plot the phase trajectory.

17. The following LTI ODEs are considered as stiff. Solve the initial value problems using ordinary ODE solver and stiff ODE solver provided in MATLAB. Meanwhile, try to solve these equations using analytical methods and verify the accuracy of the numerical results.

(i) $$\begin{cases} \dot{y}_1 = 9y_1 + 24y_2 + 5\cos t - \dfrac{1}{3}\sin t, \ y_1(0) = \dfrac{1}{3} \\ \dot{y}_2 = -24y_1 - 51y_2 - 9\cos t + \dfrac{1}{3}\sin t, \ y_2(0) = \dfrac{2}{3}, \end{cases}$$

(ii) $$\begin{cases} \dot{y}_1 = -0.1y_1 - 49.9y_2, \ y_1(0) = 1 \\ \dot{y}_2 = -50y_2, \ y_2(0) = 2 \\ \dot{y}_3 = 70y_2 - 120y_3, \ y_3(0) = 1. \end{cases}$$

18. Consider the following chemical reaction equations
$$\begin{cases} \dot{y}_1 = -0.04y_1 + 10^4y_2y_3 \\ \dot{y}_2 = 0.04y_1 - 10^4y_2y_3 - 3\times 10^7y_2^2 \\ \dot{y}_3 = 3\times 10^7y_2^2 \end{cases}$$
where the initial values are $y_1(0) = 1, y_2(0) = y_3(0) = 0$. Note that the above ODE might be regarded as stiff. Solve the initial value problem with

`ode45()` and the stiff solver `ode15s()` and check whether the obtained results are comparable or not.

19. Solve the boundary value problem in Problem 4 using numerical methods and draw the solution $y(t)$. Compare the accuracy of the results with the analytical results obtained earlier.
20. Solve the boundary value problem where $\ddot{x}+\dfrac{1}{t}\dot{x}+\left(1-\dfrac{1}{4t^2}\right)x=\sqrt{t}\cos t$, with $x(1)=1, x(6)=-0.5$.
21. For the Van der Pol equation $\ddot{y}+\mu(y^2-1)\dot{y}+y=0$, if $\mu=1$, find the numerical solutions for boundary conditions $y(0)=1, y(5)=3$. If μ is an undetermined parameter, an extra condition $\dot{y}(5)=-2$ can be used. Solve the parameter μ as well as the equation. Plot the obtained solution and verify the results.
22. Solve numerically the partial differential equations below and draw the surface plot of the solution u.
$$\begin{cases} \dfrac{\partial^2 u}{\partial x^2}+\dfrac{\partial^2 u}{\partial y^2}=0 \\ u\big|_{x=0,y>0}=1, \quad u\big|_{y=0,\ x\geqslant 0}=0 \\ x>0, \quad y>0. \end{cases}$$
23. Consider the following simple LTI ODE
$$y^{(4)}+4y^{(3)}+6\ddot{y}+4\dot{y}+y=\mathrm{e}^{-3t}+\mathrm{e}^{-5t}\sin(4t+\pi/3)$$
with the initial conditions given by $y(0)=1, \dot{y}(0)=\ddot{y}(0)=1/2, y^{(3)}(0)=0.2$. Construct the simulation model with Simulink, and find the simulation results.
24. Consider the following LTV ODE
$$y^{(4)}+4ty^{(3)}+6t^2\ddot{y}+4\dot{y}+y=\mathrm{e}^{-3t}+\mathrm{e}^{-5t}\sin(4t+\pi/3)$$
with the initial conditions $y(0)=1$, $\dot{y}(0)=\ddot{y}(0)=1/2, y^{(3)}(0)=0.2$. Construct a Simulink model, solve the ODE and plot the solution trajectories.
25. Consider the following delay differential equation
$$y^{(4)}(t)+4y^{(3)}(t-0.2)+6\ddot{y}(t-0.1)+6\ddot{y}(t)+4\dot{y}(t-0.2)+y(t-0.5)=\mathrm{e}^{-t^2}.$$
It is assumed that for $t\leqslant 0$, the above DDE has zero initial conditions. Construct a Simulink model and solve the DDE. Moreover, use the function `dde23()` to solve the same problem and compare the results from these two methods.

Chapter 8

Data Interpolation and Functional Approximation Problems

In scientific research and engineering practice, a lot of experimental data are generated. Based on the experimental data, the problems of data interpolation and function fitting may always be encountered. The existing data can be regarded as the samples, the so-called *data interpolation* is to numerically generate new data points from a discrete set of known samples. In Section 8.1, one-dimensional, two-dimensional or even high-dimensional interpolation problems are solved in MATLAB. An interpolation-based numerical integration method is also introduced. In Section 8.2, two of the most widely used splines for interpolation are introduced, the cubic spline and the B-spline. Spline function-based numerical differentiations and integrations are also introduced. The integration results are even more accurate than those presented in Chapter 3 and Section 8.1. Data interpolation problems can easily be solved by following the examples in these two sections.

The so-called *function approximation problem* is to extract the function representation from the measured sample points. Polynomial approximations and the least squares method for nonlinear function approximation will be studied. Furthermore, Padé approximations and continued fraction approximations for given functions are explored in Section 8.3. In Section 8.4, the correlation analysis of signals and experimental data are introduced. Fast Fourier transforms and filter-based de-noising and other signal processing problems are introduced.

At this point, let us note the difference between "interpolation" and "fitting." The key distinguishing feature for "interpolation" is that the obtained interpolation function passes through all given data points, while for "fitting" the obtained fitting function may not pass all the given data points.

Some problems in the chapter are also related to those topics presented in other chapters, with Chapter 10 in particular. For instance, the data fitting problems can be solved with neural networks to be discussed in Section 10.2. Data filtering and de-noising problems can also be solved with wavelet transform techniques to be introduced in Section 10.4.

For readers who wish to check the detailed explanations of various data interpolation techniques and function fitting methods, we recommend the free textbook [4] (Chapters 3, 5 and 15). We also found the online resource [37] useful and interesting (Chapter 5).

8.1 Interpolation and Data Fitting

8.1.1 One-dimensional data interpolation

Solving one-dimensional data interpolation problems

Assume that the function $f(x)$ is a one-dimensional function and the mathematical formula is not known. If at a set of N distinct points $x_1, x_2, \cdots, x_N$, the values of the function are measured as $y_1, y_2, \cdots, y_N$. The points (x_1, y_1), (x_2, y_2), $\cdots$, (x_N, y_N) are often referred to as *sample points*. The idea of using sample points for finding other unknown points is called *interpolation*.

Many interpolation functions have been provided in MATLAB, such as the one-dimensional interpolation function `interp1()`, two-dimensional `interp2()`, and the polynomial fitting function `polyfit()`.

One-dimensional interpolation function `interp1()` can be called

```
y1=interp1(x,y,x1,method)
```

where $\boldsymbol{x} = [x_1, x_2, \cdots, x_N]^{\mathrm{T}}$, and $\boldsymbol{y} = [y_1, y_2, \cdots, y_N]^{\mathrm{T}}$. The two vectors provide the information of the sample points. The argument $\boldsymbol{x}_1$ is the points of independent variable to be interpolated. It can be a scalar, a vector or a matrix. The interpolated results are returned in argument $\boldsymbol{y}_1$. The default of `method` argument is `'linear'`. Alternative options for the `method` argument can be `'nearest'`, `'cubic'` (newly changed to `'pchip'`) and `'spline'`. Usually it is recommended that the `'spline'` option be used. Extrapolations, i.e., the interpolation point is outside of the interval $[x_1, x_N]$ can also be performed with `y1=interp1(x,y,x1,method,'extrap')`.

Example 8.1 Assume the sample points are obtained from the function $f(x) = (x^2 - 3x + 5)\mathrm{e}^{-5x} \sin x$. Generate some sample points and interpolate them for a smoother curve.

Solution The sample data can be generated with the following statements, where the spacing can be selected as a large value, such as 0.12. The sample points are then obtained as shown in Figure 8.1 (a).

```
>> x=0:.12:1;
   y=(x.^2-3*x+5).*exp(-5*x).*sin(x); plot(x,y,x,y,'o')
```

It can be seen that the curve is formed by joining the adjacent points with straight lines. The curves around the sample points are not very smooth. One may select another set of data and then use `interp1()` function to calculate the interpolations to these points.

```
>> x1=0:.02:1; y0=(x1.^2-3*x1+5).*exp(-5*x1).*sin(x1);
   y1=interp1(x,y,x1); y2=interp1(x,y,x1,'cubic');
   y3=interp1(x,y,x1,'spline'); y4=interp1(x,y,x1,'nearest');
   plot(x1,[y1',y2',y3',y4'],':',x,y,'o',x1,y0)
```

Then with different methods, the interpolation results are compared with the theoretical results from the given function obtained as shown in Figure 8.1 (b).

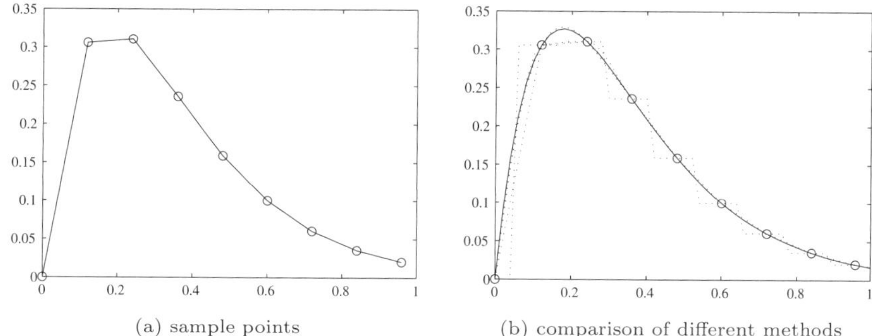

(a) sample points (b) comparison of different methods

FIGURE 8.1: Results of one-dimensional interpolation methods

It can be seen that the linear interpolation is exactly the same as the one in Figure 8.1 (a) and the quality of `'nearest'` option is very poor. The interpolation results under the `'cubic'` and `'spline'` options are much smoother. In fact, the interpolation using the `'spline'` method gives much better interpolation to the original function. Thus it is recommended to use the `'spline'` option in solving one-dimensional interpolation problems.

Example 8.2 Write a piece of code, which allows the user to draw manually a a smooth curve, interpolated by splines.

Solution In applications, the user can pick a few points with the `ginput()` function. Then interpolation can be made to obtain a smooth curve. The above idea can be implemented in the MATLAB function shown below

```
function sketcher(vis)
x=[]; y=[]; i=1; h=[]; axes(gca);
while 1, [x0,y0,but]=ginput(1);
   if but==1, x=[x,x0]; y=[y,y0];
      h(i)=line(x0,y0); set(h(i),'Marker','o'); i=i+1;
   else, break; end
end
if nargin==1, delete(h); end
xx=[x(1):(x(end)-x(1))/100: x(end)];
yy=interp1(x,y,xx,'spline'); line(xx,yy)
```

Lagrange interpolation algorithm and its application

The Lagrange interpolation algorithm is the most presented interpolation method in the interpolation part of numerical analysis textbooks, e.g. [38]. For known sample points x_i, y_i, the interpolation on $\boldsymbol{x}$ vector can be obtained as

$$\phi(\boldsymbol{x}) = \sum_{i=1}^{N} y_i \prod_{j=1, j\neq i}^{N} \frac{\boldsymbol{x} - x_j}{(x_i - x_j)}. \tag{8.1}$$

Based on the above algorithm, a MATLAB function is prepared as follows:

```
function y=lagrange(x0,y0,x)
ii=1:length(x0); y=zeros(size(x));
for i=ii
   ij=find(ii~=i); y1=1;
   for j=1:length(ij), y1=y1.*(x-x0(ij(j))); end
   y=y+y1*y0(i)/prod(x0(i)-x0(ij));
end
```

Example 8.3 Consider a well-known function $f(x) = 1/(1 + 25x^2), -1 \leqslant x \leqslant 1$, where a set of points can be evaluated as sample points. Based on the points, the interpolation points can be obtained using Lagrange interpolation method and the results are shown in Figure 8.2 (a).

```
>> x0=-1+2*[0:10]/10; y0=1./(1+25*x0.^2);
   x=-1:.01:1; y=lagrange(x0,y0,x);   % Lagrange interpolation
   ya=1./(1+25*x.^2); plot(x,ya,x,y,':')
```

From the interpolation results it can be seen that the interpolation is far from the theoretic curve. The higher the degree of the polynomials, the more serious the divergence. This phenomenon is referred to as the *Runge phenomenon*. Thus for this example, the Lagrange interpolation method failed. The function `interp1()` can be used instead to solve the problem. The spline interpolation can be obtained with the following statements and the fitting results are shown in Figure 8.2 (b). Thus the function provided in MATLAB does not have the Runge phenomenon any more, so it can be safely used.

```
>> y1=interp1(x0,y0,x,'cubic'); y2=interp1(x0,y0,x,'spline');
   plot(x,ya,x,y1,':',x,y2,'--')
```

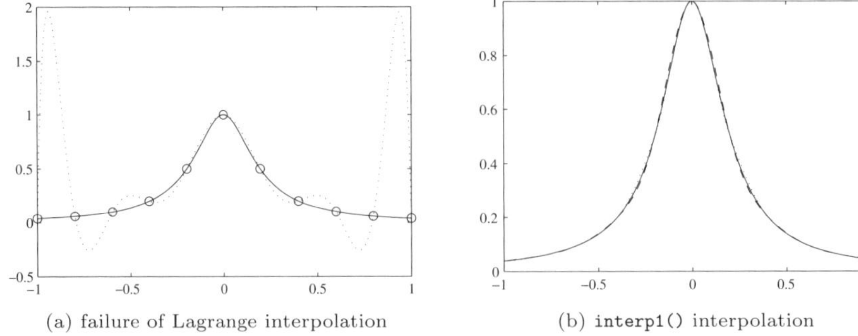

(a) failure of Lagrange interpolation (b) `interp1()` interpolation

FIGURE 8.2: Interpolation results comparison

8.1.2 Definite integral evaluation from given samples

The definite integral evaluation based on sample points has been discussed and the trapezoidal approach `trapz()` has been presented in Section 3.4.1. From Example 3.29, it is seen that if the sample points are sparsely distributed, there exist large errors in the results. If interpolation is used in the evaluation of the intermediate points, the interpolation-based numerical integration function can be written as

```
function y=quadspln(x0,y0,a,b)
f=@(x,x0,y0)interp1(x0,y0,x,'spline'); y=quadl(f,a,b,1e-8,[],x0,y0);
```

whose syntax is `I=quadspln(`$\boldsymbol{x}_0$`,`$\boldsymbol{y}_0$`,`a`,`b`)`, where $\boldsymbol{x}_0$ and $\boldsymbol{y}_0$ are vectors composed of sample points, $[a, b]$ is the integration interval. With `quadspln()` function, the definite integral can be obtained as illustrated in the following examples.

Example 8.4 Consider again the problem in Example 3.29. Use the interpolation-based method to evaluate the definite integral.

Solution From the theoretical method it is known that the integral is 2. The trapezoidal method can be applied to evaluate the integral. If the step-size is too large, the approximation accuracy is not satisfactory. Here the interpolation-based method is applied to the same sample points

```
>> x0=0:pi/30:pi; y0=sin(x0); I=quadspln(x0,y0,0,pi)
```

and the result is $I = 1.99999970244699$. It can be seen that the result is much more accurate than the one obtained by the trapezoidal method in Example 3.29. If the given sample points are even more sparsely distributed, the advantage of the interpolation-based method is much more evident.

```
>> x0=0:pi/10:pi; y0=sin(x0); I1=trapz(x0,y0)
   I2=quadspln(x0,y0,0,pi)
```

The trapezoidal result is $I_1 = 1.9835235375$, and with interpolation $I_2 = 2.00000016$.

An even more exaggerated example is as follows. Suppose that there are only 5 unevenly distributed sample points measured, the results by the use of interpolation method and trapezoidal methods can be compared

```
>> x0=[0,0.4,1 2,pi]; y0=sin(x0);  % sample points generation
   plot(x0,y0,x0,y0,'o')   % the sample points are shown in Figure 8.3 (a)
   I=quadspln(x0,y0,0,pi)   % only 1% relative error obtained
   I1=trapz(x0,y0)    % with the trapz() function, relative error reaches 7.9%!
```

and the results with these two methods are $I = 2.0191$, $I_1 = 1.84156$, respectively. In fact, even with such sparsely distributed sample points, the fitting results are still very satisfactory. The interpolation function together with the sample points are shown in Figure 8.3 (b), which are quite close to the original true function.

```
>> x=[0:0.01:pi,pi]; y0a=sin(x); y=interp1(x0,y0,x,'spline');
   plot(x0,y0,x,y,':',x,y0a,x0,y0,'o')
```

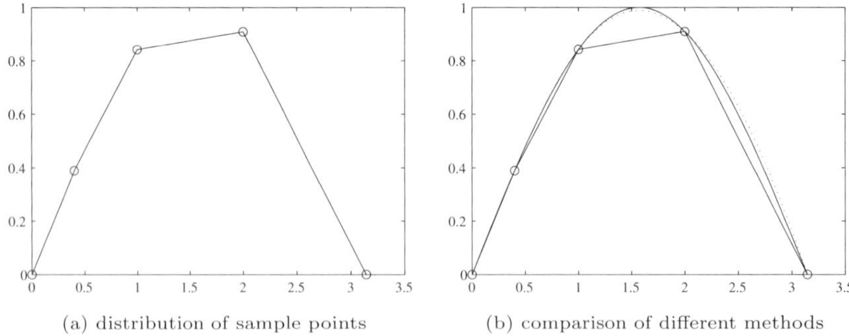

FIGURE 8.3: The integration results with only five sample points

Example 8.5 Consider again the oscillatary function in Example 3.30. Assume that 150 sample points are measured. Evaluate the definite integral with the `quadspln()` function and verify the accuracy of the results.

Solution The numerical solutions to the definite integral can be obtained from the generated sample points

```
>> x=[0:3*pi/2/200:3*pi/2]; y=cos(15*x); I=quadspln(x,y,0,3*pi/2)
```

and the result obtained is $I = 0.06667117837770$. Note that the true integral is $1/15$.

Clearly, the interpolation-based method achieves quite high accuracy. The following statements can be used to draw the original and interpolated curves as shown in Figure 8.4. The fitting results are also quite satisfactory. The difference between the original and interpolated curves are hardly seen from the figure.

```
>> x0=[0:3*pi/2/1000:3*pi/2]; y0=cos(15*x0);
   y1=interp1(x,y,x0,'spline'); plot(x,y,x0,y1,':')
```

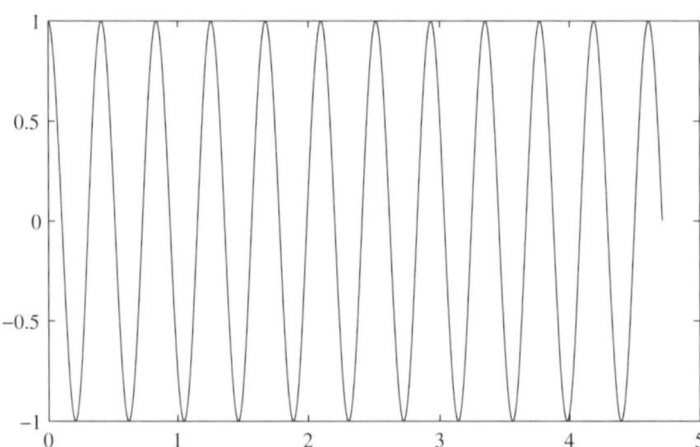

FIGURE 8.4: Original function and interpolation curves

From this example, it can be seen that if there are significant oscillations in the integrand, the accuracy of integral evaluations cannot be guaranteed if the number of sample points is large enough.

8.1.3 Two-dimensional grid data interpolation

The two-dimensional (2D) interpolation function `interp2()` is provided in MATLAB with the syntax `z1=interp2(x0,y0,z0,x1,y1,method)`, where $\boldsymbol{x}_0, \boldsymbol{y}_0, \boldsymbol{z}_0$ is the measured sample points in mesh grid form. The arguments $\boldsymbol{x}_1, \boldsymbol{y}_1$ are the points to be interpolated. They are not necessarily given in mesh grid form. They can be in any form, scalars, vectors, matrices or even multi-dimensional arrays. The returned argument $\boldsymbol{z}_1$ is the interpolation results, which is exactly the same data type as $\boldsymbol{x}_1$ or $\boldsymbol{y}_1$. The `method` options are `'linear'`, `'cubic'` and `'spline.'` Similar to the one-dimensional interpolation function, the method `'spline'` is recommended. Examples will be given in the following to illustrate the 2D grid data interpolation.

Example 8.6 From the given function $z = (x^2 - 2x)\mathrm{e}^{-x^2-y^2-xy}$, generate a set of relatively sparsely distributed mesh grid sample data. From the data, compute the whole surface using interpolation methods, and compare the results with the exact surface.

Solution The mesh grid dataa can be obtained as shown in Figure 8.5 (a) by the following MATLAB scripts. It can be seen that the surface obtained is not very smooth.

```
>> [x,y]=meshgrid(-3:.6:3,-2:.4:2); z=(x.^2-2*x).*exp(-x.^2-y.^2-x.*y);
   surf(x,y,z), axis([-3,3,-2,2,-0.7,1.5])
```

Now, select more densely distributed interpolation points in mesh grid form. The following statements can be used to evaluate the interpolation points with the interpolated surface shown in Figure 8.5 (b).

```
>> [x1,y1]=meshgrid(-3:.2:3, -2:.2:2);
   z1=interp2(x,y,z,x1,y1); surf(x1,y1,z1), axis([-3,3,-2,2,-0.7,1.5])
```

It can be seen that the interpolation results using the linear method is still rather rough. Let us try `'cubic'` and `'spline'` options and the obtained results are compared in Figure 8.6.

```
>> z1=interp2(x,y,z,x1,y1,'cubic'); z2=interp2(x,y,z,x1,y1,'spline');
   surf(x1,y1,z1), axis([-3,3,-2,2,-0.7,1.5])
   figure; surf(x1,y1,z2), axis([-3,3,-2,2,-0.7,1.5])
```

It can be seen that the interpolation results are all satisfactory, especially with the spline interpolation option. Thus the `'spline'` option is recommended.

Furthermore, since the original function is known, the exact solutions can be obtained. The following statements can be used to compute the error surface between the two interpolated matrices z_1 and z_2 and the exact z, respectively, as shown in Figure 8.7 (a), (b). It can be seen that the spline interpolation results

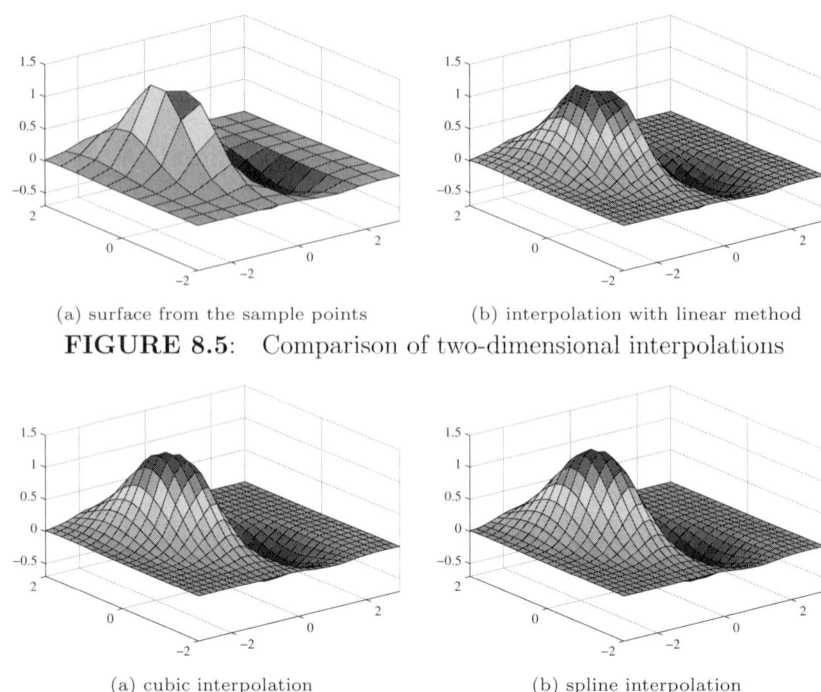

(a) surface from the sample points (b) interpolation with linear method

FIGURE 8.5: Comparison of two-dimensional interpolations

(a) cubic interpolation (b) spline interpolation

FIGURE 8.6: Comparisons of interpolation with other methods

are much more accurate than the ones by the cubic interpolation. The error surface comparison suggests again that the 'spline' option is recommended.

```
>> z=(x1.^2-2*x1).*exp(-x1.^2-y1.^2-x1.*y1);   % exact functions
   surf(x1,y1,abs(z-z1)), axis([-3,3,-2,2,0,0.08])
   figure; surf(x1,y1,abs(z-z2)), axis([-3,3,-2,2,0,0.025])
```

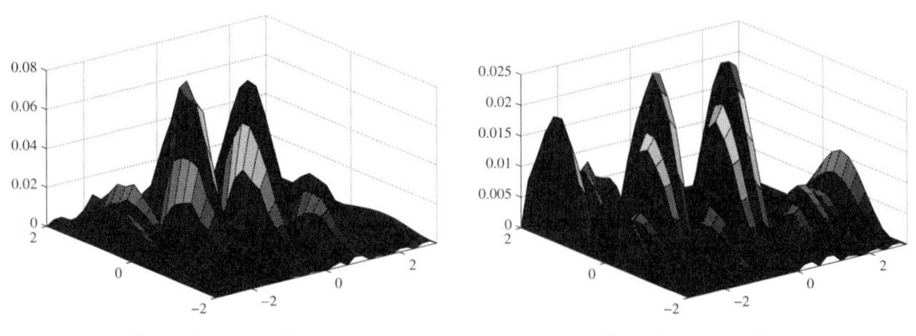

(a) cubic algorithm (b) spline algorithm

FIGURE 8.7: Error surfaces of the interpolation results

8.1.4 Two-dimensional scattered data interpolation

Through the above examples, it can be seen that the two-dimensional interpolation problems can easily be solved with the function `interp2()`. However, it should be noted that there are some restrictions. For example, only the data given in grid format can be handled by the function `interp2()`. If a scattered sample data set is provided, the `interp2()` function cannot be used. However, in many practical problems, the scattered data set (x_i, y_i, z_i) is usually provided, rather than the grid data. The general function `griddata()` can then be used instead, with the syntax `z=griddata(x0,y0,z0, x,y,'v4')`, where $\boldsymbol{x}_0, \boldsymbol{y}_0, \boldsymbol{z}_0$ are the vectors composed of the sample points. They can be the data vectors of arbitrarily distributed sample points. The arguments $\boldsymbol{x}, \boldsymbol{y}$ are the expected interpolation positions, which can be described as single point, vectors or mesh grid matrices. The returned argument $\boldsymbol{z}$ is in the same format as $\boldsymbol{x}$, representing the interpolation results. The option `'v4'` is the unnamed interpolation algorithm used in MATLAB version 4.0, which has many advantages. Apart from the `'v4'` option, other options such as `'linear'`, `'cubic'` and `'nearest'` can also be used. However, the option `'v4'` is recommended.

Example 8.7 Consider again the function $z = (x^2 - 2x)e^{-x^2-y^2-xy}$. In the rectangular region, $x \in [-3, 3]$, $y \in [-2, 2]$, a set of 200 sample points (x_i, y_i) can be selected randomly and the values z_i can be calculated. Based on the data, perform the interpolation using `griddata()` and visualize the error surface.

Solution The randomly selected 200 can be generated with the following statements, and the vectors $\boldsymbol{x}$, $\boldsymbol{y}$ and bmz can be established. Since the data set is not given in the grid format, the surface plot cannot be drawn directly from the data. Scattered points can be displayed instead as shown in Figure 8.8 (a), with the sample points distribution on the x-y plane shown in Figure 8.8 (b). It can be seen that the sample points are evenly scattered.

```
>> x=-3+6*rand(200,1); y=-2+4*rand(200,1);
   z=(x.^2-2*x).*exp(-x.^2-y.^2-x.*y);    % data generation
   plot3(x,y,z,'x'), axis([-3,3,-2,2,-0.7,1.5]), grid
   figure, plot(x,y,'x')   % two-dimensional distribution
```

The points to be interpolated can still be selected as grid format data using the method shown in Example 8.6. The interpolated data can then be obtained with the `'cubic'` and `'v4'` options, and the interpolated surfaces are shown in Figures 8.9 (a) and (b), respectively. It can be seen that the surface interpolation using the `'v4'` algorithm is much better. Some of the points with the `'cubic'` options are actually missing.

```
>> [x1,y1]=meshgrid(-3:.2:3, -2:.2:2);
   z1=griddata(x,y,z,x1,y1,'cubic'); surf(x1,y1,z1),
   axis([-3,3,-2,2,-0.7,1.5]); z2=griddata(x,y,z,x1,y1,'v4');
   figure; surf(x1,y1,z2), axis([-3,3,-2,2,-0.7,1.5])
```

Next, let us check the interpolation errors. The error surfaces using the two interpolation algorithms are obtained as shown in Figures 8.10 (a) and (b), respectively.

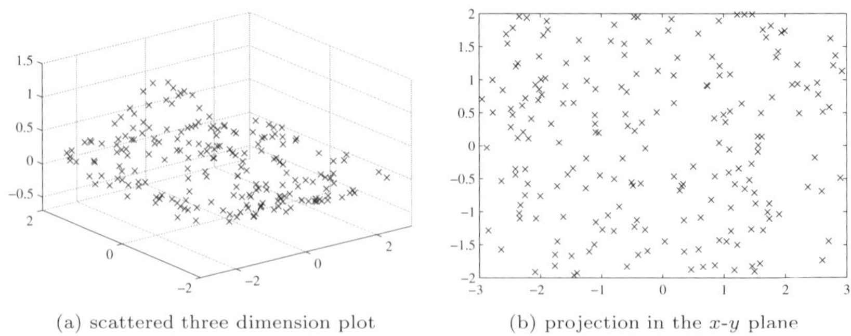

(a) scattered three dimension plot (b) projection in the x-y plane

FIGURE 8.8: Visualization of the known sample points

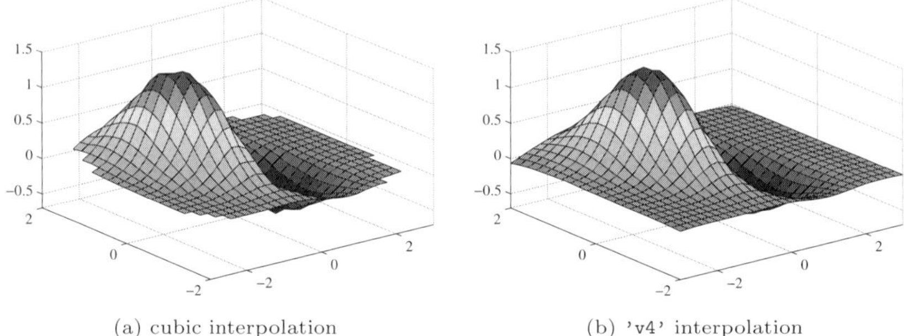

(a) cubic interpolation (b) 'v4' interpolation

FIGURE 8.9: Comparison of interpolation surfaces for a 2D function

It can be seen that the interpolation quality of the 'v4' option is much superior to the one obtained with the cubic interpolation algorithm.

```
>> z0=(x1.^2-2*x1).*exp(-x1.^2-y1.^2-x1.*y1);
   surf(x1,y1,abs(z0-z1));  axis([-3,3,-2,2,0,0.1])
   figure; surf(x1,y1,abs(z0-z2));  axis([-3,3,-2,2,0,0.1])
```

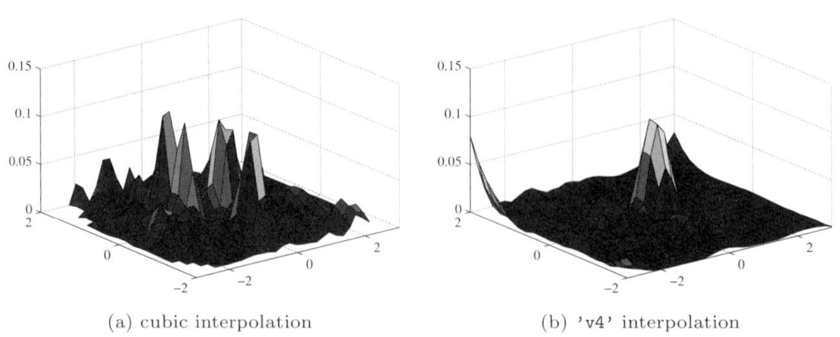

(a) cubic interpolation (b) 'v4' interpolation

FIGURE 8.10: Interpolations errors for a two-dimensional function

Example 8.8 In the previous example, the sample points in the x-y plane are evenly scattered. Now, let us remove some of the points and then perform interpolation and observe what happens.

Solution If the points within the circle centered at $(-1,-1/2)$ with a radius of 0.5 are removed from the vectors $\boldsymbol{x}$, $\boldsymbol{y}$ and $\boldsymbol{z}$, the following statements can be used to do so.

```
>> ii=find((x+1).^2+(y+0.5).^2>0.5^2); % find the points within the circle
   x=x(ii); y=y(ii); z=z(ii); plot(x,y,'x')
   t=[0:.1:2*pi,2*pi]; x0=-1+0.5*cos(t); y0=-0.5+0.5*sin(t);
   line(x0,y0)    % superimpose the circle on the samples
```

The new samples are distributed as shown in Figure 8.11 (a) with the circle superimposed. The samples within the circle have been removed. With the new set of samples, the interpolated surface is obtained as shown in Figure 8.11 (b). The quality of fittings looks satisfactory.

```
>> [x1,y1]=meshgrid(-3:.2:3, -2:.2:2); z1=griddata(x,y,z,x1,y1,'v4');
   surf(x1,y1,z1), axis([-3,3,-2,2,-0.7,1.5])
```

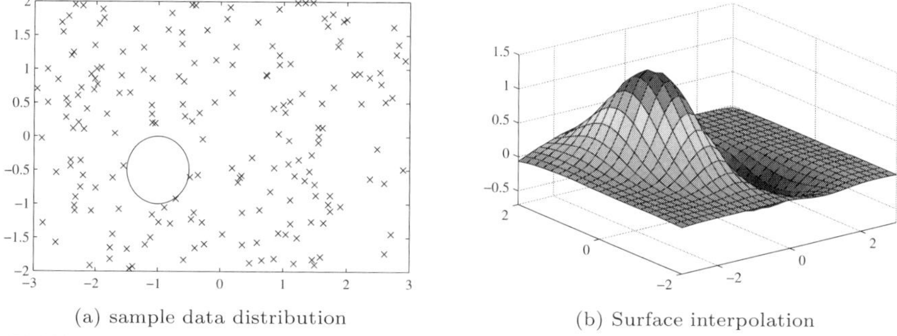

(a) sample data distribution (b) Surface interpolation

FIGURE 8.11: The new samples distribution and interpolated surface

Error surface of the interpolation is visualized in Figure 8.12 (a) where it can be observed that although some samples are removed, the fitting result is still satisfactory.

```
>> z0=(x1.^2-2*x1).*exp(-x1.^2-y1.^2-x1.*y1);
   surf(x1,y1,abs(z0-z1)), axis([-3,3,-2,2,0,0.15])
```

The error contours are also shown in Figure 8.12 (b), superimposed with the circle. It can be seen that the fitting results are satisfactory, apart from the regions where samples are removed.

```
>> contour(x1,y1,abs(z0-z1),30); hold on, plot(x,y,'x'); line(x0,y0)
```

It can be concluded that the quality of interpolation depends mainly on the distribution of measured samples. If the samples within a certain area are not sufficient, the quality of interpolation in that area cannot be obtained satisfactorily. Clearly, the more sample data, the better interpolation accuracy.

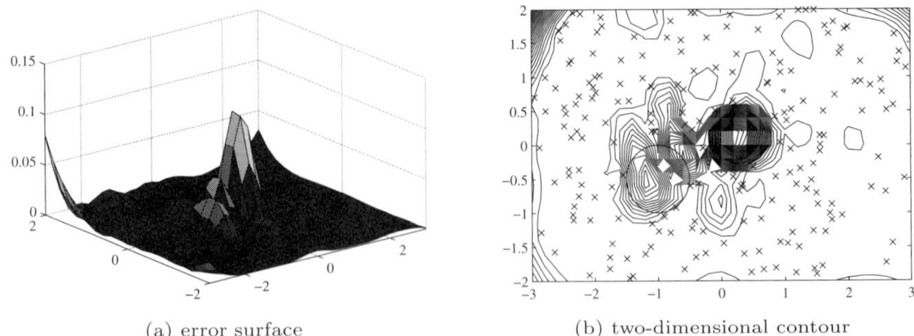

(a) error surface (b) two-dimensional contour

FIGURE 8.12: Error surface and error contour for the 2D interpolation

8.1.5 High-dimensional data interpolations

Three-dimensional mesh grid can also be generated with the previously discussed function `meshgrid()`, with `[`$\boldsymbol{x}$`,`$\boldsymbol{y}$`,`$\boldsymbol{z}$`]=meshgrid(`$\boldsymbol{x}_1$`,`$\boldsymbol{y}_1$`,`$\boldsymbol{z}_1$`)`, where $\boldsymbol{x}_1$, $\boldsymbol{y}_1$, $\boldsymbol{z}_1$ are the necessary segmentation vectors. The returned arguments $\boldsymbol{x}$, $\boldsymbol{y}$, $\boldsymbol{z}$ are in three-dimensional arrays, in grid format.

The n-dimensional grid data can be generated with the function `ndgrid()`, and the syntax of the function is `[`$\boldsymbol{x}_1,\cdots,\boldsymbol{x}_n$`]=ndgrid(`$\boldsymbol{v}_1,\cdots,\boldsymbol{v}_n$`)`, where $\boldsymbol{v}_1$, $\cdots$, $\boldsymbol{v}_n$ are the segmentation of the n-dimensional data, which are given in vectors. The returned argument $\boldsymbol{x}_1, \cdots, \boldsymbol{x}_n$ are the n-dimensional arrays in grid format.

For the samples obtained in grid format, the functions `interp3()` and `interpn()` can be used to solve interpolation problems. The syntaxes of these functions are similar to the `interp2()` function. If the scattered three-dimensional or n-dimensional data are given, the functions `griddata3()` and `griddatan()` can be used and these functions are similar to the `griddata()` function.

Example 8.9 Consider the function $V(x,y,z) = \mathrm{e}^{x^2z+y^2x+z^2y}\cos(x^2yz + z^2yx)$. Generate some samples in grid format and solve the interpolation problem.

Solution The function `meshgrid()` can be used to generate three-dimensional grids. The function values are in the variable $\boldsymbol{V}$. It can be seen that the quality of interpolation is very high

```
>> [x,y,z]=meshgrid(-1:0.2:1); [x0,y0,z0]=meshgrid(-1:0.05:1);
   V=exp(x.^2.*z+y.^2.*x+z.^2.*y).*cos(x.^2.*y.*z+z.^2.*y.*x);
   V0=exp(x0.^2.*z0+y0.^2.*x0+z0.^2.*y0).*...
       cos(x0.^2.*y0.*z0+z0.^2.*y0.*x0);
   V1=interp3(x,y,z,V,x0,y0,z0,'spline'); err=V1-V0; max(err(:))
```

with the maximum error of 0.04186.

8.2 Spline Interpolation and Numerical Calculus

The Spline Toolbox provided in MATLAB can be used to better solve the interpolation problems. Moreover, numerical differentiation and integration problems can be solved easily using this toolbox. This section can be regarded as an extension to Sections 8.1, 3.3 and 3.4.

Spline functions can be regarded as an effective approximation method. The most widely used spline functions are the cubic splines and B-splines. The creation of these splines will be shown in this section, and then numerical differentiation and integration can be solved using the Spline Toolbox.

8.2.1 Spline interpolation in MATLAB

Cubic splines and MATLAB solutions

The definition of cubic spline function is that for n sample points (x_i, y_i) $(i = 1, 2, \cdots, n)$ where $x_1 < x_2 < \cdots < x_n$, the following three conditions are to be satisfied. The function $S(x)$ is referred to as *cubic spline* function with n nodes.

(i) $S(x_i) = y_i$, $(i = 1, 2, \cdots, n)$, i.e., the function passes through all the samples.

(ii) In each interval $[x_i, x_{i+1}]$, the function $S(x)$ is given as a cubic polynomial

$$S(x) = c_{i1}(x - x_i)^3 + c_{i2}(x - x_i)^2 + c_{i3}(x - x_i) + c_{i4}. \tag{8.2}$$

(iii) $S(x)$ has continuous 1st- and 2nd-order derivatives in the whole interval $[x_1, x_n]$.

A cubic spline object can be established with the function `csapi()` provided in the Spline Toolbox. The syntax of the function is simply `S=csapi(x,y)`, where $\boldsymbol{x} = [x_1, x_2, \cdots, x_n]$, $\boldsymbol{y} = [y_1, y_2, \cdots, y_n]$ are the sample points. Thus the returned argument S is a cubic spline object, whose members include the sub-intervals, cubic function coefficients of each piecewise cubic function.

The interpolation curves can be drawn with the function `fnplt()`. For given independent variable vector $\boldsymbol{x}_p$, the interpolation results can be evaluated from the function `fnval()` with the syntax `fnplt(S)`, `y_p=fnval(S,x_p)`, where the vector $\boldsymbol{y}_p$ contains the interpolation results for the given vector $\boldsymbol{x}_p$.

Example 8.10 Compute the cubic spline interpolation results for the sparsely distributed sample points in Example 8.5.

Solution The cubic spline interpolation can be obtained and compared with the theoretical data as shown in Figure 8.13.

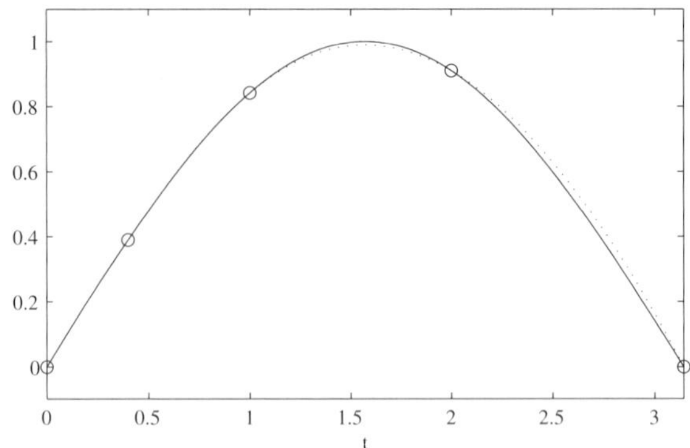

FIGURE 8.13: Interpolation with cubic splines

```
>> x0=[0,0.4,1 2,pi]; y0=sin(x0);
   sp=csapi(x0,y0), fnplt(sp,':'); hold on,
   ezplot('sin(t)',[0,pi]); plot(x0,y0,'o'); sp.coefs
```

The coefficients of the piecewise polynomials are given in Table 8.1. For instance, in the interval $(0.4, 1)$, the interpolation polynomial can be expressed as

$$S_2(x) = -0.1627(x-0.4)^3 - 0.1876(x-0.4)^2 + 0.9245(x-0.4) + 0.3894.$$

TABLE 8.1: The coefficients of the piecewise cubic functions

interval	c_0	c_1	c_2	c_3
$(0, 0.4)$	−0.16265031	0.007585654	0.99653564	0
$(0.4, 1)$	−0.16265031	−0.18759472	0.92453202	0.38941834
$(1, 2)$	0.024435717	−0.48036529	0.52375601	0.84147098
$(2, \pi)$	0.024435717	−0.40705814	−0.36366741	0.90929743

Example 8.11 Interpolate the data given in Example 8.1 using cubic splines.

Solution The following statements can be used to establish a cubic spline object-based on the given data. The obtained piecewise cubic function coefficients are shown in Table 8.2. The cubic spline object can then be used in the interpolation method.

```
>> x=0:.12:1; y=(x.^2-3*x+5).*exp(-5*x).*sin(x);
   sp=csapi(x,y); fnplt(sp)
   c=[sp.breaks(1:4)' sp.breaks(2:5)' sp.coefs(1:4,:),...
      sp.breaks(5:8)' sp.breaks(6:9)' sp.coefs(5:8,:) ];
```

The function `csapi()` can be used to establish the cubic spline object for multivariable data in the grid format, with S=`csapi(`$\{\boldsymbol{x}_1,\boldsymbol{x}_2,\cdots,\boldsymbol{x}_n\},\boldsymbol{z}$`)`, where the variables $\boldsymbol{x}_i$ and $\boldsymbol{z}$ are the grid points, and the cubic spline object is returned in S.

TABLE 8.2: Coefficients of piecewise cubic spline interpolation

piecewise interval	coefficients of cubic polynomials c_1	c_2	c_3	c_4	piecewise interval	coefficients of cubic polynomials c_1	c_2	c_3	c_4
$(0, 0.12)$	24.7396	−19.359	4.5151	0	$(0.48, 0.6)$	−0.2404	0.7652	−0.5776	0.1588
$(0.12, 0.24)$	24.7396	−10.4526	0.9377	0.3058	$(0.6, 0.72)$	−0.4774	0.6787	−0.4043	0.1001
$(0.24, 0.36)$	4.5071	−1.5463	−0.5022	0.3105	$(0.72, 0.84)$	−0.4559	0.5068	−0.2621	0.0605
$(0.36, 0.48)$	1.9139	0.07623	−0.6786	0.2358	$(0.84, 0.96)$	−0.4559	0.3427	−0.1601	0.03557

Example 8.12 Interpolate the grid data in Example 8.7 with cubic splines and draw the interpolation surface.

Solution The following statements can be used to establish the cubic spline object `sp`. The surface obtained is shown in Figure 8.14, which is the same as the one obtained with the `interp2()` function.

```
>> x0=-3:.6:3; y0=-2:.4:2; [x,y]=ndgrid(x0,y0);
   z=(x.^2-2*x).*exp(-x.^2-y.^2-x.*y);
   sp=csapi({x0,y0},z); fnplt(sp);
```

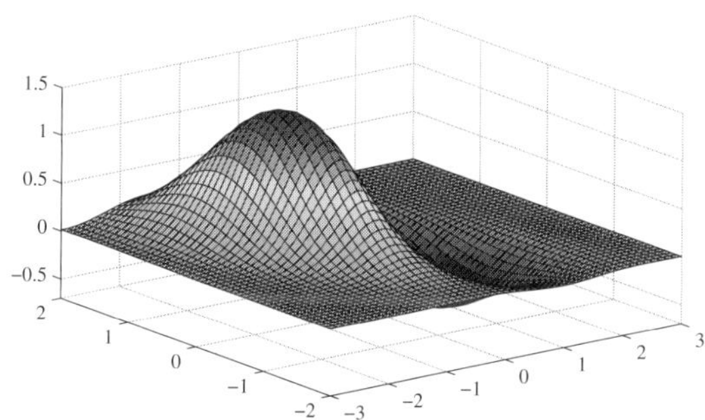

FIGURE 8.14: Interpolation results of two-dimensional function

It should be noted that the matrices $\boldsymbol{x}$ and $\boldsymbol{y}$ are obtained with `ndgrid()` rather than `meshgrid()`. Be careful that the arrangement of the matrices is different in this example.

B-spline and its MATLAB functions

B-spline is another type of commonly used splines. The function `spapi()` for defining the B-spline object is introduced. If the samples are given in vectors $\boldsymbol{x}$ and $\boldsymbol{y}$, the following statements can be used to define a B-spline object S with the syntax `S=spapi(k,x,y)`, where k is the degree of B-spline. Normally one may select k=4 or 5 to ensure good interpolation results. For specific problems, a suitable increase in k may improve the interpolation results.

Example 8.13 Interpolate the functions given in Examples 8.10 and 8.11 with B-splines. Compare with the results obtained using cubic splines.

Solution For the Example 8.10, the following statements can be used to interpolate the data with the results shown in Figure 8.15 (a). It can be seen that one can hardly distinguish the interpolation curve from the theoretical one.

```
>> x0=[0,0.4,1 2,pi]; y0=sin(x0); ezplot('sin(t)',[0,pi]); hold on
   sp1=csapi(x0,y0); fnplt(sp1,'--'); % cubic spline interpolation
   sp2=spapi(5,x0,y0); fnplt(sp2,':') % B-spline with degree k = 5
```

From the above observations, the B-spline interpolation is far better than the cubic spline for this example. For the data in Example 8.11, the B-spline interpolation results are shown in Figure 8.15 (b). Again, here, the B-spline interpolation is much better than the cubic spline interpolation.

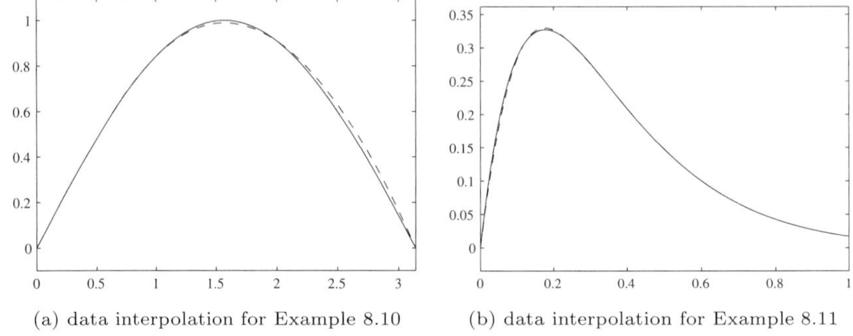

(a) data interpolation for Example 8.10 (b) data interpolation for Example 8.11

FIGURE 8.15: The interpolation curves with spline interpolations

```
>> x=0:.12:1; y=(x.^2-3*x+5).*exp(-5*x).*sin(x);
   ezplot('(x^2-3*x+5)*exp(-5*x)*sin(x)',[0,1]), hold on
   sp1=csapi(x,y); fnplt(sp1,'--'); sp2=spapi(5,x,y); fnplt(sp2,':')
```

8.2.2 Numerical differentiation and integration with splines

It has been shown that splines can be used to evaluate numerical integrals and even when the samples are sparsely distributed, good results can still be obtained. Compared with the algorithms in Sections 3.3 and 3.4, the spline-based algorithm has its own advantages. The spline-based integration is defined as $F(x) = \int_{x_0}^{x} f(t)\,\mathrm{d}t$, where x_0 is the pre-specified boundary. Clearly, the definite integral can be evaluated from $I = F(b) - F(a)$.

Numerical differentiation

The spline function-based numerical differentiation to the given sample points can be calculated from the function `fnder()`

S_d=fnder(S,k)	% kth order derivative of S
S_d=fnder(S,[$k_1,\cdots,k_n$])	% partial derivatives for multivariable functions

Example 8.14 Consider the sample points in Example 8.11. Compute the numerical differentiations to the samples with cubic function or B-spline functions and compare the results with the theoretical ones.

Solution From the generated samples, the cubic and B-spline data objects can be established. Then the derivatives can be obtained with `fnder()` function, and the curves can be shown in Figure 8.16.

```
>> syms x; f=(x^2-3*x+5)*exp(-5*x)*sin(x); ezplot(diff(f),[0,1]),
   hold on, x=0:.12:1; y=(x.^2-3*x+5).*exp(-5*x).*sin(x);
   sp1=csapi(x,y); dsp1=fnder(sp1,1); fnplt(dsp1,'--')
   sp2=spapi(5,x,y); dsp2=fnder(sp2,1);
   fnplt(dsp2,':'); axis([0,1,-0.8,5])
```

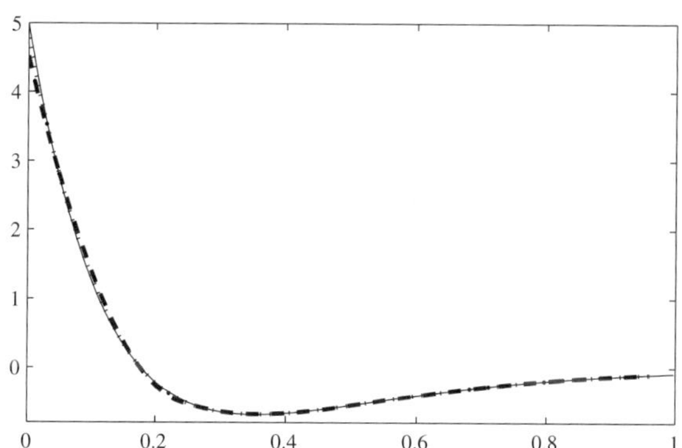

FIGURE 8.16: Numerical differentiations based on spline interpolations

From Figure 8.16, many other theoretical curves can also be displayed. It can be seen that the numerical differentiations with B-spline is very satisfactory. Piecewise cubic polynomials may also give very good results. Since the samples are extremely sparsely distributed, the method in Section 3.3 cannot yield good results.

Example 8.15 Fit the $\partial^2 z/(\partial x \partial y)$ surface with the data obtained in Example 8.12. Compare the surface with the exact results.

Solution The following statements can be used to generate the data. With the B-spline fitting, the numerical differentiation can be evaluated and the surface is then shown in Figure 8.17 (a).

```
>> x0=-3:.3:3; y0=-2:.2:2; [x,y]=ndgrid(x0,y0);
   z=(x.^2-2*x).*exp(-x.^2-y.^2-x.*y);
   sp=spapi({5,5},{x0,y0},z); dspxy=fnder(sp,[1,1]); fnplt(dspxy)
```

The following statements can be used to evaluate the exact partial derivatives theoretically and the surface is obtained as shown in Figure 8.17 (b). It can be seen that the results are almost the same, which indicates that the numerical method is reliable.

```
>> syms x y; z=(x^2-2*x)*exp(-x^2-y^2-x*y);
   ezsurf(diff(diff(z,x),y),[-3 3],[-2 2])
```

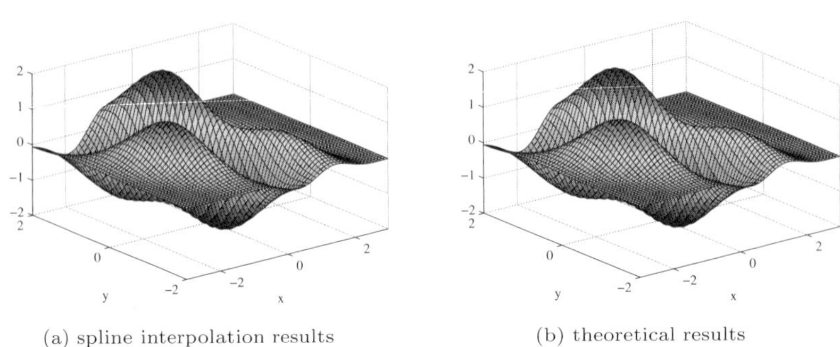

(a) spline interpolation results (b) theoretical results

FIGURE 8.17: The second-order partial derivative surface

Numerical integration

The cubic and B-splines can be used to approximate the integrand that is defined by a given data set. From the spline interpolation, the numerical integration can be obtained. The method to be introduced here is different from the `quadspln()` function discussed in Section 8.1.2. The function `quadspln()` can only be used to evaluate the definite integral, while the function `fnint()` can be used to obtain an approximate integration function. Of course, it can also be used to evaluate definite integrals, $\boldsymbol{f}_i$`=fnint(`S`)`, where $\boldsymbol{f}_i$ are vectors which return the integration at each $\boldsymbol{x}$ point. The indefinite integral can be obtained by adding a constant to the results. If the definite integration over the $[a, b]$ interval is required, we can use I`=diff(fnint(`S`,`$[a,b]$`))`.

Example 8.16 Consider again the sparsely distributed samples in Example 8.4. Compute the definite and indefinite integrals using spline interpolations.

Solution The two spline objects can be established and with the function `fnint()`, the integration function can be evaluated and also the definite integral can be obtained

```
>> x=[0,0.4,1 2,pi]; y=sin(x);
   sp1=csapi(x,y); a=fnint(sp1,1); xx=fnval(a,[0,pi]); p1=xx(2)-xx(1)
   sp2=spapi(5,x,y); b=fnint(sp2,1); xx=fnval(b,[0,pi]); p2=xx(2)-xx(1)
```

and the definite integrals by the two spline functions are obtained as $p_1 = 2.01905235$, and $p_2 = 1.99994177102$, respectively. It can be seen that the results with cubic spline are the same as the results obtained in Example 8.4. With the B-splines, much better results can be obtained.

The approximate primitive function can also be obtained from sample data. For instance, the approximate primitive function can be obtained with B-splines and it is displayed together with the true one as shown in Figure 8.18.

```
>> ezplot('-cos(t)+2',[0,pi]); hold on
   fnplt(a,'--'); fnplt(b,':')
```

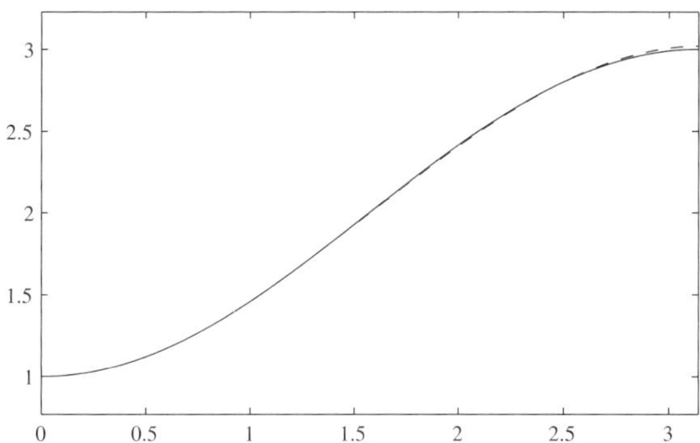

FIGURE 8.18: The approximate primitive function from spline interpolation and the true function

8.3 Data Modeling

8.3.1 Polynomial fitting

The Lagrange interpolation is a type of polynomial fitting. The objective of the polynomial fitting is to find a set of coefficients $a_i, i = 1, 2, \cdots, n+1$, such that the polynomial

$$\psi(x) = a_1 x^n + a_2 x^{n-1} + \cdots + a_n x + a_{n+1} \tag{8.3}$$

can be used to fit the original data in least squares sense. The difference between the polynomial fitting and the interpolation is that, in polynomial fitting, the samples are not necessarily on the fitting curve. A MATLAB function `polyfit()` can be used to solve polynomial fitting problems, with the syntax `p=polyfit(`$\boldsymbol{x}$`,`$\boldsymbol{y}$`,`n`)`, where the arguments $\boldsymbol{x}$ and $\boldsymbol{y}$ are the vectors of sample points. The argument n is the selected degree of polynomial fitting. The returned argument $\boldsymbol{p}$ stores the coefficients of the polynomial in descending order. The function `poly2sym()` can be used to convert the results into symbolic polynomials. Also the function `polyval()` can be used to evaluate the values of polynomials. The following examples illustrate the use of polynomial fitting.

Example 8.17 Consider the sample points in Example 8.1. Solve the polynomial fitting problems for different degrees and find a suitable degree.

Solution Using the following statements, the fitting curve is shown in Figure 8.19 (a)

```
>> x0=0:.1:1; y0=(x0.^2-3*x0+5).*exp(-5*x0).*sin(x0);
   p3=polyfit(x0,y0,3); vpa(poly2sym(p3),10)
   x=0:.01:1; ya=(x.^2-3*x+5).*exp(-5*x).*sin(x);
   y1=polyval(p3,x); plot(x,y1,x,ya,x0,y0,'o')
```

where the cubic function is $p_3(x) = 2.84x^3 - 4.7898x^2 + 1.9432x + 0.0598$. From the fitting curve, it can be seen that the fitting quality is very poor. Thus to address the fitting accuracy problem, it is natural to increase the degree of the polynomial. For different degrees such as 4, 5 and 8, the fitting curves can be obtained as shown in Figure 8.19 (b).

```
>> p4=polyfit(x0,y0,4); y2=polyval(p4,x);
   p5=polyfit(x0,y0,5); y3=polyval(p5,x);
   p8=polyfit(x0,y0,8); y4=polyval(p8,x);
   plot(x,ya,x0,y0,'o',x,y2,x,y3,x,y4)
```

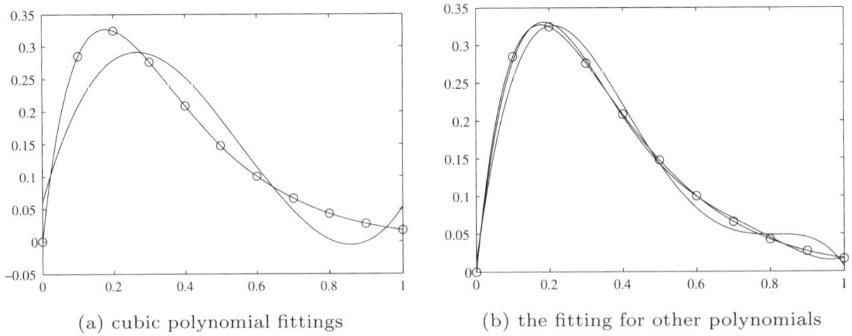

(a) cubic polynomial fittings (b) the fitting for other polynomials

FIGURE 8.19: Fitting results of polynomials of different degrees

From the fitting results, it can be seen that when $n \geqslant 8$, the fitting quality is satisfactory. The eighth degree polynomial obtained can be obtained from

```
>> vpa(poly2sym(p8),5)
```

that is, $p_8(x) = -8.2586x^8 + 43.566x^7 - 101.98x^6 + 140.22x^5 - 125.29x^4 + 74.450x^3 - 27.672x^2 + 4.9869x + .42037\times 10^{-6}$.

Polynomial fitting to a given function is in fact related to the Taylor series expansion. However, to evaluate the Taylor series, it is required that the original function be given. This is not realistic for many practical problems. For this example, since the original function is known, the Taylor series can be used to get a truncated polynomial

```
>> syms x; y=(x^2-3*x+5)*exp(-5*x)*sin(x);
   P=vpa(taylor(y,9),5)
```

which reads

$$P = 5x - 28x^2 + 77.667x^3 - 142x^4 + 192.17x^5 - 204.96x^6 + 179.13x^7 - 131.67x^8.$$

Comparing the results of truncated Taylor series expansion with the polynomial fitting, it can be seen that the two polynomials are significantly different. This means that the polynomial approximation to a given function may not be unique. Although the mathematical forms of the two polynomials are completely different, the fitting curves could be very close within a specific interval.

Example 8.18 Consider again the function in Example 8.3. Observe the polynomial fittings to the original function.

Solution Polynomial fittings are not always accurate. Consider the samples in Example 8.3. Using different degrees n, the polynomial fitting can be obtained, with fitting results shown in Figure 8.20 (a).

```
>> x0=-1+2*[0:10]/10; y0=1./(1+25*x0.^2);
   x=-1:.01:1; ya=1./(1+25*x.^2);
   p3=polyfit(x0,y0,3); y1=polyval(p3,x);
   p5=polyfit(x0,y0,5); y2=polyval(p5,x);
   p8=polyfit(x0,y0,8); y3=polyval(p8,x);
   p10=polyfit(x0,y0,10); y4=polyval(p10,x);
   plot(x,ya,x,y1,x,y2,'-.',x,y3,'--',x,y4,':')
```

In fact, truncated Taylor series expansion to this function is even poorer. The following statements can be used to find the Taylor series for the original function and the fitting is shown in Figure 8.20 (b). It can be seen that the Taylor series fitting is erroneous for the example.

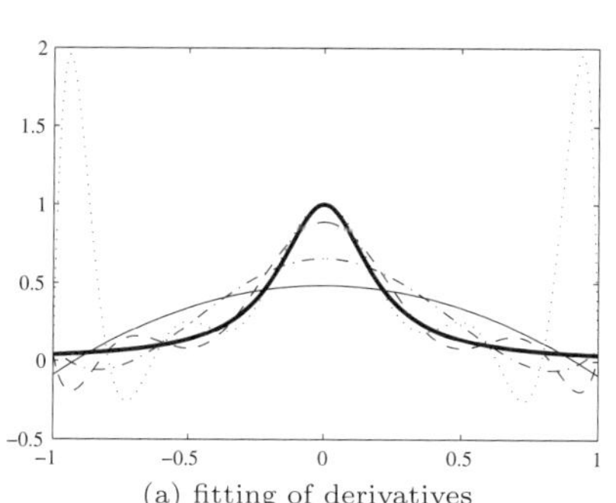

(a) fitting of derivatives

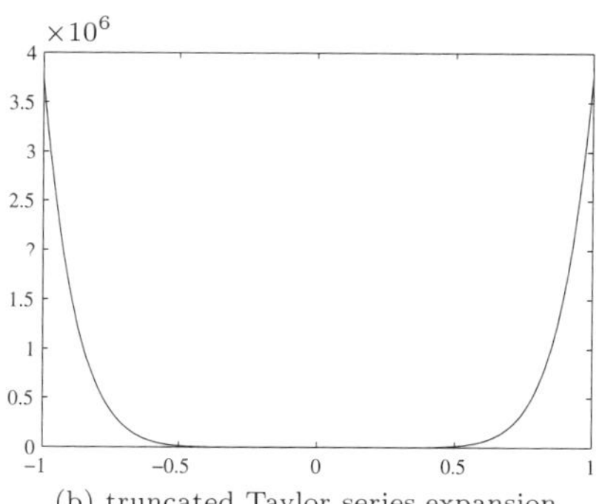

(b) truncated Taylor series expansion

FIGURE 8.20: Polynomial fitting and the truncated Taylor series expansion

```
>> syms x; y=1/(1+25*x^2); p=taylor(y,x,10)
   x1=-1:0.01:1; ya=1./(1+25*x1.^2); y1=subs(p,x,x1); plot(x1,y1)
```

the polynomial obtained is $p(x) = 1 - 25x^2 + 625x^4 - 15625x^6 + 390625x^8$.

8.3.2 Approximation by continued fraction expansions

Continued fraction is often regarded as an effective way in approximating certain functions. The continued fraction expansion to a given function $f(x)$ can be expressed by

$$f(x) = \cfrac{1}{f_1(x) + \cfrac{1}{f_2(x) + \cfrac{1}{f_3(x) + \cfrac{1}{f_4(x) + \cfrac{1}{\cdots}}}}} \tag{8.4}$$

and the most widely used continued fraction is in Cauer II form[39], where

$$f(x) = \cfrac{\alpha_1}{\beta_1 + \cfrac{\alpha_2 x}{\beta_2 + \cfrac{\alpha_3 x}{\beta_3 + \cfrac{\alpha_4 x}{\beta_4 + \cfrac{\alpha_5 x}{\cdots}}}}}. \tag{8.5}$$

Continued fraction evaluation is not supported in MATLAB. However, the `cfrac()` function in Maple can be used instead. The `numtheory` package should be loaded using the `with()` function. The continued fraction of Cauer II form can be obtained with

```
maple('with(numtheory):')                  % load the numtheory package
f=maple(['cfe:=cfrac(' fun ',x,n)']);     % continued fraction in cfe
```

where the continued fraction to the given function *fun* can be obtained. The independent variable is x and the number of terms required is n. The partial fraction expansion `cfe` can be obtained, and f is returned back to MATLAB workspace. If continued fraction expansion to a number is required, the argument x should be omitted.

If the first n continued fraction terms are used, one can use the functions `thenumer()` and `thedenom()` in Maple to evaluate the rational approximation, with the syntaxes

```
p=maple('nthnumer','cfe',n);   % extract the numerator in variable cfe
q=maple('nthdenom','cfe',n);   % extract the denominator
```

Example 8.19 Let us first observe the continued fraction expansion-based approximation to a constant. The irrational number π can be approximated with a

20 staged continued fraction. Find a suitable degree such that good approximation can be obtained.

Solution The continued fractions to π can be directly obtained using the following statements[1]

```
>> maple('with(numtheory):'); f=maple(['cfe:=cfrac(pi,20)'])
```

which can be written as

$$\pi \approx 3 + \cfrac{1}{7 + \cfrac{1}{15 + \cfrac{1}{1 + \cfrac{1}{292 + \cfrac{1}{1 + \cdots}}}}}$$

where the value 292 is very large compared with other values. Thus the rest of the terms can be truncated and a reasonably high-precision approximation can be ensured. The numerator and denominator after truncation can be obtained as

```
>> n=maple('nthnumer','cfe',4); d=maple('nthdenom','cfe',4);
   [vpa(n),vpa(d)]
```

which is $\dfrac{103993}{33102} \approx 3.141592653012$ with four stages of continued fractions. It can be seen that the value obtained is very close to π.

Example 8.20 Establish the first 10-stage continued fractions for $f(x) = \dfrac{e^{-x}\sin x}{(x+1)^3}$. Obtain the rational approximation to the given function.

Solution The first 10-stage continued fractions can be obtained by

```
>> syms x; fun='sin(x)*exp(-x)/(x+1)^3'; % fun should be given as a string
   maple('with(numtheory):'); f=maple(['cfe:=cfrac(' fun ',x,10)'])
```

which can be written as

$$f(x) = \cfrac{x}{1 + \cfrac{4x}{1 - \cfrac{5x}{3 + \cfrac{43x}{20 - \cfrac{337x}{43 + \cfrac{28274x}{1685 - \cfrac{66157779x}{395836 - \cfrac{9881300005x}{512851 + \cfrac{140501598188444x}{158335371 - \cdots}}}}}}}}}.$$

The following statements can be used to evaluate the eighth- and tenth-degree rational approximation with the continued fraction using the following MATLAB scripts:

[1] In MATLAB 2008a, `pi` in the following sentence should be changed to `Pi`.

```
>> n=collect(maple('nthnumer','cfe',8),x);
   d=collect(maple('nthdenom','cfe',8),x); [n,d]=numden(n/d); G=n/d
   n=collect(maple('nthnumer','cfe',10),x);
   d=collect(maple('nthdenom','cfe',10),x); [n,d]=numden(n/d); G1=n/d
```

which gives

$$f_8(x) \approx 10\frac{x\left(845713x^3 - 4973560x^2 + 11841438x - 10769871\right)}{5864273x^4 - 83147900x^3 - 294069480x^2 - 312380460x - 107698710}.$$

The following statements can be used to examine the fitting over the interval $(0, 0.5)$, where the original function $f(x)$ and the fitting function $f_8(x)$ are compared in Figure 8.21 (a). It can be seen that the fitting quality is fairly satisfactory. If $n = 10$, one cannot distinguish the difference between the two curves. However, if the interval is increased to $(0, 5)$, the fitting results shown in Figure 8.21 (b) suggest that the more stages, the better fitting quality.

```
>> ezplot(fun,[0,2]), hold on; ezplot(G,[0,2]); ezplot(G1,[0,2])
   figure; ezplot(fun,[0,5]), hold on; ezplot(G,[0,5]); ezplot(G1,[0,5])
```

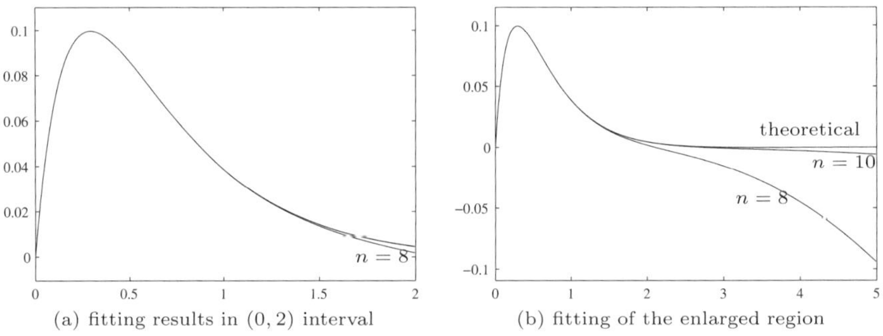

(a) fitting results in (0, 2) interval (b) fitting of the enlarged region

FIGURE 8.21: Comparison of continued fraction expansion-based function approximation

8.3.3 Padé rational approximations

Assume that the power series expansion of a given function $f(s)$ is

$$f(s) = c_0 + c_1 s + c_2 s^2 + c_3 s^3 + \cdots = \sum_{i=0}^{\infty} c_i s^i \tag{8.6}$$

and assume that the r/mth degree Padé approximation is expressed by

$$G_m^r(s) = \frac{\beta_{r+1}s^r + \beta_r s^{r-1} + \cdots + \beta_1}{\alpha_{m+1}s^m + \alpha_m s^{m-1} + \cdots + \alpha_1} = \frac{\sum_{i=1}^{r+1} \beta_i s^{i-1}}{\sum_{i=1}^{m+1} \alpha_i s^{i-1}} \tag{8.7}$$

where $\alpha_1 = 1$, $\beta_1 = c_1$. Let $\sum_{i=0}^{\infty} c_i s^i = G_m^r(s)$. The following equation can be obtained:

$$\sum_{i=1}^{m+1} \alpha_i s^{i-1} \sum_{i=0}^{\infty} c_i s^i = \sum_{i=1}^{r+1} \beta_i s^{i-1}. \tag{8.8}$$

Equating the terms with the same power of s, the coefficients α_i, $i = 2, \cdots, m+1$ and β_i, $i = 2, \cdots, k+1$ can be obtained from

$$\boldsymbol{W}\boldsymbol{x} = \boldsymbol{w}, \quad \boldsymbol{v} = \boldsymbol{V}\boldsymbol{y} \tag{8.9}$$

where

$$\begin{aligned} \boldsymbol{x} &= [\alpha_2, \alpha_3, \cdots, \alpha_{m+1}]^{\mathrm{T}}, \quad \boldsymbol{w} = [-c_{r+2}, -c_{r+3}, \cdots, -c_{m+r+1}]^{\mathrm{T}} \\ \boldsymbol{v} &= [\beta_2 - c_2, \beta_3 - c_3, \cdots, \beta_{r+1} - c_{r+1}]^{\mathrm{T}}, \quad \boldsymbol{y} = [\alpha_2, \alpha_3, \cdots, \alpha_{r+1}]^{\mathrm{T}} \end{aligned} \tag{8.10}$$

and

$$\boldsymbol{W} = \begin{bmatrix} c_{r+1} & c_r & \cdots & 0 & \cdots & 0 \\ c_{r+2} & c_{r+1} & \cdots & c_1 & \cdots & 0 \\ \vdots & \vdots & \cdots & \vdots & \ddots & \vdots \\ c_{r+m} & c_{r+m-1} & \cdots & c_{m-1} & \cdots & c_{r+1} \end{bmatrix} \tag{8.11}$$

$$\boldsymbol{V} = \begin{bmatrix} c_1 & 0 & 0 & \cdots & 0 \\ c_2 & c_1 & 0 & \cdots & 0 \\ \vdots & \vdots & \vdots & \ddots & \vdots \\ c_r & c_{r-1} & c_{r-2} & \cdots & c_1 \end{bmatrix}. \tag{8.12}$$

It can be shown[40] that, if the degree of numerator is one less than that of the denominator, the Padé approximation is equivalent to the Cauer II form of the continued fraction expansion. A MATLAB function `padefcn()` can be prepared to compute the Padé rational approximation to a given $f(x)$ function. The function is

```
function [nP,dP]=padefcn(c,r,m)
w=-c(r+2:m+r+1)'; vv=[c(r+1:-1:1)', zeros(m-1-r,1)];
W=rot90(hankel(c(m+r:-1:r+1),vv)); V=rot90(hankel(c(r:-1:1)));
x=[1 (W\w)']; y=[1 x(2:r+1)*V'+c(2:r+1)];
dP=x(m+1:-1:1)/x(m+1); nP=y(r+1:-1:1)/x(m+1);
```

Example 8.21 Find the rational function approximation to the function $f(x) = \mathrm{e}^{-2x}$.

Solution Let us select the degree of the numerator as 0, and select different degrees for the denominator. The Padé rational approximation can be obtained by the following MATLAB scripts and the fitting results are given in Figure 8.22.

```
>> syms x; c=taylor(exp(-2*x),10);
   c=sym2poly(c); c=c(end:-1:1); x=0:0.01:8; nd=[3:7];
```

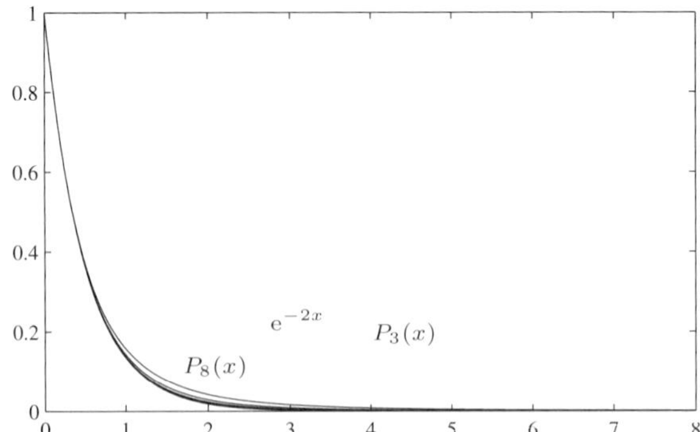

FIGURE 8.22: Original curve and fitting curves

```
xx=[0,2,2+eps,8]; yy=[0,0,1,1]; plot(xx,yy); hold on
for i=1:length(nd)
   [n,d]=padefcn(c,0,nd(i));
   y=polyval(n,x)./polyval(d,x); plot(x,y)
end
```

It can be seen from Figure 8.22 that third-degree Padé approximation gives a reasonably good fitting. If the degree is increased, the quality of fitting may also increase. The eighth-degree fitting to the original function is actually very satisfactory, which reads

$$P_8(s) = \frac{157.5}{x^8 + 4x^7 + 14x^6 + 42x^5 + 105x^4 + 210x^3 + 315x^2 + 315x + 157.5}.$$

8.3.4 Curve fitting by linear combination of basis functions

Assume that the original function is composed of linear combination of a set of known functions $f_1(x), f_2(x), \cdots, f_n(x)$

$$g(x) = c_1 f_1(x) + c_2 f_2(x) + c_3 f_3(x) + \cdots + c_n f_n(x) \tag{8.13}$$

where $c_1, c_2, \cdots, c_n$ are undetermined coefficients. The measured data points obtained are $(x_1, y_1), (x_2, y_2), \cdots, (x_M, y_M)$. The following linear equations can be established

$$\boldsymbol{A}\boldsymbol{c} = \boldsymbol{y} \tag{8.14}$$

where

$$\boldsymbol{A} = \begin{bmatrix} f_1(x_1) & f_2(x_1) & \cdots & f_m(x_1) \\ f_1(x_2) & f_2(x_2) & \cdots & f_m(x_2) \\ \vdots & \vdots & \ddots & \vdots \\ f_1(x_M) & f_2(x_M) & \cdots & f_m(x_M) \end{bmatrix}, \quad \boldsymbol{y} = \begin{bmatrix} y_1 \\ y_2 \\ \vdots \\ y_M \end{bmatrix} \tag{8.15}$$

and $\boldsymbol{c} = [c_1, c_2, \cdots, c_n]^{\mathrm{T}}$. The least squares solution of the problem is $\boldsymbol{c} = \boldsymbol{A} \backslash \boldsymbol{y}$.

Example 8.22 The measured data points (x_i, y_i) are given in the table below. Assume that the original function is given in the form $y(x) = c_1 + c_2 e^{-3x} + c_3 \cos(-2x) e^{-4x} + c_4 x^2$. Compute the undetermined coefficients c_i with least squares method.

x_i	0	0.2	0.4	0.7	0.9	0.92	0.99	1.2	1.4	1.48	1.5
y_i	2.88	2.2576	1.9683	1.9258	2.0862	2.109	2.1979	2.5409	2.9627	3.155	3.2052

Solution The undetermined coefficients c_i can be obtained from the following:

```
>> x=[0,0.2,0.4,0.7,0.9,0.92,0.99,1.2,1.4,1.48,1.5]';
   y=[2.88;2.2576;1.9683;1.9258;2.0862;2.109;2.1979;2.5409;...
      2.9627;3.155;3.2052];
   A=[ones(size(x)) exp(-3*x), cos(-2*x).*exp(-4*x) x.^2];
   c=A\y; c1=c', x0=[0:0.01:1.5]';
   A1=[ones(size(x0)),exp(-3*x0),cos(-2*x0).*exp(-4*x0),x0.^2];
   y1=A1*c; plot(x0,y1,x,y,'x')
```

and the result is $\boldsymbol{c}^{\mathrm{T}} = [1.22002081341, 2.33972067466, -0.67973291883, 0.86999835216]$. The fitting curve is shown in Figure 8.23 with the given samples. It can be seen that the fitting is successful.

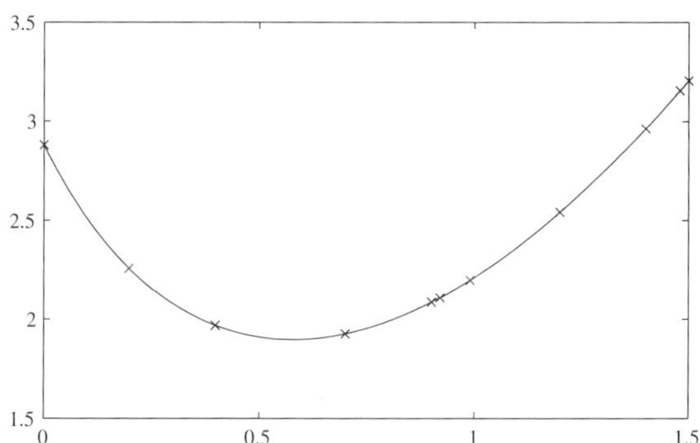

FIGURE 8.23: Original data and the fitting curve

Example 8.23 Given the measured data below, find a suitable fitting function.

x_i	1.1052	1.2214	1.3499	1.4918	1.6487	1.8221	2.0138	2.2255	2.4596	2.7183	3.6693
y_i	0.6795	0.6006	0.5309	0.4693	0.4148	0.3666	0.3241	0.2865	0.2532	0.2238	0.1546

Solution The measured data can be shown in Figure 8.24 (a).

```
>> x=[1.1052,1.2214,1.3499,1.4918,1.6487,1.8221,2.0138,...
      2.2255,2.4596,2.7183,3.6693];
   y=[0.6795,0.6006,0.5309,0.4693,0.4148,0.3666,0.3241,...
      0.2864,0.2532,0.2238,0.1546];
   plot(x,y,x,y,'*')
```

By inspection, it is hard to determine a possible form of the fitting function. One may perform a nonlinear transformation to the original data, and see whether they can be described by linear functions. For instance, the logarithmic transformation to both $\boldsymbol{x}, \boldsymbol{y}$ can be attempted as shown in Figure 8.24 (b), and it can be observed that it is linear.

```
>> x1=log(x); y1=log(y);  plot(x1,y1,x1,y1,'*')
```

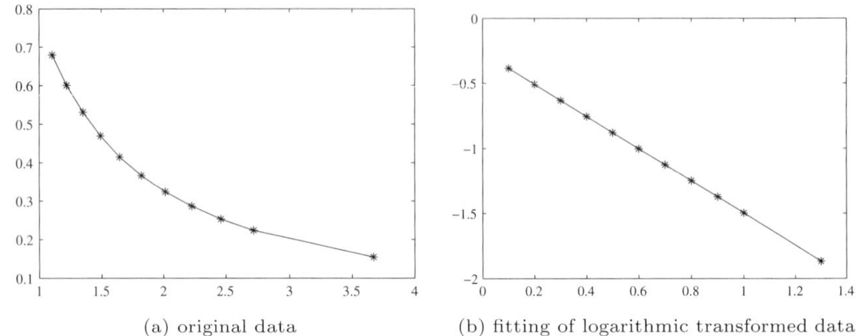

(a) original data (b) fitting of logarithmic transformed data

FIGURE 8.24: Original data and fitting results for the transformed data

The linear fitting function can be tested where $\ln y = a \ln x + b$, i.e., $y = \mathrm{e}^b x^a$. The coefficients a, b and e^b can be obtained with the following statements

```
>> A=[x1' ones(size(x1'))]; c=[A\y1']', d=exp(c(2))
```

then $\boldsymbol{c} = [-1.23389448522593, -0.26303708610220]^{\mathrm{T}}$, $\mathrm{e}^b = 0.76871338819924$. The fitting function is then $y(x) = 0.76871338819924x^{-1.23389448522593}$.

Example 8.24 Polynomial fitting can be regarded as a special example of the linear combination of basis function fittings. In this case, the basis functions can be written as $f_i(x) = x^{n+1-i}$, $i = 1, 2, \cdots, n$. Solve the polynomial fitting problem in Example 8.17 and observe the fitting results.

Solution From the above algorithm, the sample data can be generated and the polynomial fitting can be performed which yields exactly the same results as in Example 8.17.

```
>> x=[0:.1:2]'; y=(x.^2-3*x+5).*exp(-5*x).*sin(x); n=7; A=[];
   for i=1:n+1, A(:,i)=x.^(n+1-i); end, c=A\y
```

that is

$$-0.715x^8+6.62x^7-25.9x^6+55.6x^5-71.2x^4+55.0x^3-24.16x^2+4.75x+0.00073.$$

8.3.5 Least squares curve fitting

Given a set of data $x_i, y_i, i = 1, 2, \cdots, N$, and given the original function, referred to as the *prototype function*, $\hat{y}(x) = f(\boldsymbol{a}, x)$, where $\boldsymbol{a}$ is the vector of undetermined coefficients, the objective of the least squares approximation is to find the undetermined coefficients which minimize the objective function

$$J = \min_{\boldsymbol{a}} \sum_{i=1}^{N} [y_i - \hat{y}(x_i)]^2 = \min_{\boldsymbol{a}} \sum_{i=1}^{N} [y_i - f(\boldsymbol{a}, x_i)]^2. \tag{8.16}$$

The `lsqcurvefit()` function provided in the Optimization Toolbox can be used to solve the least squares curve fitting problems. The syntax of the function is `[a,J_m]=lsqcurvefit(fun,a_0,x,y)`, where *fun* is the MATLAB description to the prototype function. It can either be an M-function, an anonymous or an inline function. The argument $\boldsymbol{a}_0$ is a vector containing the initial guess of $\boldsymbol{a}$. The vectors $\boldsymbol{x}$ and $\boldsymbol{y}$ store respectively the input and output data. The undetermined coefficients are returned in the $\boldsymbol{a}$ vector, and the objective function is in $J_{\rm m}$.

Example 8.25 The sample data set is generated with the following statements

```
>> x=0:.1:10; y=0.12*exp(-0.213*x)+0.54*exp(-0.17*x).*sin(1.23*x);
```

and assume that the prototype function satisfies $y(x) = a_1 {\rm e}^{-a_2 x} + a_3 {\rm e}^{-a_4 x} \sin(a_5 x)$, where a_i are the undetermined coefficients. The least squares curve fitting algorithm can be used to find the coefficients. Compute the coefficients which minimize the objective function.

Solution The prototype function can first be expressed by an anonymous function, then the following statements can be used to evaluate the undetermined coefficients

```
>> f=@(a,x)a(1)*exp(-a(2)*x)+a(3)*exp(-a(4)*x).*sin(a(5)*x);
   [xx,res]=lsqcurvefit(f,[1,1,1,1,1],x,y); xx',res
```

and the estimated vector $\boldsymbol{a} = [0.11971712065965, 0.21246, 0.54040, 0.17019, 1.23003]^{\rm T}$, with the minimized objective function 7.16368×10^{-7}.

It can be seen from the results that the accuracy of fitting is not very high, compared with the known exact values $\boldsymbol{a} = [0.12, 0.213, 0.54, 0.17, 1.23]^{\rm T}$. In order to further increase the accuracy of fitting, the optimization options should be modified such that a better estimation of the parameters can be obtained

```
>> ff=optimset; ff.TolFun=1e-20; ff.TolX=1e-15;
   [a,res]=lsqcurvefit(f,[1,1,1,1,1],x,y,[],[],ff);
   x1=0:0.01:10; y1=f(xx,x1); plot(x1,y1,x,y,'o')
```

where the empty matrices are used to represent the lower- and upper-bounds of vector $\boldsymbol{a}$. Since there are no other constraints, the undetermined coefficients can then be estimated as $\boldsymbol{a}^{\rm T} = [0.12, 0.213, 0.54, 0.17, 1.23]$, and the error is $9.503499130199283 \times 10^{-21}$. It can be seen that after modifying the options, the accuracy is significantly improved, The sample points and the fitting curves are shown in Figure 8.25.

Example 8.26 The measured data points are given below and also the prototype function is known to be $y(x) = ax + bx^2{\rm e}^{-cx} + d$. Compute using least squares method the undetermined constants a, b, c, d.

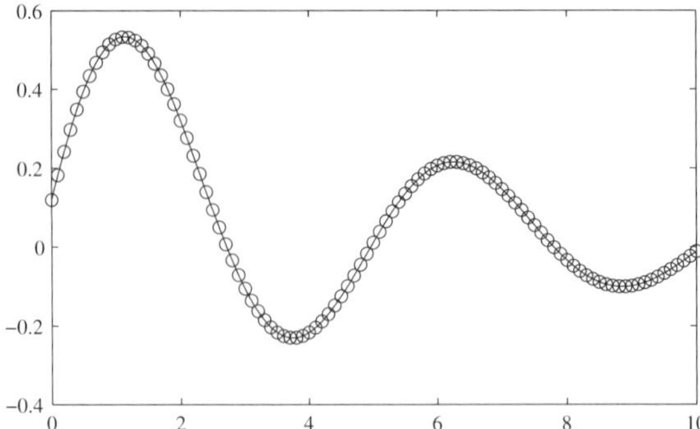

FIGURE 8.25: Comparisons of fitting results

x_i	0.1	0.2	0.3	0.4	0.5	0.6	0.7	0.8	0.9	1
y_i	2.3201	2.6470	2.9707	3.2885	3.6008	3.9090	4.2147	4.5191	4.8232	5.1275

Solution The following statements can be entered in MATLAB workspace

```
>> x=0.1:0.1:1;
   y=[2.3201,2.6470,2.9707,3.2885,3.6008,3.9090,...
      4.2147,4.5191,4.8232,5.1275];
```

Let $a_1 = a, a_2 = b, a_3 = c, a_4 = d$, the prototype function can be rewritten as $y(x) = a_1x + a_2x^2\mathrm{e}^{-a_3x} + a_4$. Thus an anonymous function can be written for the prototype function, and the following statement can be used to evaluate the coefficients

```
>> f=@(a,x)a(1)*x+a(2)*x.^2 .*exp(-a(3)*x)+a(4);
   a=lsqcurvefit(f,[1;2;2;3],x,y), y1=f(a,x); plot(x,y,x,y1,'o')
```

with $\boldsymbol{a} = [2.458711, 2.448946, 1.4465565, 2.073377]^{\mathrm{T}}$. The sample points and the fitted curve are shown in Figure 8.26 and it can be seen that the fitting is satisfactory.

8.4 Signal Analysis and Digital Signal Processing

8.4.1 Correlation analysis

Correlation analysis can be used to characterize both stochastic and deterministic signals. Consider a deterministic signal $x(t)$. The auto-correlation function is defined as

$$R_{xx}(\tau) = \lim_{T\to\infty}\frac{1}{T}\int_0^T x(t)x(t+\tau)\,\mathrm{d}t, \quad \tau \geqslant 0 \tag{8.17}$$

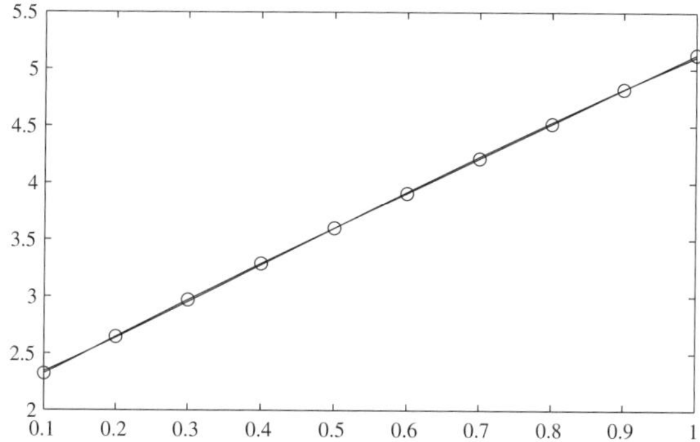

FIGURE 8.26: Comparison of fitting results

and the correlation function is an even function, i.e., $R_{xx}(-\tau) = R_{xx}(\tau)$. The cross-correlation function of two signals $x(t)$ and $y(t)$ is defined as

$$R_{xy}(\tau) = \lim_{T\to\infty} \frac{1}{T}\int_0^T x(t)y(t+\tau)\,\mathrm{d}t, \quad \tau \geqslant 0. \tag{8.18}$$

Clearly, when the functions are all given, the auto-correlation and cross-correlation functions can be evaluated by using Symbolic Math Toolbox. In some cases, it is possible to obtain analytical solutions.

Example 8.27 Consider the signal $x(t) = A_1\cos(\omega_1 t + \theta_1) + A_2\cos(\omega_2 t + \theta_2)$. Compute its auto-correlation function.

Solution It is known that the symbolic variables should be declared first. Then the function $x(t)$ can be defined and the following statements can be used to evaluate the auto-correlation function.

```
>> syms A1 A2 w1 w2 t1 t2 t tau T; x=A1*cos(w1*t+t1)+A2*cos(w2*t+t2);
   Rxx=simple(limit(int(x*subs(x,t,t+tau),t,0,T)/T,T,inf))
```

The analytical solution can be obtained as

$$R_{xx}(\tau) = \frac{1}{2}A_1^2\cos(\omega_1\tau) + \frac{1}{2}A_2^2\cos(\omega_2\tau).$$

Given two sets of data $x_i, y_i, (i = 1, 2, \cdots, n)$, the following formula can be used to evaluate the correlation coefficients

$$r_{xy} = \frac{\sqrt{\sum(x_i - \bar{x})(y_i - \bar{y})}}{\sqrt{\sum(x_i - \bar{x})^2}\sqrt{\sum(y_i - \bar{y})^2}}. \tag{8.19}$$

The MATLAB function `corrcoef()` can be used to obtain the correlation coefficients $\boldsymbol{R}$ of vectors $\boldsymbol{x}$ and $\boldsymbol{y}$ using the following syntaxes:

`R=corrcoef(x,y)` or `R=corrcoef([x,y])`

Example 8.28 Generate sample points from the functions $y_1 = te^{-4t}\sin 3t$ and $y_2 = te^{-4t}\cos 3t$, and then compute the correlation coefficient matrix.

Solution With the following statements, the correlation matrix can be obtained

```
>> x=0:0.01:5; y1=x.*exp(-4*x).*sin(3*x); y2=x.*exp(-4*x).*cos(3*x);
   R=corrcoef(y1,y2)
```

and $\boldsymbol{R} = \begin{bmatrix} 1 & 0.477585851214039 \\ 0.477585851214039 & 1 \end{bmatrix}$.

Consider now the sampled data $x_i, y_i, (i = 1, 2, \cdots, n)$. The auto-correlation function of the discrete signal x_i can be evaluated from

$$c_{xx}(k) = \frac{1}{N}\sum_{l=1}^{n-[k]-1} x(l)x(k+l), \quad 0 \leqslant k \leqslant m-1 \tag{8.20}$$

where $m < n$. The auto-correlation function obtained is also an even function. Similarly, the cross-correlation function is defined as

$$c_{xy}(k) = \frac{1}{N}\sum_{l=1}^{n-[k]-1} x(l)y(k+l), \quad 0 \leqslant k \leqslant m-1. \tag{8.21}$$

Similarities of two discrete signals can be studied with the help of correlation functions. The auto-correlation function and cross-correlation function can both be evaluated with the use of function `xcorr()`. The syntaxes are

C_{xx}=`xcorr(`$\boldsymbol{x}$`,`N`)` and C_{xy}=`xcorr(`$\boldsymbol{x}$`,`$\boldsymbol{y}$`,`N`)`

where N is the maximum value of k and it can be omitted.

Example 8.29 Consider again the data generated from Example 8.28. Compute numerically the auto-correlation, cross-correlation functions and compare the results with theoretical curves.

Solution Specify a time vector in the interval $t \in (0, 5)$. The output signals can be generated in vectors $\boldsymbol{x}$ and $\boldsymbol{y}$. The auto-correlation and cross-correlation functions can be obtained as shown in Figures 8.27 (a) and (b), respectively.

```
>> t=0:0.01:5; x=t.*exp(-4*t).*sin(3*t); y=t.*exp(-4*t).*cos(3*t);
   N=150; c1=xcorr(x,N); x1=[-N:N]; stem(x1,c1)
   figure; c1=xcorr(x,y,N); x1=[-N:N]; stem(x1,c1)
```

8.4.2 Fast Fourier transforms

The Fourier transform for discrete sequences $x_i, i = 1, 2, \cdots, N$ is fundamental to digital signal processing. The discrete Fourier transform is defined as

$$X(k) = \sum_{i=1}^{N} x_i e^{-2\pi j(k-1)(i-1)/N}, \quad \text{where } 1 \leqslant k \leqslant N \tag{8.22}$$

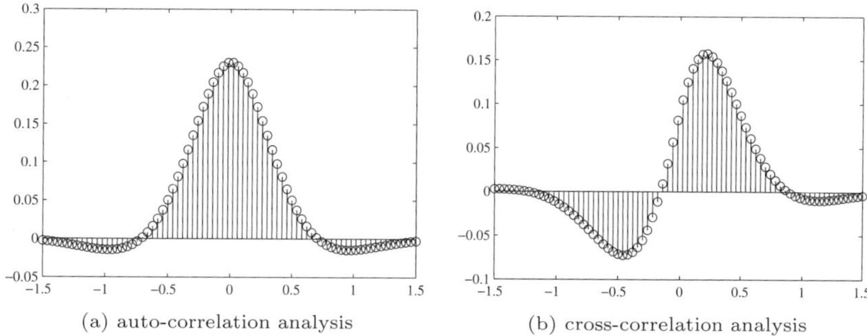

(a) auto-correlation analysis (b) cross-correlation analysis

FIGURE 8.27: Correlation function analysis

and its inverse transform is defined as

$$x(k) = \frac{1}{N}\sum_{i=1}^{N} X(i)\mathrm{e}^{2\pi \mathrm{j}(k-1)(i-1)/N}, \quad \text{where } 1 \leqslant k \leqslant N. \tag{8.23}$$

Fast Fourier transform (FFT) technique is the most effective and most practical way in solving Fourier transform. A built-in function `fft()` has been provided such that `f=fft(x)`. A useful feature `fft()` is that the length of the vector does not have to be constrained as 2^n. However, setting length of 2^n indeed makes the computation much faster.

For inverse FFT, the function `ifft()` can be used such that `x̂=ifft(f)`.

Example 8.30 Given a signal $x(t) = 12\sin(2\pi t+\pi/4)+5\cos(8\pi t)$ and the step-size selected as h, draw the relationship between the frequency versus FFT magnitude. Observe whether the original signal can be using inverse FFT.

Solution For different sampling periods h, the L time t_i and the function x_i can be obtained. The frequency points are$f_0 = 1/h, 2f_0, 3f_0, \cdots$. The following statements can be used

```
>> h=0.01; t=0:h:1; x=12*sin(2*pi*t+pi/4)+5*cos(2*pi*4*t); X=fft(x);
   f=t/h; plot(f(1:floor(length(f)/2)),abs(X(1:floor(length(f)/2))))
   xlim([0,10])
```

and the relationship between frequency and FFT magnitude is shown in Figure 8.28 (a). Here only half of the data are used to avoid aliasing phenomenon. It can be seen that there are two peaks in the magnitude plots, and the corresponding frequencies are 1Hz and 4Hz, which are the same as appeared in the original signal.

Inverse FFT can be evaluated by

```
>> ix=real(ifft(X)); plot(t,x,t,ix,':'); norm(x-ix)
```

and it can be seen that the error in restoring the original function by inverse FFT is as small as 5.6263×10^{-14}. The function restored is shown in Figure 8.28 (b), together with the original function. It can be seen that they are almost the same.

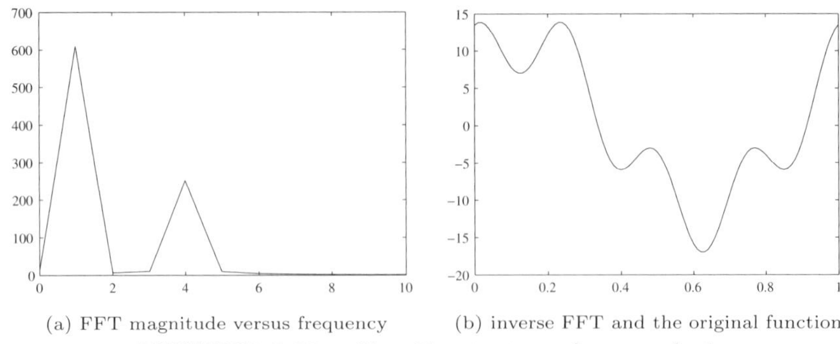

(a) FFT magnitude versus frequency (b) inverse FFT and the original function

FIGURE 8.28: Fast Fourier transform analysis

Two-dimensional and even high-dimensional FFT and their inverses can be solved with the MATLAB functions `fft2()`, `ifft2()`, `fftn()` and `ifftn()`, respectively.

8.4.3 Filtering techniques and filter design

Example 8.31 To illustrate the concept of filtering techniques, the function $y(x) = \mathrm{e}^{-x}\sin(5x)$ is studied. Suppose a set of function data is generated, disturbed by the white noise signal with the zero mean and the variance of 0.05. Draw the corrupted signal first.

Solution The following statements can be used and the noisy signal is shown in Figure 8.29.

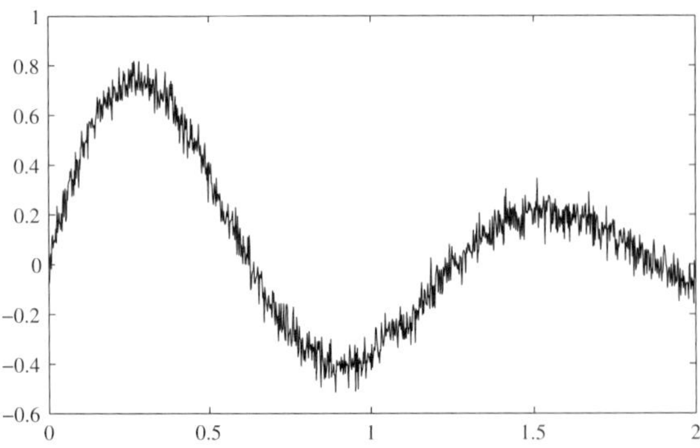

FIGURE 8.29: Data corrupted by noises

```
>> x=0:0.002:2; y=exp(-x).*sin(5*x);
   r=0.05*randn(size(x)); y1=y+r; plot(x,y1)
```

For the corrupted signal shown in Figure 8.29, a noise elimination technique should be applied. For instance, filters can be used to filter out the noise to extract the actual signal.

Linear filters — general models

Linear filters can generally be written as

$$H(z)=\frac{b_1+b_2z^{-1}+b_3z^{-2}+\cdots+b_{n+1}z^{-n}}{1+a_1z^{-1}+a_2z^{-2}+\cdots+a_mz^{-m}}. \tag{8.24}$$

Assume that the input signal is $x(n)$, the filtered output signal can be expressed using the following difference equation:

$$y(k)=-a_1y(k-1)-\cdots-a_my(k-m)+b_1x(k)+b_2x(k-1)+\cdots+b_{n+1}x(k-n). \tag{8.25}$$

With different combinations of n and m, three types of filters can be defined as

(i) **FIR filter** The finite impulse response (FIR) filter requires $m=0$ in (8.24). The variable $\boldsymbol{a}$ is then a scalar. This filter is also known as the *moving average* (MA) filter. The vector $\boldsymbol{b}$ can be used to define the filter.

(ii) **IIR filter** The infinite impulse response (IIR) filter, also known as *auto-regressive* (AR) or all-pole filter required $n=0$, i.e., $\boldsymbol{b}$ is a scalar and the vector $\boldsymbol{a}$ can be used to represent the filter. An advantage of FIR is that it is always stable[41].

(iii) **ARMA filter** The auto-regressive moving average (ARMA) filter is also known as *general IIR* filter, where n and m are not zero. The vectors $\boldsymbol{a}$ and $\boldsymbol{b}$ can be used to express the filter.

Assume that the filter is given by vectors $\boldsymbol{a}$ and $\boldsymbol{b}$, and the signal to be filtered is given by vector $\boldsymbol{x}$. The function `filter()` can be used to evaluate the filtered signal vector $\boldsymbol{y}$ where `y=filter(b,a,x)`.

Filters can be classified into low-pass filters, high-pass filters, band-stop and band-pass filters. It is known from the filter names that a low-pass filter allows the signal with low frequency to pass the filter directly with little or no attenuation while the signal's high-frequency content will be filtered out or attenuated. High-pass filters filter out the low-frequency signal components. A band-pass filter allows the signal components of certain frequency range to pass directly, while those of other frequencies are filtered out. In real applications, these filters are used depending on the purpose of the signal processing. For example, the noise in Example 8.31 is a high-frequency noise, thus a low-pass filter is needed, i.e., the magnitude is set to 1 or nearly 1 for low-frequency signals, while to 0 for high-frequency noises. In this way, the noise may be eliminated. If the filter is known, the function `freqz()` can be used to analyze the magnitude of the filter, such that `[h,w]=freqz(b,a,N)`, where N is the number of points to be analyzed. The returned argument $\boldsymbol{h}$ is the complex gain, and $\boldsymbol{w}$ is the frequency vector. The complex gain includes the information of magnitudes and phases. If only the magnitude information is required, `plot(w,abs(h))` can be used to visualize the filter frequency response.

Example 8.32 Assume that a filter is given by

$$H(z)=\frac{1.2296\times 10^{-6}(1+z^{-1})^7}{\begin{array}{c}(1-0.7265z^{-1})(1-1.488z^{-1}+0.5644z^{-2})\\(1-1.595z^{-1}+0.6769z^{-2})(1-1.78z^{-1}+0.8713z^{-2})\end{array}}.$$

Compute the complex gain and observe the filtered signals.

Solution The vectors $\boldsymbol{b}$ and $\boldsymbol{a}$ of the filter can be entered into MATLAB, where the function `conv()` can be used to perform polynomial multiplications. The following statements can also draw the gain-frequency curve as shown in Figure 8.30 (a). It can be seen from the curve that, for low frequencies, the gain is close to 1, which means that no filtering action is made over these frequencies. For high frequencies, the gain approaches to 0, which means that the high-frequency noise can be removed or attenuated from the signal.

```
>> b=1.2296e-6*conv([1 4 6 4 1],[1 3 3 1]); a=conv([1,-0.7265],...
     conv([1,-1.488,0.5644],conv([1,-1.595,0.6769],[1,-1.78,0.8713])));
   x=0:0.002:2; y=exp(-x).*sin(5*x); r=0.05*randn(size(x)); y1=y+r;
   [h,w]=freqz(b,a,100); plot(w,abs(h))     % draw magnitude
   figure; y2=filter(b,a,y1);  plot(x,y1,x,y2) % result after filtering
```

It can be seen from the results shown in Figure 8.30 (b) that, in the filtered signal, the noise is successfully removed. Note that there exist small delays in the filtered signal. In real-time or online filtering, this delay is unavoidable. However, when performing off-line filtering, it is possible to have zero delay by using the so-called "zero-phase filtering" which can be done by the MATLAB function `filtfilt()` with the same syntax as `filter()`.

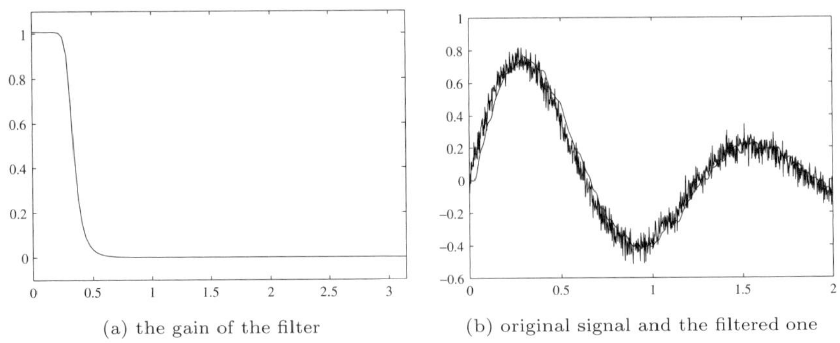

(a) the gain of the filter (b) original signal and the filtered one

FIGURE 8.30: Filter gain and filter results

Filter design using MATLAB

It has been shown from the previous examples that the noisy signal can be filtered with a properly designed filter. There are various algorithms and relevant toolboxes in MATLAB for filter design and simulation problems. The most widely used filters are the Butterworth filters and the Chebyshev filters. They can be designed with the `butter()`, `cheby1()` (Chebyshev type I) and `cheby2()` (Chebyshev type II) functions. The syntaxes of these functions are

`[`b`,`a`]=butter(`n`,`$\omega_{\rm n}$`)`, `[`b`,`a`]=cheby1(`n`,`r`,`$\omega_{\rm n}$`)`, `[`b`,`a`]=cheby2(`n`,`r`,`$\omega_{\rm n}$`)`

where n is the order of the filter, which can either be selected by the user, or be a relevant function, for instance, `buttord()`. The argument $\omega_{\rm n}$ is the normalized frequency, which defines the ratio of the filter frequency and the Nyquist frequency of the signal. Assume that there are N sampling points, and step-size is Δt. The fundamental frequency can then be calculated as $f_0 = 1/\Delta t$ Hz. Then the Nyquist frequency is defined as $f_0/2$.

Example 8.33 Consider the signal in Example 8.31. Design Butterworth filters for different combinations of orders and the values of $w_{\rm n}$, and compare the filtering results.

Solution Again $\omega_{\rm n} = 0.1$ can be selected. With the following statements, the Butterworth filters of different orders can be designed, and the gain and filter results are shown in Figures 8.31 (a) and (b), respectively. It can be seen that with the increment of order n, the filter curve becomes smoother, and the delay also increases.

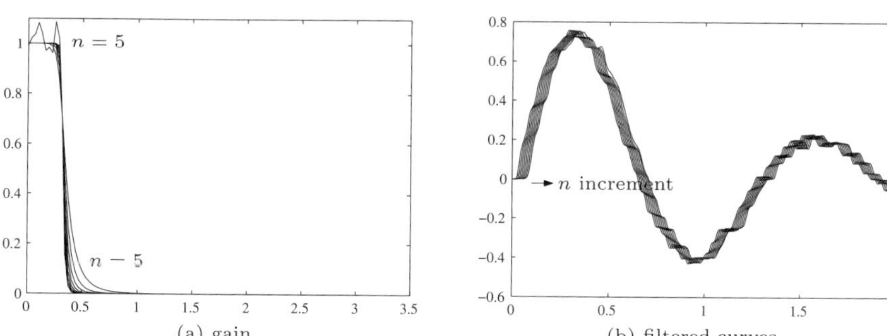

(a) gain (b) filtered curves

FIGURE 8.31: Butterworth filters with different orders

```
>> f1=figure; f2=figure;
   for n=5:2:20
      figure(f1); [b,a]=butter(n,0.1);
      y2=filter(b,a,y1); plot(x,y2); hold on
      figure(f2); [h,w]=freqz(b,a,100); plot(w,abs(h)); hold on
   end
```

If one selects the 7th order filter structure, different values of ω_n can be tested for Butterworth filters. The gain and filter results can be obtained as shown in Figures 8.32 (a) and (b), respectively. It can be seen that when ω_n increases, the delay will decrease and filtering results may get worse. The large value of ω_n may not provide any benefit.

```
>> for wn=0.1:0.1:0.7
      figure(f1); [b,a]=butter(7,wn);
      y2=filter(b,a,y1); plot(x,y2); hold on
      figure(f2); [h,w]=freqz(b,a,100); plot(w,abs(h)); hold on
   end
```

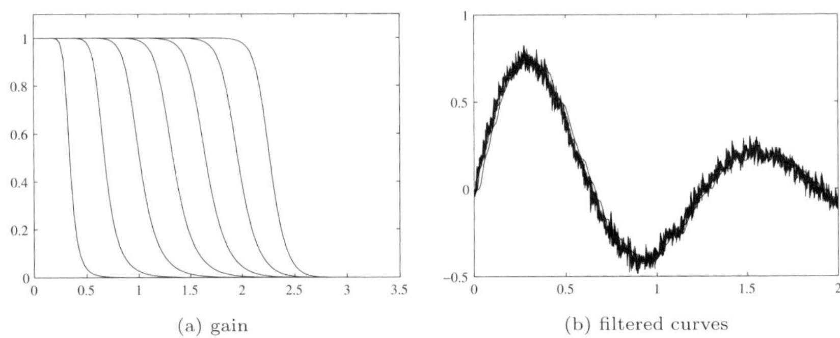

(a) gain (b) filtered curves

FIGURE 8.32: Butterworth filters with different frequencies

If high-pass filters are required, the simplest way is to design it with the formula $1 - H(z^{-1})$, where $H(z^{-1})$ is the low-pass filter designed using the methods shown earlier. Also the function `butter()` can be used to design high-pass and band-pass filters with the syntaxes

```
[b,a]=butter(n,wn,'high')   % high-pass filter
[b,a]=butter(n,[w1,w2])     % band-path filter
```

Exercises

1. Generate a sparsely distributed data from the following functions. Use one-dimensional interpolation method to smooth the curves, with different methods. Compare the interpolation results with the theoretical curves.

   (i) $y(t) = t^2 e^{-5t} \sin t$, where $t \in (0, 2)$,

   (ii) $y(t) = \sin(10t^2 + 3)$, for $t \in (0, 3)$.

2. Generate a set of mesh grid data and randomly distribute the data from the prototype function $f(x, y) = \dfrac{1}{3x^3 + y} e^{-x^2 - y^4} \sin(xy^2 + x^2 y)$. Fit the original

3D surface with two-dimensional interpolation methods and compare the results with the theoretical ones.

3. Assume that a set of data is given as shown below. Interpolate the data into a smooth curve in the interval $x \in (-2, 4.9)$. Compare the advantages and disadvantages of the algorithms.

x_i	-2	-1.7	-1.4	-1.1	-0.8	-0.5	-0.2	0.1	0.4	0.7	1	1.3
y_i	0.1029	0.1174	0.1316	0.1448	0.1566	0.1662	0.1733	0.1775	0.1785	0.1764	0.1711	0.1630
x_i	1.6	1.9	2.2	2.5	2.8	3.1	3.4	3.7	4	4.3	4.6	4.9
y_i	0.1526	0.1402	0.1266	0.1122	0.0977	0.0835	0.0702	0.0579	0.0469	0.0373	0.0291	0.0224

4. Assume that a set of measured data is given in a file c8pdat.dat. Draw the 3D surface using interpolation methods.

5. Assume that a set of measured data is given in a file c8pdat3.dat, whose 1~3 columns are the coordinates of x, y, z, and the fourth column saves the measured function value $V(x, y, z)$. Perform three-dimensional interpolation from the data.

6. Generate a set of data from the function $f(x) = \dfrac{\sqrt{1+x} - \sqrt{x-1}}{\sqrt{2+x} + \sqrt{x-1}}$, for $x = 3 : 0.4 : 8$. The cubic splines and B-splines can be used to perform data interpolation tasks. From the fitted splines, take the second-order derivatives and compare the results with the theoretical curves.

7. Assume that the measured data are given below. Draw the 3D surface plot for (x, y) within the rectangular intervals $(0.1, 0.1) \sim (1.1, 1.1)$.

y_i	x_1	x_2	x_3	x_4	x_5	x_6	x_7	x_8	x_9	x_{10}	x_{11}
0	0.1	0.2	0.3	0.4	0.5	0.6	0.7	0.8	0.9	1	1.1
0.1	0.8304	0.8272	0.824	0.8209	0.8182	0.8161	0.8148	0.8146	0.8157	0.8185	0.823
0.2	0.8317	0.8324	0.8358	0.842	0.8512	0.8637	0.8797	0.8993	0.9226	0.9495	0.9801
0.3	0.8358	0.8434	0.8563	0.8746	0.8986	0.9284	0.9637	1.0045	1.0502	1.1	1.1529
0.4	0.8428	0.8601	0.8853	0.9186	0.9598	1.0086	1.0642	1.1253	1.1903	1.2569	1.3222
0.5	0.8526	0.8825	0.9228	0.9734	1.0336	1.1019	1.1763	1.254	1.3308	1.4017	1.4605
0.6	0.8653	0.9104	0.9684	1.0383	1.118	1.2045	1.2937	1.3793	1.4539	1.5086	1.5335
0.7	0.8807	0.9439	1.0217	1.1117	1.2102	1.311	1.4063	1.4859	1.5377	1.5484	1.5052
0.8	0.899	0.9827	1.082	1.1922	1.3061	1.4138	1.5021	1.5555	1.5572	1.4915	1.346
0.9	0.92	1.0266	1.1482	1.2768	1.4005	1.5034	1.5661	1.5678	1.4888	1.3156	1.0454
1	0.9438	1.0752	1.2191	1.3624	1.4866	1.5684	1.5821	1.5032	1.315	1.0155	0.6247
1.1	0.9702	1.1278	1.2929	1.4448	1.5564	1.5964	1.5341	1.3473	1.0321	0.6126	0.1476

8. For the measured data samples (x_i, y_i) given below, piecewise cubic polynomial splines can be used and find the coefficients of each polynomial.

x_i	1	2	3	4	5	6	7	8	9	10
y_i	244.0	221.0	208.0	208.0	211.5	216.0	219.0	221.0	221.5	220.0

9. The one-dimensional and two-dimensional data given in Exercises 3 and 7 can be used for cubic splines and B-splines interpolation. Find the derivatives of the related interpolated functions.

10. Consider again the data in Exercise 3. Polynomial fitting can be used to model the data. Select a suitable degree such that good approximation by polynomials can be achieved. Compare the results with interpolation methods.

11. Consider again the data in Exercise 3. Assume that the prototype of the function for the data is $y(x) = \dfrac{1}{\sqrt{2\pi}\sigma}\mathrm{e}^{-(x-\mu)^2/2\sigma^2}$. The values of the parameters μ and σ are not known. Use least squares curve fitting methods to see whether suitable μ and σ can be identified. Observe the fitting results.

12. Express the constant e in terms of continued fractions. Observe how many continued fraction stages are expected to get a suitable approximation.

13. Find good approximations to the functions given below using continued fraction expansions and Padé approximations. Observe the fitting results obtained and find suitable degrees of the rational functions.

    (i) $f(x) = \mathrm{e}^{-2x}\sin 5x$, (ii) $f(x) = \dfrac{x^3+7x^2+24x+24}{x^4+10x^3+35x^2+50x+24}\mathrm{e}^{-3x}$

14. Assume that the data in Exercise 7 satisfies a prototype function of $z(x,y) = a\sin(x^2y) + b\cos(y^2x) + cx^2 + dxy + e$. Identify the values of a, b, c, d, e with least squares method. Verify the identification results.

15. Given a signal $f(t) = \mathrm{e}^{-3t}\cos(2t+\pi/3)+\mathrm{e}^{-2t}\cos(t+\pi/4)$, evaluate the formula of the auto-correlation function of the signal. Generate a sequence of randomly distributed data and verify the results in a numerical way.

16. Evaluate the auto-correlation function for the Gaussian distribution function defined as $f(t) = \dfrac{1}{\sqrt{2\pi}3}\mathrm{e}^{-t^2/3^2}$. Generate a sequence of signals and compare them using Gaussian distributed data to check whether the resulted description is close to the theoretical result.

17. Assume that the noisy signal can be generated with the following statements

    ```
    >> t=0:0.005:5; y=15*exp(-t).*sin(2*t);
       r=0.3*randn(size(y)); y1=y+r;
    ```

    Find the Nyquist frequency of the signal. Based on the Nyquist frequency, design an eighth-order Butterworth filter which can be used to effectively filter out the noise while having a relatively small delay due to filtering.

18. High-pass filters can be used to filter out information with low-frequencies, and retain the high-frequency details. Design a high-pass filter for the data shown in Exercise 17. From the obtained high frequency noise information, compare the statistical behavior of the noise signal obtained.

Chapter 9

Probability and Mathematical Statistics Problems

The theory of probability and mathematical statistics is a very important branch in experimental sciences. The solutions to probability and mathematical statistics problems sometimes could be quite involved. The traditional statistics usually relies heavily on lookup-tables. A very comprehensive statistics toolbox is provided in MATLAB which contains a lot of handy functions for solving related probability and mathematical statistics problems. In Section 9.1, the basic concepts of *probability density function* (PDF) and *cumulative distribution function* (CDF) are introduced. Given probability distributions, various probability related example problems are solved using Statistics Toolbox. In this section, pseudo-random number generators of different distributions, such as uniform distribution, normal distribution, Poisson distribution, etc., are introduced and demonstrated. In Section 9.2, how to compute statistical quantities, such as mean, variance, moments and covariances is covered for both univariate and multivariate distributions together with an introduction to Monte Carlo method and its applications. The parametric estimation and interval estimation problems, together with their MATLAB implementations, are given in Section 9.3 where multi-variable linear regression and least squares data fitting problems are also presented. In Section 9.4, the hypothesis test problems including mean value test, normality test and given distribution test, are discussed. The variance analysis problems and applications are presented in Section 9.5 with detailed example solutions in MATLAB.

For readers who wish to check the detailed explanations of probability and statistics, we recommend the open source textbook [42] for probability and the online e-textbook [43] for statistics.

9.1 Probability Distributions and Pseudo-Random Number Generators

9.1.1 Introduction to PDFs and CDFs

The PDF for a continuous random variable is often denoted as $p(x)$, where

$$p(x) \geqslant 0, \text{ and } \int_{-\infty}^{\infty} p(x)\,\mathrm{d}x = 1. \tag{9.1}$$

The CDF is then defined as

$$F(x) = \int_{-\infty}^{x} p(\tau)\,\mathrm{d}\tau. \tag{9.2}$$

The CDF $F(x)$ is defined as the probability of an event when $\xi \leqslant x$ happens for the random variable ξ. This function is a monotonic increasing function which satisfies

$$0 \leqslant F(x) \leqslant 1, \text{ and } F(-\infty) = 0, \; F(\infty) = 1. \tag{9.3}$$

Given the probability $f_i = F(x_i)$, to determine the value of x_i, lookup-table methods can be used. Since $F(x_i)$ is monotonic, x_i can easily be found. This problem is also referred to as the *inverse distribution problem*. In fact, functions provided in the Statistics Toolbox can easily, accurately and straightforwardly solve this type of inverse problems. In this section, some of the commonly encountered probability distributions will be summarized and visualized.

9.1.2 PDFs/CDFs of commonly used distributions

Normal distribution

The PDF of normal distribution is defined as

$$p_{\rm n}(x) = \frac{1}{\sqrt{2\pi}\sigma} \mathrm{e}^{-\frac{(x-\mu)^2}{2\sigma^2}} \tag{9.4}$$

where μ and σ^2 are respectively the mean and variance of the normal distribution. The functions `normpdf()` and `normcdf()` provided in the Statistics Toolbox can be used to evaluate the PDFs and CDFs of such a distribution. Also if the distribution value $\boldsymbol{F}$ is known, the function `norminv()` can be used to evaluate the corresponding $\boldsymbol{x}$. The syntaxes of these functions are

```
y=normpdf(x,μ,σ),     F=normcdf(x,μ,σ),     x=norminv(F,μ,σ)
```

where $\boldsymbol{x}$ is a vector of given points, $\boldsymbol{y}$ vector contains the values of PDF at $\boldsymbol{x}$. The vector $\boldsymbol{F}$ returns the CDFs at $\boldsymbol{x}$. Clearly, the normal distribution is a function of the mean μ and the standard deviation σ, which is visualized by the following example.

Example 9.1 For different combinations of (μ, σ^2) such as $(-1, 1), (0, 0.1), (0, 1), (0, 10), (1, 1)$, draw the PDFs and CDFs of normal distributions.

Solution With MATLAB Statistics Toolbox, these density functions can be drawn easily and accurately for any combination of μ and σ^2. Thus the use of computer mathematics languages can help the readers to solve effectively the probability and statistics problems.

The problem can easily be solved with MATLAB. One can assign a vector $\boldsymbol{x}$ composed of points in the interval $(-5, 5)$ and two other vectors defined to represent the possible combinations of the variables μ and σ^2. Then the functions `normpdf()` and `normcdf()` can be used to draw the corresponding PDFs and CDFs, as shown in Figures 9.1 (a) and (b), respectively.

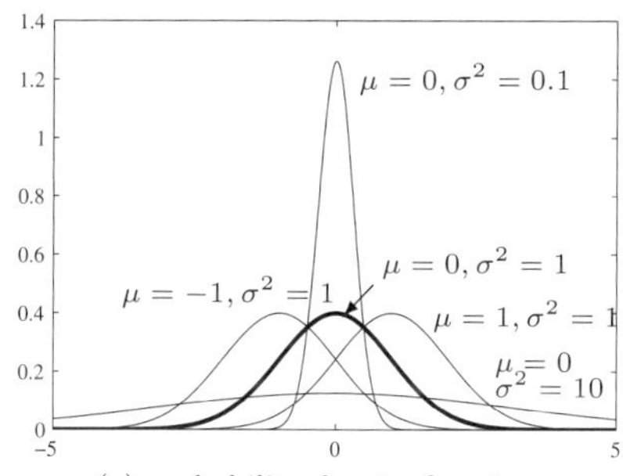

(a) probability density functions

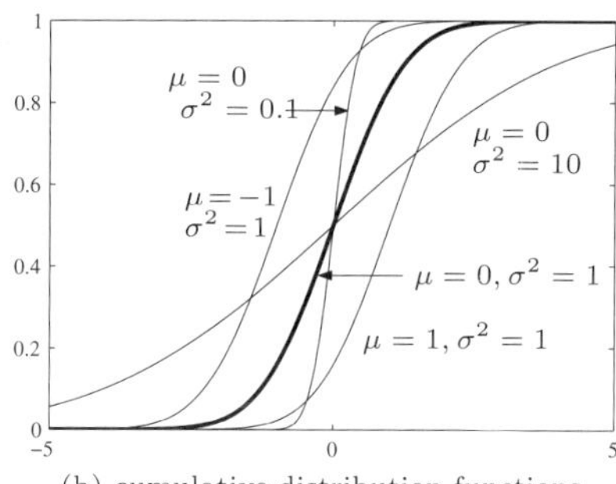

(b) cumulative distribution functions

FIGURE 9.1: PDFs and CDFs of normal distribution

```
>> x=[-5:.02:5]'; y1=[]; y2=[];
   mu=[-1,0,0,0,1]; sig=[1,0.1,1,10,1]; sig=sqrt(sig);
   for i=1:length(mu)
      y1=[y1,normpdf(x,mu(i),sig(i))]; y2=[y2,normcdf(x,mu(i),sig(i))];
   end
   plot(x,y1), figure; plot(x,y2)
```

From the PDF curves, it can be observed that if σ^2 remains the same, the shapes of the PDFs are exactly the same as if they are obtained by translating the curves according to μ. If the values of σ are different, the shapes are also different. The smaller the value of σ^2, the narrower the curve of PDFs.

The combination of $\mu = 0, \sigma^2 = 1$ is referred to as the *standard normal distribution*, which is usually denoted by $N(0, 1)$.

Many similar functions are provided in Statistics Toolbox. The functions with suffixes of `pdf`, `cdf` and `inv` are used to indicate the probability density function, cumulative distribution function and inverse distribution function respectively. The related functions are summarized in Table 9.1. Other suffixes of the function include `rnd`, `stat` and `fit`, meaning random number generator, statistics analysis and parametric estimation.

Poisson distribution

The Poisson distribution is different from the continuous distribution functions discussed earlier. It requires that vector $\boldsymbol{x}$ is composed of positive integers. The PDF of the Poisson distribution is defined as

$$p_{\rm P}(x) = \frac{\lambda^x}{x!}{\rm e}^{-\lambda x}, \ x = 0, 1, 2, 3, \cdots \tag{9.5}$$

where λ is a positive integer.

TABLE 9.1: Keywords and function names in Statistics Toolbox

keyword	distribution	parameters	keyword	distribution	parameters
`beta`	β distribution	a, b	`bino`	binomial distribution	n, p
`chi2`	χ^2 distribution	k	`ev`	extreme value	μ, σ
`exp`	exponential	λ	`f`	F distribution	p, q
`gam`	Γ-distribution	a, λ	`geo`	geometric	p
`hyge`	hypergeometric	m, p, n	`logn`	lognormal distribution	μ, σ
`mvn`	multivariate normal	$\boldsymbol{\mu}, \boldsymbol{\sigma}$	`nbin`	negative binomial	ν_1, ν_2, δ
`ncf`	noncentral F	k, δ	`nct`	noncentral T	k, δ
`ncx2`	noncentral χ^2	k, δ	`norm`	normal distribution	μ, σ
`poiss`	Poisson distribution	λ	`rayl`	Rayleigh distribution	b
`t`	T-distribution	k	`unif`	uniform distribution	a, b
`wbl`	Weibull distribution	a, b			

Poisson distribution is a function of positive integer parameter λ. The related functions `poisspdf()`, `poisscdf()` and `poissinv()` are provided in Statistics Toolbox to evaluate the PDFs, CDFs and inverse distribution function, respectively. The syntaxes of the function are

```
y=poisspdf(x,λ),    F=poisscdf(x,λ),    x=poissinv(F,λ)
```

where the vector $\boldsymbol{x}$ contains given integers, which can be specified with `x=0:k`. The values in $\boldsymbol{y}$ are the PDFs corresponding to the points in the vector $\boldsymbol{x}$. The vector $\boldsymbol{F}$ returns the CDFs corresponding to vector $\boldsymbol{x}$.

Example 9.2 For the parameter λ selected as $\lambda = 1, 2, 5, 10$, draw the PDFs and CDFs of the Poisson distributions.

Solution One can create a vector of $\boldsymbol{x}$, and then for different values of λ, the functions `poisspdf()` and `poisscdf()` can be used in a loop structure. The PDFs and CDFs can be obtained as shown in Figures 9.2 (a), (b), respectively.

```
>> x=[0:15]'; y1=[]; y2=[]; lam1=[1,2,5,10];
   for i=1:length(lam1)
      y1=[y1,poisspdf(x,lam1(i))]; y2=[y2,poisscdf(x,lam1(i))];
   end
   plot(x,y1), figure; plot(x,y2)
```

Rayleigh distribution

The PDF of Rayleigh distribution is defined as

$$p_{\rm r}(x) = \begin{cases} \dfrac{x}{b^2}\mathrm{e}^{-\frac{x^2}{2b^2}} & x \geqslant 0 \\ 0 & x < 0 \end{cases} \tag{9.6}$$

which contains a parameter b. The functions `raylpdf()`, `raylcdf()` and `raylinv()` provided in the Statistics Toolbox can be used to evaluate the relevant functions. The syntaxes of these functions are

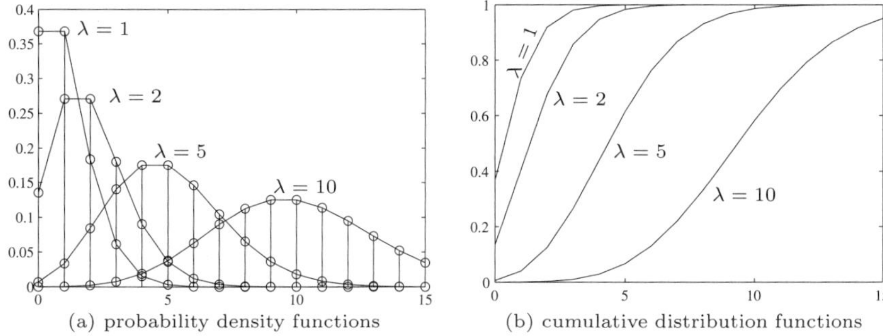

FIGURE 9.2: PDFs and CDFs of the Poisson distribution

`y=raylpdf(x,b),   F=raylcdf(x,b),   x=raylinv(F,b)`

Example 9.3 Draw the PDFs and CDFs of the Rayleigh distribution for $b = 0.5, 1, 3, 5$.

Solution Assign a vector $\boldsymbol{x}$ in the interval $(-0.1, 5)$, and also a parameter vector $\boldsymbol{b}_1$, the `raylpdf()` and `raylcdf()` functions can be used to draw PDFs and CDFs, as shown in Figures 9.3 (a) and 9.3 (b), respectively.

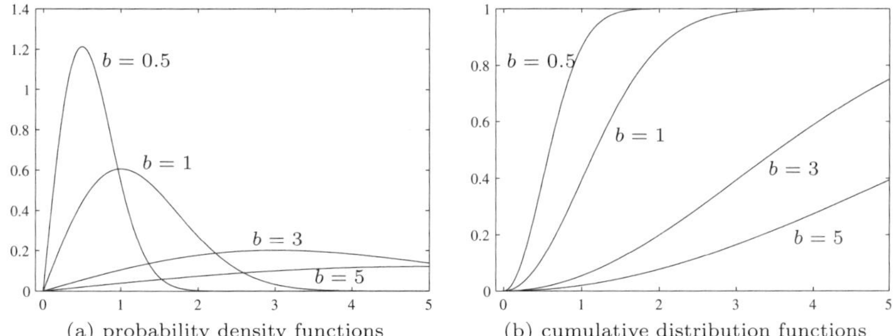

FIGURE 9.3: PDFs and CDFs of the Rayleigh distribution

```
>> x=[-eps:-0.02:-0.05,0:0.02:5]; x=sort(x');
   b1=[.5,1,3,5]; y1=[]; y2=[];
   for i=1:length(b1)
      y1=[y1,raylpdf(x,b1(i))]; y2=[y2,raylcdf(x,b1(i))];
   end
   plot(x,y1), figure; plot(x,y2)
```

Γ distribution

The PDF of the Γ distribution is defined as

$$p_\Gamma(x) = \begin{cases} \dfrac{\lambda^a x^{a-1}}{\Gamma(a)} \mathrm{e}^{-\lambda x} & x \geqslant 0 \\ 0 & x < 0 \end{cases} \tag{9.7}$$

where $\Gamma(a) = \int_0^\infty x^{a-1}e^{-x}dx$. Note that the $\Gamma(a)$ function satisfies $\Gamma(a) = a\Gamma(a-1)$, $\Gamma(1) = 1$ and $\Gamma(1/2) = \pi$. For other the values of a, $\Gamma(a)$ can be evaluated through integrals. In particular, for integers n, one has $\Gamma(n) = (n-1)!$. They can also be evaluated from the function `gamma()`. For instance, the value of $\Gamma(\pi)$ can be evaluated with `gamma(pi)` which gives 2.28803779534003. The function `gamma()` is also provided in the Symbolic Math Toolbox to evaluate the value through arbitrary precision with `vpa(gamma(sym(pi)))`.

The PDF of Γ distribution is a function of the parameters a and λ. The functions `gampdf()`, `gamcdf()` and `gaminv()` can be used with the following syntaxes

```
y=gampdf(x,a,λ),    F=gamcdf(x,a,λ),    x=gaminv(F,a,λ)
```

Example 9.4 For the parameter combinations (a,λ) as $(1,1)$, $(1,0.5)$, $(2,1)$, $(1,2)$, $(3,1)$, draw the PDFs and CDFs of the Γ distribution.

Solution A vector $\boldsymbol{x}$ can be established over the interval $(-0.5, 5)$, two variable vectors are also defined as `a1` and `lam1`, then the functions `gampdf()` and `gamcdf()` can be used to plot the PDFs and CDFs as shown in Figures 9.4 (a) and (b), respectively.

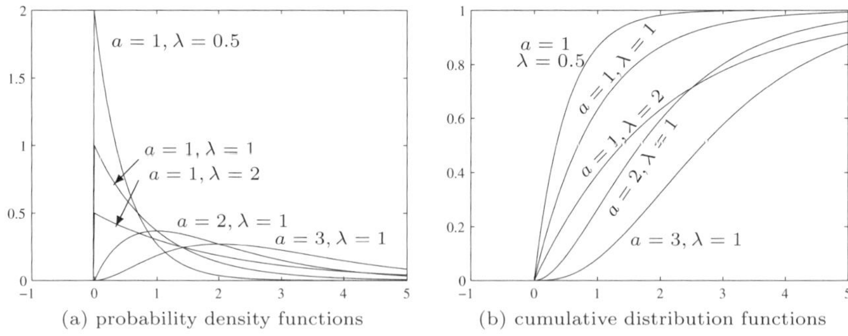

(a) probability density functions (b) cumulative distribution functions

FIGURE 9.4: PDFs and CDFs of the Γ distribution

```
>> x=[-0.5:.02:5]'; y1=[]; y2=[]; a1=[1,1,2,1,3]; lam1=[1,0.5,1,2,1];
   for i=1:length(a1)
      y1=[y1,gampdf(x,a1(i),lam1(i))]; y2=[y2,gamcdf(x,a1(i),lam1(i))];
   end
   plot(x,y1), figure; plot(x,y2)
```

There is a minor problem in the PDF curves. Since a step-size of 0.02 is selected to form the $\boldsymbol{x}$ vector, the PDF will jump from 0 to the maximum value. To avoid this, the $\boldsymbol{x}$ vector generating statement should be modified to

```
>> x=[eps, 0:0.02:5]; x=sort(x');
```

χ^2 distribution

The PDF of the χ^2 distribution is defined as

$$p_{\chi^2}(x) = \begin{cases} \dfrac{1}{2^{\frac{k}{2}}\Gamma\left(\dfrac{k}{2}\right)} x^{\frac{k}{2}-1} e^{-\frac{x}{2}} & x \geqslant 0 \\ 0 & x < 0 \end{cases} \tag{9.8}$$

where the parameter k is a positive integer. It can also be seen that the χ^2 function is a special Γ distribution, with $a = k/2$ and $\lambda = 1/2$. The PDF of the χ^2 distribution is also a function of k. The functions `chi2pdf()`, `chi2cdf()` and `chi2inv()` can be used to evaluate relevant functions for χ^2 distribution. The syntaxes of the functions are

```
y=chi2pdf(x,k),    F=chi2cdf(x,k),    x=chi2inv(F,k)
```

Example 9.5 For the parameter k selected as $1, 2, 3, 4, 5$, draw the PDFs and CDFs of χ^2 distributions.

Solution Select a vector $\boldsymbol{x}$ within the interval $(-0.05, 1)$, and also define a vector $\boldsymbol{k}_1$. Call the functions `chi2pdf()` and `chi2cdf()` directly. The PDFs and CDFs can be drawn as shown in Figures 9.5 (a) and (b), respectively.

```
>> x=[-eps:-0.02:-0.05,0:0.02:2]; x=sort(x');
   k1=[1,2,3,4,5]; y1=[]; y2=[];
   for i=1:length(k1)
      y1=[y1,chi2pdf(x,k1(i))]; y2=[y2,chi2cdf(x,k1(i))];
   end
   plot(x,y1), figure; plot(x,y2)
```

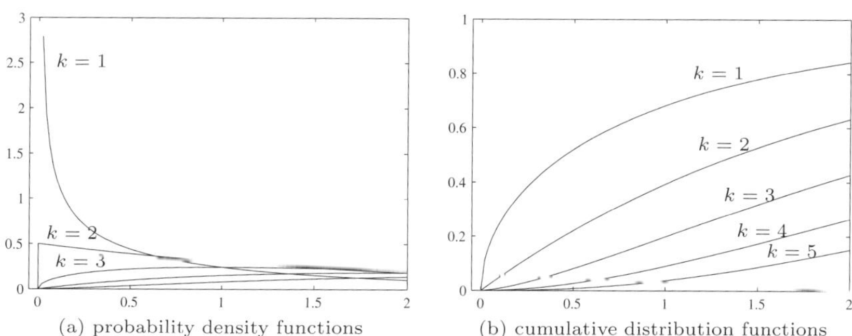

(a) probability density functions (b) cumulative distribution functions

FIGURE 9.5: PDFs and CDFs of the χ^2 distribution

T distribution

The PDF of the T distribution is defined as

$$p_{\mathrm{T}}(x) = \frac{\Gamma\left(\dfrac{k+1}{2}\right)}{\sqrt{k\pi}\,\Gamma\left(\dfrac{k}{2}\right)}\left(1 + \frac{x^2}{k}\right)^{-\frac{k+1}{2}} \tag{9.9}$$

with a positive integer parameter k. The functions `tpdf()`, `tcdf()` and `tinv()` can be used to evaluate relevant functions, with

```
y=tpdf(x,k),    F=tcdf(x,k),    x=tinv(F,k)
```

Example 9.6 Draw the PDFs and CDFs of T distribution for $k = 1, 2, 5, 10$.

Solution Again an $\boldsymbol{x}$ vector is defined over the interval $(-5, 5)$. Another vector $\boldsymbol{k}_1$ can also be established and with the following statements, the PDFs and CDFs can be obtained as shown in Figures 9.6 (a) and (b), respectively.

```
>> x=[-5:0.02:5]'; k1=[1,2,5,10]; y1=[]; y2=[];
   for i=1:length(k1)
      y1=[y1,tpdf(x,k1(i))]; y2=[y2,tcdf(x,k1(i))];
   end
   plot(x,y1), figure; plot(x,y2)
```

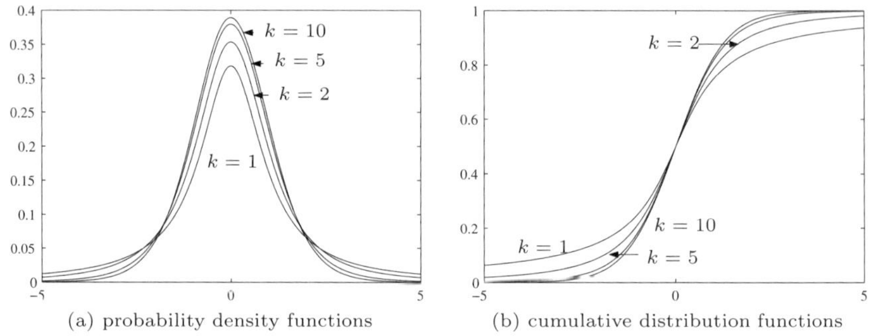

(a) probability density functions (b) cumulative distribution functions

FIGURE 9.6: PDFs and CDFs of the T distribution

F distribution

The PDF of the F distribution is defined as

$$p_{\mathrm{F}}(x) = \begin{cases} \dfrac{\Gamma\left(\dfrac{p+q}{2}\right)}{\Gamma\left(\dfrac{p}{2}\right)\Gamma\left(\dfrac{q}{2}\right)} p^{\frac{p}{2}} q^{\frac{q}{2}} x^{\frac{p}{2}-1} (p+qx)^{-\frac{p+q}{2}} & x \geqslant 0 \\ 0 & x < 0 \end{cases} \tag{9.10}$$

and the PDF is a function of parameters p and q, with p and q positive integers. The functions `fpdf()`, `fcdf()` and `finv()` in the Statistics Toolbox can be used. The syntaxes of the functions are

```
y=fpdf(x,a,b),    F=fcdf(x,p,q),    x=finv(F,p,q)
```

Example 9.7 Draw the PDFs and CDFs for the T distributions with the (p, q) pairs as $(1, 1), (2, 1), (3, 1), (3, 2), (4, 1)$.

Solution Define a vector $\boldsymbol{x}$ in the interval $(-0.1, 1)$, and two other vectors $\boldsymbol{p}_1$ and $\boldsymbol{q}_1$ can be defined for the given pairs. The PDFs and CDFs can be obtained with the functions `fpdf()` and `fcdf()`, as shown in Figures 9.7 (a) and (b), respectively.

```
>> x=[-eps:-0.02:-0.05,0:0.02:1]; x=sort(x');
   p1=[1 2 3 3 4]; q1=[1 1 1 2 1]; y1=[]; y2=[];
   for i=1:length(p1)
      y1=[y1,fpdf(x,p1(i),q1(i))]; y2=[y2,fcdf(x,p1(i),q1(i))];
   end
   plot(x,y1), figure; plot(x,y2)
```

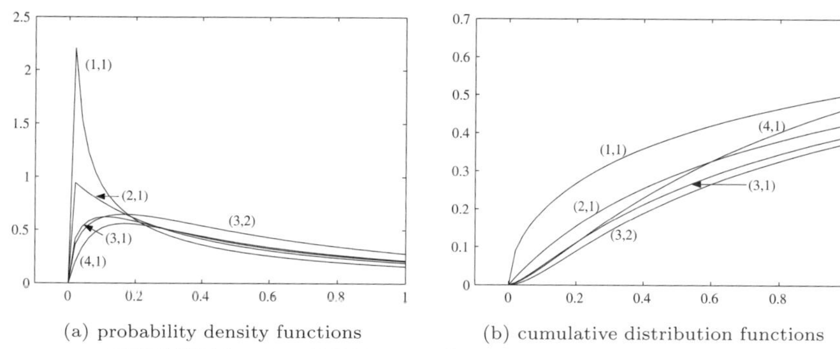

(a) probability density functions (b) cumulative distribution functions

FIGURE 9.7: PDFs and CDFs of the F distribution

9.1.3 Solving probability problems

It has been shown earlier that the definition of the CDF $F(x)$ is the probability of a random variable ξ falling in the interval $(-\infty, x)$. For instance, the probability of the random number ξ falling in the interval $[x_1, x_2]$ can be represented by $P[x_1 \leqslant \xi \leqslant x_2]$. The following formulas can be used to evaluate the probabilities

$$\begin{cases} P[\xi \leqslant x] = F(x), & \text{\% probability of } \xi \leqslant x \\ P[x_1 \leqslant \xi \leqslant x_2] = F(x_2) - F(x_1), & \text{\% probability of } x_1 \leqslant \xi \leqslant x_2 \\ P[\xi \geqslant x] = 1 - F(x), & \text{\% probability of } \xi \geqslant x \end{cases} \tag{9.11}$$

where the CDFs are given.

Example 9.8 Assume that a random variable x satisfies Rayleigh distribution, with $b = 1$. Compute the probabilities when x falls in the intervals $[0.2, 2]$ and $[1, \infty)$.

Solution With the definitions of probability, the probabilities of the random variable x falling in the intervals $(-\infty, 0.2]$ and $(-\infty, 2]$ can easily be found. Thus the probability of variable x falling into the interval $[0.2, 2]$ can be evaluated from

```
>> b=1; p1=raylcdf(0.2,b); p2=raylcdf(2,b); P1=p2-p1
```

and the probability is $P_1 = 0.844863$. Also the probability of the variable x falling in the interval $(1, \infty]$ can be evaluated from

```
>> p1=raylcdf(1,b); P2=1-p1
```

and the probability can be found as $P_2 = 0.606531$.

Example 9.9 Assume that the joint PDF for the variables ξ and η is given by $p(x,y) = \begin{cases} x^2 + \dfrac{xy}{3}, & 0 \leqslant x \leqslant 1,\ 0 \leqslant y \leqslant 2 \\ 0, & \text{otherwise} \end{cases}$. Determine $P(\xi < 1/2, \eta < 1/2)$.

Solution From the given joint PDF $p(x,y)$, the probability $P(\xi < x_0, \eta < y_0)$ can be evaluated with the following statement

```
>> syms x y; f=x^2+x*y/3; P=int(int(f,x,0,1/2),y,0,1/2)
```

and it can be found that the probability $P = 5/192$. In the original problem, $p(x,y)$ is zero when the independent variables x and/or y are zero. Thus, in the integrals, only the first quadrant is considered.

9.1.4 Random numbers and pseudo-random numbers

In scientific research and statistical analysis, random data are always expected. There are two ways of generating random numbers. One is to generate the random numbers by specific electronic devices, which is referred to as the *physical method.* The other method is to generate random numbers using mathematical algorithms. The random numbers generated by this method are calculated from corresponding mathematical formulae and the random numbers generated are referred to as *pseudo-random numbers.*

There are two advantages to using pseudo-random numbers. One is that the random numbers are repeatable for the same random number seed, which makes repetitive experiments possible. The other advantage is that the distributions of the random numbers generated can be specified as needed. For instance, random numbers satisfying uniform distribution, normal distribution or Poisson distribution, etc. can be selected by the user.

Two functions, `rand()` and `randn()`, have been introduced in Section 4.1 for generating uniformly distributed and normally distributed random matrices. Apart from these two types of random numbers, other types of random numbers are summarized in Table 9.1. In particular, several commonly used pseudo-random number generators are list below.

```
A=gamrnd(a,λ,n,m)   % generates an n × m random matrix satisfying Γ
B=chi2rnd(k,n,m)    % generates pseudo-random numbers satisfying χ^2
C=trnd(k,n,m)       % T distribution
D=frnd(p,q,n,m)     % F distribution
E=raylrnd(b,n,m)    % Rayleigh distribution
```

Example 9.10 Generate a 30000×1 vector with numbers satisfying Rayleigh distribution for $b = 1$. Verify the PDF of the random numbers with histograms.

Solution From the above list, the function `raylrnd()` can be used to generate a vector containing 30000×1 random values satisfying Rayleigh distribution. Assigning a vector `xx`, the function `hist()` can be used to evaluate the numbers in each bin defined in vector `xx`. The PDF can be approximated with the `bar()` function. Also the exact result can be obtained for comparison as shown in Figure 9.8. It can be seen that the generated data correspond to the Rayleigh distribution very well.

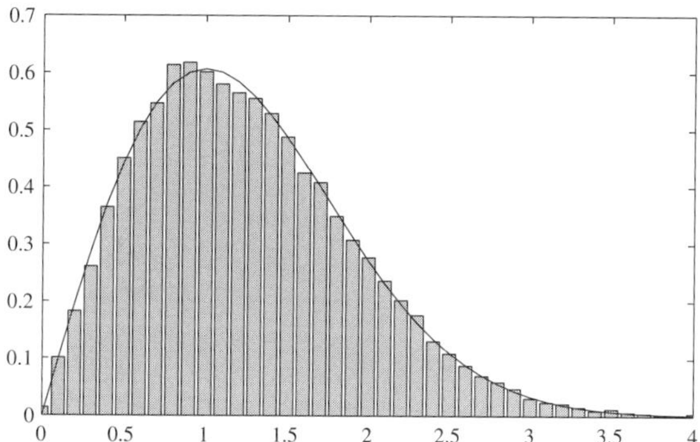

FIGURE 9.8: Distribution evaluation with Rayleigh functions

```
>> b=1; p=raylrnd(1,30000,1);
   xx=0:.1:4; yy=hist(p,xx); yy=yy/(30000*0.1);
   bar(xx,yy), y=raylpdf(xx,1); line(xx,y)
```

9.2 Statistics

9.2.1 Mean and variance of random variables

Assume that the probability density function of a continuous random variable x is given by $p(x)$. The following formulae can be used to evaluate the mathematical expectation $\mathrm{E}[x]$ and the variance $\mathrm{D}[x]$, respectively:

$$\mathrm{E}[x] = \int_{-\infty}^{\infty} xp(x)\,\mathrm{d}x, \quad \mathrm{D}[x] = \int_{-\infty}^{\infty} (x - \mathrm{E}[x])^2\, p(x)\,\mathrm{d}x. \tag{9.12}$$

Clearly, these two important statistical quantities or statistics can be evaluated through the direct integration method.

Example 9.11 Calculate the mean and variance of Γ distribution with $a > 0, \lambda > 0$ using direct integration method.

Solution The following statements can be used to evaluate the mean and variance of Γ distribution:

```
>> syms x; syms a lam positive
   p=lam^a*x^(a-1)/gamma(a)*exp(-lam*x); m=int(x*p,x,0,inf)
   s=simple(int((x-1/lam*a)^2*p,x,0,inf))
```

The results obtained are $m = \dfrac{a}{\lambda}$ and $s = \dfrac{a}{\lambda^2}$.

For a set of measured data $x_1, x_2, x_3, \cdots, x_n$, the mean and variance are defined as

$$\bar{x} = \frac{1}{n}\sum_{i=1}^{n} x_i, \quad \hat{s}_x^2 = \frac{1}{n}\sum_{i=1}^{n}(x_i - \bar{x})^2. \tag{9.13}$$

It can be shown that the variance $\hat{s}_x^2$ thus defined is biased. An unbiased variance is defined as

$$s_x^2 = \frac{1}{n-1}\sum_{i=1}^{n}(x_i - \bar{x})^2 \tag{9.14}$$

which is referred to as the *standard deviation*, with $s_x \geqslant 0$.

Let a set of measured data be expressed by vector $\boldsymbol{x} = [x_1, x_2, x_3, \cdots, x_n]^{\mathrm{T}}$. The MATLAB functions `mean()`, `var()` and `std()` can be used to evaluate the mean, variance and standard deviation, respectively, with the following syntaxes: `m=mean(x), s2=var(x), s=std(x)`. These three functions can also be used to process the data given in matrices on a column basis. Also these functions can be used to measure the properties of all the elements in a matrix or multi-dimensional array with the command `m=mean(x(:))`.

Example 9.12 Generate a set of 30000 normally distributed random numbers with the mean and standard deviation given by 0.5 and 1.5, respectively. If the total number of points is reduced, what may happen?

Solution The following statements can be used to generate the random numbers. The mean, variance and deviation can be evaluated from

```
>> p=normrnd(0.5,1.5,30000,1); [mean(p), var(p), std(p)]
```

and it can be seen that $\mu = 0.48792268249878$, $\sigma^2 = 2.27484783652451$ and $\bar{s} = 1.50825987$, pretty close to the original exact values. If one reduces the number of points, for instance to 300 points, the following statements can be used and it can be seen that the statistics become $\mu = 0.47448749620068$, $\sigma^2 = 1.91182304623802$, $\bar{s} = 1.38268689378254$, which deviates far away from the expected values.

```
>> p=normrnd(0.5,1.5,300,1); [mean(p), var(p), std(p)]
```

The commonly encountered distribution functions have been summarized earlier. If the distribution is given, the functions such as `normstat()` and `gamstat()` can be used directly to evaluate the mean and variance of the normal and Γ distributions. Similar to the suffix `pdf`, the suffix `stat` is used to indicate the statistics properties. For instance, the syntax of `gamstat()` is $[\mu,\sigma^2]$`=gamstat(`a,λ`)`, where the returned arguments are the mean and variance of the given distribution.

Example 9.13 Evaluate the mean and variance of Rayleigh distribution with b=0.45.

Solution Since Rayleigh distribution is considered, the function `raylstat()` should be used and the mean and variance of the signal can be evaluated as

```
>> [m,s]=raylstat(0.45)
```

with $m = 0.56399136179198$, $s = 0.08691374382403$.

9.2.2 Moments of random variables

Suppose that x is a continuous variable and $p(x)$ is its PDF, its rth *raw moment* and rth *central moment* are defined respectively as

$$\nu_r = \int_{-\infty}^{\infty} x^r p(x)\,\mathrm{d}x, \quad \mu_r = \int_{-\infty}^{\infty} (x-\mu)^r p(x)\,\mathrm{d}x \tag{9.15}$$

and it can be seen that $\nu_1 = \mathrm{E}[x]$, $\mu_2 = \mathrm{D}[x]$.

Example 9.14 Consider the raw moment and central moment problems for Γ distributions in Example 9.11. Summarize the general formula from the first few quantities.

Solution The raw moment of the signal can easily be evaluated from

```
>> syms x; syms a lam positive; p=lam^a*x^(a-1)/gamma(a)*exp(-lam*x);
   for n=1:5, m=int(x^n*p,x,0,inf), end
```

and the first five raw moments are
$\dfrac{a}{\lambda}$, $\dfrac{a}{\lambda^2}(a+1)$, $\dfrac{a}{\lambda^3}(a+1)(a+2)$, $\dfrac{a}{\lambda^4}(a+1)(a+2)(s+3)$, $\dfrac{a}{\lambda^5}(a+1)(a+2)(a+3)(a+4)$

which can be summarized as $\nu_k = \dfrac{1}{\lambda^k}a(a+1)(a+2)\cdots(a+k-1) = \dfrac{1}{\lambda^k}\prod_{i=0}^{k-1}(a+i)$.

In fact these results can also be obtained from

```
>> syms k; m=simple(int((x)^k*p,x,0,inf))
```

with $m = \lambda^{-k}\Gamma(k+a)/\Gamma(a)$, which is exactly the same as the one summarized above.

Similarly, the central moments of the random variable can be evaluated from

```
>> for n=1:7, s=simple(int((x-1/lam*a)^n*p,x,0,inf)), end
```

The first seven central moments obtained are

$$0,\ \frac{a}{\lambda^2},\ \frac{2a}{\lambda^3},\ \frac{3a(a+2)}{\lambda^4},\ \frac{4a(5a+6)}{\lambda^5},\ \frac{5a(3a^2+26a+24)}{\lambda^6},\ \frac{6a(35a^2+154a+120)}{\lambda^7}.$$

No further compact formulae, however, can be summarized from the above results.

For a set of random samples $x_1, x_2, \cdots, x_n$, the rth raw moment and central moment are defined as

$$A_r = \frac{1}{n}\sum_{i=1}^{n} x_i^r, \quad B_r = \frac{1}{n}\sum_{i=1}^{n}(x_i-\bar{x})^r. \tag{9.16}$$

The `moment()` function provided in the Statistics Toolbox can be used to evaluate the central moments of all given order. There is no such function to evaluate the raw moment. In fact, according to the definition, the rth raw moment and central moment can be evaluated from

`A_r=sum(x.^r)/length(x)` and `B_r=moment(x,r)`

Example 9.15 Consider again the random numbers generated earlier. The following statements can be used to evaluate the moments of different orders

```
>> A=[]; B=[]; p=normrnd(0.5,1.5,30000,1); n=1:5;
   for r=n,
      A=[A, sum(p.^r)/length(p)]; B=[B,moment(p,r)];
   end
```

and the moments obtained are

$$\boldsymbol{A} = [0.508053449, 2.5154715759, 3.5456642980, 18.8911497429, 40.791198267]$$

$$\boldsymbol{B} = [0, 2.257353268872, -0.026041939834, 15.381462246655, -1.208702122593]$$

Actually, the analytical method can be used to evaluate the exact values of the moments, with the following statements

```
>> syms x; A1=[]; B1=[]; p=1/(sqrt(2*pi)*1.5)*exp(-(x-0.5)^2/(2*1.5^2));
   for i=1:5
      A1=[A1,vpa(int(x^i*p,x,-inf,inf),12)];
      B1=[B1,vpa(int((x-0.5)^i*p,x,-inf,inf),12)];
   end
```

with $\boldsymbol{A}_1 = [0.5, 2.5, 3.5, 18.625, 40.8125]$, and $\boldsymbol{B}_1 = [0, 2.25, 0, 15.1875, 0]$, which agree very well with the above numerical results.

9.2.3 Covariance analysis of multivariate random variables

Assume that the data pairs (x_1, y_1), (x_2, y_2), (x_3, y_3), $\cdots$, (x_n, y_n) are samples of a function with two variables (x, y). The covariance s_{xy} and correlation coefficient η of the two variables is defined as

$$s_{xy} = \frac{1}{n-1}\sum_{i=1}^{n}(x_i - \bar{x})(y_i - \bar{y}), \quad \eta = \frac{s_{xy}}{s_x s_y} \tag{9.17}$$

and from the above definition, a matrix $\boldsymbol{C}$ can also be defined

$$\boldsymbol{C} = \begin{bmatrix} c_{xx} & c_{xy} \\ c_{yx} & c_{yy} \end{bmatrix} \tag{9.18}$$

where $c_{xx} = \sigma_x^2$, $c_{xy} = c_{yx} = s_{xy}$, and $c_{yy} = \sigma_y^2$. The matrix $\boldsymbol{C}$ is referred to as the *covariance matrix*.

The definition of the covariance matrix with more variables can also be extended from the above definition. The function `cov()` can also be used to evaluate the covariance matrix, with the syntax `C=cov(X)`, where each column in matrix $\boldsymbol{X}$ is regarded as the samples of a random signal. If $\boldsymbol{X}$ is a vector, however, the variance will be returned instead.

Example 9.16 Generate four sets of standard normally distributed random data points. Evaluate the covariance matrix for the four signals.

Solution The random signals can be generated with the `randn()` function, such that it has four columns, each one containing 30000 elements. This means that four signals are defined with each one containing 30000 samples. The covariance matrix can then be obtained, and theoretically, it should be an identity matrix.

```
>> p=randn(30000,4); C=cov(p)
```

The covariance matrix can be found as

$$C = \begin{bmatrix} 1.006377779 & 0.001258868464 & 0.004726044354 & -0.000519010644 \\ 0.001258868464 & 1.003955935 & -0.000945450079 & 0.004802025815 \\ 0.004726044354 & -0.000945450079 & 1.011043483 & -0.01189535769 \\ -0.0005190106435 & 0.004802025815 & -0.01189535769 & 0.9947556372 \end{bmatrix}.$$

9.2.4 Joint PDFs and CDFs of multivariate normal distributions

Assume that a set of n random variables $\xi_1, \xi_2, \cdots, \xi_n$ satisfy normal distribution. Also it is assumed that their means are $\mu_1, \mu_2, \cdots, \mu_n$, from which a mean vector $\boldsymbol{\mu}$ can be defined. The covariance matrix can be represented as $\boldsymbol{\Sigma}^2$. Thus the joint probability density function is defined as

$$p(x_1, x_2, \cdots, x_n) = \frac{1}{\sqrt{2\pi}} \boldsymbol{\Sigma}^{-1} \mathrm{e}^{-\frac{1}{2}(\boldsymbol{x}-\boldsymbol{\mu})^{\mathrm{T}} \boldsymbol{\Sigma}^{-2} (\boldsymbol{x}-\boldsymbol{\mu})} \tag{9.19}$$

where $\boldsymbol{x} = [x_1, x_2, \cdots, x_n]^{\mathrm{T}}$.

The function `mvnpdf()` is provided in Statistics Toolbox to evaluate the joint PDF, with the following syntax `p=mvnpdf(X,μ,Σ^2)`, where $\boldsymbol{X}$ is a matrix with n columns, each representing a random variable.

Example 9.17 If the mean vector and covariance matrix are defined as $\boldsymbol{\mu} = [-1, 2]^{\mathrm{T}}$, $\boldsymbol{\Sigma}^2 = [1, 1; 1, 3]$, draw the joint PDF. If the covariance matrix is a diagonal matrix, draw the surface of new joint PDF.

Solution The random matrix with two columns can be generated from the grid data obtained with the `meshgrid()` function. The function `mvnpdf()` can be used to evaluate the joint PDF as a column vector. The function `reshape()` can be used to restore the grid matrix. The surface of the joint PDF can be obtained as shown in Figure 9.9 (a).

```
>> mu1=[-1,2]; Sigma2=[1 1; 1 3]; % mean vector and covariance matrix
   [X,Y]=meshgrid(-3:0.1:1,-2:0.1:4); xy=[X(:) Y(:)];  % grid data
   p=mvnpdf(xy,mu1,Sigma2); P=reshape(p,size(X));      % joint pdf
   surf(X,Y,P)                   % the surface of the joint pdf
```

If diagonal covariance matrix is studied, the following statements can then be introduced and the joint PDF can be obtained as shown in Figure 9.9 (b).

```
>> Sigma2=diag(diag(Sigma2)); % eliminate the non-diagonal elements
   p=mvnpdf(xy,mu1,Sigma2); P=reshape(p,size(X)); surf(X,Y,P)
```

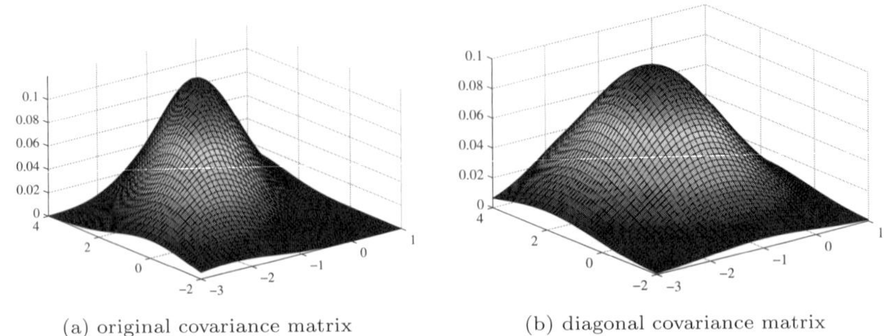

(a) original covariance matrix (b) diagonal covariance matrix

FIGURE 9.9: The joint PDF surface

The `mvnrnd()` function provided in the Statistics Toolbox can be used to generate multivariate pseudo-random numbers of the normal distribution, with the syntax $\boldsymbol{R}$=`mvnrnd(`μ`,`Σ^2`,`m`,`n`)`, where the m sets of random numbers will be returned in an $m \times n$ matrix $\boldsymbol{R}$, with each column corresponding to a random signal.

Example 9.18 Observe the two types of normally distributed random numbers with specifications in Example 9.17.

Solution The following statements can be used to generate the two sets of random numbers, with each set containing 2000 samples. The distribution of the samples on the x-y plane are obtained as shown in Figures 9.10 (a) and (b), respectively. It can be seen that when the covariance matrix is diagonal, there is no relationship between the two variables. Thus the variables are independent.

```
>> mu1=[-1,2]; Sigma2=[1 1; 1 3];
   R1=mvnrnd(mu1,Sigma2,2000); plot(R1(:,1),R1(:,2),'o')
   Sigma2=diag(diag(Sigma2));
   figure; R2=mvnrnd(mu1,Sigma2,2000); plot(R2(:,1),R2(:,2),'o')
```

9.2.5 Monte Carlo solutions to mathematical problems

Buffon's Needle is an old example of the applications of probability. The so-called *Monte Carlo method* is to find the values of uncertain variables using a large number of random quantities, based on random distributions.

Consider the plot shown in Figure 9.11 (a). Generate N sets of two random numbers x_i and y_i satisfying uniform distribution in the interval $[0, 1]$. Assume

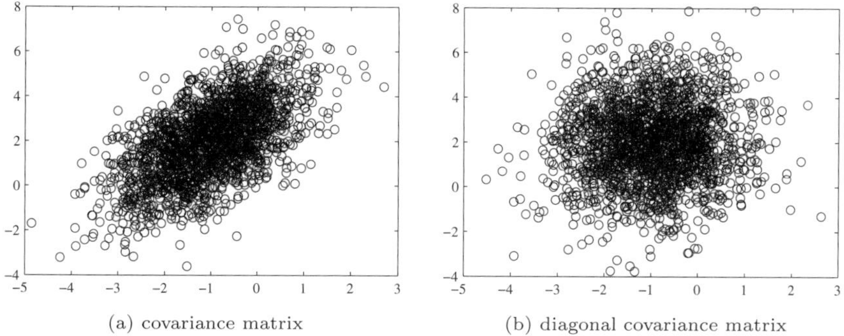

FIGURE 9.10: Distribution of two-dimensional random variables

that the number of points falling in the quarter circle, i.e., $x_i^2 + y_i^2 \leqslant 1$, is N_1. Since the area of the quarter circle is $\pi/4$, while the area of the square is 1, it can be seen that $N_1/N \approx \pi/4$, from which it is found that $\pi \approx 4N_1/N$. If N is large enough, the value of π can be approximated using such a formula.

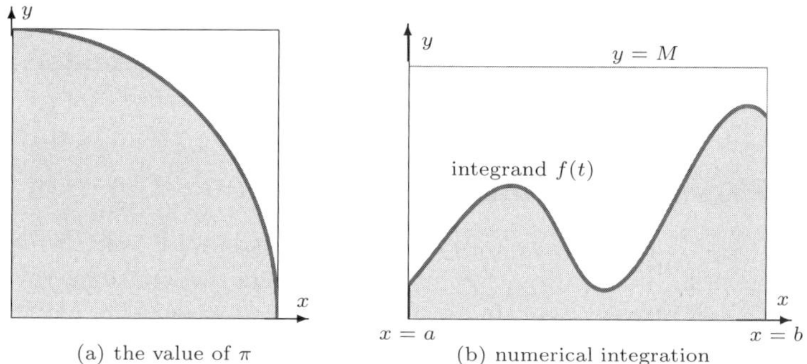

FIGURE 9.11: Applications of Monte Carlo method

Example 9.19 Use Monte Carlo method to evaluate approximately the value π.

Solution The following MATLAB statements can be used

```
>> N=100000; x=rand(1,N); y=rand(1,N);
   i=(x.^2+y.^2)<=1; N1=sum(i);  p=N1/N*4
```

and it is found that $\pi \approx 3.1448$. If the value of N further increases, the precision may increase as well. However, never expect the exact value of π to be obtained this way.

Consider also the integration problem $\displaystyle\int_a^b f(x)\,\mathrm{d}x$ shown in Figure 9.11 (b). Again generate N sets of two random numbers x_i and y_i satisfying the uniform distribution in the intervals $[a, b]$ and $(0, M)$, respectively. For a value of x_i, the numbers of points satisfying $f(x_i) \leqslant y_i$ is counted as N_1. It is found that

$\dfrac{N_1}{N} \approx \dfrac{1}{M(b-a)}\displaystyle\int_a^b f(x)\,dx$, from which the integration can be found that[44]

$$\int_a^b f(x)\,dx \approx \frac{M(b-a)N_1}{N}. \tag{9.20}$$

Example 9.20 Evaluate $\displaystyle\int_1^3 \left[1+e^{-0.2x}\sin(x+0.5)\right] dx$ using Monte Carlo method.

Solution The following statements can be used, and the integration can be evaluated numerically and analytically such that $p = 2.73432$, $I = 2.74393442001018$.

```
>> f=@(x)1+exp(-0.2*x).*sin(x+0.5); a=1; b=3; M=2; N=100000;
   x=a+(b-a)*rand(N,1); y=M*rand(N,1); i=y<=f(x);
   N1=sum(i); p=M*N1*(b-a)/N,
   syms x; I=vpa(int(1+exp(-0.2*x)*sin(x+0.5),x,a,b))
```

It should be noted that the evaluation is only possible when $f(x) \geqslant 0$, otherwise, modified representations should be used.

9.3 Statistical Analysis

9.3.1 Parametric estimation and interval estimation

If a set of data $\boldsymbol{x} = [x_1, x_2, \cdots, x_n]^{\mathrm{T}}$ is measured and it is known that they satisfy a certain distribution, for instance, normal distribution, the function `normfit()` can be used to evaluate the mean μ and variance σ^2 via maximum likelihood method. Also the confidence intervals $\Delta\mu$ and $\Delta\sigma^2$ can be obtained, with the syntax

`[`μ`,`σ^2`,`$\Delta\mu$`,`$\Delta\sigma^2$`]=normfit(`$\boldsymbol{x}$`,`P_{ci}`)`

where the argument P_{ci} is the confidence level, for instance, 0.95 for 95% of confidence. The function `norminv()` can then be used to evaluate the relevant needed quantities. It can be predicted that the bigger the value of P_{ci}, the narrower the confidence interval, and the more accurate the parameters estimated.

Similar to the PDFs, other functions for evaluating the means and variances for other distributions can also be made. For instance, the mean and variance of Γ distribution can be estimated by using `gamfit()`, and those for Rayleigh distribution by using `raylfit()`, as summarized in Table 9.1.

Example 9.21 Generate a set of random numbers with Γ distribution, where $a = 1.5$ and $\lambda = 3$. From the generated data, the parameters can be estimated with different confidence levels, and compare the results.

Solution First, generate a set of 30000 random data. Select the confidence levels at 90%,92%, 95%, 98%. The following statements can be used to estimate the parameters with the given confidence levels

```
>> p=gamrnd(1.5,3,30000,1); Pv=[0.9,0.92,0.95,0.98]; A=[];
   for i=1:length(Pv)
      [a,b]=gamfit(p,Pv(i)); A=[A; Pv(i),a(1),b(:,1)',a(2),b(:,2)'];
   end
```

The results are obtained and shown in Table 9.2. It can be seen that the estimated parameters are not affected with the confidence levels. However, the confidence intervals can be different. Normally, a confidence level of 95% should be used.

TABLE 9.2: Parameter estimation results

confidence	a estimation results			λ estimation results		
level	$\hat{a}$	$a_{\min}$	$a_{\max}$	$\hat{\lambda}$	$\lambda_{\min}$	$\lambda_{\max}$
90%	1.506500556	1.505099132	1.507901979	2.991117941	2.987797191	2.994438691
92%	1.506500556	1.505380481	1.507620631	2.991117941	2.988463861	2.993772021
95%	1.506500556	1.505801226	1.507199886	2.991117941	2.98946084	2.992775042
98%	1.506500556	1.506220978	1.506780134	2.991117941	2.990455465	2.991780417

Now consider the impact of the size of random number group on the estimation, selecting respectively 300, 3000, 30000, 300000, 3000000 random numbers. The confidence level of 95% is assumed; thus the estimation intervals may also change and the variations are shown in Table 9.3.

TABLE 9.3: Parameter estimation results

number of	a estimation results			λ estimation results		
values	$\hat{a}$	$a_{\min}$	$a_{\max}$	$\hat{\lambda}$	$\lambda_{\min}$	$\lambda_{\max}$
300	1.548677954	1.540991679	1.55636423	2.91172985	2.896265076	2.927194623
3000	1.476057561	1.473908973	1.47820615	3.040589493	3.035438607	3.04574038
30000	1.503327455	1.502624743	1.504030167	2.976242793	2.974591027	2.977894559
300000	1.509546583	1.509323617	1.50976955	2.984774009	2.984252596	2.985295421
3000000	1.498005677	1.497935817	1.498075536	3.006048895	3.005882725	3.006215065

```
>> num=[300,3000,30000,300000,3000000]; A=[];
   for i=1:length(num)
      p=gamrnd(1.5,3,num(i),1);
      [a,b]=gamfit(p,0.95); A=[A;num(i),a(1),b(:,1)',a(2),b(:,2)'];
   end
```

It can be seen from the table that, when the number of random samples is small, the estimated parameters may not be satisfactory. Normally 30000 samples are acceptable.

9.3.2 Multivariable linear regression and interval estimation

Assume that the output signal y is the linear combination of n input signals $x_1, x_2, \cdots, x_n$ such that

$$y = a_1 x_1 + a_2 x_2 + a_3 x_3 + \cdots + a_n x_n \tag{9.21}$$

where $a_1, a_2, \cdots, a_n$ are undetermined constants. Assume that m groups of experiments are made with the measured data arranged in the following format:

$$\begin{array}{cccccc} x_{11} & x_{12} & \cdots & x_{1,n} & \Longrightarrow & y_1 \\ x_{21} & x_{22} & \cdots & x_{2,n} & \Longrightarrow & y_2 \\ \vdots & \vdots & \ddots & \vdots & \cdots & \vdots \\ x_{m1} & x_{m2} & \cdots & x_{m,n} & \Longrightarrow & y_m. \end{array} \tag{9.22}$$

The following matrix equation can then be set up

$$\boldsymbol{y} = \boldsymbol{X}\boldsymbol{a} + \boldsymbol{\varepsilon} \tag{9.23}$$

where $\boldsymbol{a} = [a_1, a_2, \cdots, a_n]^{\mathrm{T}}$ is the vector with undetermined constants. If the original data was obtained in experiments, there exist errors in each equation in (9.21) denoted by $\boldsymbol{\varepsilon} = [\varepsilon_1, \varepsilon_2, \cdots, \varepsilon_m]^{\mathrm{T}}$. Also the vector $\boldsymbol{y} = [y_1, y_2, \cdots, y_m]^{\mathrm{T}}$ is the observed values, and the matrix $\boldsymbol{X}$ is composed of observed independent variables such that

$$\boldsymbol{X} = \begin{bmatrix} x_{11} & x_{12} & \cdots & x_{1,n} \\ x_{21} & x_{22} & \cdots & x_{2,n} \\ \vdots & \vdots & \ddots & \vdots \\ x_{m1} & x_{m2} & \cdots & x_{m,n} \end{bmatrix}. \tag{9.24}$$

Assume that the objective function for the problem is to have the minimized sum of the squared errors, i.e., $J = \min \boldsymbol{\varepsilon}^{\mathrm{T}}\boldsymbol{\varepsilon}$, then the undetermined constant vector $\boldsymbol{a}$ of the linear regression model can be obtained from

$$\hat{\boldsymbol{a}} = (\boldsymbol{X}^{\mathrm{T}}\boldsymbol{X})^{-1}\boldsymbol{X}^{\mathrm{T}}\boldsymbol{y}. \tag{9.25}$$

From the knowledge of linear algebra illustrated in Chapter 4, the least squares solution to the above equation can be obtained from `a=X\y` or more formally `a=inv(X'*X)*X'*y`.

A multivariable linear regression function `regress()` for parameter estimation and confidence interval estimation is provided in MATLAB, and the syntax of the function is `[â,a_ci]=regress(y,X,α)`, where $1-\alpha$ is the confidence level specified by the user.

Example 9.22 Assume that the linear regression model is $y = x_1 - 1.232x_2 + 2.23x_3 + 2x_4 + 4x_5 + 3.792x_6$. With 120 measured data points x_i, the output vector $\boldsymbol{y}$ can be computed first. Based on this information, the undetermined parameters a_i can be estimated, with given confidence intervals.

Solution Linear regression can be used to process the measured data. The following statements can be used to construct the matrix $\boldsymbol{X}$ and the vector $\boldsymbol{y}$. The least squares method can be used to estimate the undetermined coefficients $\boldsymbol{a}$ given a confidence level

```
>> a=[1 -1.232 2.23 2 4,3.792]'; X=randn(120,6); y=X*a;
   [a,aint]=regress(y,X,0.02)
```

The estimated parameter and the confidence interval can be obtained as

$$\boldsymbol{a} = \begin{bmatrix} 1 \\ -1.232 \\ 2.23 \\ 2 \\ 4 \\ 3.792 \end{bmatrix}, \quad \boldsymbol{a}_{\rm int} = \begin{bmatrix} 1 & 1 \\ -1.232 & -1.232 \\ 2.23 & 2.23 \\ 2 & 2 \\ 4 & 4 \\ 3.792 & 3.792 \end{bmatrix}.$$

It can be seen that there is no error in the estimated results, and the confidence interval is 100%. If the samples are corrupted by noises, for instance, normally distributed noise $N(0, 0.5)$ can be added to the samples within the interval. The following statements can then be used to perform linear regression analysis. The estimated parameters and their confidence intervals can be obtained and shown in Figure 9.12 (a), with the help of the `errorbar()` function.

```
>> yhat=y+sqrt(0.5)*randn(120,1); [a,aint]=regress(yhat,X,0.02)
   errorbar(1:6,a,aint(:,1)-a,aint(:,2)-a)
```

The parameters and the intervals are obtained as

$$\boldsymbol{a} = \begin{bmatrix} 1.03887888369425 \\ -1.20360949260729 \\ 2.19454416841833 \\ 1.89146235051598 \\ 34.0628445587228 \\ 8.70540411609337 \end{bmatrix}, \quad \boldsymbol{a}_{\rm int} = \begin{bmatrix} 0.92959508774995 & 1.14816267963855 \\ -1.313314026578 & 1.09390495863658 \\ 2.07413724843311 & 2.31495108840356 \\ 1.77199410730299 & 2.01093059372896 \\ 33.9418338141614 & 34.1838553032841 \\ 8.59651534574785 & 8.8142928864389 \end{bmatrix}.$$

Reducing the variance to 0.1, the estimated parameters are obtained using the following statements as shown in Figure 9.12 (b). It can be seen that the estimation results are even more accurate.

```
>> yhat=y+sqrt(0.1)*randn(120,1); [a,aint]=regress(yhat,X,0.02);
   errorbar(1:6,a,aint(:,1)-a,aint(:,2)-a)
```

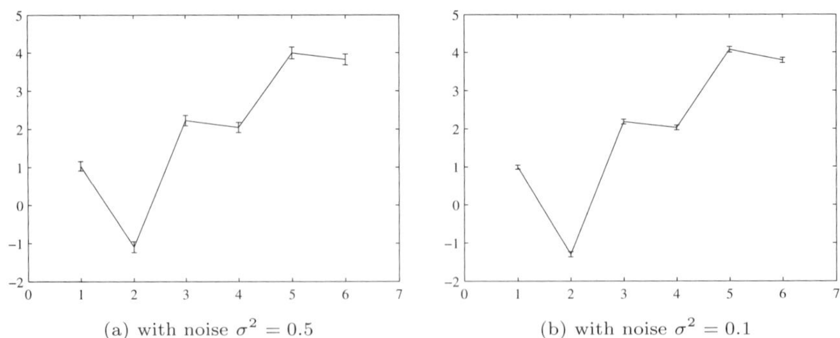

(a) with noise $\sigma^2 = 0.5$ (b) with noise $\sigma^2 = 0.1$

FIGURE 9.12: Parameter estimations with confidence intervals

9.3.3 Nonlinear least squares parametric and interval estimations

Assume that a set of data $x_i, y_i, i = 1, 2, \cdots, N$ is measured satisfying a given prototype functional relationship $\hat{y}(x) = f(\boldsymbol{a}, x)$, where $\boldsymbol{a}$ is the vector containing undetermined coefficients. Since the measured data is corrupted with noises, the original function can be written as $\hat{y}(x) = f(\boldsymbol{a}, x) + \varepsilon$, where ε represents the *residual error*. An objective function can be introduced such that

$$I = \min_{\boldsymbol{a}} \sum_{i=1}^{N} [y_i - \hat{y}(x_i)]^2 = \min_{\boldsymbol{a}} \sum_{i=1}^{N} [y_i - f(\boldsymbol{a}, x_i)]^2. \tag{9.26}$$

Minimizing the above objective function, the undetermined parameters $\boldsymbol{a}$ can be estimated. Substituting the estimated $\boldsymbol{a}$ back to the prototype function, the residue error $\varepsilon_i = y_i - f(\boldsymbol{a}, x_i)$ can be obtained. Similar to the least squares estimation shown in Section 8.3.5, the least squares fitting with Gauss-Newton's algorithm is implemented in the function `nlinfit()`. The Jacobian vector $\boldsymbol{j}_i$ of the residual error with respect to $\boldsymbol{a}$ can also be obtained. Confidence interval estimation can also be obtained with the function `nlparci()`, and the results can be obtained with 95% confidence level. The syntax of the function is

```
[a,r,J]=nlinfit(x,y,fun,a0) % least squares estimation
c=nlparci(a,r,J)            % confidence interval with 95% confidence
```

where $\boldsymbol{x}$ and $\boldsymbol{y}$ contain the measured data. The function *fun* represents the prototype function, which can either be described by an M-function, an anonymous function or an inline function. The initial values $\boldsymbol{a}_0$ of the estimated parameters should also be specified. It can be seen that the input arguments are the same as the `lsqcurvefit()` function. The returned argument $\boldsymbol{a}$ vector contains the estimated parameters, and $\boldsymbol{r}$ returns the residue error vector for the estimation. Matrix $\boldsymbol{J}$ is the Jacobian and based on the information, the estimation of confidence intervals $\boldsymbol{c}$ can be found. The parametric estimation and confidence interval estimation are illustrated by the following examples.

Example 9.23 The parametric estimation can be performed to the measured data in Example 8.25 with least squares fitting method. The 95% confidence level is assumed. Solve the parametric estimation and confidence interval estimation problems.

Solution Assume that the prototype function is in the form of $y(x) = a_1 e^{-a_2 x} + a_3 e^{-a_4 x} \sin(a_5 x)$, where the parameters a_i are the undetermined coefficients. An anonymous function can be used to describe the prototype function and the undetermined constants can be evaluated by using `nlinfit()` function

```
>> f=@(a,x)a(1)*exp(-a(2)*x)+a(3)*exp(-a(4)*x).*sin(a(5)*x);
   x=0:0.1:10; y=f([0.12,0.213,0.54,0.17,1.23],x);
   [a,r,j]=nlinfit(x,y,f,[1;1;1;1;1]); ci=nlparci(a,r,j)
```

and it is found that

$$\boldsymbol{a} = \begin{bmatrix} 0.11999999763418 \\ 0.21299999458274 \\ 0.54000000196818 \\ 0.17000000068705 \\ 1.22999999996315 \end{bmatrix}, \quad c_i = \begin{bmatrix} 0.11999999712512 & 0.11999999814323 \\ 0.21299999340801 & 0.21299999575747 \\ 0.54000000124534 & 0.54000000269101 \\ 0.17000000036077 & 0.17000000101332 \\ 1.22999999978603 & 1.23000000014028 \end{bmatrix}.$$

It can be seen that the results obtained are more accurate than the default results obtained with `lsqcurvefit()`. However, more accurate results are not possible. The associated `nlparci()` function can be used to get the confidence intervals for the parameters, with 95% of confidence level.

If the original samples y_i are corrupted with noises uniformly distributed in the interval $[0, 0.02]$, the following statements can be used to estimate the parameters and confidence intervals for the new samples

```
>> y=f([0.12,0.213,0.54,0.17,1.23],x)+0.02*rand(size(x));
   [a,r,j]=nlinfit(x,y,f,[1;1;1;1;1]); ci=nlparci(a,r,j)
   errorbar(1:5,a,ci(:,1)-a,ci(:,2)-a)
```

and it can be found that

$$\boldsymbol{a} = \begin{bmatrix} 0.12281531581639 \\ 0.17072641296744 \\ 0.55113088779121 \\ 0.17347639675132 \\ 1.2291686258648 \end{bmatrix}, \quad c_i = \begin{bmatrix} 0.11857720435195 & 0.12705342728083 \\ 0.16221631527879 & 0.17923651065609 \\ 0.54465309442893 & 0.55760868115349 \\ 0.17055714192171 & 0.17639565158094 \\ 1.22755955648343 & 1.23077769524618 \end{bmatrix}.$$

The estimated parameters and their confidence intervals are shown in Figure 9.13.

Example 9.24 Assume that the prototype function is

$$f(\boldsymbol{a}, \boldsymbol{x}) = (a_1 x_1^3 + a_2)\sin(a_3 x_2 x_3) + (a_4 x_2^3 + a_5 x_2 + a_6).$$

Solve the multivariable nonlinear regression problem for a set of noisy data with the function `nlinfit()`.

Solution Assume that the initial values of a_i are all 1's. The following statements can be used to define the function f, and to generate a data set $\boldsymbol{X}$ serving as the observed data.

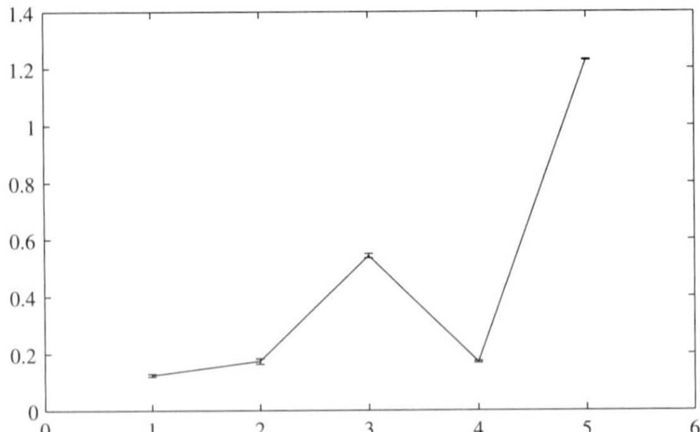

FIGURE 9.13: Estimated parameters and confidence intervals

```
>> a=[1;1;1;1;1;1]';
   f=@(a,x)(a(1)*x(:,1).^3+a(2)).*sin(a(3)*x(:,2).*x(:,3))+...
     (a(4)*x(:,3).^3+a(5)*x(:,3)+a(6));
   X=randn(120,4); y=f(a,X)+sqrt(0.2)*randn(120,1);
```

With the observed data, the following statements can be used to estimate a_i, and the results can be shown in Figure 9.14 (a) where we can observe that the fitting is satisfactory.

```
>> [ahat,r,j]=nlinfit(X,y,f,[0;2;3;2;1;2]); ci=nlparci(ahat,r,j);
   y1=f(ahat,X); plot([y y1])
```

The estimated parameters and the confidence intervals are

$$\boldsymbol{a}=\begin{bmatrix}1.04839959073146\\1.01882085899938\\0.98446778587739\\0.99107092667601\\1.02519403669663\\1.05040136072101\end{bmatrix},\quad \boldsymbol{c}_i=\begin{bmatrix}0.968935454453576 & 1.12786372692717\\0.88884528207121 & 1.14879643592754\\0.89167073268603 & 1.07726483906874\\0.96675355288178 & 1.01538830047025\\0.9270517440877 & 1.12333632930555\\0.99419821923406 & 1.10660450220796\end{bmatrix}.$$

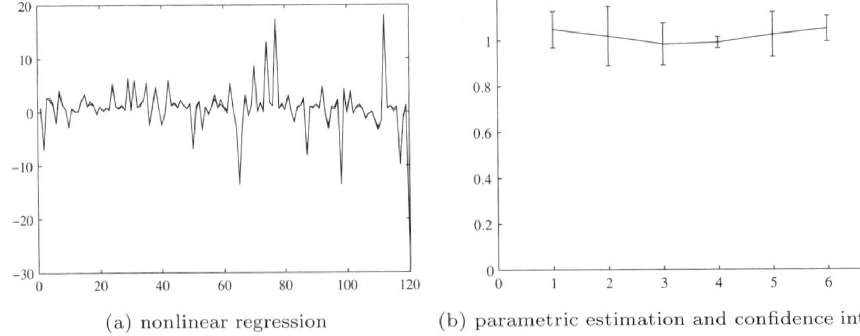

(a) nonlinear regression (b) parametric estimation and confidence interval

FIGURE 9.14: Nonlinear regression for multivariable functions

The results are exactly the same as the ones obtained previously. With the `nlparci()` function, the confidence interval can also be found. The confidence intervals with 95% confidence level are visualized in Figure 9.14 (b).

```
>> errorbar(1:6,ahat,ci(:,1)-ahat,ci(:,2)-ahat)
```

9.4 Statistic Hypothesis Tests

9.4.1 Basic concept and procedures for statistic hypothesis test

For a given data set, we can assume certain statistic properties, for instance, certain distribution. How to check the assumption made on the data set is also known as *hypothesis test.* Hypothesis test is very important in statistics. For instance, someone may propose a hypothesis that the average lifetime of a certain light bulb product is over 3000 hours. How to check whether this hypothesis is correct? The exact way to check it is to turn all the bulbs on, and measure the lifetime of all the bulbs. This is of course not a feasible solution to the test. In statistics, one may randomly select a certain number of bulbs for the hypothesis test. The following example is used to illustrate the procedures of hypothesis test.

Example 9.25 Suppose it is known that the average strength of a product is $\mu_0 = 9.94$kg. Now a new manufacturing technique is adopted. Two hundred pieces of the product were selected randomly and the average strength is measured as $\bar{x} = 9.73$kg, with a standard deviation $s = 1.62$kg. Check whether the new manufacturing technique will affect the average strength of the product.

Solution To solve this type of hypothesis test problem, the following procedures can be followed:

(i) Establish two hypotheses

$$\begin{cases} \mathscr{H}_0 : \mu = \mu_0 & \text{\% no significant change} \\ \mathscr{H}_1 : \text{reject the hypothesis } \mathscr{H}_0 \end{cases}$$

(ii) Define a statistical variable u

$$u = \frac{\sqrt{n}(\bar{x} - \mu_0)}{s} \tag{9.27}$$

satisfying the standard normal distribution $N(0, 1)$. For this example, the statistical variable u can be computed from

```
>> n=200; mu0=9.94; xbar=9.73; s=1.62; u=sqrt(n)*(mu0-xbar)/s
```

with $u = 1.83323980307623$.

(iii) Assign a significance level. Note that statistical method cannot be used to draw any definite conclusion. Any decision on the hypothesis, accept or eject, may have errors in statistical sense. Thus the introduction of the significance level α establishes the probability of wrong acceptance. Of course, one cannot assign $\alpha = 0$. Normally a significance level of $\alpha = 5\%$ or $\alpha = 2\%$ can be used. In other words, one can accept or reject a hypothesis with 95% (or 98%) confidence level.

(iv) For a given significance level α, compute the inverse normal distribution function $K_{\alpha/2}$ such that

$$\int_{-K_{\alpha/2}}^{K_{\alpha/2}} \frac{1}{\sqrt{2\pi}} \mathrm{e}^{-\frac{x^2}{2}} \mathrm{d}x < 1 - \alpha. \tag{9.28}$$

Given α, the value of $K_{\alpha/2}$ can easily be obtained with the following statements

```
>> alpha=[0.01:0.01:0.05];
   K=norminv(1-alpha/2,0,1); [alpha' K']
```

and the (α_i, K_i) pairs are obtained as $(0.01, 2.5758)$, $(0.02, 2.3263)$, $(0.03, 2.1701)$, $(0.04, 2.0537)$, $(0.05, 1.96)$.

(v) The value of u in (9.27) can be obtained and $|u| < K_{\alpha/2}$. Thus the hypothesis $\mathscr{H}_0$ cannot be rejected.

Further increasing α, for instance, $\alpha = 0.07, 0.09$, the H becomes 1. Thus the hypothesis can be rejected. However, since the confidence level is decreased, the probability of error conclusion is large, which means that the results may not be reliable.

9.4.2 Solving hypothesis test problems in MATLAB

MATLAB solution to a simple hypothesis test has been shown in the previous example. In fact, many MATLAB functions are provided in the Statistics Toolbox to solve different hypothesis test problems, for instance, mean value test for normal distributions, normality test and arbitrary given distribution test. In this subsection, hypothesis tests based on the Statistics Toolbox are presented.

Hypothesis test of the mean of normal distribution

Given a set of data satisfying the normal distribution with its standard deviation σ known. The hypothesis $\mathscr{H}_0$ is made to claim that the mean of the distribution is μ. This hypothesis test can be performed through Z test with the function `ztest()`, where `[`H`,`s`,`$\mu_{\rm ci}$`]=ztest(`$\boldsymbol{X}$`,`μ`,`σ`,`α`)`, with H the test result. When $H = 0$, it means that the hypothesis $\mathscr{H}_0$ cannot be rejected, otherwise the hypothesis should be rejected, with a confidence level of α. The value of s is the significance level, and $\mu_{\rm ci}$ is the confidence interval of the mean.

If the standard deviation for the normal distribution is not known, the T test can be used to perform hypothesis test on the means. The MATLAB function `ttest()` can be used to perform such a test, with `[`H`,`s`,`$\mu_{\rm ci}$`]=ttest(`$\boldsymbol{X}$`,`μ`,`α`)`.

Example 9.26 Generate a set of random numbers satisfying the normal distribution. Perform hypothesis test on the mean of the random numbers.

Solution Let us first generate 400 random numbers satisfying $N(1,2^2)$. Since the standard deviation is 2, the hypothesis $\mathscr{H}_0 : \mu = 1$ can be established. With the following MATLAB statements

```
>> r=normrnd(1,2,400,1); [H,p,ci]=ztest(r,1,2,0.02)
```

it can be found that $H = 0, p = 0.43594320476, \boldsymbol{c}_i = [0.8453, 1.3105]$. Since $H = 0$, the hypothesis cannot be rejected. In other words, the hypothesis can be accepted with 98% of confidence.

If the hypothesis is changed to $\mathscr{H}_0 : \mu = 0.5$, the following statements can be used and the value of H is 1. In this case, the hypothesis $\mathscr{H}_0$ should be rejected. The confidence interval for the mean value remains the same.

```
>> [H,p,ci]=ztest(r,0.5,2,0.02)
```

It is found that $H = 1, p = 7.5118\times 10^{-9}, \boldsymbol{c}_i = [0.845, 1.3105]$.

If the standard deviation is not known, the T test can be used to test the $\mathscr{H}_0 : \mu = 1$ hypothesis by using the following MATLAB statement

```
>> [H,p,ci]=ttest(r,1,0.02)
```

and the results obtained are $H = 0, p = 0.4517, \boldsymbol{c}_i = [0.8363534005, 1.3194589956]$, which means that the hypothesis can be accepted with 98% of confidence.

Hypothesis test of normality

Testing whether a set of random numbers satisfies the normal distribution can be performed with MATLAB directly. Two functions, `jbtest()` and `lillietest()` are provided in the Statistics Toolbox, which implements the Jarque-Bera and Lilliefors hypothesis test algorithms, respectively[45]. They can be used to test whether the random samples are normal or not. The syntaxes of the two functions are

```
[H,s]=jbtest(X,α)         % Jarque-Bera test
[H,s]=lillietest(X,α)     % Lilliefors test
```

Example 9.27 A batch of light bulbs produced by a factory may have different lumen levels. The lumen here is regarded as a random variable ξ. It is assumed that ξ satisfies the normal distribution $N(\mu,\sigma^2)$. Now take 120 samples randomly from the products and the lumen levels of the samples are given in Table 9.4. Verify whether the normal distribution assumption is acceptable.

Solution The data should be entered into MATLAB workspace first and the function `jbtest()` can be used

```
>> X=[216,203,197,208,206,209,206,208,202,203,206,213,218,207,208,...
      202,194,203,213,211,193,213,208,208,204,206,204,206,208,209,...
      213,203,206,207,196,201,208,207,213,208,210,208,211,211,214,...
      220,211,203,216,224,211,209,218,214,219,211,208,221,211,218,...
```

TABLE 9.4: The lumen levels of the test samples

216	203	197	208	206	209	206	208	202	203	206	213	218	207	208	202	194	203	213	211
193	213	208	208	204	206	204	206	208	209	213	203	206	207	196	201	208	207	213	208
210	208	211	211	214	220	211	203	216	224	211	209	218	214	219	211	208	221	211	218
218	190	219	211	208	199	214	207	207	214	206	217	214	201	212	213	211	212	216	206
210	216	204	221	208	209	214	214	199	204	211	201	216	211	209	208	209	202	211	207
202	205	206	216	206	213	206	207	200	198	200	202	203	208	216	206	222	213	209	219

```
        218,190,219,211,208,199,214,207,207,214,206,217,214,201,212,...
        213,211,212,216,206,210,216,204,221,208,209,214,214,199,204,...
        211,201,216,211,209,208,209,202,211,207,202,205,206,216,206,...
        213,206,207,200,198,200,202,203,208,216,206,222,213,209,219];
    [H,p]=jbtest(X,0.05)
```

The obtained results are $H = 0$, $p = 0.7281$, which means that the experimental data satisfies a normal distribution.

Having shown that the distribution is normal, the mean and variance with confidence intervals can be estimated with the use of the `normfit()` function

```
>> [mu1,sig1,mu_ci,sig_ci]=normfit(X,0.05); mu=[mu1,mu_ci']
   sig=[sig1, sig_ci']
```

and the estimated parameters are $\mu = 208.8167, \sigma^2 = 6.3232$, and the corresponding confidence intervals are $(207.6737, 209.9596)$, and $(5.6118, 7.2428)$, respectively.

Example 9.28 Generate a set of random numbers satisfying Γ distribution. Verify whether the same random numbers also satisfy a normal distribution using the hypothesis test method. Of course, we know that it does not satisfy normal distribution and the result of hypothesis test should be 1.

Solution The following statements can be used to generate a set of pseudo-random numbers satisfying the Γ distribution. The function `jbtest()` can then be used and the following results can be obtained.

```
>> r=gamrnd(1,3,400,1); [H,p,c,d]=jbtest(r,0.05)
```

It can be found that $H = 1$, $p = 0$, i.e., the hypothesis $\mathscr{H}_0$ should be rejected.

Kolmogorov-Smirnov test for other distributions

The Jarque-Bera and Lilliefors hypothesis test algorithms can only be used to test whether the set of data to be tested satisfies normal distribution or not. They cannot be used for testing other distributions. The Kolmogorov-Smirnov test algorithm is an effective algorithm which can be used to test whether it satisfies arbitrarily given distributions. The `kstest()` function can be used `[H,s]=kstest(X,cdffun,α)`, where the matrix *cdffun* is composed of two columns, one for the independent variable, and the other for the CDF of the target distribution. By establishing the *cdffun* matrix, the existing function or the user-defined functions can be used. Thus this MATLAB function can be used to test arbitrary distribution function.

Example 9.29 Test whether the random numbers generated in Example 9.28 satisfy the Γ distribution.

Solution Assume that the data satisfy Γ distribution. Then the function `gamfit()` can be used to estimate the two parameters a and λ

```
>> r=gamrnd(1,3,400,1); alam=gamfit(r)
```

and $\hat{a} = 1.0456$ and $\hat{\lambda} = 3.2868$. The CDF of Γ distribution is computed by `gamcdf(sort(r),alam(1),alam(2))`. The results can be given in the `kstest()` function so that the hypothesis test can be performed

```
>> r=sort(r);
   [H,p]=kstest(r,[r gamcdf(r,alam(1),alam(2))],0.05)
```

and for this example, $H = 0$ and $p = 0.8772$. Thus the hypothesis can be accepted which means that the samples satisfy Γ distribution, with $\hat{a} = 1.0456$ and $\hat{\lambda} = 3.2868$, which validates the distribution test problem. In fact, the samples are generated by Γ distribution with $a = 1$ and $\lambda = 3$.

9.5 Analysis of Variance and Its Computation

Analysis of variance, also known as *ANOVA*, is a statistical analysis approach proposed by the British statistician R. A. Fischer. It can be used in many fields, such as in medical research, scientific tests and quality control.

Since the classification of the test samples are different, the method of variance analysis may also be different. Commonly used classification methods include one-way method, two-way method and n-way method. The analysis of variance method and its solutions in MATLAB will be demonstrated via several examples.

9.5.1 One-way ANOVA

One-way analysis of variance is to check whether the external factor has a significant impact on the observations. Assume that for a certain disease, N types of medicines can be tested. The target of the analysis is determining whether the medicines have the same healing effect. In mathematics words, whether each group has the same mean.

To do the test, the patients are divided randomly into N groups, with m patients each. The healing time can be measured and denoted as $y_{i,j}$, where i is the group number, $i = 1, 2, \cdots, N$, and j is the patient number, $j = 1, 2, \cdots, m$. Using the colon notation in MATLAB, the measured data in the ith group for all patients is $y_{i,:}$, and the data for the jth patient for all groups are $y_{:,j}$. Also $\bar{y}_{i,:}$ denotes the average healing time in the ith group, while $\bar{y}_{:,:}$ for the average healing time for all the patients. The standard variance analysis table can be constructed as shown in Table 9.5. Based on the table, certain necessary rules can be extracted.

TABLE 9.5: Variance analysis table

source	sum of squares	DOF	mean squares	F	probability p
factor	$\text{SSA}=\sum_i n_i\bar{y}_{i,:}^2-N\bar{y}_{:,:}^2$	$I-1$	$\text{MSSA}=\dfrac{\text{SSA}}{I-1}$	$\dfrac{\text{MSSA}}{\text{MSSE}}$	$p=P(F_{I-1,N-I}>c)$
error	$\text{SSE}=\sum_i\sum_k y_{i,k}^2-\sum_i n_i\bar{y}_{i,:}^2$	$N-I$	$\text{MSSE}=\dfrac{\text{SSE}}{N-I}$		
total	$\text{SST}=\sum_i\sum_k y_{i,k}^2-N\bar{y}_{:,:}^2$	$N-1$			

Hypothesis tests can also be used in the analysis of variance, that is, to propose $\mathscr{H}_0$ assuming that all the average observed measures are the same. The most important quantities are the last two columns of the table, the statistic F and the probability p, which can be evaluated from the inverse distribution function. If the value of the probability $p<\alpha$, with α the selected confidence level, the hypothesis $\mathscr{H}_0$ should be rejected. Otherwise, the hypothesis cannot be rejected.

A function `anova1()` provided in the Statistics Toolbox can be used to perform one-way analysis of variance for the test data. The syntax of the function is `[p,tab,stats]=anova1(X)`, where $\boldsymbol{X}$ is the data set to be analyzed. The experimental data should be given as an $m\times n$ matrix, with each column corresponding to the measured data of the same group. The probability p, variance analysis table `tab` in the format of Table 9.5 are returned. The other returned argument `stats` is the structured data containing the statistic quantities. Two windows will also be automatically opened, one displaying the table shown in Table 9.5 and the other showing the boxed plot.

Example 9.30 Suppose that there are five medicines to cure a certain disease. Assume that there are 30 patients, randomly divided in 5 test groups, with 6 patients in each test group. For each test group, only one medicine is used. The healing time for each patient is recorded as shown in Table 9.6. Conclude whether there is significant difference in the effect. (Data and example source, Reference [46].)

Solution From the given table, matrix $\boldsymbol{A}$ can be established in MATLAB and the mean value of each column is obtained as $[7.5, 5, 4.3333, 5.1667, 6.1667]$. The analysis of variance is then carried out

```
>> A=[5,4,6,7,9; 8,6,4,4,3; 7,6,4,6,5; 7,3,5,6,7; 10,5,4,3,7; 8,6,3,5,6];
   mean(A), [p,tbl,stats]=anova1(A)
```

TABLE 9.6: Experimental data on healing time (days)

patient number	m 1	m 2	m 3	m 4	m 5	patient number	m 1	m 2	m 3	m 4	m 5
1	5	4	6	7	9	2	8	6	4	4	3
3	7	6	4	6	5	4	7	3	5	6	7
5	10	5	4	3	7	6	8	6	3	5	6

and the probability is returned as $p = 0.01359$. Two windows are opened automatically by function `anova1()`, as shown respectively in Figures 9.15 (a) and (b). The value of probability is $p = 0.0136 < \alpha$, where $\alpha = 0.02$ or 0.05. Thus the hypothesis is to be rejected, i.e., it can be concluded that there is no significant difference among the medicines. The results are exactly the same as the one obtained by SAS program[47]. In fact, from the boxed plot obtained, the healing time of medicine 3 is significantly shorter than medicine 1.

ANOVA Table

Source	SS	df	MS	F	Prob>F
Columns	36.4667	4	9.11667	3.9	0.0136
Error	58.5	25	2.34		
Total	94.9667	29			

(a) variance table

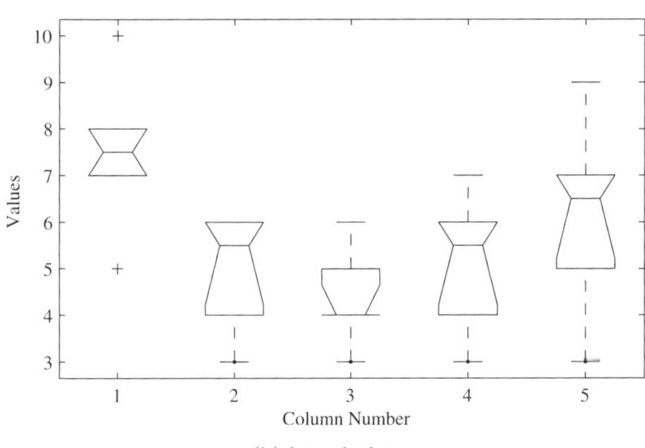

(b) boxed plot

FIGURE 9.15: Results of one-way analysis of variance

9.5.2 Two-way ANOVA

If there are two factors which may affect the statistical properties of a certain phenomenon, the concept of two-way variance analysis can be used. The measured data $y_{i,j,k}$ are expressed as a three-dimensional array $y(i,j,k)$ in MATLAB.

Based on the two-way analysis, three hypotheses are introduced:

$$\begin{cases} \mathscr{H}_1: \ \alpha_1 = \alpha_2 = \cdots = \alpha_I, \ \alpha_i \text{ is the effect when factor 1 acts alone} \\ \mathscr{H}_2: \ \beta_1 = \beta_2 = \cdots = \beta_J, \ \beta_j \text{ is the effect when factor 2 acts alone} \\ \mathscr{H}_3: \ \gamma_1 = \gamma_2 = \cdots = \gamma_{IJ}, \ \gamma_k \text{ is the effect when both factors act.} \end{cases} \tag{9.29}$$

For the two-way analysis of variance, the ANOVA table constructed is shown in Table 9.7, where the interactive SSAE effect can be evaluated from (9.30).

TABLE 9.7: Table of two-way analysis of variance

source	square of sums	DOF	mean squared error	F	p
factor A	$\text{SSA}=JK\sum_i \bar{y}_{i,:,:}^2 - IJK\bar{y}_{:,:,:}^2$	$I-1$	$\text{MSSA}=\dfrac{\text{SSA}}{I-1}$	$\dfrac{\text{MSSA}}{\text{MSSE}}$	p_A
factor B	$\text{SSB}=IK\sum_i \bar{y}_{:,j,:}^2 - IJK\bar{y}_{:,:,:}^2$	$J-1$	$\text{MSSB}=\dfrac{\text{SSB}}{J-1}$	$\dfrac{\text{MSSB}}{\text{MSSE}}$	p_B
interaction	SSAB see (9.30)	$(I-1)(J-1)$	$\text{MSSAB}=\dfrac{\text{SSAB}}{(I-1)(J-1)}$	$\dfrac{\text{MSSAB}}{\text{MSSE}}$	p_{AB}
errors	$\text{SSE}=\sum_{ijk} y_{i,j,k}^2 - K\sum_i\sum_j \bar{y}_{i,j,:}^2$	$IJ(K-1)$	$\text{MSSE}=\dfrac{\text{SSE}}{IJ(K-1)}$		
total	$\text{SST}=\sum_{ijk} y_{i,j,k}^2 - IJK\bar{y}_{:,:,:}^2$	$IJK-1$			

$$\text{SSAB} = K\sum_{ij}\bar{y}_{i,j,:}^2 - JK\sum_i \bar{y}_{i,:,:}^2 - IK\sum_j \bar{y}_{:,j,:}^2 + IJK\bar{y}_{:,:,:}^2. \tag{9.30}$$

The definitions of the three probabilities are

$$\begin{cases} p_A = P\left(F_{[I-1,IJ(K-1)]} > c_1\right) & \text{if } p_A < c_1 \text{ reject hypothesis } \mathscr{H}_1 \\ p_B = P\left(F_{[J-1,IJ(K-1)]} > c_2\right) & \text{if } p_B < c_2 \text{ reject hypothesis } \mathscr{H}_2 \\ p_{AB} = P\left(F_{[(I-1)(J-1),IJ(K-1)]} > c_3\right) & \text{if } p_{AB} < c_3 \text{ reject hypothesis } \mathscr{H}_3 \end{cases} \tag{9.31}$$

Two-way analysis of variance problems can be solved with the `anova2()` function `[p,tab,stats]=anova2(X)`, which is similar to `anova1()`.

Example 9.31 Suppose that there are three species of pines planted in four different living conditions. Select randomly five samples in each group of subjects for measurement. The chest radii of the trees are measured and the data are given in Table 9.8. Perform two-way analysis of variance on the data and check whether there exists significant differences (Data source, book [46]).

Solution The data in the table can be entered into MATLAB workspace. Calling the function `anova2()`, the results are obtained as shown in Figure 9.16, which are exactly the same as the ones in Reference [46].

TABLE 9.8: Measured data for the pines

pine species	living conditions																			
	1					2					3					4				
1	23	15	26	13	21	25	20	21	16	18	21	17	16	24	27	14	17	19	20	24
2	28	22	25	19	26	30	26	26	20	28	19	24	19	25	29	17	21	18	26	23
3	18	10	12	22	13	15	21	22	14	12	23	25	19	13	22	16	12	23	22	19

```
>> B=[23,15,26,13,21,25,20,21,16,18,21,17,16,24,27,14,17,19,20,24;
      28,22,25,19,26,30,26,26,20,28,19,24,19,25,29,17,21,18,26,23;
      18,10,12,22,13,15,21,22,14,12,23,25,19,13,22,16,12,23,22,19];
   anova2(B',5);
```

ANOVA Table

Source	SS	df	MS	F	Prob>F
Columns	355.6	2	177.8	9.68	0.0003
Rows	49.65	3	16.55	0.9	0.4478
Interaction	106.4	6	17.733	0.97	0.4588
Error	882	48	18.375		
Total	1393.65	59			

FIGURE 9.16: Results of two-way analysis of variance

It can be seen from the results that, since p_{A} is very small, the hypothesis $\mathscr{H}_1$ should be rejected. It can be concluded that factor A has a significant impact upon the phenomenon observed, i.e., the tree species has significant impact on the chest radius of the trees. The other two hypothesis cannot be rejected.

9.5.3 n-way ANOVA

Similar to two-way analysis of variance presented earlier, three-way and even n-way analysis of variance are supported in the Statistics Toolbox with the function `manova1()`. Interested readers may refer to References [47, 48].

Exercises

1. The PDF of Rayleigh distribution is given by $p_{\mathrm{r}}(x) = \begin{cases} \dfrac{x}{b^2}\mathrm{e}^{-\frac{x^2}{2b^2}} & x \geqslant 0 \\ 0 & x < 0 \end{cases}$.

   Derive analytically the CDF, mean, variance, central moment and raw moments of the distribution. Generate a pseudo-random sequence satisfying Rayleigh distribution and verify numerically whether the numerical generation is correct or not.

2. Assume that between locations A and B, there are six sets of traffic lights. The probability of a red light at each set of traffic lights is the same, with $p = 1/3$. Suppose that the number of red traffic lights on the road for locations A to B satisfies a binomial distribution $B(6, p)$. Find the probability of the event that one meets only once the red traffic light from A to B, varying the value of p to plot the probability functions.
3. Assume that in a foreign language examination, the randomly selected samples indicate that the scores satisfy approximately the normal distribution, with a mean value of 72. The number of those scores higher than 96 is 2.3% of all the number of students. Find the probability of a student whose score is between 60 and 80.
4. Generate 30000 pseudo-random numbers satisfying normal distribution of $N(0.5, 1.4^2)$. Find the mean and standard deviation of the data set. Observe the histogram of the data to see whether they agree with theoretical distribution. Change the width of the bins and see what may happen.
5. Assume that a set of data was measured as shown below. Use MATLAB to perform the following hypothesis tests:
   (i) Assume that the data satisfies normal distribution with a standard deviation of 1.5. Test whether the mean value of the data is 0.5.
   (ii) If the standard deviation is not known, test whether the mean is still 0.5.
   (iii) Test whether the distribution of the data is normal.

−1.7908	0.3238	4.6927	−2.3586	−0.0940	2.8943	5.3067	3.1634	−3.2812	−3.4389
0.0903	2.5006	−0.6758	−3.2431	−3.2440	−0.0521	−0.0796	0.4653	−0.7905	2.0690
3.9223	5.6959	0.7327	−1.083	−1.8152	−2.9145	−2.6714	1.7065	0.0819	2.3258
0.4135	1.6804	−1.3172	1.132	1.047	0.5219	4.4827	−1.112	0.5201	1.9318
3.2618	0.4735	2.031	−0.7177	−2.3273	0.6606	1.2325	−0.9750	2.3831	3.4477
−1.0665	2.5546	−4.8203	−2.5004	−0.2812	1.2122	−2.0178	1.2073	−1.1251	1.236
0.5169	0.6259	2.7278	2.9135	−1.6181	1.6246	1.8958	0.7403	−1.1234	−1.0142
−1.2615	−1.9909	0.9925	−1.1022	−2.1428	3.3757	3.357	4.6585	0.04734	0.1640
1.8206	1.5924	1.0887	0.47461	−1.7976	−0.7326	−1.5161	−0.1190	0.4540	−5.0103
−0.0652	0.48874	3.2303	0.49816	−0.40375	1.0868	0.80414	5.4782	1.1275	1.5649
1.5803	−0.1215	−0.118	−0.0612	0.8908	0.4704	0.1872	3.8942	2.8812	0.7631
2.0033	3.372	0.2005	1.3923	0.23873	−0.80559	−2.1176	−3.8764	1.8988	−0.8300

6. Suppose that tests have been made on a group of randomly selected fuses, and it is found that the burn-out currents of the fuses are 10.4, 10.2, 12.0, 11.3, 10.7, 10.6, 10.9, 10.8, 10.2, 12.1 A. Suppose that these random values satisfy normal distribution. Find the burn-out current and its confidence interval under the confidence level $\alpha \leqslant 0.05$.
7. Assume that the boiling points under certain atmospheric pressures are tested with the multiple measured data 113.53, 120.25, 106.02, 101.05, 116.46, 110.33, 103.95, 109.29, 93.93, 118.67°C. Check whether they satisfy normal distribution under the confidence level of $\alpha \leqslant 0.05$.
8. Assume that 12 samples are obtained for a random variable as 9.78, 9.17, 10.06, 10.14, 9.43, 10.60, 10.59, 9.98, 10.16, 10.09, 9.91, 10.36. Find the deviation of the data and its confidence interval.

9. Twenty patients suffering from insomnia are divided randomly into groups A and B, with ten patients each. They were given medicines A and B respectively. The extended sleeping hours are measured as shown below. Determine whether there are significant differences in the healing effect.

A	1.9	0.8	1.1	0.1	−0.1	4.4	5.5	1.6	4.6	3.4
B	0.7	−1.6	−0.2	−1.2	−0.1	3.4	3.7	0.8	0	2

10. For a prototype linear function $y = a_1x_1 + a_2x_2 + a_3x_3 + a_4x_4 + a_5x_5$, with five independent variables x_1, x_2, x_3, x_4, x_5 and one output y, the following data are obtained. Find the values a_i and their confidence intervals using linear regression method.

x_1	8.11	9.25	7.63	7.89	12.94	10.11	7.57	9.92	7.74	7.3	9.48	11.91
x_2	2.13	2.66	0.83	1.54	1.74	0.79	0.68	2.93	2.01	1.35	2.81	2.23
x_3	3.98	−0.68	1.42	−0.96	−0.28	3.37	4.58	2.15	2.66	3.69	1	−0.98
x_4	6.55	6.85	6.25	5.34	6.85	7.2	6.12	6.07	5.51	6.6	6.15	6.43
x_5	5.92	7.54	5.39	4.65	6.47	5.1	6.04	5.37	6.54	6.55	5.8	3.95
y	27.676	38.774	23.314	23.828	35.154	21.779	25.516	29.845	32.642	28.443	31.5	23.554

11. Assume that a set of measured data x_i and y_i are given below, and the prototype function is $f(x) = a_1\mathrm{e}^{-a_2x}\cos(a_3x + \pi/3) + a_4\mathrm{e}^{-a_5x}\cos(a_6x + \pi/4)$. Estimate the values of a_i and their confidence intervals.

x	1.027	1.319	1.204	0.684	0.984	0.864	0.795	0.753	1.058	0.914	1.011	0.926
y	8.8797	5.9644	7.1057	8.6905	9.2509	9.9224	9.8899	9.6364	8.5883	9.7277	9.023	9.6605

12. For a prototype function $y = a_1\mathrm{e}^{-a_2x_1}(a_3x_2 + x_3) + a_4x_4(a_5x_5 + 1)$, assume that the measured data are given below. Estimate the values of a_i and their confidence intervals.

x_1	8.11	9.25	7.63	7.89	12.94	10.11	7.57	9.92	7.74	7.3	9.48	11.91
x_2	2.13	2.66	0.83	1.54	1.74	0.79	0.68	2.93	2.01	1.35	2.81	2.23
x_3	3.98	−0.68	1.42	−0.96	−0.28	3.37	4.58	2.15	2.66	3.69	1	−0.98
x_4	6.55	6.85	6.25	5.34	6.85	7.2	6.12	6.07	5.51	6.6	6.15	6.43
x_5	5.92	7.54	5.39	4.65	6.47	5.1	6.04	5.37	6.54	6.55	5.8	3.95
y	8.19	7.68	5.42	3.98	5.99	6.19	7.09	6.83	7.23	8.06	6.74	4.24

13. Assume the measured data given below satisfies the following prototype function $y(t) = c_1\mathrm{e}^{-5t}\sin(c_2t) + (c_3t^2 + c_4t^3)\mathrm{e}^{-3t}$. Find from the data the parameters c_i's and their confidence interval.

t	0	0.1	0.2	0.3	0.4	0.5	0.6	0.7	0.8	0.9
y	0	0.1456	0.2266	0.2796	0.3187	0.3479	0.3677	0.3777	0.3782	0.37
t	1	1.1	1.2	1.3	1.4	1.5	1.6	1.7	1.8	1.9
y	0.3546	0.3335	0.3085	0.2812	0.253	0.225	0.198	0.1726	0.1492	0.1279
t	2	2.1	2.2	2.3	2.4	2.5	2.6	2.7	2.8	2.9
y	0.109	0.0922	0.0776	0.065	0.0541	0.0449	0.0371	0.0305	0.025	0.0204

14. Assume that 12 sample plants are randomly selected from areas A and B. The iron element content in μg/g is measured as shown below. Assume that the iron element content in the plant satisfies a normal distribution and the variance of the distribution is not affected by the area. Test whether the distribution of the iron element content is the same.

area A	11.5	18.6	7.6	18.2	11.4	16.5	19.2	10.1	11.2	9	14	15.3
area B	16.2	15.2	12.3	9.7	10.2	19.5	17	12	18	9	19	10

15. Assume that random variables A and B are sampled below. Check whether they have significant statistic differences.

A	10.42	10.48	7.98	8.52	12.16	9.74	10.78	10.18	8.73	8.88	10.89	8.1
B	12.94	12.68	11.01	11.68	10.57	9.36	13.18	11.38	12.39	12.28	12.03	10.8

16. Suppose that five different dyeing techniques are tested for the same cloth. Different dyeing techniques and different machines are tested randomly, and the percentage of washing shrinkage are given below. Judge whether the dyeing techniques have significant effect on the washing shrinkage.

machine	dyeing techniques					machine	dyeing techniques				
number	1	2	3	4	5	number	1	2	3	4	5
1	4.3	6.1	6.5	9.3	9.5	2	7.8	7.3	8.3	8.7	8.8
3	3.2	4.2	8.6	7.2	11.4	4	6.5	4.2	8.2	10.1	7.8

17. Assume that the heights of randomly selected Year-5 pupils in three schools are measured in the table below. Check whether there are significant differences in the heights in these three schools. ($\alpha = 0.05$)

school	measured height data					
1	128.1	134.1	133.1	138.9	140.8	127.4
2	150.3	147.9	136.8	126	150.7	155.8
3	140.6	143.1	144.5	143.7	148.5	146.4

18. The table below recorded the daily output of three operators on four different machines. Check the following
(i) whether there are significant differences in the skill of the operators
(ii) whether there are significant differences in the machines
(iii) whether the interaction is significant ($\alpha = 0.05$)

machine	operator number									machine	operator number								
number	1			2			3			number	1			2			3		
M_1	15	15	17	19	19	16	16	18	21	M_3	15	17	16	18	17	16	18	18	18
M_2	17	17	17	15	15	15	19	22	22	M_4	18	20	22	15	16	17	17	17	17

Chapter 10

Nontraditional Solution Methods

In the previous chapters, traditional branches of advanced applied mathematics have been summarized and our focus was on computer aided solutions to those problems. Over the last few decades, many new applied math topics emerged which are referred to as *nontraditional mathematics* in this book. For instance, fuzzy logic and fuzzy inference are presented and used for imitating imprecise human thinking and linguistic behaviors. The artificial neural networks are established based on the mathematical model imitating the neural network of biological systems. The genetic algorithm-based optimization procedures are proposed based on the principles of survival of the fittest. These new branches of applied mathematics are promising areas of research in science and engineering offering important tools for real-life problems. In Section 10.1, classical set theory, fuzzy set and fuzzy inference are presented with their implementations in MATLAB. Section 10.2 introduces artificial neural network in general and feedforward neural network in particular with back-propagation algorithms, where MATLAB solutions for network construction, training and generalization are presented using data fitting problems for illustration. In Section 10.3, evolution type optimization algorithms, including genetic algorithms and particle swarm optimization methods, are introduced. The global solutions to optimization problems are also explored. Wavelet-based methods and solutions are given in Section 10.4 using signal de-noising as an application example. Fundamental introduction to rough set and rough set-based attribute reduction is given in Section 10.5 with real-life examples. In Section 10.6, a comprehensive introduction to fractional-order calculus and numerical solutions to fractional-order ordinary differential equations is given, where from a programming point of view, the design and application of the classes and objects dedicated for fractional-order calculus are demonstrated thoroughly.

It should be noted that only very brief introductions to the mathematical background and theoretical description are given. Our focus, again, is on the solutions of the related math problems using MATLAB.

10.1 Fuzzy Logic and Fuzzy Inference

10.1.1 Classical set theory and fuzzy sets

MATLAB solutions to enumerable set problems

Set theory is the foundation of modern mathematics. The so-called *set* is a collections of objects, with each object defined as a *member* of the set. If the object a is a member of set A, it is denoted as $a \in A$, and we call it "a belongs to set A." If b is not a member of the set A, it is denoted as $b \notin A$. A set is called an *enumerable set* if all its members can be enumerated. In MATLAB, enumerable sets can be represented by vectors or cell arrays.

Example 10.1 The following MATLAB statements can be used to define enumerable sets

```
>> A=[1 2 3 5 6 7 9 3 4 11]  % set of digits, repeated members are allowed
   B={1 2 3 5 6 7 9 3 4 11}  % the above set can also be described by cells
   C={'ssa','jsjhs','su','whi','kjshd','kshk'}  % set of strings
```

The set operations provided in MATLAB are summarized in Table 10.1 with brief explanations. These functions can be nested to establish complicated set operations. Unfortunately, these functions are not applicable to symbolic variables.

TABLE 10.1: Set operations under MATLAB

operations	MATLAB functions	descriptions to set operations
set union	A=`union`(B,C)	Union of sets B and C, such that $A = B \bigcup C$, where the returned results are sorted
difference	A=`setdiff`(B,C)	The difference of the sets B and C, denoted as $A = B \backslash C$. This operation removes the members in set C from set B. The remaining members, after sorting, are returned
intersection	A=`intersect`(B,C)	The sorted results of the intersect of sets B and C, where $A = B \bigcap C$
exclusive or	A=`setxor`(B,C)	Exclusive or operation of the sets B and C, i.e., removes set $B \bigcap C$ from set $B \bigcup C$, mathematically $A = (B \bigcup C) \backslash (B \bigcap C)$. Sorted results are returned
unique	A=`unique`(B)	Find the non-repeating elements in set B, i.e., to delete the extra repeated members with returned set sorted
belong to	`key=ismember`(a,B)	Check where a belongs to set B, `key`=$a \in B$, returns 1 or 0. If a is a set, `key` is a vector of 0's and 1's

Example 10.2 For the three sets $A = \{1, 4, 5, 8, 7, 3\}$, $B = \{2, 4, 6, 8, 10\}$ and $C = \{1, 7, 4, 2, 7, 9, 8\}$, perform the relevant set operations. Verify the commutative law $(A \bigcup B) \bigcap C = (A \bigcap C) \bigcup (B \bigcap C)$.

Solution The three sets can be expressed in MATLAB easily and the related functions can be called to perform set operations

```
>> A=[1,4,5,8,7,3]; B=[2,4,6,8,10]; C=[1,7,4,2,7,9,8]; % define sets
   D=unique(C), E=union(A,B), F=intersect(A,B)  % set operations
```

and the unique set D is found such that $D = [1, 2, 4, 7, 8, 9]$, where the extra repeated set member 7 is removed, and the result is sorted. The union and intersection are also found as $E = [1, 2, 3, 4, 5, 6, 7, 8, 10]$, and $F = [4, 8]$.

The following statement is used to verify the commutative law where the set difference is taken between the sets at the left-hand side and the right-hand side. The result is an empty matrix, indicating that the commutative law holds for the given sets.

```
>> G=setdiff(intersect(union(A,B),C),...
     union(intersect(A,C),intersect(B,C)))
```

The function `ismember()` can be called such that

```
>> H=ismember(A,B), I=A(ismember(A,B))
```

and it is found that $H = [0, 1, 0, 1, 0, 0]$, indicating the second and fourth members in set A, i.e., members 4 and 8, belong to set B. These members can be extracted to form set I, which is the same as the intersection of sets A and B, i.e., $I = A \bigcap B$.

Example 10.3 Consider the sets A and B containing strings
$A = \{$`'skhsak','ssd','ssfa'`$\}$, and $B = \{$`'sdsd','ssd','sssf'`$\}$.
Find the union and intersection of A and B. If $C=\{$`'jsg','sjjfs','ssd'`$\}$, verify the associative law such that $\left(A \bigcap B\right) \bigcup \left(C \bigcap B\right) = \left(A \bigcup C\right) \bigcap B$.

Solution The set containing strings can be expressed by cell objects. The set operations can be calculated easily by the following scripts

```
>> A={'skhsak','ssd','ssfa'}; B={'sdsd','ssd','sssf'};
   F=union(A,B), D=intersect(A,B),
```

and the union and intersection of sets A and B are found
$F = \{$`'sdsd','skhsak','ssd','ssfa','sssf'`$\}$, $D = \{$`'ssd'`$\}$.

When set C is specified, the associative law can be verified with the following statements, and the result should be empty.

```
>> C={'jsg','sjjfs','ssd'};
   E=setdiff(union(intersect(A,B),intersect(C,B)),...
         intersect(union(A,C),B))
```

Subset and set inclusion are the important concepts in set theory. The so-called *set inclusion* means that if all the members in set A belong to set B, it is said that A is included in set B, denoted by $A \subseteq B$, or if set B includes set A, A is also called a *subset* of set B. If $B \backslash A$ is not empty, the set inclusion is referred to as *strict inclusion*, denoted by $A \subset B$. Set A is then referred to as a *proper subset* of B. Set inclusion functions are not directly provided in MATLAB, but the following statements can be used to check the set inclusion and strict inclusion.

```
key=all(ismember(A,B)) % key= 1 if A⊆B, all elements in A belong to B
key=all(ismember(A,B))&(length(setdiff(B,A))>0) % key= 1 then A⊂B
```

Example 10.4 Consider the sets E, F in Example 10.2, check whether $F \subset E$ is true. Verify the reflexive law of set A, i.e., $A \subseteq A$.

Solution The following statements can be used to confirm the inclusion set relationship, with key=1, meaning F is a subset of E.

```
>> A=[1,4,5,8,7,3]; B=[2,4,6,8,10];
   E=union(A,B); F=intersect(A,B); key=all(ismember(F,E))
```

In fact, $F = A \bigcup B$ and $E = A \bigcap B$, thus $E \subset F$. It can also be verified that $A \subseteq A$, i.e., the reflexive law, with key=0, key1=1, meaning A is not a proper subset of A. However, A is a subset of set A.

```
>> key=all(ismember(A,A)) & (length(setdiff(A,A))>0)
   key1=all(ismember(A,A))  % A⊆A is satisfied
```

Example 10.5 Let us consider the verification of the well-known *Goldbach's conjecture*, which states that any even integer greater than 2 can be expressed by the sum of two prime numbers. It is one of the oldest unsolved problems in number theory which dates back to the time of Euler.

Solution For finite even integers, the full set of all the possible sums of two prime numbers within a given range can be constructed as set c. One can then check whether the terms in the set of even integers c_1 not belonging to c is empty. The following statements can be used to verify the conjecture for even integers in $[4, 2000]$.

```
>> iA=1:1040; iA=iA(isprime(iA)); c=[];
   for i=iA, c=[c i+iA]; end, c=unique(c);
   c1=4:2:2000; c2=ismember(c1,c); key=c1(c2==0)
```

It can be seen that the set key is an empty one, which verifies the conjecture for small integer numbers. It should be noted that a larger c should be constructed. With the parallel computers, the even integers c up to 10^{19} have been verified with no exception[49].

Fuzzy sets

From the classical set theory, it is seen that an event a either belongs to set A, or does not belong to set A. There is no other relationship. In modern science and engineering, the fuzzy concepts might be useful, which may require that the event a belongs to set A to some extent. This is the fundamental motivation of fuzzy set theory.

The concept of fuzzy set was proposed by Professor Lotfi A. Zadeh in 1965[50]. Currently the concept of fuzzy logic has been applied to almost all research and application areas. For example, fuzzy control is an attractive and promising research topic in control engineering.

Actually, fuzzy concept has been used earlier in the book. For instance, in the variable-step computation, it has been stated that "when the error is large $\cdots$." The word "large" is a fuzzy description. However, fuzzy procedures were not used there.

In real world situations, precision is not everything. Professor Zadeh points out that "As complexity rises, precise statements lose meaning, and meaningful statements lose precision."

10.1.2 Membership function and fuzzification

Membership function is a very important concept in fuzzy set theory. It states the "grade of membership" of an element a belonging to set A, denoted as $\mu_A(a)$. The membership function of classical set is either 1 or 0, indicating a either belongs to A or does not. However, for fuzzy sets, the value of the membership function is between 0 and 1. Some of the commonly used membership functions are summarized below.

Bell-shaped membership functions

The mathematical description to bell-shaped membership function is

$$f(x) = \frac{1}{1 + \left|\dfrac{x-c}{a}\right|^{2b}} \tag{10.1}$$

and function `gbellmf()` in Fuzzy Logic Toolbox can be used to evaluate the membership function with `y=gbellmf(x,[a,b,c])`, where $\boldsymbol{x}$ is the values of independent variable $\boldsymbol{x}$, and $\boldsymbol{y}$ contains the values of membership functions.

Example 10.6 The following statements can be used to draw the bell-shaped membership functions for different combinations of parameters a, b, c, in Figure 10.1.

```
>> x=[0:0.05:10]'; y=[]; a0=1:5; b=2; c=3;
   for a=a0, y=[y gbellmf(x,[a,b,c])]; end
   y1=[]; a=1; b0=1:4; c=3; for b=b0, y1=[y1 gbellmf(x,[a,b,c])]; end
   y2=[]; a=2; b=2; c0=1:4; for c=c0, y2=[y2 gbellmf(x,[a,b,c])]; end
   plot(x,y); figure; plot(x,y1); figure; plot(x,y2)
```

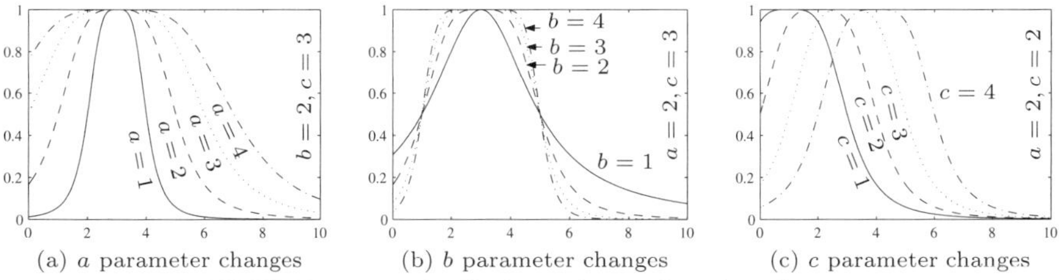

(a) a parameter changes (b) b parameter changes (c) c parameter changes

FIGURE 10.1: Bell-shaped membership function

It can be seen from the curves that the shapes of the membership function can be modified by changing respectively the parameters a, b, c. One may use these parameters to shape the membership functions depending on the application requirements.

Gaussian membership functions

The mathematical description to Gaussian membership function is given by

$$f(x) = \mathrm{e}^{-\frac{(x-c)^2}{2\sigma^2}} \tag{10.2}$$

and function `gaussmf()` in Fuzzy Logic Toolbox can be used to evaluate the membership function with `y=gaussmf(x,[σ,c])`, where $\boldsymbol{x}$ is the values of independent variable $\boldsymbol{x}$, and $\boldsymbol{y}$ contains the values of membership function.

Example 10.7 For different values of c and σ, the Gaussian membership function can be drawn as shown in Figure 10.2. The shape of the function is the same as the normal PDF studied in Chapter 9. It can be seen that when the values of c change, the shape of the membership function is unchanged. Only translations are made.

```
>> x=[0:0.05:10]'; y=[]; c0=1:4; s=3; y1=[]; c=5; sig0=1:4;
   for c=c0, y=[y gaussmf(x,[s,c])]; end
   for sig=sig0, y1=[y1 gaussmf(x,[sig,c])]; end;
   plot(x,y); figure; plot(x,y1)
```

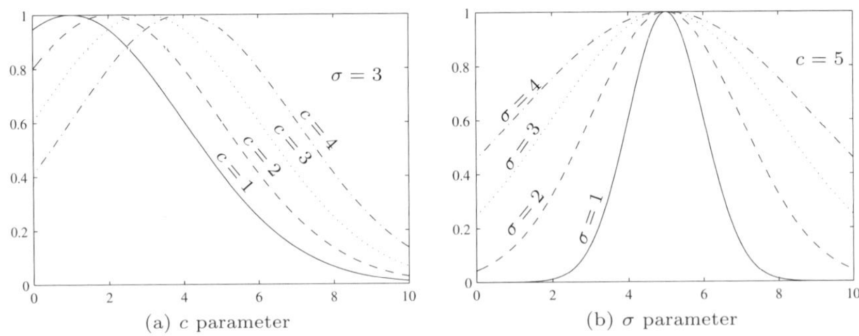

(a) c parameter (b) σ parameter

FIGURE 10.2: Gaussian membership function

Sigmoid membership functions

The mathematical description to sigmoid membership function is given by

$$f(x) = \frac{1}{1+\mathrm{e}^{-a(x-c)}} \tag{10.3}$$

which can be evaluated with the MATLAB function `y=sigmf(x,[a,c])`.

Example 10.8 The shapes of sigmoid function for different parameters of a and c are visualized in Figure 10.3. When c varies, the curve may translate to left or right with the shape of the curve unchanged.

```
>> x=[0:0.05:10]'; y=[]; c0=1:4; a=3;
   for c=c0, y=[y sigmf(x,[a,c])]; end
   y1=[]; c=5; a0=1:2:7; for a=a0, y1=[y1 sigmf(x,[a,c])]; end;
   plot(x,y); figure; plot(x,y1)
```

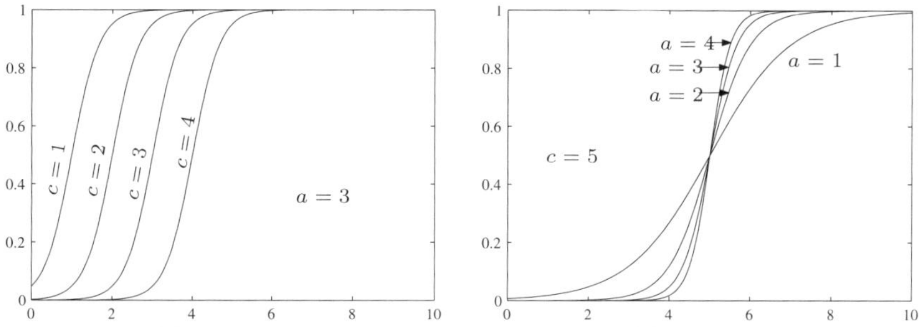

(a) changing c (b) changing a

FIGURE 10.3: Sigmoid membership function

10.1.3 An interactive membership function editor

A graphical user interface is provided for membership function manipulations in the Fuzzy Logic Toolbox: the command `mfedit` can be used to open the interface shown in Figure 10.4. The prototypes of three membership functions are given and one can edit the membership function with mouse drag and click.

If one more fuzzy set, or membership function is to be added, the menu item Edit → Add custom MF can be selected and a dialog box is shown in Figure 10.5 (a). The membership function can then be introduced as shown in Figure 10.5 (b).

10.1.4 Building fuzzy inference systems

The `newfis()` function provided in the Fuzzy Logic Toolbox can be used to construct the data structure of the fuzzy inference system (FIS). The syntax of the function is `fis=newfis(`*name*`)`, where the string variable *name* is used to describe the name of the FIS. The structured variable `fis` can be established. The properties in the FIS include the fuzzification, fuzzy inference and defuzzification, etc. These properties can also be used in the definition of `newfis()` function, or, they can be defined later. Having defined the FIS object `fis`, the function `addvar()` can be used to define input and output variables to the object, with the following syntaxes

```
fis=addvar(fis,'input',iname,vi)     % add an input variable iname
fis=addvar(fis,'output',oname,vo)    % add an output variable oname
```

where $\boldsymbol{v}_{\rm i}$ and $\boldsymbol{v}_{\rm o}$ are the ranges, i.e., the *universes*, of the input and output variables. They are described as row vectors of the minimum and maximum

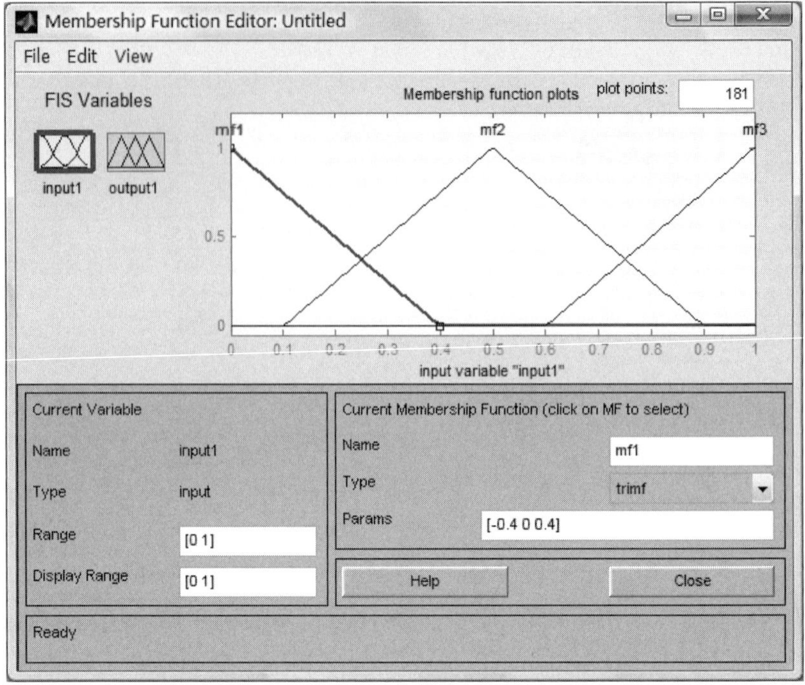

FIGURE 10.4: Editing interface for membership functions

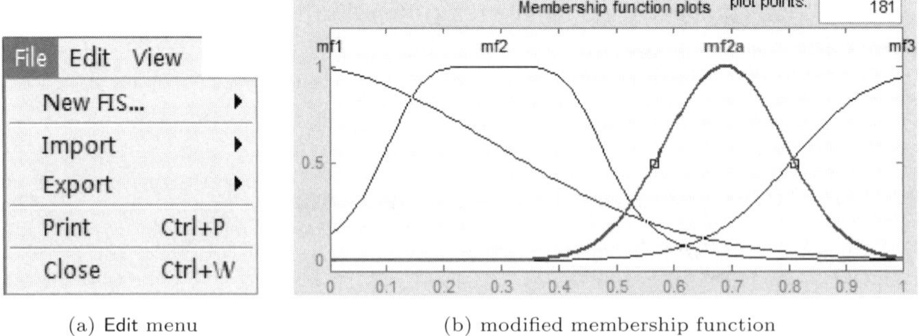

(a) Edit menu (b) modified membership function

FIGURE 10.5: Modifications of membership functions

of the variable. The input and output variables can also be defined with the graphical user interface `fuzzy`. The membership functions can be specified with the `addmf()` function, or directly, with the `mfedit()` interface.

Example 10.9 Suppose that there are two input variables, $\mathtt{ip}_1$ and $\mathtt{ip}_2$, and one output variable `op`. Assume that the universe of the input $\mathtt{ip}_1$ is $(-3, 3)$, with three membership functions, all selected as bell-shaped function. The universe for input signal $\mathtt{ip}_2$ is $(-5, 5)$, also three Gaussian-type membership functions are defined. The universe of the output `op` is $(-2, 2)$, whose membership function is sigmoidal functions. The framework of the fuzzy inference system can be established. Also the graphical user interface `fuzzy()` can be used to edit the FIS.

```
>> fff=newfis('c10mfis');        % set up a new fuzzy inference model
   fff=addvar(fff,'input','ip1',[-3,3]);   % define input 1
   fff=addvar(fff,'input','ip2',[-5,5]);   % define input 1
   fff=addvar(fff,'output','op',[-2,2]);   % define the output
   fuzzy(fff)             % edit using graphical interface fuzzy()
```

The fuzzy inference system modification interface can be opened with the function `fuzzy()`, as shown in Figure 10.6.

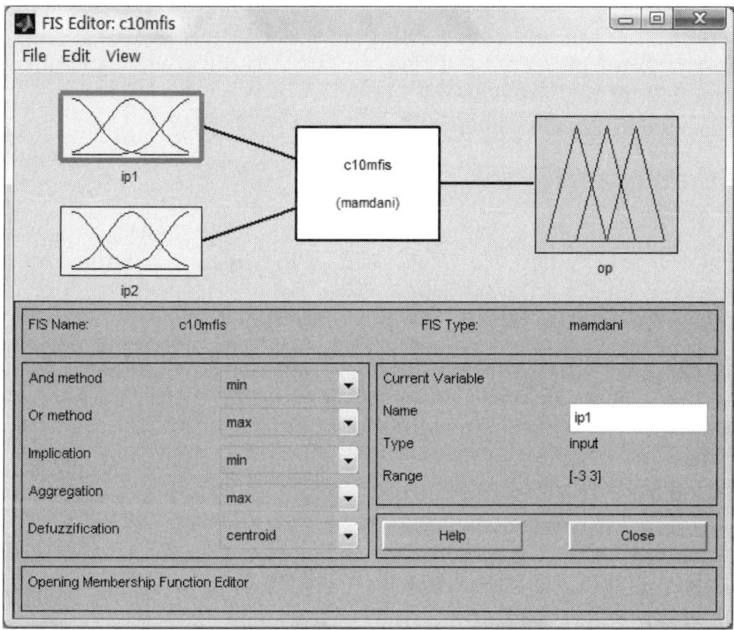

FIGURE 10.6: Graphical user interface of fuzzy inference system

The Edit → Membership functions menu item can be selected in the interface, and the membership function editing interface will be displayed as shown in Figure 10.4. The icon labeled $\mathtt{ip}_1$ in the interface can be selected, and Edit → Add MFs menu item can be selected, a dialog box shown in Figure 10.7 (a) opens. The user can define the membership functions in the dialog box. For instance, the output membership function, after editing procedure, can be obtained as shown in Figure 10.7 (b).

10.1.5 Fuzzy rules and fuzzy inference

Fuzzification

If three fuzzy sets or membership functions are defined for a certain signal, the physical meanings of the three sets may be "very small," "medium" and

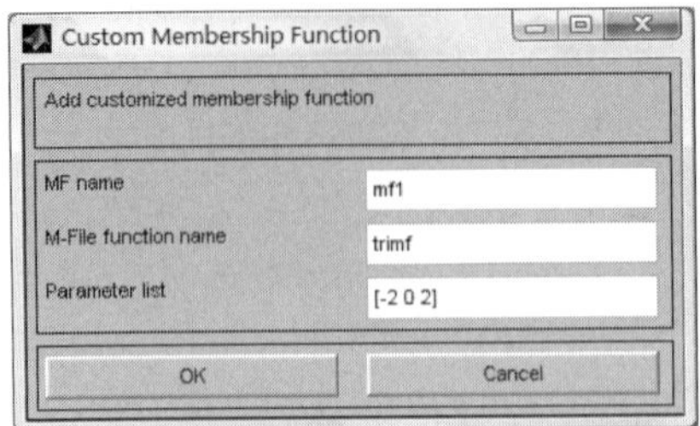

(a) membership function dialog box

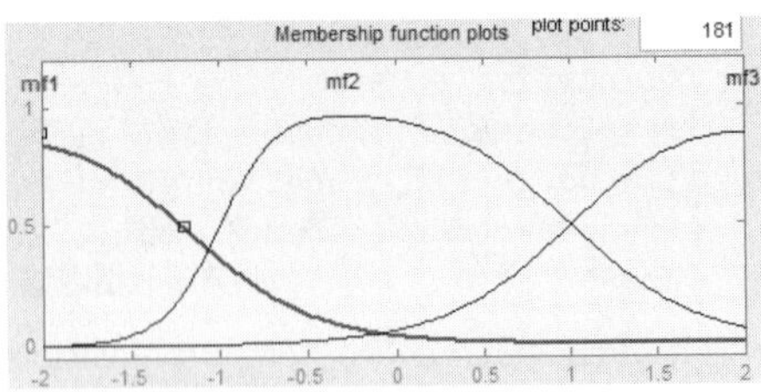

(b) modified membership function for output

FIGURE 10.7: Edited results of membership functions

"very large." If five sets are used, the physical meanings can also be defined as "very small," "small," "medium," "large" and "very large." An exact signal can be fuzzified into a fuzzy signal.

Fuzzy rules and inference

If all the input signals are fuzzified, the `if` – `else` clauses can be used to represent the fuzzy inference relationship. For instance, if the input signal $\mathtt{ip}_1$ is "very small," and input $\mathtt{ip}_2$ is "very large," then set the output signal `op` to "very large." The inference rule can be represented by

`if` $\mathtt{ip}_1$ is " very small" `and` $\mathtt{ip}_2$ is "very large," `then op=`"very large"

Fuzzy inference rule can be established with the `ruleedit()` interface, or by the menu item Edit → Rules in the `mfedit()` interface. A typical rule editing interface is shown in Figure 10.8, and the user can add all rules to the rule library, with the Add rule button. One may also delete certain rules by clicking the Delete rule button.

The rules can further be modified by clicking the Change Rule button. When the modification of the rules is completed, the Close button can be used to close the relevant windows. The fuzzy inference can be displayed as 3D surfaces with the View → Surface button, as shown in Figure 10.9, indicating the map from input signals to outputs.

Fuzzy rules can also be expressed by vectors. Multiple rules can be expressed by matrices comprising different vectors. The matrix is referred to as a *fuzzy rule matrix*. In each row of the matrix, there are $m+n+2$ elements, where m and n represent the numbers of input and output variables. The first m elements in the vector correspond to the sequence numbers of the input fuzzy sets, while the next n elements correspond to the numbers of the output sets. The $m+n+1$th element expresses the weighting, while the last element represents the logic relationship, such that 1 represents "and" and 2 represents "or." For instance, the third rule in Example 10.8 can be expressed by vector $[3, 2, 1, 1, 2]$.

From these rule vectors, a matrix $\boldsymbol{R}$ consisting of all the rule vectors can be established. The following command can be used to add a rule matrix to a `fis` object: `fis=addrule(fis,`$\boldsymbol{R}$`)`.

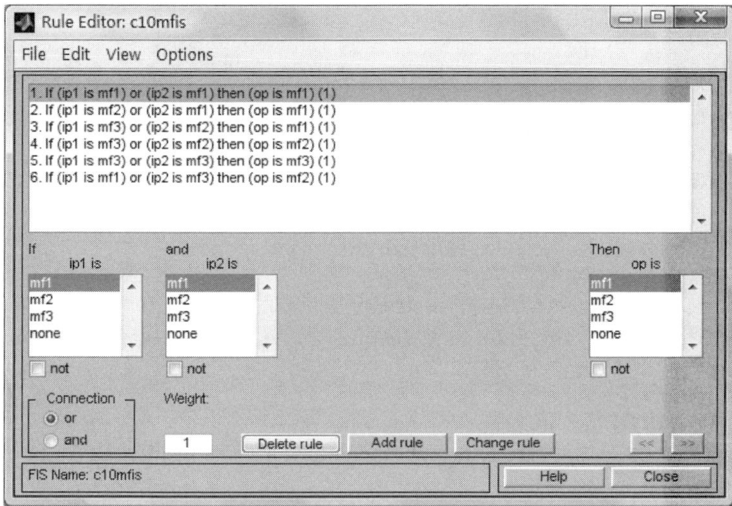

FIGURE 10.8: Edit box of fuzzy inference rules

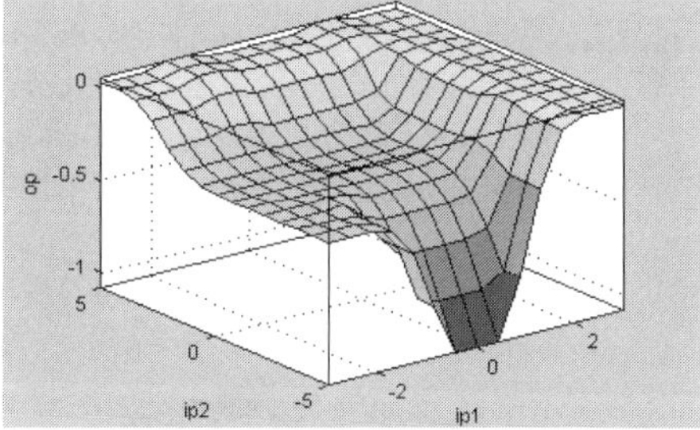

FIGURE 10.9: 3D surface

Defuzzification

Through fuzzy inference, a fuzzy output variable `op` can be generated. In practice, the exact or crispy value of the obtained fuzzy output signal is needed. The process of converting a fuzzy signal to the exact signal is referred to as *defuzzification*. Various defuzzification algorithms are supported in the Fuzzy Logic Toolbox, and they can be selected from the dialog box in Figure 10.6, i.e., from the list box shown in Figure 10.10.

The fuzzy inference data thus created can be saved to a file by the File → Export → To Disk menu item, with the file extension of `.fis`. For instance, the fuzzy inference system created above can be saved into file c10mfis.fis. The process can be saved too with the `writefis()` function. The menu item File → Export → To Workspace saves the FIS into MATLAB workspace.

Fuzzy inference problems can be solved by `y=evalfis(X,fis)` function where $\boldsymbol{X}$ is a matrix, whose columns contain the exact values of each input signal. The function `evalfis()` can be used to define the fuzzy inference

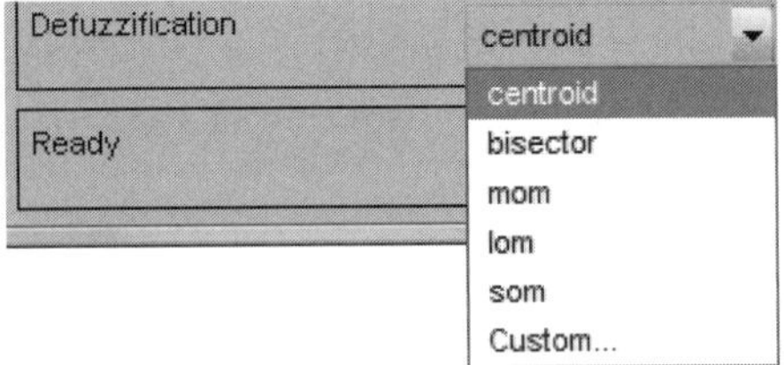

FIGURE 10.10: FIS system

system, to perform fuzzification first, then to perform fuzzy inference and finally to defuzzify the inference results. The exact output is returned in argument $\boldsymbol{y}$.

Example 10.10 Assume that the fuzzy inference system is the one defined in the last example. Draw the 3D surface of the fuzzy inference results.

Solution The following statements can be used to import the fuzzy inference system, draw mesh grids in the interested region on the x-y plane, and call `evalfis()` function to evaluate the z values. The 3D surface can then be obtained as shown in Figure 10.11.

```
>> fff=readfis('c10mfis.fis');  % read in the fuzzy inference system
   [x,y]=meshgrid(-3:.2:3,-5:.2:5);              % create mesh grids
   x1=x(:); y1=y(:); z1=evalfis([x1 y1],fff); % fuzzy inference
   z=reshape(z1,size(x)); surf(x,y,z)           % surface plot
```

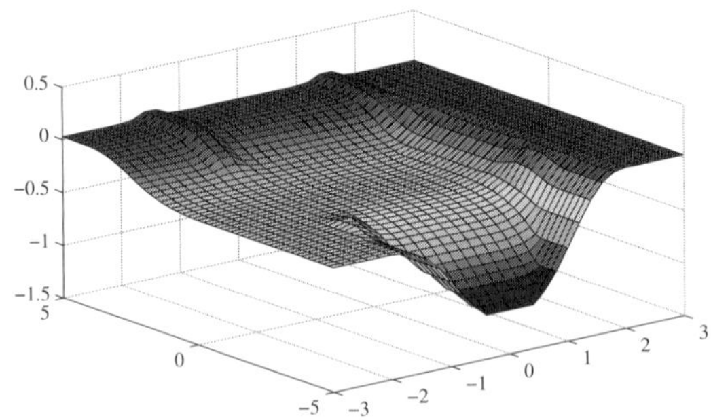

FIGURE 10.11: Output surface obtained from fuzzy inference system

10.2 Neural Network and Its Applications in Data Fitting Problems

Artificial neural networks (ANN) were originated from studying and understanding the behavior of complicated neural networks of living creatures. In human brains, there are about 10^{11} interlinked units, known as *neurons*. Each neuron has about 10^4 links with other neurons[51]. When the mathematical representations of artificial neurons are established, the interconnected neurons can be used to construct artificial neural networks. However, due to the limitations of the status of computers today, neural networks as complicated as human brains cannot be built so far.

In this section, an introduction to mathematical descriptions of artificial neurons and neural networks are given first, followed by the descriptions of MATLAB solutions to the neural network-based problems. The neural networks graphical user interface is demonstrated in detail.

10.2.1 Fundamentals of neural networks

Concept and structure of neural networks

The structure of an artificial neuron is shown in Figure 10.12, where $x_1, x_2, \cdots, x_n$ are input signals. The weighted sum of them, plus the threshold b, forms linear function u_i. The signal is further processed by the nonlinear transfer function $f(u_i)$ to generate the output signal y.

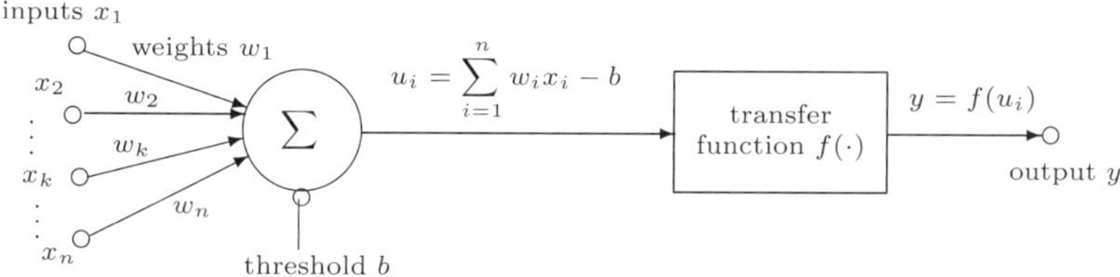

FIGURE 10.12: Basic structure of an artificial neuron

In this artificial neuron model, the weights w_i and the transfer function or activation function $f(\cdot)$ are the two important elements. The weights can be considered as the intensities of the input signals which can be determined by repeated training from the samples. Normally the transfer function should be selected as a monotonic function such that the inverse function uniquely exists. Commonly used transfer functions are sigmoidal function and logarithmic sigmoidal function, expressed respectively by

$$\text{sigmoid function} \quad f(x) = \frac{2}{1+\mathrm{e}^{-2x}} - 1 = \frac{1-\mathrm{e}^{-2x}}{1+\mathrm{e}^{-2x}} \tag{10.4}$$

$$\text{logarithmic sigmoid function} \quad f(x) = \frac{1}{1+\mathrm{e}^{-x}}. \tag{10.5}$$

The simple saturation function and step function can also be used as transfer functions. The shapes of some typical transfer functions are shown through the following examples.

Example 10.11 Draw some of the commonly used activation functions.

Solution The following MATLAB statements can be used to draw the sigmoid functions, as shown in Figure 10.13.

```
>> x=-2:0.01:2; y=tansig(x); plot(x,y)
```

The `logsig()` function can be used to replace the `tansig()` function to draw the logarithmic sigmoid function. Also other MATLAB functions can be used to draw the commonly used transfer functions, as shown in Figure 10.13.

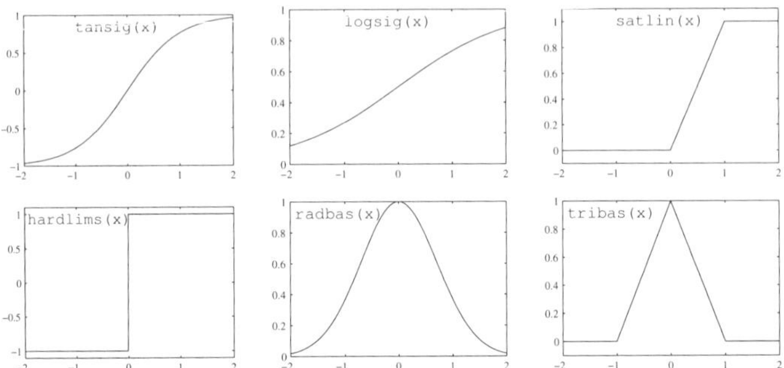

FIGURE 10.13: Different curves of the commonly used transfer functions

Part of the artificial neurons can be connected together to form a network, known as the *artificial neural network*. The word "artificial" is usually dropped and *neural network* is commonly used. With different ways of connections, different types of neural networks can be established. In this section, only the feedforward structure will be presented. Since in the training procedure, the error is propagated in the reversed direction, this network type is often referred to as the *back-propagation* (or BP) neural network. The typical structure of a BP neural network is shown in Figure 10.14. In the network, there are input layer, several intermediate layers known as *hidden layers*, and the output layer. In this book, the last hidden layer is in fact the output layer.

It is quite easy to construct a feedforward BP network with the Neural Network Toolbox of MATLAB. The function `newff()` can be used with the syntax

`net=newff(`$[\boldsymbol{x}_{\rm m},\boldsymbol{x}_{\rm M}]$`,`$[h_1,h_2,\cdots,h_k]$`,`$\{f_1,f_2,\cdots,f_k\}$`)`

where $\boldsymbol{x}_{\rm m}$ and $\boldsymbol{x}_{\rm M}$ are column vectors, containing the minimum and maximum values of the sample data. If the samples are given, the functions `min()` and `max()` can be used to extract the boundaries. The user may also select the number of nodes in each layer, and the number of hidden layers. The third argument in the function should be expressed by a cell object, whose elements are strings, indicating the transfer functions used in each layer. The neural network object `net` is then created, whose important properties are listed in

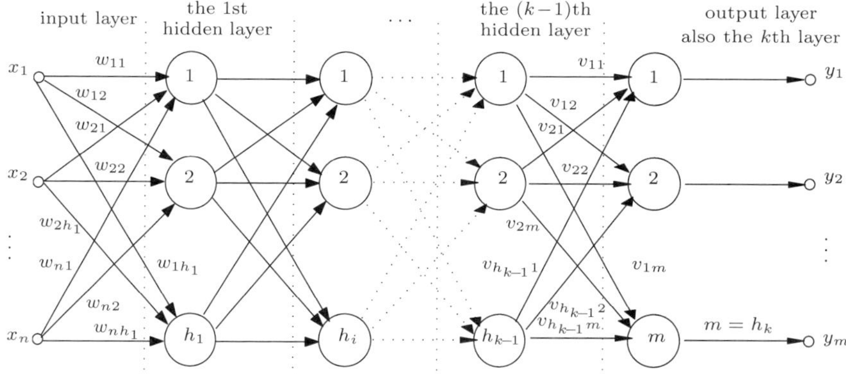

FIGURE 10.14: Basic structure of neural network

Table 10.2. Examples are given below to show the creation of neural network objects.

TABLE 10.2: Common properties of neural network

property names	data type	descriptions	defaults
`net.IW`	cell	input and hidden layer weighting, where `net.IW{1}` stores the weighting of the input layer, while `net.IW{i+1}` for the ith hidden layer	random
`net.numInputs`	integer	number of inputs, it can be calculated automatically from the sizes of $\boldsymbol{x}_{\rm m}$ or $\boldsymbol{x}_{\rm M}$	
`net.numLayers`	integer	number of hidden layers, which can be determined automatically by the `newff()` function call	
`net.LW`	cell	output layer weighting matrix	random
`net.trainParam.epochs`	integer	maximum training steps, when the error criterion is met, the training stops	100
`net.trainParam.lr`	double	learning rate	0.01
`net.trainParam.goal`	double	training error criterion, if the error is smaller than this value, the training stops	0
`net.trainFcn`	string	training algorithms, the selectable options are `'traincgf'` (conjugate gradient with Fletcher-Reeves updates), `'train'` (batch training), `'traingdm'` (gradient descent with momentum), `'trainlm'` (Levenberg-Marquardt algorithm)	`'train'`

Example 10.12 Assume that there are two inputs whose ranges are respectively $[0, 1]$ and $(-1, 5)$. There is one output signal. Establish a feedforward neural network object with the `newff()` function.

Solution Consider a feedforward network, with two hidden layers, and in the first hidden layer, there are 8 nodes, with logarithmic sigmoid transfer function. The second layer is in fact the output layer, with one node whose transfer function can be selected as a sigmoid function. The following statements can be used to establish the neural network object in MATLAB workspace

```
>> net=newff([0,1; -1,5],[8,1],{'tansig','logsig'});
```

Now assume that in the new network, there are three hidden layers. In the first layer there are 4 nodes with linear transfer function. In the second layer, there are 8 nodes with `logsig()` transfer function. In the third layer, there is one node having `tansig()` transfer function. The following statements can be used to establish the neural network model.

```
>> net=newff([0,1; -1,5],[4 6 1],{'purelin','logsig','tansig'});
```

Apart from the structures of the neural network, the other parameters can be set, for instance, one may use the following commands

`net.trainParam.epochs=300` or `net.trainFcn='trainlm'`

Training and generalization of neural networks

If the neural network model `net` has been established, the function `train()` can be used to train the parameters in the network

`[net,tr,`Y_1`,`E`]=train(net,`X`,`Y`)`

where the variable X is an $n \times M$ matrix, with n the number of input signals, and M the number of samples for training. The variable Y is an $m \times M$ matrix, with m the number of outputs. The variables X and Y store respectively the inputs and outputs data of the samples. From the function call, the network can be trained, and the returned variable `net` is the trained object. The argument `tr` is a structure variable with training information, where `tr.epochs` returns the number of epochs, meaning steps in the training process of an artificial neural network, and `tr.perf` returns the objective function values in each training step. The arguments Y_1 and E matrices are the output and model errors of the network, respectively, and the training error can be visualized by `plotperf(tr)`.

If the training up to the maximum epochs cannot find a satisfactory network, a warning message will be given. The training results can be used as the initial weights, and one can continue the training process until a satisfactory network is obtained. If the satisfactory network still can not be obtained, there might be problems in the network structure and a new structure should be tested.

After training, the neural network can be used as a computation unit. When the input signals other than the ones in the samples are provided, the output can be evaluated from the unit. This process is also known as the *simulation* or *generalization* of neural networks. This process can be used to solve data fitting problems. For the input signals in matrix X_1, the output can be evaluated using the function `sim()` where Y_1`=sim(net,`X_1`)`.

Example 10.13 For the problem in Example 8.25, use neural network to fit the data.

Solution The following statements can be given to load the sample data. A two-layer feedforward network structure can be selected. Assume that the first layer comes with 5 nodes, and the second layer has one node. The network parameters are trained for generalization, with the training error shown in Figure 10.15 (a) and the generalization results shown in Figure 10.15 (b). It can be observed that the training is satisfactory, and almost no difference can be observed from the true results.

```
>> x=0:.5:10; x0=[0:0.1:10];
   y=0.12*exp(-0.213*x)+0.54*exp(-0.17*x).*sin(1.23*x);
   y0=0.12*exp(-0.213*x0)+0.54*exp(-0.17*x0).*sin(1.23*x0);
   net=newff([0,10],[5,1],'tansig','tansig');
   net.trainParam.epochs=1000;    % set maximum epochs
   net=train(net,x,y);            % training network
   figure; y1=sim(net,x0); plot(x,y,'o',x0,y0,x0,y1,':');
```

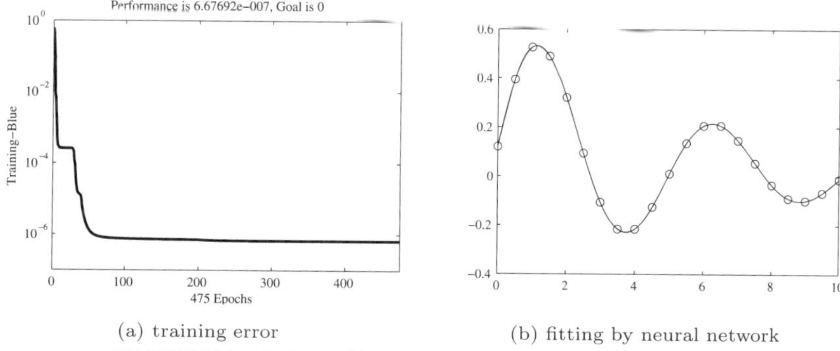

(a) training error (b) fitting by neural network

FIGURE 10.15: Data fitting with neural network

The following statements can be used to extract the weights of the trained neural network

```
>> w1=net.IW{1}, w2=net.LW{2,1} % weights for hidden layer and output
```

with

input to hidden layer: $\boldsymbol{w}_1^{\mathrm{T}} = [0.5131, 0.3705, -0.3047, -5.0318, 1.0329]$

hidden layer to output: $\boldsymbol{w}_2 = [0.7837, 13.4970, 15.5915, -2.1293, 1.0486]$.

Selecting different training algorithms, the training errors for each algorithm are compared in Figure 10.16 (a).

```
>> net=newff([0,10],[5,1],{'tansig','tansig'});
   net.trainParam.epochs=100;
   net.trainFcn='trainlm'; [net,b1]=train(net,x,y);
   net=newff([0,10],[5,1],{'tansig','tansig'});
   net.trainParam.epochs=100;
   net.trainFcn='traincgf'; [net,b2]=train(net,x,y);
   net=newff([0,10],[5,1],{'tansig','tansig'});
   net.trainParam.epochs=100;
   net.trainFcn='traingdx'; [net,b3]=train(net,x,y);
```

```
semilogy(b1.epoch,b1.perf); hold on
semilogy(b2.epoch,b2.perf,'--'); semilogy(b3.epoch,b3.perf,':')
```

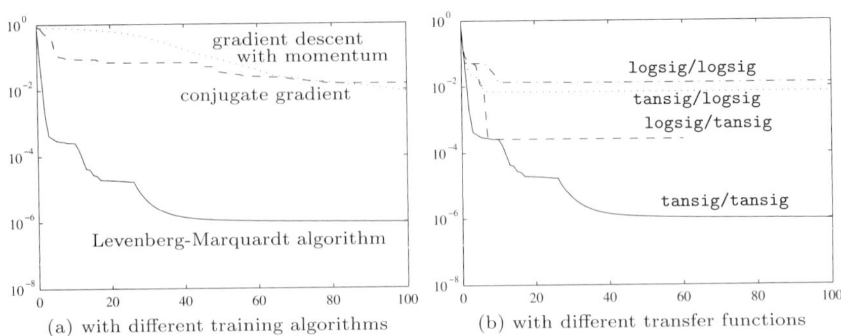

(a) with different training algorithms (b) with different transfer functions

FIGURE 10.16: Control parameters in neural network and their effects

It can be seen from the training results that the Levenberg-Marquardt algorithm behaves the best among the three algorithms tested. The transfer functions combinations are also tested and the objective functions for different transfer functions are shown in Figure 10.16 (b) and it is found that the `'tansig'/'tansig'` combination is much superior to other combinations in this example. This combination is recommended for other applications as well.

```
>> net=newff([0,10],[5,1],{'tansig','logsig'});
   net.trainParam.epochs=100;
   net.trainFcn='trainlm'; [net,b2]=train(net,x,y);
   net=newff([0,10],[5,1],{'logsig','tansig'}); [net,b3]=train(net,x,y);
   net=newff([0,10],[5,1],{'logsig','logsig'}); [net,b4]=train(net,x,y);
   semilogy(b1.epoch,b1.perf); hold on; semilogy(b2.epoch,b2.perf,'--')
   semilogy(b3.epoch,b3.perf,':'); semilogy(b4.epoch,b4.perf,'-.')
```

The number of hidden layer nodes can further be increased, for instance, to 15. The new network can be created and trained, and the training error reaches an extremely small value for only 50 epochs, as shown in Figure 10.17 (a). The curve fitting, i.e., generalization, results are shown in Figure 10.17 (b).

```
>> net=newff([0,10],[15,1],{'tansig','tansig'});
   net.trainParam.epochs=100;
   net.trainFcn='trainlm'; [net,b2]=train(net,x,y);
   figure; y1=sim(net,x0); plot(x0,y0,x0,y1,x,y,'o')
```

It can be seen from the generalization results that the curve fitting results are very poor, albeit on the samples, the fittings are good. This means that due to the increase of hidden layer nodes, the generalization process goes wrong. However, to date there is no universally accepted method on how to assign reasonable numbers of nodes. The node numbers and number of layers can only be assigned by trial-and-error methods.

Example 10.14 Fitting the two-dimensional data with neural network for the data given in Example 8.6.

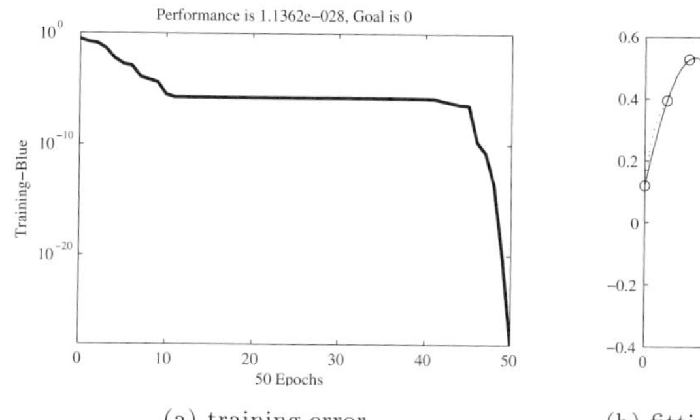

(a) training error

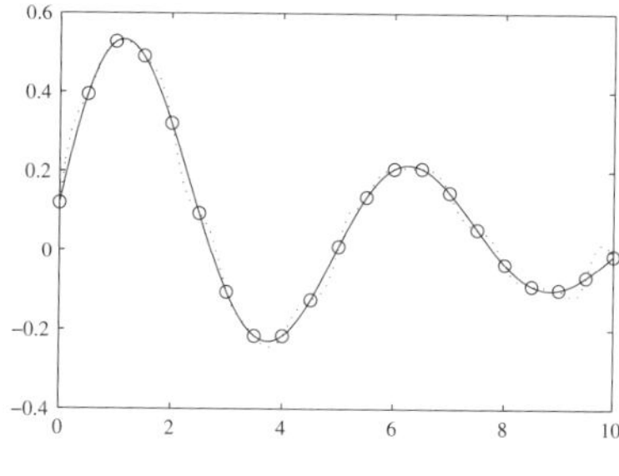

(b) fitting results with neural network

FIGURE 10.17: Fitting results with 15 nodes in the hidden layer

Solution Three hidden layer network structure can be tested first for the sample data. Assume that there are 10 nodes each in layers 1 and 2, with sigmoid transfer functions. In the third layer of course one should use one node, the same as the number of outputs. The following statements can be used to train the network, and the generalization results, represented as a 3D surface, are shown in Figure 10.18 (a).

```
>> [x,y]=meshgrid(-3:.6:3, -2:.4:2); x=x(:)'; y=y(:)';
   z=(x.^2-2*x).*exp(-x.^2-y.^2-x.*y);
   net=newff([-3 3; -2  2],[10,10,1],{'tansig','tansig','tansig'});
   net.trainParam.epochs=1000; net.trainFcn='trainlm';
   [net,b]=train(net,[x; y],z);
   [x2,y2]=meshgrid(-3:.1:3, -2:.1:2); x1=x2(:)'; y1=y2(:)';
   z1=sim(net,[x1; y1]); z2=reshape(z1,size(x2)); surf(x2,y2,z2)
```

It can be seen that the generalization is not satisfactory, and there exist oscillations at several points. Now select 20 nodes in the second layer. From the new generalization results shown in Figure 10.18 (b), it can be seen that even though the number of nodes increased, the generalization becomes even worse. This is because insufficient sample points are not adequate for good training and generalization for the data fitting problem.

```
>> net=newff([-3 3; -2  2],[10,20,1],{'tansig','tansig','tansig'});
   [net,b]=train(net,[x; y],z);
   z1=sim(net,[x1; y1]); z2=reshape(z1,size(x2)); surf(x2,y2,z2)
```

Now assume that more sample points are known. The network structures used earlier can be tried again with the fitting surfaces shown respectively in Figures 10.19 (a) and (b). It can be seen that the fitting results are improved. However, with surprise, the NN fitting results are far poorer than the interpolation results in Chapter 8.

```
>> [x,y]=meshgrid(-3:.2:3, -2:.2:2); x=x(:)'; y=y(:)';
   z=(x.^2-2*x).*exp(-x.^2-y.^2-x.*y);
   net=newff([-3 3; -2  2],[10,10,1],{'tansig','tansig','tansig'});
   net.trainParam.epochs=100; net.trainFcn='trainlm';
   net=train(net,[x; y],z);
   [x1,y1]=meshgrid(-3:.1:3, -2:.1:2); a=x1;  x1=x2(:)'; y1=y2(:)';
   z1=sim(net,[x1; y1]); z2=reshape(z1,size(a)); surf(x2,y2,z2)
```

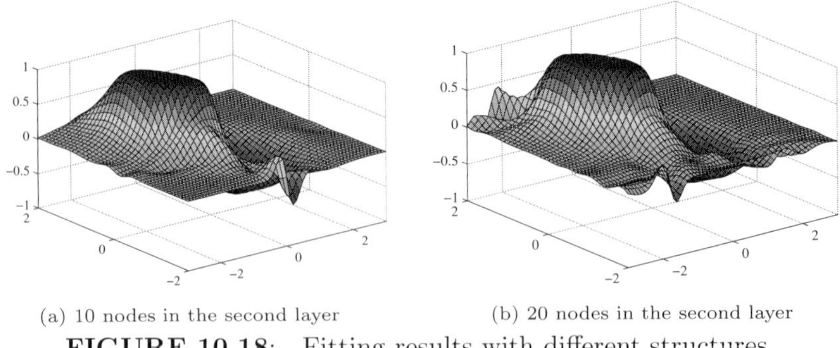

(a) 10 nodes in the second layer (b) 20 nodes in the second layer

FIGURE 10.18: Fitting results with different structures

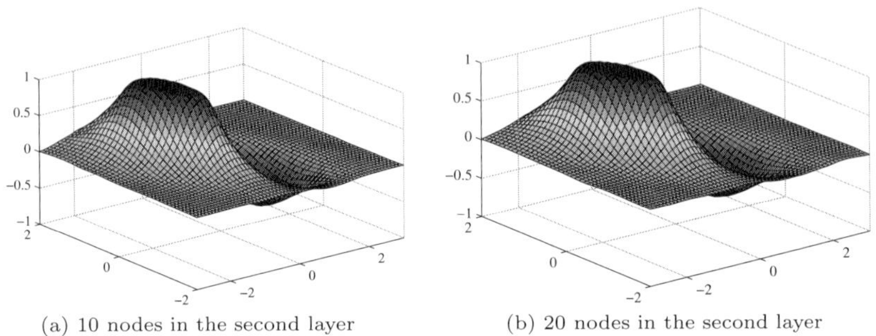

(a) 10 nodes in the second layer (b) 20 nodes in the second layer

FIGURE 10.19: Fitting with modified sample data set

```
net=newff([-3 3; -2 2],[10,20,1],{'tansig','tansig','tansig'});
net=train(net,[x; y],z); figure;
z1=sim(net,[x1; y1]); z2=reshape(z1,size(a)); surf(x2,y2,z2)
```

In neural network research, people always quote a theoretical result, i.e., a three-layer network can approximate any given continuous function with any specified accuracy. However, for this continuous function, we found that no matter how we assign the numbers of nodes, transfer function and training algorithm, the fitting precision was never lower than 10^{-3}.

10.2.2 Graphical user interface for neural networks

A graphical user interface is provided in the Neural Network Toolbox. The command `nntool` can be used to start the GUI shown in Figure 10.20. The interface can be used to create neural network, then train and simulate the network. The example below is given to show the use of the interface.

Example 10.15 Consider the problem in Example 10.13. Use the `nntool` interface to solve the same problem.

Solution The following statements can be used to import data into MATLAB workspace. Then `nntool` can be used to load the interface shown in Figure 10.20. The interface can be used in neural network data fitting.

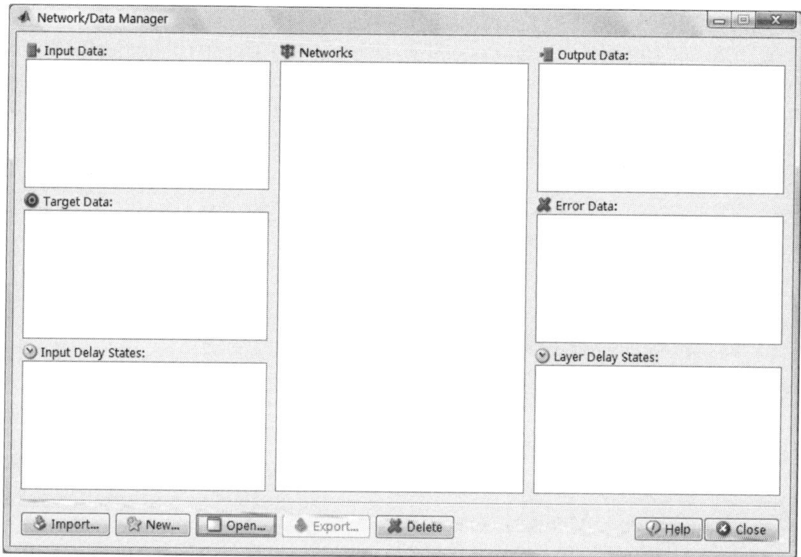

FIGURE 10.20: Graphical user interface for neural network

```
>> x=0:.5:10; x0=[0:0.1:10];
   y=0.12*exp(-0.213*x)+0.54*exp(-0.17*x).*sin(1.23*x);
   y0=0.12*exp(-0.213*x0)+0.54*exp(-0.17*x0).*sin(1.23*x0);
   nntool      % start the user interface of neural networks
```

Click the Import button to import the samples and the dialog box shown in Figure 10.21 appears. Select the variables `x` and `xx` as input variables, and `y` as target variables, by selecting the combinations in the interface.

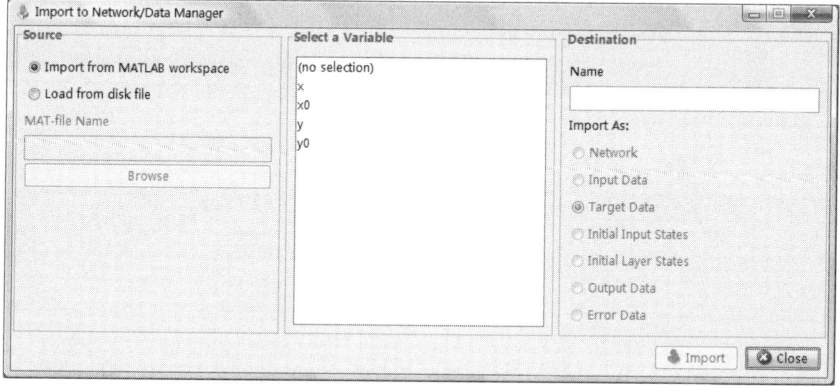

FIGURE 10.21: Data input interface

Click the New Network button to select the structure of neural network, with the graphical user interface shown in Figure 10.22 (a). The default (Feedforward Backprop) network structure can be established, with (Number of layers) set to 2.

The node numbers in each layer can also be set, where the number for layer 1 is 8. One may also set the transfer functions by selecting the Transfer Function list box to Logsig. Click the Create button, and the neural network structure can be created.

The network structure can be displayed by clicking the View button, as shown in Figure 10.22 (b).

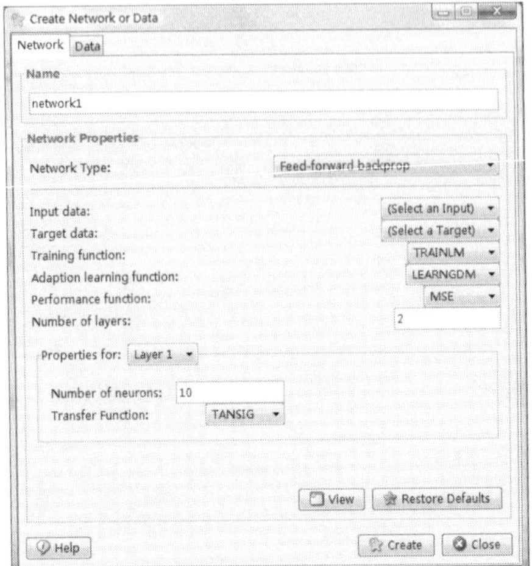

(a) neural network structure setting

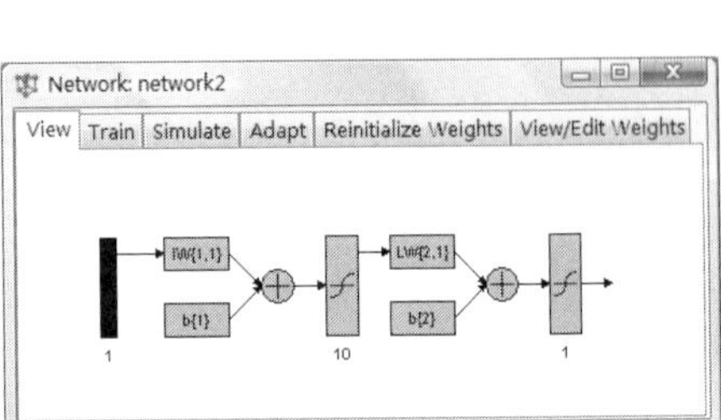

(b) display of neural network

FIGURE 10.22: Structure setting and display of neural network

The network can be trained by clicking the Train button, and the dialog box shown in Figure 10.23 is displayed. The training data can be specified first and training parameters can be set by clicking the Training Parameters tag to show the dialog box shown in Figure 10.24. The network can then be trained by clicking the Train button, and the training curve is shown in Figure 10.15 (a). If the termination conditions are satisfied, the training stops and the required parameters can be obtained. The trained network can be exported back to the environment with the Export button. The generalization results are obtained with the following statements, as shown in Figure 10.15 (b).

```
>> y1=sim(network1,x0); plot(x,y,'o',x0,y0,x0,y1,':')
```

10.3 Evolution Algorithms and Their Applications in Optimization Problems

10.3.1 Basic idea of genetic algorithms

Genetic algorithm (GA) is a class of evolution-based methods following the law of "survival of the fittest." [24] The method was first proposed by Professor

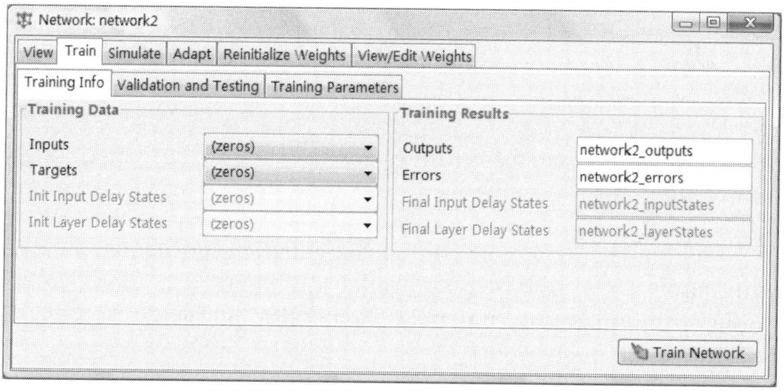

FIGURE 10.23: Dialog box for neural network training

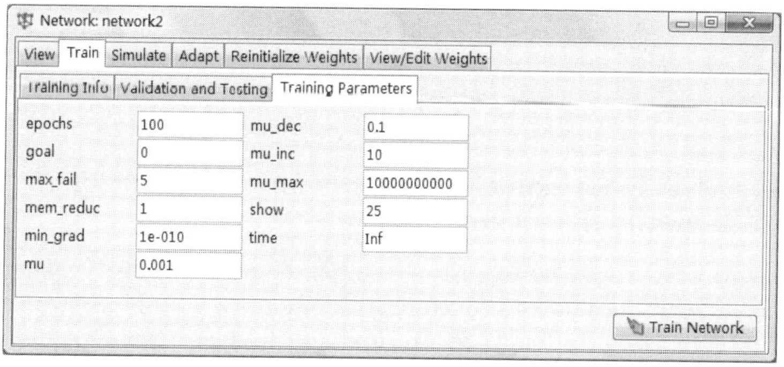

FIGURE 10.24: Training parameter setting dialog box

John Holland of Michigan University in 1975. The main idea of the method is to search from a population consisting of randomly distributed individuals. The individuals are encoded, regarded as genes with chromosomes, in a certain way. The population evolves generation by generation through reproduction, crossover and mutation, until individuals with the best fitness function are found. Another similar method is the particle swarm optimization (PSO) method, which imitates the coordinated motion in flocks of birds searching for food. In this section, MATLAB solutions to optimization problems using genetic algorithm and particle swarm optimization methods are introduced.

The general procedures of simple genetic algorithm are as follows:

(i) Select an initial population $\boldsymbol{P}_0$ with N individuals. Evaluate the objective functions for all the individuals. The initial population $\boldsymbol{P}_0$ can be established randomly.

(ii) Set the generation to $i = 1$, which means the first generation.

(iii) Compute the values of selective functions, i.e., select some individuals in a probabilistic way from the current population.

(iv) Create the population of the next generation $\boldsymbol{P}_{i+1}$, by reproduction, crossover and mutation.

(v) Set $i = i + 1$. If the termination conditions are not satisfied, go to (iii) to continue evolution.

Compared with traditional optimization methods, the genetic algorithms have mainly the following differences[22]:

(i) In searching the optimum points, the genetic algorithms allow searches from many initial points in a parallel way. Thus it is more likely to find global optimum points than with the traditional methods, which initiate search from a single point.

(ii) Genetic algorithms do not depend on the gradient information of the objective functions. Only the fitness functions, i.e., objective functions are necessary in optimum points search.

(iii) Genetic algorithms evaluate and select the objective function in a probabilistic way rather than a deterministic way. Thus there are slight differences among each run.

10.3.2 MATLAB solutions to optimization problems with genetic algorithms

There are several genetic algorithm toolboxes under MATLAB. The GADS (Genetic Algorithm and Direct Search) Toolbox by The MathWorks is the official toolbox, and its functions are updated in each release of MATLAB. Apart from that, the GAOT[52] (Genetic Algorithm Optimization Toolbox) written by Christopher Houck, Jeffery Joines and Michael Kay, Northern Carolina State University and the GA Toolbox[22] written by Peter Fleming and Andrew Chipperfield of Sheffield University are among the most commonly used free toolboxes.

For instance, in the GAOT toolbox, the function `gaopt()`[1] can be used directly to solve optimization problems. The benefit of such a function is that the intermediate search results are also available. In GAOT Toolbox, three selective functions are provided, such as `roulette()`, `normGeomSelect()` and `tournSelect()`, with `normGeomSelect()` the default.

In this section, the two toolboxes GADS and GAOT are explored for optimization problems.

Applications of GAOT Toolbox in optimization

The `gaopt()` function in GAOT Toolbox is illustrated first. `gaopt()` can be called in the following formats:

---

[1]The original name of the function is `ga()`, however it is the same as the one provided in GADS of The MathWorks, which may cause conflict. Thus it is recommended that the function is renamed as `gaopt()` and this name is used throughout this book.
The toolbox can be downloaded from: `http://www.ise.ncsu.edu/mirage/GAToolBox/gaot/`

```
[a,b,c]=gaopt(bound,fun)                      % the simplest form
[x,b,c]=gaopt(bound,fun,p,v,P0,fun1,n)        % with more arguments
```

where `bound=`$[\boldsymbol{x}_{\rm m}, \boldsymbol{x}_{\rm M}]$ stores the lower-bound $\boldsymbol{x}_{\rm m}$ and upper-bound $\boldsymbol{x}_{\rm M}$ of the decision variables $\boldsymbol{x}$. The argument *fun* is the file name string of M-function describing the objective function. It should be noted that the description of objective function is different from the other optimization routines. Examples will be given later to demonstrate the syntax of the objective function. The returned argument $\boldsymbol{a}$ is composed of the solution $\boldsymbol{x}$ and the value of the objective function $f_{\rm opt}$. The argument `b` stores the information of the final population. The argument `c` returns the intermediate search results.

In the second syntax, the argument $\boldsymbol{p}$ is the additional parameters in the objective function, $\boldsymbol{v}$ is the display precision control vector, $\boldsymbol{P}_0$ is the initial population, and *fun1* is the extra function name, with default of `'maxGenTerm'`, the maximum allowed generation, and n is the number of generations. Of course, other functions are allowed, for instance, selective function, and mutation function, etc. Details can be found from Reference [52].

Example 10.16 Now consider a simple function $f(x) = x\sin(10\pi x) + 2$, $x \in (-1, 2)$. Find the maximum value of $f(x)$, and the value of x.

Solution The objective function within the specified interval is drawn as shown in Figure 10.25, with the following simple MATLAB statements. Clearly, $f(x)$ is oscillatory with many extreme points in the interval.

```
>> ezplot('x*sin(10*pi*x)+2',[-1,2])
```

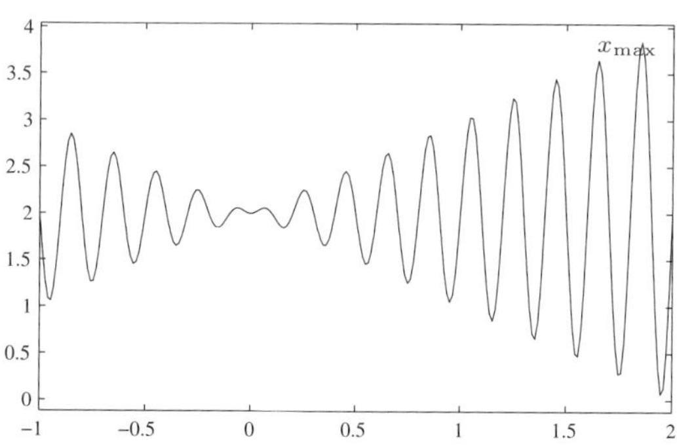

FIGURE 10.25: Objective function curve

Optimization functions provided in the Optimization Toolbox, like all traditional optimization approaches, need the initial values for searching the optimum. Different initial values may yield different optimization results. For this example, the following statements are tested and the search results for different initial values are given in Table 10.3.

```
>> f=@(x)-x.*sin(10*pi*x)-2; v=[];
   for x0=[-1:0.8:1.5,1.5:0.1:2]
```

TABLE 10.3: Optimal solutions from different search points x_0

x_0	x_1	$f(x_1)$	x_0	x_1	$f(x_1)$	x_0	x_1	$f(x_1)$
−1	−1	−2	1.4	1.4506983	−3.4503492	1.7	1.2508097	−3.2504050
−0.2	−0.65155382	−2.6507777	1.5	0.25396846	−2.2519973	1.8	1.8505475	−3.8502738
0.6	0.65155379	−2.6507777	1.6	1.6506138	−3.6503069	1.9	0.4522327	−2.4511207

TABLE 10.4: Intermediate search results using genetic algorithm

gen.	x	$f(x)$	gen.	x	$f(x)$	gen.	x	$f(x)$
1	1.833411	3.590014	6	1.85168	3.849101	11	1.850281	3.850209
2	1.647955	3.644557	8	1.851659	3.849144	12	1.85054	3.850274
3	1.858199	3.796899	9	1.851091	3.850004	100	1.850547	3.850274

```
    x1=fmincon(f,x0,[],[],[],[],-1,2); v=[v; x0,x1,f(x1)];
end
```

So, from this example, when the initial value is selected randomly, it may be very hard to find the global optimum point.

To solve the problem with genetic algorithm, an objective function should be expressed by an M-function

```
function [sol,y]=c10mga1(sol,options)
x=sol(1); y=x.*sin(10*pi*x)+2;
```

It can be seen that the objective function description is different from the ones required by traditional optimization algorithms in the following sense: (i) the maximum function is specified; (ii) the input and returned arguments are differently defined. With the use of `gaopt()` function, the following statements can be given.

```
>> [a,b,c,d]=gaopt([-1,2],'c10mga1'); x=a(1), f=a(2)
```

where the global result returned in a is $x^* = 1.8505$, with the optimum value $f(x^*) = 3.8503$. The intermediate results are given in Table 10.4, with the abbreviation *gen.* for the generation number. It can be seen that from generation 12, the results obtained are acceptable for this example.

Example 10.17 Find the minimum value of the function $f(x) = (x_1 + x_2)^2 + 5(x_3 - x_4)^2 + (x_2 - 2x_3)^4 + 10(x_1 - x_4)^4$ using genetic algorithm.

Solution It is obvious that the global optimum solution to the problem is $x_1 = x_2 = x_3 = x_4 = 0$. In order to use genetic algorithm, the M-function below should be established. It should be noted that here, the maximum objective function is expressed.

```
function [sol,f]=c10mga3(sol,options)
x=sol(1:4);
f=-(x(1)+x(2))^2-5*(x(3)-x(4))^2-(x(2)-2*x(3))^4-10*(x(1)-x(4))^4;
```

Selecting the ranges for x_i such that $-1 \leqslant x_i \leqslant 1$, $i = 1, 2, 3, 4$, the function `gaopt()` can be called to solve this optimization problem, with the intermediate results shown in Table 10.5.

TABLE 10.5: Intermediate search results using genetic algorithm

gen.	x_1	x_2	x_3	x_4	$f(x_1, x_2, x_3, x_4)$
1	0.053034683	0.40724952	0.13840848	−0.01682375	−0.33287436
5	0.063661288	0.084587914	0.042225122	0.096858012	−0.036913734
10	0.061948198	−0.00062694769	0.089349763	0.074719112	−0.0058649774
14	0.061948198	−0.035582875	0.089349763	0.074719112	−0.003874042
21	0.061948198	−0.039947967	0.089349763	0.087316197	−0.0027943115
28	0.030899005	−0.028464643	0.089349763	0.087316197	−0.0019697798
32	0.022521945	−0.0085565429	0.089313262	0.087316197	−0.0016188571
37	0.016334305	−0.0076107093	0.089291541	0.087316197	−0.0015513478
41	0.016334305	−0.0082057068	−0.02253066	−0.02328211	-9.5374641×10^{-5}
44	0.014834811	−0.0082619779	−0.016819586	−0.017633545	-5.8042763×10^{-5}
70	0.014761402	−0.013351969	−0.017506533	−0.017633545	-1.3300401×10^{-5}
77	0.014761402	−0.013966665	−0.01184689	−0.01104842	-8.2658502×10^{-6}
85	0.013772082	−0.013965442	−0.011333	−0.01107414	-4.1891828×10^{-6}
100	0.013121641	−0.01320957	−0.010778171	−0.010696633	-3.264236×10^{-6}

```
>> [a,b,c,d]=gaopt([-1,1; -1 1; -1 1; -1 1],'c10mga3'); x=a(1:4)
```

It is found from the solutions that $x_1 = 0.0131$, $x_2 = -0.0132$, $x_3 = -0.0108$, $x_4 = -0.0107$, with some errors in the solutions.

Similar to the procedures used in the previous example, here 2000 generations are tested. The intermediate results are shown in Table 10.6. It can be found that the results are $\boldsymbol{x}^* = [-0.0032, 0.0032, -0.0014 - 0.0014]$, which are more accurate than the ones obtained in the previous example.

```
>> xmM=[-ones(4,1),ones(4,1)];
   [a,b,c,d]=gaopt(xmM,'c10mga3',[],[],[],'maxGenTerm',2000);
```

Now consider again the unconstrained optimization problem in Chapter 6. The solution to such a problem can be solved with the following statements

```
>> f=@(x)(x(1)+x(2))^2+5*(x(3)-x(4))^2+(x(2)-2*x(3))^4+10*(x(1)-x(4))^4;
   ff=optimset; ff.MaxIter=10000; ff.TolX=1e-7;
   x=fminsearch(f,10*ones(4,1),ff)
```

with the solution $\boldsymbol{x}^{\mathrm{T}} = [0.0304, -0.0304, -0.7534, -0.7534]\times10^{-6}$. It can be seen that the accuracy of traditional methods is much higher than that of the genetic algorithm-based methods.

Solving constrained problems

It has been shown that the ordinary genetic algorithm-based methods can only be used in solving unconstrained problems. For constrained optimization problems, it is possible to convert it into unconstrained problems.

Example 10.18 Solve the following linear programming problem with genetic algorithm.

TABLE 10.6: Intermediate search results for more generations

gen.	x_1	x_2	x_3	x_4	$f(x_1,x_2,x_3,x_4)$
1	0.52943125	−0.18554265	0.17102598	0.42105685	−0.50969811
4	0.30026929	−0.15235023	0.0046551376	0.073501054	−0.072705948
9	0.25969559	−0.24700356	0.0046551376	0.073501054	−0.040194938
21	0.18429634	−0.17440885	0.051175939	0.073501054	−0.009963743
32	0.18448868	−0.17115196	0.055131714	0.062716218	−0.0089361333
40	0.15139802	−0.17115196	0.055131714	0.060727809	−0.0074944505
60	0.12157312	−0.14269825	0.0069953616	0.011040366	−0.0026235247
127	0.11836763	−0.11881963	0.0087307561	0.01355787	−0.0016683703
147	−0.055877859	0.060775121	0.0097285242	0.014063368	−0.00036014666
252	−0.044906673	0.04604043	0.0078889251	0.0078620251	-7.9664135×10^{-5}
562	−0.025653189	0.024757092	0.0070596265	0.007048969	-1.2253193×10^{-5}
765	−0.024463789	0.024672404	0.0070538013	0.007039069	-9.9062381×10^{-6}
841	−0.0026816486	0.003508872	−0.001446297	−0.0014440959	-6.8602555×10^{-7}
2000	−0.0031643672	0.0031641119	−0.0013999645	−0.0014000418	-1.3622025×10^{-9}

$$\min_{\boldsymbol{x} \text{ s.t.} \begin{cases} -2x_1+x_2+x_3\leqslant 9 \\ -x_1+x_2\geqslant -4 \\ 4x_1-2x_2-3x_3=-6 \\ x_{1,2}\leqslant 0, x_3\geqslant 0 \end{cases}} (x_1+2x_2+3x_3)$$

Solution From the equality constraint, it is found that $x_3 = (6+4x_1-2x_2)/3$. Substituting x_3 into the original problem, problems with two variables can be reformulated. The following M-function can be written to describe the objective function

```
function [sol,y]=c10mga4(sol,options)
x=sol(1:2); x=x(:); x(3)=(6+4*x(1)-2*x(2))/3;
y1=[-2 1 1]*x; y2=[-1 1 0]*x;
if (y1>9 | y2<-4 | x(3)<0), y=-100; else, y=-[1 2 3]*x; end
```

where x_3 can be calculated first from x_1 and x_2. The constraints are checked with $\boldsymbol{x}$. For the points where the constraints are not satisfied, the objective functions are set to -100 as penalties. The constrained optimization problem can now be solved using the genetic algorithm. The intermediate results are listed in Table 10.7.

```
>> [a,b,c]=gaopt([-200 0; -200 0],'c10mga4',[],[],[],'maxGenTerm',1000)
```

The obtained results are $x_1 = -6.9983, x_2 = -10.9967$, and it is found from $x_3 = (6+4x_1-2x_2)/3$ that $x_3 = 3.2583\times10^{-5}$.

In fact, the linear programming function can be used to find a more precise solution, where $\boldsymbol{x}^{\mathrm{T}} = [-6.99999999999967, -10.99999999999935, 0]$.

```
>> f=[1 2 3]; A=[-2 1 1; 1 -1 0]; B=[9; 4]; Aeq=[4 -2 -3]; Beq=-6;
   x=linprog(f,A,B,Aeq,Beq,[-inf;-inf;0],[0;0;inf]);
```

TABLE 10.7: Intermediate search results using genetic algorithm

gen.	x_1	x_2	$f(\boldsymbol{x})$	gen.	x_1	x_2	$f(\boldsymbol{x})$
1	−186.28915	−196.6231	−100	596	−6.9076459	−10.824295	28.53823
58	−1.0379947	0	−0.81002637	797	−6.9372398	−10.909409	28.686199
63	−1.2028576	0	0.01428822	841	−6.9796692	−10.968332	28.898346
114	−1.3578999	−2.0200479	0.78949953	921	−6.9909303	−10.982501	28.954652
141	−6.1888743	−9.9596522	24.944372	995	−6.9982728	−10.996638	28.991364
230	−6.6159194	−10.232137	27.079597	999	−6.998303	−10.996655	28.991515
481	−6.7766327	−10.561651	27.883163	1000	−6.998303	−10.996655	28.991515

10.3.3 Particle swarm optimizations

Particle swarm optimization (PSO) is a class of evolution-based optimization algorithm initially proposed in Reference [53]. The algorithm is motivated by the phenomenon in nature where birds are seeking food. It is a useful algorithm in finding global solutions to optimization problems.

Assume that within a certain area, there is a piece of food (global optimum point), and there are a flock of randomly distributed birds (or particles). Each particle has its personal best value $p_{i,\mathrm{b}}$, and the swarm has its best value g_b up to now. The position and speed of each particle can be updated with the formula

$$v_i(k+1) = \phi(k)v_i(k) + \alpha_1\gamma_{1i}(k)[p_{i,\mathrm{b}} - x_i(k)] + \alpha_2\gamma_{2i}(k)[g_\mathrm{b} - x_i(k)] \quad (10.6)$$

$$x_i(k+1) = x_i(k) + v_i(k+1) \quad (10.7)$$

where γ_{1i} and γ_{2i} are uniformly distributed random numbers in the interval $[0, 1]$. The argument $\phi(k)$ is the momentum function, while α_1 and α_2 are acceleration constants.

A particle swarm optimization toolbox(PSOt)[54] developed by Brian Birge can be used to solve optimization problems. The toolbox is provided on the companion CD. The main function of the toolbox, `pso_Trelea_vectorized()`, can be used to perform the optimization process, whose syntax is

```
[sol,t,y]=pso_Trelea_vectorized(fun,n,vM,[xm,xM],key,options)
```

where *fun* is used to describe the objective function and it can be given as an M-function. The argument n is the number of variables to be optimized. These two arguments must be provided in solving the optimization problems. The other arguments are optional, with v_M the maximum allowed speed, default 4; $\boldsymbol{x}_\mathrm{m}$ and $\boldsymbol{x}_\mathrm{M}$ the boundaries of the variable $\boldsymbol{x}$, defaults ± 100; `key` the type of optimization, default 0 for minimization and 1 for maximization; `options` contains the control variables. The returned argument `sol` is an $(n+1)\times 1$ column vector, whose first n elements form the solution vector $\boldsymbol{x}$, and the $(n+1)$th element the objective function. The returned arguments $\boldsymbol{t}$ and $\boldsymbol{y}$ are the recorded intermediate objective functions. In the PSO function, Trelea algorithm[55] is implemented, and some functions in the Neural Network Toolbox are used for particle swarm optimization. The intermediate results can also be shown in a graphics window. The Clerc algorithm[56] can also be used in the toolbox.

The "vectorized" version of the objective function is allowed, which means that the simultaneous computation for a swarm of particles is evaluated. The computational efficiency is much higher than the ordinary nonvectorized version. Note that the argument $\boldsymbol{x}$ can be a matrix, whose ith column corresponds to the values of the ith particle, thus it should be expressed by $\boldsymbol{x}(:,i)$. Examples are given below to demonstrate the particle swarm optimization method.

Example 10.19 Solve the unconstrained optimization problem in Example 10.17 by the PSO method.

Solution The objective function for the PSO solver can be expressed by an M-function

```
function f=c10mpso1(x)
f=(x(:,1)+x(:,2)).^2 + 5*(x(:,3)-x(:,4)).^2 +...
  (x(:,2)-2*x(:,3)).^4 + 10*(x(:,1)-x(:,4)).^4;
```

It should be noted that vectorized descriptions and dot operations are used in describing the objective function. The objective function similar to the ideas used in the genetic algorithm example can be written. The problem can then be used to solve the problem

```
>> x=pso_Trelea_vectorized('c10mpso1',4); x(1:4)
```

and the approximate solution $x = [-0.75, 0.75, -0.38, -0.38]^{\mathrm{T}} \times 10^{-3}$ can be obtained. The solution process can be visualized in Figure 10.26.

Example 10.20 Consider again the constrained optimization problem in Example 10.18. Find the solution with the particle swarm optimization method.

Solution Similar to the M-function for the genetic algorithm, another M-function can be written for the objective function with the constraints. It should be noted that vectorized format should be used.

```
function y=c10mpso4(x)
x1=x(:,1); x2=x(:,2); x3=(6+4*x1-2*x2)/3; x=[x x3]';
y1=[-2 1 1]*x; y2=[-1 1 0]*x; y=[1 2 3]*x;
ii=find(y1>9|y2<-4|x3'<0); y(ii)=100; y=y(:);
```

With the above MATLAB statements, the accurate solution $\boldsymbol{x}^{\mathrm{T}} = [-7, -11, 0]$ can be obtained.

```
>> [x,tr]=pso_Trelea_vectorized('c10mpso4',2);
   x=[x(1:2); (6+4*x(1)-2*x(2))/3]
```

10.3.4 Solving optimization problems with GADS Toolbox

The `ga()` function provided in the GADS Toolbox offers an alternative way in solving optimization problems using genetic algorithms. The function is

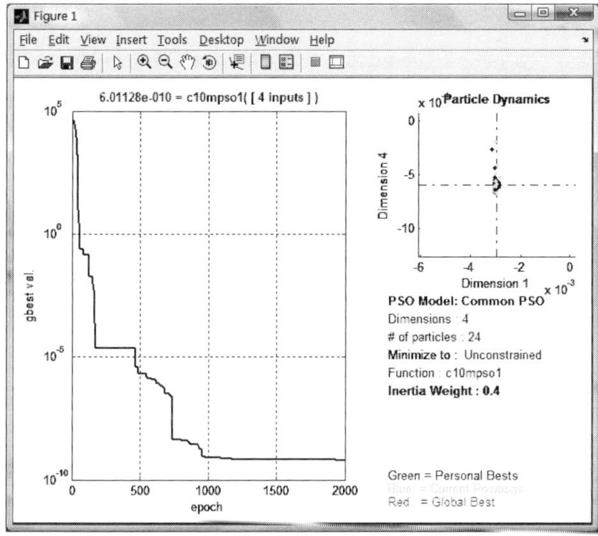

FIGURE 10.26: Visualization of the particle swarm optimization process

TABLE 10.8: Commonly used control properties in GADS Toolbox

property name	explanation to the options
`Generations`	maximum allowed generations, with a default 100
`InitialPopulation`	initial population matrix, the default population is created by random numbers
`PopulationSize`	the number of individuals in the population, with default 20
`SelectionFcn`	setting of the selective function, and the default `@selectionstochunif`, and other functions `@selectionremainder`, `@selectionuniform`, `@selection roulette`, `@selectiontournament`
`TolFun`	similar to `optimset` properties, the termination condition for the objective function. Also `TolX`, `TolCon` can be set

quite similar to other traditional optimization functions and it can be used to solve constrained problems as well. The syntax of the function is

```
[x,f,flag,out]=ga(fun,n,opts)          % simplest unconstrained problem
[x,f,flag,out]=ga(fun,n,A,B,Aeq,Beq,xm,xM,CFun,opts) % constrained
```

where *fun* is a MATLAB description of the objective function, whose format is consistent to the ones in the Optimization Toolbox. The argument n is the number of variables to be optimized. The `opts` argument contains the control properties which can be set by the function `gaoptimset()`, similar to `optimset()` function in Chapter 6. The commonly used control properties are listed in Table 10.8.

The search results are returned in the vector $\boldsymbol{x}$. The other input and returned arguments are the same as the functions in the Optimization Toolbox. In particular, if `flag` is larger than 0, successful solutions to the problem are achieved.

A graphical user interface, `gatool()`, is also provided in the GADS Toolbox. The interface is shown in Figure 10.27. The properties and settings can be modified in a visual way. The Start button can be used to start the searching process using genetic algorithm to find the optimum point.

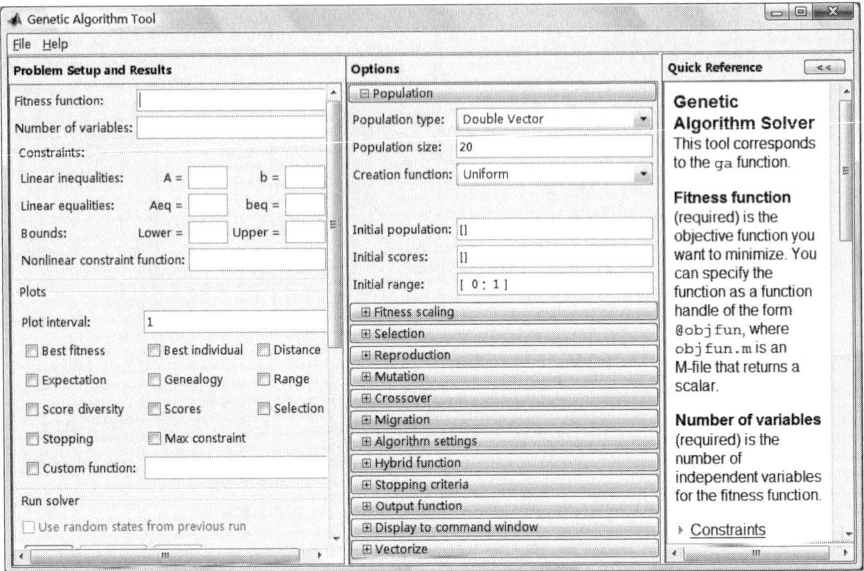

FIGURE 10.27: Graphical interface of GADS Toolbox

Example 10.21 Solve the problems in Examples 10.17 and 10.18 using GADS Toolbox.

Solution The objective function can be expressed by an anonymous function. The function `ga()` can then be used to solve the optimization problem.

```
>> f=@(x)(x(1)+x(2))^2+5*(x(3)-x(4))^2+(x(2)-2*x(3))^4+10*(x(1)-x(4))^4;
   x=ga(f,4)
```

The solution found is $\boldsymbol{x}^{\mathrm{T}} = [-0.0620, 0.0489, -0.0226, 0.022]$. One may further set control properties to increase the accuracy of the solution

```
>> ff=gaoptimset; ff.Generations=2000; ff.PopulationSize=80;
   ff.SelectionFcn=@selectiontournament;
   ff.TolX=1e-6; ff.TolX=1e-10; x=ga(f,4,ff)
```

and the improved solution is obtained as $\boldsymbol{x}^{\mathrm{T}} = [0.0176, 0.0145, 0.0171, 0.0231]$.

Consider the problem in Example 10.18. If the following statements are used, the solution may not be successful due to the equality constraint.

```
>> f=@(x)[1 2 3]*x(:); A=[-2 1 1; 1 -1 0]; B=[9; 4];
   Aeq=[4 -2 -3]; Beq=-6; x=ga(f,3,A,B,Aeq,Beq)
```

With the objective function defined for PSO, c10mpso4.m, the ga() function can be used and the solution obtained is $\boldsymbol{x} = [-6.9041, -10.8267, 0.0123]$, which is, for this example, not quite accurate.

```
>> x=ga(@c10mpso4,2); x=[x, (6+4*x(1)-2*x(2))/3]
```

10.3.5 Towards accurate global minimum solutions

In general, it can be concluded that the traditional optimization algorithms may find a solution with good accuracy, however the global optimality cannot be ensured. For oscillatory surfaces, local minima are inevitable. The evolution algorithms can likely find the global optimal solutions, since multiple initial points are used. However, the accuracy of the solution may not be satisfactorily high. A combination of the two types of optimization algorithms can be made, for instance, with the use of evolution algorithms, a less accurate global solution can be found first. This solution can then be used as an initial search point for the traditional optimization algorithms in order to find the accurate global optimum solutions.

Example 10.22 Solve accurately the optimization problem

$$\min_{(x,y)\ \text{s.t.} \begin{cases} -1\leqslant x\leqslant 3 \\ -3\leqslant y\leqslant 3 \end{cases}} \sin(3xy) + (x-0.1)(y-1) + x^2 + y^2.$$

Solution The surface plot of the objective function can be obtained as shown in Figure 10.28. It can be seen that the surface is oscillatory, which makes the global optimization a difficult task.

```
>> [x,y]=meshgrid(-1:0.1:3,-3:0.1:3);
   z=sin(3*x.*y+2)+(x-0.1).*(y-1)+x.^2+y.^2; surf(x,y,z);
```

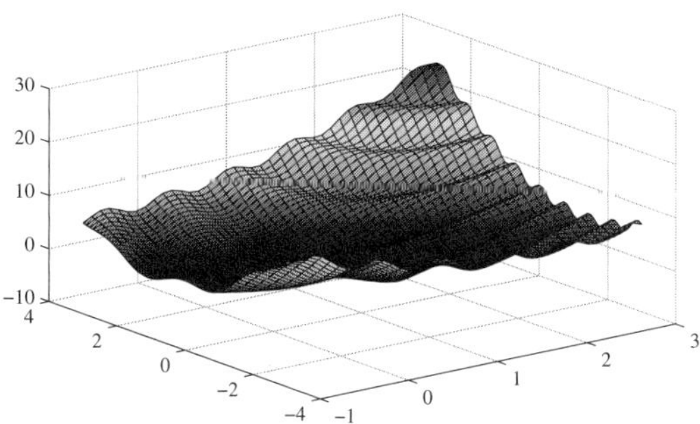

FIGURE 10.28: Surface of the objective function

If the traditional nonlinear programming technique is used, using an anonymous function to describe the objective function, the following statements can be used

```
>> f=@(x)sin(3*x(1)*x(2)+2)+(x(1)-0.1)*(x(2)-1)+x(1)^2+x(2)^2;
   x0=rand(1,2); x=fmincon(f,x0,[],[],[],[],[-1;-3],[3;3]), f(x)
```

and the optimum point is $\boldsymbol{x}^{\mathrm{T}} = [0.9299, 0.6577]$, with the objective function value 0.3742.

Using GAOT, the objective function can be written as

```
function [sol,y]=c10mga6(sol,options)
x=sol(1:2); y=-sin(3*x(1)*x(2)+2)-(x(1)-0.1)*(x(2)-1)-x(1)^2-x(2)^2;
```

Within the selected search boundaries, the problem can be solved again to find the solution, which is $x_1 = 1.225588050491944$, $x_2 = -0.918235615244950$. The objective function in this case is -0.795033762096730. It is clear that the solution by traditional optimization method is a poor local minimum.

```
>> xmM=[-1 3; -3 3];
   [a,b,c,d]=gaopt(xmM,'c10mga6'); x=a(1:2), f0=-a(3)
```

Using the results from applying the genetic algorithm as the initial search points, the following statements can be used to find a more accurate solution which is $x_1 = 1.225533324173843$ and $x_2 = -0.918286942316160$, and the objective function becomes -0.795033772873523, which is slightly better than the genetic algorithm result.

```
>> ff=optimset; ff.TolX=1e-10; ff.TolFun=1e-20;
   x=fmincon(f,a(1:2),[],[],[],[],[-1;-3],[3;3],[],ff), f(x)
```

10.4 Wavelet Transform and Its Applications in Data Processing

Fourier transform is a very important technique in signal processing. However, the ordinary Fourier transform has certain limitations. For instance, it maps the time-domain representation into the frequency-domain representation, where features in the original domain may get lost. Therefore, for many types of signals, the processed results may not be satisfactory. For stationary signals, Fourier transform may be very useful. However, for non-stationary and transient signals with sudden changes, the Fourier transform may not be suitable, since the features of the sudden changes may be neglected. Thus other types of transforms, for instance, short-time Fourier transform should be used. Moreover, the wavelet transform technique emerged in the 1980s is a very powerful tool used widely, in particular, in signal and image processing.

10.4.1 Wavelet transform and waveforms of wavelet bases

A wavelet is a function with zero mean which can be translated and scaled to form a family of functions. In this section, fundamental concepts of continuous and discrete wavelet transforms are introduced. Some of the commonly used wavelet bases are introduced.

Continuous wavelet transform

The continuous wavelet transformation formula is given by

$$\mathscr{W}_{a,b}^{\psi}[f(t)] = \frac{1}{\sqrt{|a|}} \int_{-\infty}^{\infty} f(t)\overline{\psi_{a,b}(t)}\,\mathrm{d}t = W_{\psi}(a,b) \tag{10.8}$$

where

$$\psi_{a,b}(t) = \psi\left(\frac{t-b}{a}\right), \quad \text{and} \quad \int_{-\infty}^{\infty} \psi(t)\,\mathrm{d}t = 0 \tag{10.9}$$

and the function $\psi(t)$ is referred to as *wavelet basis*, and $\psi_{a,b}(t)$ is generated by translation and scaling from the wavelet basis.

Example 10.23 Consider the "Mexican hat" wavelet basis given by $\psi(t) = \dfrac{1-t^2}{\sqrt{2\pi}}\mathrm{e}^{-\frac{t^2}{2}}$. Draw the wavelet bases for different values of a and b.

Solution From the given mathematical formula of the wavelet basis, the function `ezplot()` can be used to draw the relevant curves. The translation and scaling operations can be carried out with the `subs()` function. The curves for different values of a and b can be obtained as shown in Figures 10.29 (a) and (b).

```
>> syms t; f=(1-t^2)*exp(-t^2/2)/sqrt(2*pi);
   ezplot(f,-4,4), hold on;  % wavelet basis plot
   ezplot(subs(f,t,t-1),-4,4); ezplot(subs(f,t,t+1),-4,4) % translation
   figure; ezplot(f,-4,4), hold on;
   ezplot(subs(f,t,t/2),-4,4); ezplot(subs(f,t,2*t),-4,4) % scaling
```

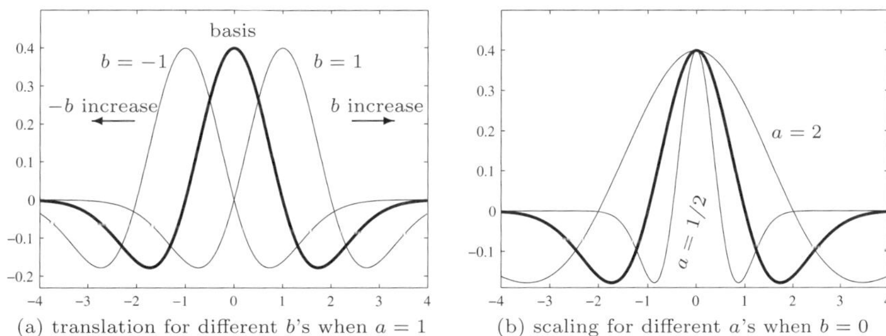

(a) translation for different b's when $a=1$ (b) scaling for different a's when $b=0$

FIGURE 10.29: Waveforms of Mexican hat basis for different a and b

From the above example, it is seen that the parameter b can translate the wavelet basis to the right or left while a expands or shrinks the waveform accordingly. If $a < 1$, the waveform of the basis will be shrunken to form the new wavelet signal. Wavelet analysis is performed by computing the coefficients for different combinations of a and b. The wavelet signals multiplied by the corresponding coefficients sum up to reconstruct the original signal. Similar to other integral transform problems, the inverse wavelet transforms are also defined:

$$f(t) = \frac{1}{C_\psi} \int_{-\infty}^{\infty} \int_{-\infty}^{\infty} W_\psi(a, b)\psi_{a,b}(t)\, \mathrm{d}a\mathrm{d}b \tag{10.10}$$

where

$$C_\psi = \int_{-\infty}^{\infty} \frac{|\hat{\psi}(\omega)|^2}{|\omega|}\, \mathrm{d}\omega. \tag{10.11}$$

The coefficients of continuous wavelet transforms can be evaluated with the `cwt()` function, with the following syntaxes

```
Z=cwt(y,a,fun)           % wavelet coefficients matrix Z
Z=cwt(y,a,fun,'plot')    % absolute value plots
```

where the *fun* is the name of wavelet basis to be discussed more later. Here, the wavelet basis name for Mexican hat wavelet basis is denoted by `'mexh'`.

Example 10.24 Perform continuous wavelet transform for the function $f(t) = \sin t^2$, and visualize its coefficients.

Solution The data in the interval $t \in [0, 2\pi]$ can be generated directly with the following statements and the time domain function is shown in Figure 10.30 (a).

```
>> t=0:0.03:2*pi; y=sin(t.^2); plot(t,y)
```

Taking the option `'mexh'` wavelet basis as the template, the wavelet coefficients $W_\psi(a, b)$ can be obtained as shown in Figure 10.30 (b).

```
>> a=1:32; Z=cwt(y,a,'mexh','plot');
```

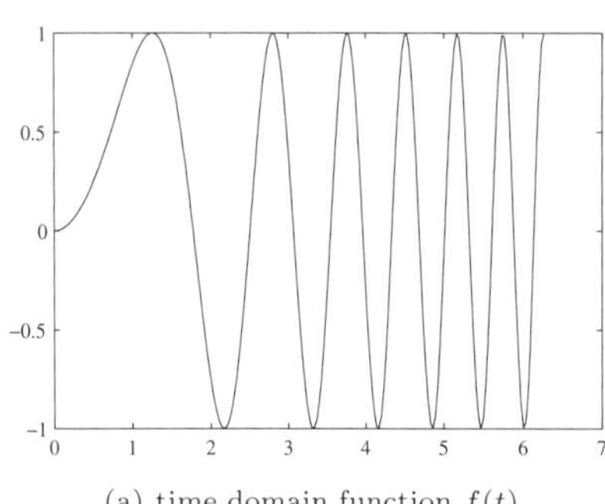

(a) time domain function $f(t)$ (b) coefficients of continuous wavelet transform

FIGURE 10.30: Continuous wavelet transform

The following statements can be used to draw the surface plot of the wavelet coefficients, as shown in Figure 10.31.

```
>> surf(t,a,Z); shading flat; axis([0 2*pi,0,32,min(Z(:)) max(Z(:))])
```

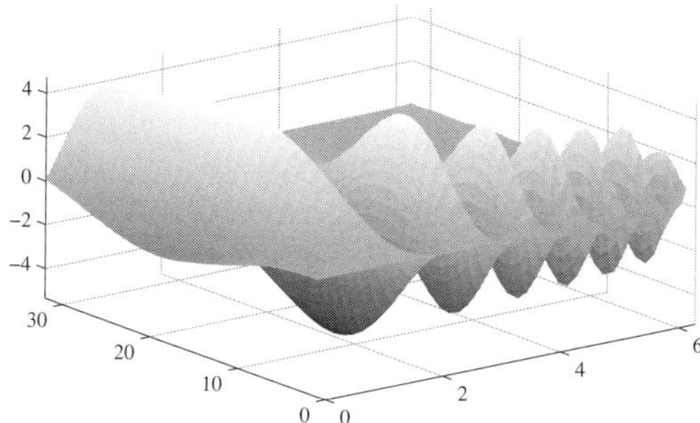

FIGURE 10.31: Surface plot of continuous wavelet transform coefficients

Discrete wavelet transform

If the signal $f(t)$ is expressed by a discrete sequence $f(k)$, and the wavelet basis $\psi(t)$ is selected, the translated and scaled wavelet function can be written as $\psi_{a,b}(t) = \sqrt{2}\psi\left(2^m t - n\right)$, whose discrete form can be written as $\psi_{a,b}(k) = \sqrt{2}\psi\left(2^m k - n\right)$. The wavelet transform of the discrete signal can be written as

$$\mathscr{W}_{n,m}^{\psi}[f(k)] = \sqrt{2}\sum_{k} f(k)\overline{\psi\left(2^m k - n\right)}\mathrm{d}t = W_{\psi}(m,n) \tag{10.12}$$

and the inverse discrete wavelet transform is expressed as

$$f(k) = \sum_{m}\sum_{n} W_{n,m}(k)\psi_{m,n}(k). \tag{10.13}$$

The function `dwt()` provided in the Wavelet Toolbox can be used for discrete wavelet transform, with the syntax `[`C_{a}`,`C_{d}`]=dwt(`x`,`*fun*`)`, where x is the original data, *fun* is the selected wavelet basis, which can be `'mexh'` or other functions to be discussed more later. The returned argument C_{a} can be used to describe the approximated coefficients of the original function, while the argument C_{d} returns the detailed coefficients. Normally the argument C_{a} corresponds to the low- and mid-frequency information, while C_{d} corresponds to the high-frequency information. The lengths of vectors C_{a} and C_{d} are half of the length of the original vector x.

The vectors C_{a} and C_{d} can be used to perform inverse wavelet transform with the function $\hat{x}$`=idwt(`C_{a}`,`C_{d}`,`*fun*`)` to restore the original vector $\hat{x}$.

Example 10.25 Generate a signal sequence corrupted by noises. Perform discrete wavelet transform to the sequence. Perform also inverse wavelet transform to the sequence, and see whether the original function can be restored.

Solution The white noise with a standard deviation of 0.1 is added to the original signal $f(t) = \sin t^2$ given in Example 10.24. The corrupted data are shown in Figure 10.32. It can be seen that the noise level is quite visible. Using discrete wavelet transform, the waveforms of approximated and detailed coefficients are shown in the same window. It can be seen that the resulted signal removes noises to some extent. For better filtering results, more steps in wavelet transform should be used.

```
>> x=0:0.002:2*pi; y=sin(x.^2); r=0.1*randn(size(x));
   y1=y+r; subplot(211); plot(x,y1)   % draw the original signal
   [cA,cD]=dwt(y1,'db4');             % discrete wavelet transform
   subplot(223), plot(cA); subplot(224), plot(cD)
```

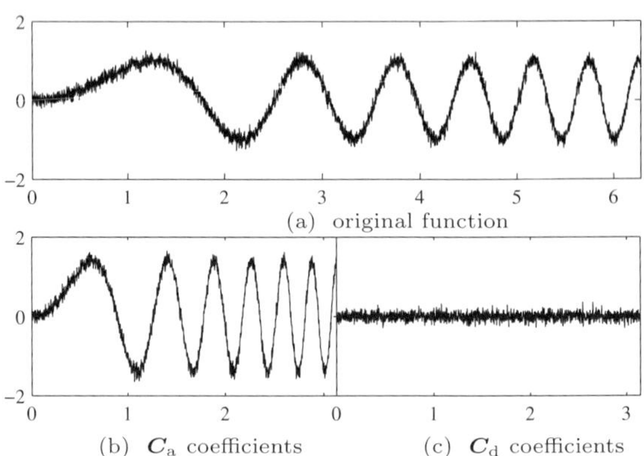

FIGURE 10.32: Discrete wavelet transform of a given function

The obtained C_a and C_d vectors can be used to perform the inverse discrete wavelet transform. Comparing the transformed sequence and the original sequence, it can be seen that the error is very small, with an error norm of 6.16×10^{-12}.

```
>> y2=idwt(cA,cD,'db4'); norm(y1-y2)
```

Wavelet bases provided in the Wavelet Toolbox

A large amount of wavelet bases such as Haar basis, Daubechies basis, Mexican hat basis, Bior basis, etc. can be generated and tested by `wavemngr()` function. The syntax of the function is `wavemngr('read',1)`. For instance, Haar wavelet basis is denoted as `'haar'`. Daubechies families are denoted as `'db1'`, `'db2'`, etc. Bior families are `'bior1.3'`, `'bior2.4'`, etc. Mexican hat wavelet basis can be denoted as `'mexh'`. The wavelet basis function can be evaluated with the `wavefun()` function. The function can be called as

```
[ψ,x]=wavefun(fun,n)          % Gaussian, Mexican hat wavelet bases
[φ,ψ,x]=wavefun(fun,n)        % Daubechies, Symlets orthogonal families
[φ1,ψ1,φ2,ψ2,x]=wavefun(fun,n)    % Bior basis
```

where n is the number of iterations computed, default is 8. The argument ψ is the wavelet basis, and ϕ is the wavelet derivatives. In Bior wavelet bases, the vectors ϕ_1 and ψ_1 are used for wavelet computation, while the vectors ϕ_2 and ψ_2 are used for wavelet reconstruction.

Example 10.26 Select Daubechies 6 wavelet basis (`'db6'`) to draw the wavelet waveforms of different iterations.

Solution The iteration numbers 2, 4, 6, 8 can be used in the function call to draw the wavelets as shown in Figure 10.33. It can be seen that when the iteration is 8, smooth waveform can be achieved. Thus normally, one should select the iteration number to 8.

```
>> [a,y,x]=wavefun('db6',2); subplot(141), plot(x,y)
   [a,y,x]=wavefun('db6',4); subplot(142), plot(x,y)
   [a,y,x]=wavefun('db6',6); subplot(143), plot(x,y)
   [a,y,x]=wavefun('db6',8); subplot(144), plot(x,y)
```

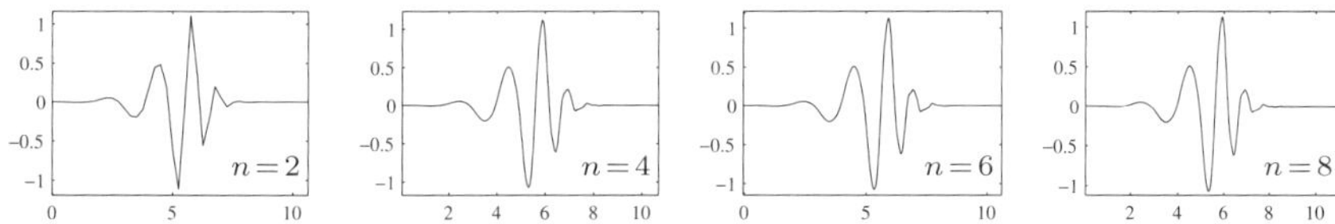

FIGURE 10.33: The waveforms of Daubechies 6 for different iterations

Example 10.27 Show the waveforms of some commonly used wavelet bases.

Solution The following statements can be used to draw wavelets of some selected wavelet bases, as shown in Figure 10.34, where `'db1'` is in fact the Haar basis. Daubechies wavelet families are smooth when `'db6'` is used. For Symlets wavelet families, the function `'sym6'` is smooth.

```
>> subplot(5,4,1), [a,y,x]=wavefun('db1'); plot(x,y), % other db bases
   subplot(5,4,9), [a,y,x]=wavefun('sym2'); plot(x,y),% other sym bases
   subplot(5,4,13), [a,y,x]=wavefun('coif2'); plot(x,y)
   subplot(5,4,15), [y,x]=wavefun('gaus2'); plot(x,y)
   subplot(5,4,17), [a,y,b,c,x]=wavefun('bior1.3'); plot(x,y)
```

10.4.2 Wavelet transform in signal processing problems

Similar to the Fourier transform technique, wavelet transform is also very useful in signal processing and two-dimensional signal processing, i.e., image processing. Some unique characteristics of wavelet transform are not offered in conventional frequency domain analysis. Here, wavelet transform-based signal decomposition and reconstruction in MATLAB are demonstrated by a de-noising example in this section.

Wavelet transform can be used to perform signal decomposition, that is, to express a given signal S by two parts, $\boldsymbol{C}_{\mathrm{a}_1}$ and $\boldsymbol{C}_{\mathrm{d}_1}$. The lengths of the two vectors are half the length of S. The vector $\boldsymbol{C}_{\mathrm{a}_1}$ preserves the low- and mid-frequency information or "approxiamte" information, while $\boldsymbol{C}_{\mathrm{d}_1}$ retains the high-frequency, or "detailed" information. From the viewpoint of de-noising signal processing task, the latter high-frequency content can be regarded as noise information. The vector $\boldsymbol{C}_{\mathrm{a}_1}$ can further be decomposed with wavelet

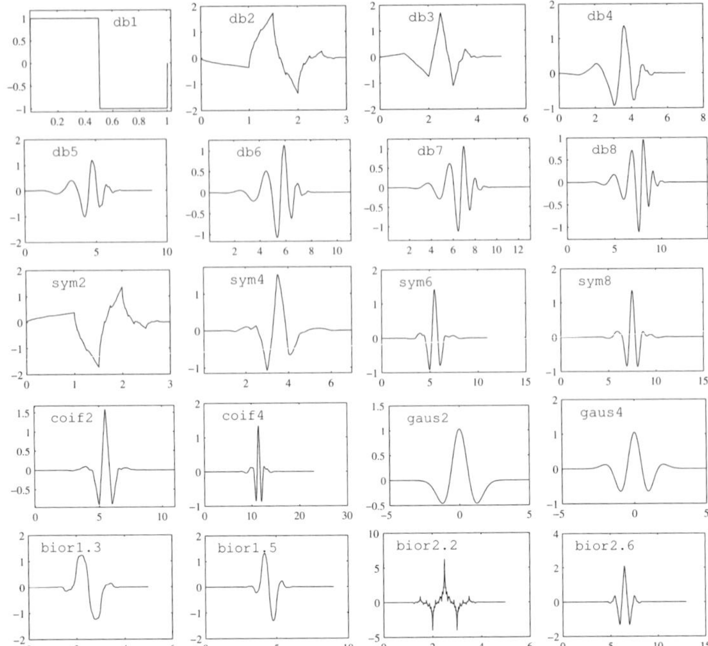

FIGURE 10.34: Waveforms of commonly used wavelet bases

transform into C_{a_2} and C_{d_2}. Further decomposition leads to C_{a_3} and C_{d_3}, and so on. The decomposition process is shown in Figure 10.35 (a). The vectors C_{a_i} are referred to as the *approximate coefficients* and C_{d_i} is the *detailed coefficients*.

The function `wavedec()` provided in Wavelet Toolbox can be used in one-dimensional wavelet decomposition, with the syntax `[C,L]=wavedec(x,n,fun)`, where x is the original signal, n is the level of decomposition. *fun* is the name of the wavelet basis, for instance `'db6'`. After the decomposition, the two vectors C and L are returned, as shown in Figure 10.35 (b). The original vector C is decomposed into a vector whose length is the same as that of x. The short sub-vectors are joined together and the lengths of the sub-vectors are indexed in the vector L.

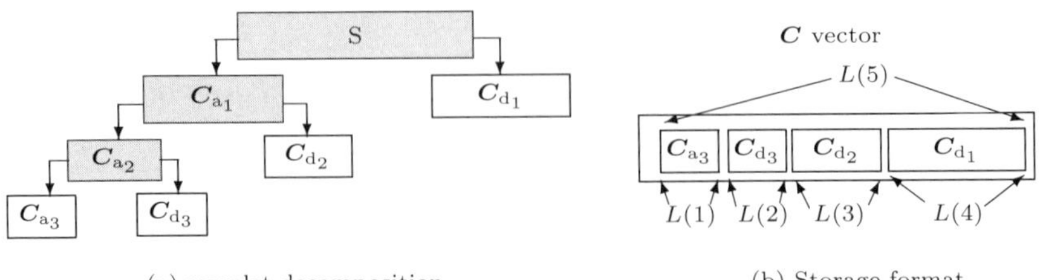

FIGURE 10.35: Illustrations of wavelet decomposition

The approximate coefficients $\boldsymbol{C}_{\rm a}$ and detailed coefficients $\boldsymbol{C}_{\rm d}$ after decomposition can be extracted from the vectors $\boldsymbol{C}$ and $\boldsymbol{L}$, with the functions `appcoef()` and `detcoef()`, respectively, by the syntaxes

```
Ca_i=appcoef(C,L,fun,n);   % extract approximate coefficients
Cd_i=detcoef(C,L,i);       % extract detailed coefficients
```

where n is the level of extraction.

Some of the noises can be filtered out from the approximate coefficients and detailed coefficients, by reconstructing the signal with the function `wrcoef()`, whose syntax is `x̂=wrcoef(type,C,L,fun,n)`, where *type* can be set to `'a'` or `'d'`, representing approximate coefficients and detailed coefficients, respectively. If `'a'` is selected, the reconstructed signal may filter out certain high-frequency noises.

Example 10.28 For the data given in Example 10.25, three-level wavelet decomposition can be performed. Compare the de-noise effects for different wavelet bases.

Solution The noisy signal in Example 10.25 shown in Figure 10.32 (a) can still be used for de-noising tests.

```
>> x=0:0.002:2*pi; y=sin(x.^2); r=0.1*randn(size(x)); y1=y+r; plot(x,y1)
```

For three-level wavelet decomposition, the following statements can be used to get the waveforms of the sub-vectors as shown in Figure 10.36, where the widths of the axes have been rearranged manually. It can be seen that part of the noise is filtered out in each decomposition level. Thus the final $\boldsymbol{C}_{\rm a_3}$ contains less noise information.

```
>> [C,L]=wavedec(y1,3,'db6');
   cA3=C(1:L(1)); subplot(141), plot(cA3)
   dA3=C(L(1)+1:sum(L([1 2]))); subplot(142), plot(dA3)
   dA2=C(sum(L(1:2))+1:sum(L(1:3))); subplot(143), plot(dA2)
   dA1=C(sum(L(1:3))+1:sum(L(1:4))); subplot(144), plot(dA1)
```

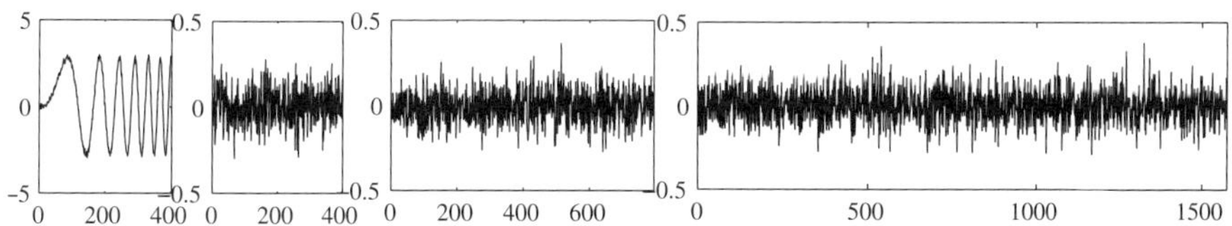

FIGURE 10.36: Wavelet decomposition illustrations

The wavelet basis `'db6'` can be used and the approximate coefficients can be obtained as shown in Figure 10.37 (a). If `'db2'` is used instead, the de-noising result shown in Figure 10.37 (b) can be obtained. The two wavelet bases do not have significant differences in de-noising effect for this example.

```
>> A3=wrcoef('a',C,L,'db6',3); plot(A3); figure
   [C,L]=wavedec(y1,3,'db2'); A3=wrcoef('a',C,L,'db2',3); plot(A3)
```

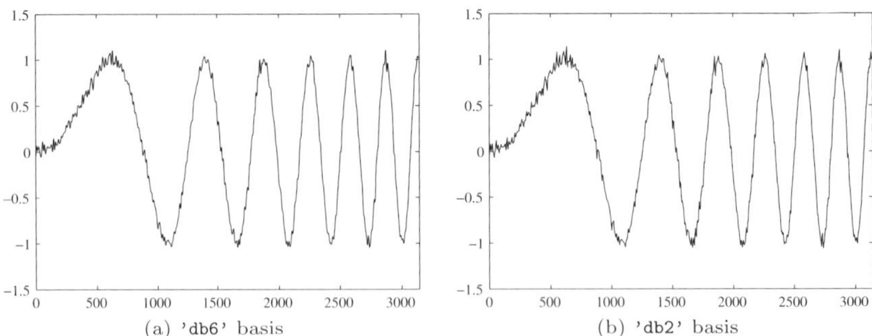

(a) 'db6' basis (b) 'db2' basis

FIGURE 10.37: De-noising effects under different wavelet bases

Two commonly used wavelet bases, 'bior2.6' and 'coif4' can also be tried for the de-noising purposes, and the results are still the same as the one shown in Figure 10.37 (a). It is seen that for this example, there is no significant difference in the de-noising effect.

```
>> [C,L]=wavedec(y1,3,'bior2.6'); A3=wrcoef('a',C,L,'bior2.6',3);
   [C,L]=wavedec(y1,3,'coif4'); A3a=wrcoef('a',C,L,'coif4',3);
   plot(A3), figure, plot(A3a)
```

Example 10.29 Reconsider the digital filtering problem in Example 8.32. Filter the corrupted signal using wavelet method and compare the results.

Solution The filter in Example 8.32 can be used again for the noisy signal. The 4-level wavelet filter with 'db6' basis is used for the same signal. The de-noising effects are compared in Figure 10.38. It can be seen that there exists delay in the filtering result in Chapter 8 while the wavelet-based filtering contributes no delay and the de-noising effect is much better.

```
>> b=1.2296e-6*conv([1 4 6 4 1],[1 3 3 1]); a=conv([1,-0.7265],...
       conv([1,-1.488,0.5644],conv([1,-1.595,0.6769],[1,-1.78,0.8713])));
   x=0:0.002:2; y=exp(-x).*sin(5*x); r=0.05*randn(size(x)); y1=y+r;
   y2=filter(b,a,y1);  % filter used in Example 8.32
   [C,L]=wavedec(y1,4,'db6'); A4=wrcoef('a',C,L,'db6',4);
   plot(x,y,x,y2,x,A4)  % comparisons of the two filters
```

10.4.3 Graphical user interface in wavelets

An easy-to-use graphical user interface is provided in the Wavelet Toolbox, which can be used for solving one-dimensional and two-dimensional wavelet transform problems. The command `wavemenu` is used to start the interface, and the window shown in Figure 10.39 appears. If a one-dimensional wavelet transform problem is to be solved, the Wavelet 1-D button should be clicked. The interface guides the user to import data file, select wavelet basis and perform wavelet analysis for the data. Details of using the interface will not be given in this book. The interested readers are suggested to consult the

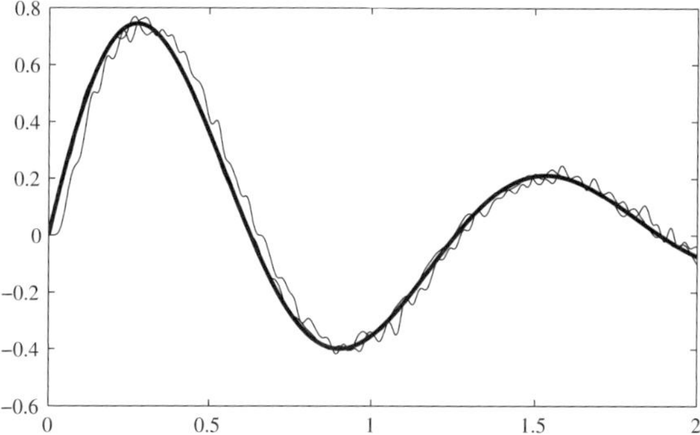

FIGURE 10.38: De-noising effect comparisons for given signal

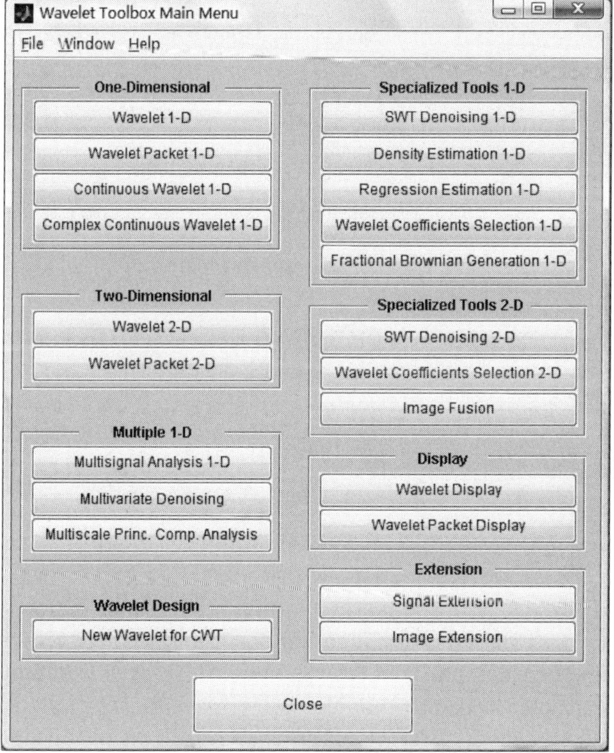

FIGURE 10.39: Graphical user interface for wavelet analysis

users' manual of Wavelet Toolbox and play with the demos offered in the toolbox.

10.5 Rough Set Theory and Its Applications

10.5.1 Introduction to rough set theory

Rough set theory

The idea of rough set was publicized by Professor Zdzisław Pawlak, a Polish mathematician, in 1982, for the development of automatic rule generation systems and soft-computing problems. In the earlier 1990s, the researchers began to realize the importance of rough sets. In 1991, Pawlak published a research monograph, which established the mathematical basis of rough sets[57]. Rough set theory provides a new mathematical framework for processing imprecise and incomplete information, and finding hidden rules from huge amounts of data.

It is widely recognized that rough set theory is the foundation for data mining, knowledge discovery and information reduction problems.

Fundamental concepts in rough sets

Assume that $X, Y \in U$ and R is the equivalent relationship defined on the universe U. The *lower-approximation set* of the set X on R is defined as

$$\underline{\mathscr{R}}(X) = \bigcup\{Y \in U/R:\ Y \subseteq X\} \tag{10.14}$$

where $\underline{\mathscr{R}}(X)$ is the maximum set of entities that for sure belong to the set X, also referred to as *positive region*, denoted by $\mathrm{POS}(X)$.

Similarly, the *upper-approximation set* of the set X on R is defined as

$$\overline{\mathscr{R}}(X) = \bigcup\{Y \in U/R:\ Y \cap R \neq \phi\} \tag{10.15}$$

where ϕ is an empty set. $\overline{\mathscr{R}}(X)$ is the minimum set of entities which possibly belong to set X.

Based on the above definitions, the boundary set can be defined as $\mathrm{BND}(X) = \overline{\mathscr{R}}(X) - \underline{\mathscr{R}}(X)$. If $\mathrm{BND}(X)$ is an empty set, the set X reduces to a crisp set on R. If on the other hand, $\mathrm{BND}(X)$ is non-empty, X is a rough set on R.

Suppose that the decision table can be represented by matrix $\boldsymbol{S}$ in MATLAB. One can extract its first m columns to represent conditional attributes C, and the rest of the columns for decisional attributes D. The upper-approximation set $\overline{\mathscr{R}}(X)$ and lower-approximation set $\underline{\mathscr{R}}(X)$ can be obtained by calling the `rslower()` and `rsupper()` functions, respectively, such that

```
Sl=rslower(X,a,S)     % find the lower-approximation set Sl
Su=rsupper(X,a,S)     % find the upper-approximation set Su
Sd=setdiff(Su,Sl)     % find the boundary set Sd
```

and the boundary set can be obtained by calling the `setdiff()` function in MATLAB. The contents of `rslower()` and `rsupper()` functions are listed below

```
function w=rslower(y,a,T)
z=ind(a,T); w=[]; [p,q]=size(z);
for u=1:p,
   zz=setdiff(z(u,:),0); if ismember(zz,y), w=cat(2,w,zz); end,
end
```

```
function w=rsupper(y,a,T)
z=ind(a,T); w=[]; [p,q]=size(z);
for u=1:p
   zz=setdiff(z(u,:),0); zzz=intersect(zz,y);
   [zp,zq]=size(zzz); if zq~=0, w=cat(2,w,zz); end
end
```

and the common supporting function `ind()` is used to evaluate the indiscernibility relationship to be defined later.

Example 10.30 For the toy set $U = \{x_1, x_2, x_3, x_4, x_5, x_6, x_7\}$. Assume that they have three attributes, namely, "color R_1," "shape R_2" and "size R_3." For the color attribute, it is assumed that $R_1 = \{0, 1, 2\}$, representing "red," "yellow" or "green." The shape attribute $R_2 = \{0, 1, 2\}$, denoting "square," "round" and "triangular." The size attribute $R_3 = \{0, 1\}$, indicating "large" and "small."

For the relationship in the attributes, the red toys can be described as $\{x_1, x_2, x_7\}$, green ones $\{x_3, x_4\}$ and blue ones $\{x_5, x_6\}$. It can then be written as $U|R_1 = \{\{x_1, x_2, x_7\}, \{x_3, x_4\}, \{x_5, x_6\}\}$.

Information decision system

The information decision system T can be expressed as a quadruple $T = (U, A, C, D)$, where U is the set of entities, i.e., the *universe*. A is the attribute set. If set A can further be divided into conditional attribute set C and decisional attribute set D, i.e., $C \cup D = A$ and $C \cap D = \phi$, the information system is referred to as a *decision system* or a *decision table.*

In rough set theory, a decision table can be used to describe the entities in a given universe. A decision table is a two-dimensional table with each row corresponding to an entity, and each column corresponding to a certain attribute. Again the attributes can be classified as conditional attributes and decisional attributes. The entities in the universe can be classified into different decisional classes due to its different conditional attributes. Table 10.9 is a typical example of a decision table. The universe $U = \{x_1, x_2, \cdots\}$ is a set of entities, $C = \{s_1, \cdots, s_m\}$ is a conditional attribute set, while $D = \{d_1, \cdots, d_k\}$ is a decisional attribute set. The term f_{ij} represents the jth conditional attribute of the ith entity, and g_{ij} represents the jth decisional attribute of the ith entity.

TABLE 10.9: Decision table

U	C				D		
	s_1	s_2	$\cdots$	s_m	d_1	$\cdots$	d_k
x_1	f_{11}	f_{12}	$\cdots$	f_{1m}	g_{11}	$\cdots$	g_{1k}
x_2	f_{21}	f_{22}	$\cdots$	f_{2m}	g_{21}	$\cdots$	g_{2k}
$\vdots$	$\vdots$	$\vdots$	$\vdots$	$\vdots$	$\vdots$	$\vdots$	$\vdots$
x_n	f_{n1}	f_{n2}	$\cdots$	f_{nm}	g_{n1}	$\cdots$	g_{nk}

TABLE 10.10: Example 10.31

	C properties				D
U	a	b	c	d	sales
1	1	0	1	1	1
2	1	0	0	0	1
3	0	0	1	0	0
4	1	1	0	1	0
5	1	1	1	2	2
6	2	1	0	2	2
7	2	2	0	2	2

TABLE 10.11: Rotated decision table

Universe X	x_1	x_2	x_3	x_4	x_5	x_6	x_7	x_8	x_9	x_{10}
U/R_1 relation	0	0	0	0	1	1	1	1	2	2
U/R_2 relation	0	0	0	1	1	1	1	2	2	2

Example 10.31 Consider the toy problem in Example 10.30. Assume that there are four relevant attributes, "a" for color, "b" for shape, "c" for size and "d" for price. It is known that $x_{1,2,4,5}$ are yellow, x_3 is red and $x_{6,7}$ are green. Also $x_{1,2,3}$ are square, $x_{4,5,6}$ are round and x_7 is triangular; $x_{1,3,5}$ are large while the rest is small; $x_{2,3}$ are cheap ones, $x_{1,4}$ are medium priced and $x_{5,6,7}$ are expensive ones. From sales status, $x_{3,4}$ are good, $x_{1,2}$ are average and $x_{5,6,7}$ are poor. Construct the decision table.

Solution Assume that for the color attribute, $\{0, 1, 2\}$ can be used to describe "red," "yellow" and "green;" for shape, $\{0, 1, 2\}$ for "square," "round" and "triangular;" for size, $\{0, 1\}$ for "small" and "large;" for price, $\{0, 1, 2\}$ for "low," "medium" and "high," and for sales, the decision attribute, $\{0, 1, 2\}$ for "good," "average" and "poor." The decision table can easily be constructed as shown in Table 10.10. One of the applications of rough set theory is that it can be used to check which of the conditional attributes contributes the most to the sales attribute, and which has little impact.

Example 10.32 Let the universe $U = \{x_1, x_2, x_3, x_4, x_5, x_6, x_7, x_8, x_9, x_{10}\}$, with relations $R = \{R_1, R_2\}$, and

$U/R_1 = \{\{x_1, x_2, x_3, x_4\}, \{x_5, x_6, x_7, x_8\}, \{x_9, x_{10}\}\}$,

$U/R_2 = \{\{x_1, x_2, x_3\}, \{x_4, x_5, x_6, x_7\}, \{x_8, x_9, x_{10}\}\}$

If $X = \{x_1, x_2, x_3, x_4, x_5\}$, find the upper- and lower-approximation sets of X.

Solution From the given conditions, the three subsets for U/R_1 and U/R_2 can be represented as $\{0, 1, 2\}$. Thus the decision table in Table 10.11 can be established in $\boldsymbol{S}$. Here the decision table is rotated for neat type-setting purposes. If $X = \{1, 2, 3, 4, 5\}$, the first two columns can be extracted for vector a. The upper- and lower-approximation sets of X can be obtained by using

```
>> S=[0,0; 0,0; 0,0; 0,1; 1,1; 1,1; 1,1; 1,2; 2,2; 2,2];
   X=[1,2,3,4,5]; a=[1,2]; S1=rslower(X,a,S) % the three sets
   S2=rsupper(X,a,S), Sd=setdiff(S2,S1)
```

The sets are obtained as $S_1 = [1, 2, 3, 4]$, $S_2 = [1, 2, 3, 4, 5, 6, 7]$, and $S_\mathrm{d} = [5, 6, 7]$.

It can be seen from the decision table that, the sets for $\{U/R_1, U/R_2\}$ are $\{0,0\}$, $\{0,1\}$, $\{1,1\}$, $\{1,2\}$, and $\{2,2\}$. For the selected entities $X = \{1,2,3,4,5\}$, those involved with $\{U/R_1, U/R_2\}$ are only $\{0,0\}$, $\{0,1\}$ and $\{1,1\}$. Thus those belonging to set X are $\{1,2,3,4\}$, i.e., $\{x_1, x_2, x_3, x_4\}$. The lower-approximation set is then $\{x_1, x_2, x_3, x_4\}$. Similarly, the set with the entities which may probably belong to X is the upper-approximation set of X. Since x_6, x_7 and x_5 are all $\{1,1\}$, thus apart from the entities in the lower-approximation set, the upper-approximation set also includes x_5, x_6, x_7. The latter three entities belong to the boundary set. Besides, since the boundary set is non-empty, it belongs to a rough set.

In the information system, for each subset $R \subseteq A$, the indiscernibility relationship is defined as

$$\mathrm{IND}(R) = \{(x,y) \in U \times U: \ r \in R: \ r(x) = r(y)\}. \tag{10.16}$$

The MATLAB implementation of the function is listed below

```
function aa=ind(a,x)
[p,q]=size(x); [ap,aq]=size(a); z=1:q;
tt=setdiff(z,a); x(:,tt(size(tt,2):-1:1))=-1;
for r=q:-1:1, if x(1,r)==-1, x(:,r)=[]; end, end
for i=1:p, v(i)=x(i,:)*10.^(aq-[1:aq]'); end
y=v'; [yy,I]=sort(y); y=[yy I];
[b,k,l]=unique(yy); y=[l I]; m=max(l); aa=zeros(m,p);
for ii=1:m, for j=1:p, if l(j)==ii, aa(ii,j)=I(j); end, end, end
```

10.5.2 Data processing problem solutions using rough sets

Reductions in rough set

Nowadays people live in the information explosive era. It is easier and easier to get more and more information. However, a huge amount of unprocessed information causes "data disasters." To tackle this kind of problem, the so-called *data mining* techniques should be used.

The aim of the reduction of information system is simply deleting and neglecting the unnecessary and redundant information, without affecting the original decision. New decision rules can be generated with the reduced information set.

Attribute reduction means finding the minimum set of conditional attributes without affecting the decisional attributes. One decision table may simultaneously have several reductions. The set of attributes which is common to all the reductions is referred to as the *core set* of R, denoted as $\mathrm{Core}(R)$.

Details of the attribute reduction algorithms are not given here. Only a few MATLAB-based functions, such as `redu()` and `core()`, are introduced with their usage explained. The Rough Set Data Analysis (RSDA) Toolbox with the related functions can be obtained from the aforementioned book. For

TABLE 10.12: Truth table of LCD

code	C attributes							D	code	C attributes							D
X	a	b	c	d	e	f	g	value	X	a	b	c	d	e	f	g	value
0	1	1	1	1	1	1	0	0	5	1	0	1	1	0	1	1	5
1	0	1	1	0	0	0	0	1	6	1	0	1	1	1	1	1	6
2	1	1	0	1	1	0	1	2	7	1	1	1	0	0	0	0	7
3	1	1	1	1	0	0	1	3	8	1	1	1	1	1	1	1	8
4	0	1	1	0	0	1	1	4	9	1	1	1	1	0	1	1	9

details of the algorithms and implementations, please refer to Reference [58]. The updated version of the Toolbox is provided with the companion CD.

Assume again that the decision table is described by matrix $\boldsymbol{S}$, whose $\boldsymbol{c}$ columns are stored in matrix $\boldsymbol{C}$ for conditional attributes, and the $\boldsymbol{d}$ columns are the decisional attributes in matrix $\boldsymbol{D}$. Thus the minimum set of attributes can be extracted by the `redu()` function, and the core set can be extracted by using the function `core()`, with the following syntaxes

```
y=redu(c,d,S)   % attribute reduction, find the minimum set from C to D
y=core(c,d,S)   % extract the core set from C to D
```

Application examples of rough set in information reduction

With the use of the attribute reduction concepts, and in particular, the relevant MATLAB functions, two illustrative examples are demonstrated in the following[58].

Example 10.33 LCDs are often used to show the 10 digits, from 0~9, with a seven-segment display unit. The seven-segment display units are shown in Figure 10.40 (a). If a segment is lighted, it is denoted as 1, otherwise 0. The truth table

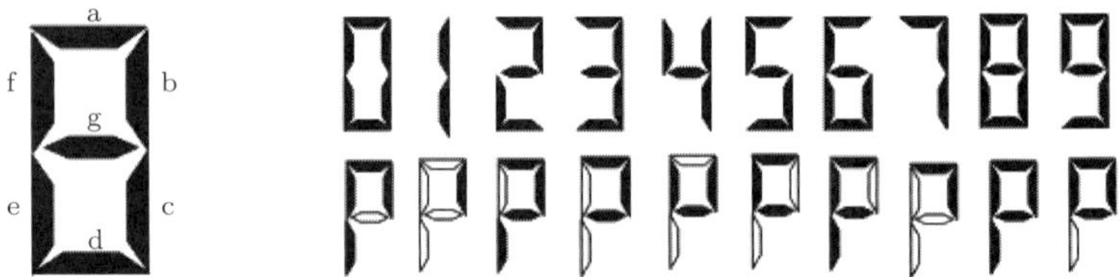

(a) seven-segment display (b) reduced display

FIGURE 10.40: LCD displays and reduction results

for the encoding system is given in Table 10.12. Find the unnecessary segments by attribute reduction using rough set theory.

Solution For digit recognition by human vision, it is necessary to use a seven-segment LCD display. If one segment is missing, the recognition of digits may become very difficult. However, sometimes it may not be necessary to have all the seven segments in place in order to recognize digit by computer vision. If one does find a certain segment which is not necessary, the LCD display may be reduced or further simplified. With the use of rough set techniques, reduction can be tried to check whether certain segments can be reduced, without affecting the recognition of digits. The following commands can be given

```
>> C=[1,1,1,1,1,1,0; 0,1,1,0,0,0,0; 1,1,0,1,1,0,1; 1,1,1,1,0,0,1;
      0,1,1,0,0,1,1; 1,0,1,1,0,1,1; 1,0,1,1,1,1,1; 1,1,1,0,0,0,0;
      1,1,1,1,1,1,1; 1,1,1,1,0,1,1];
   D=[0:9]'; X=[C D]; c=1:7; d=8;
   Y=core(c,d,X) % where columns 1-7 are C properties, 8 for D
```

with the reduction result $\boldsymbol{Y} = [1, 2, 5, 6, 7]$, which means that the segments a, b, e, f and g in the LCD are irreducible. The segments 3 and 4, i.e., the segments c and d in Figure 10.40 (a) can be removed without affecting the recognition of digits by computers. The corresponding mapping patterns are shown in Figure 10.40 (b). However, for human visual recognition, the above reduction may be useless. On the other hand, for the recognition by machines, this kind of reduction is no doubt useful, since less information is needed. The reduction idea shown in this example is promising for character recognition via machine vision.

Example 10.34 The outbreak of severe acute respiratory syndrome (SARS) in 2003 caused terrors globally. The accurate diagnosis of SARS is very difficult. Some data collected from newspapers and magazines are given in Table 10.13. All the 12 attributes are analyzed with rough set theory to see whether some of the attributes are redundant for the diagnosis purpose. The major attributes can be found from the 12 attributes. *It should be noted that some data may not be accurate, and the entries are far from complete. Thus the conclusion here is not usable for clinical diagnosis.*

TABLE 10.13: Collected incomplete data for diagnosing SARS

universe	C properties												D
U	c_1	c_2	c_3	c_4	c_5	c_6	c_7	c_8	c_9	c_{10}	c_{11}	c_{12}	SARS
1	1	1	1	1	0	0	0	0	1	1	0	1	1
2	0	0	0	0	0	0	0	0	0	0	0	0	0
3	1	0	1	0	0	0	0	0	0	1	0	0	0
4	0	0	0	1	1	1	1	0	1	0	1	1	0
5	1	0	0	1	1	1	1	1	0	1	1	0	0
6	0	1	0	1	1	1	1	1	1	0	0	1	0
7	1	0	0	0	1	1	1	0	0	1	1	1	0
8	1	1	1	1	0	0	0	0	1	1	0	1	1
9	1	0	1	1	1	0	0	0	1	1	0	1	1
10	1	1	1	1	0	0	0	0	1	1	0	1	1
11	1	0	1	1	1	0	0	0	1	1	0	1	1
12	1	0	1	1	1	0	0	0	1	1	0	1	1

c_1 — hacking cough, c_2 — breath difficulties, c_3 — blood test, c_4 — temperature $\geqslant 38°$C
c_5 — X-ray test abnormal, c_6 — with phlegm, c_7 — high white blood cells, c_8 — chill
c_9 — ache in muscles, c_{10} — hypodynamia, c_{11} — pleurodynia, c_{12} — headache

Solution From the data given in the table, the reduction of attributes with rough sets can be used to find which attributes contribute much more towards the diagnosis of SARS

```
>> D=[1; 0; 0; 0; 0; 0; 0; 1; 1; 1; 1; 1];
   C=[1,1,1,1,0,0,0,0,1,1,0,1; 0,0,0,0,0,0,0,0,0,0,0,0;
      1,0,1,0,0,0,0,0,0,1,0,0; 0,0,0,1,1,1,1,0,1,0,1,1;
      1,0,0,1,1,1,1,1,0,1,1,0; 0,1,0,1,1,1,1,1,1,0,0,1;
      1,0,0,0,1,1,1,0,0,1,1,1; 1,1,1,1,0,0,0,0,1,1,0,1;
      1,0,1,1,1,0,0,0,1,1,0,1; 1,1,1,1,0,0,0,0,1,1,0,1;
      1,0,1,1,1,0,0,0,1,1,0,1; 1,0,1,1,1,0,0,0,1,1,0,1];
   Y=redu(1:12,13,[C D])
```

with $\boldsymbol{Y} = [3, 4]$, indicating that attributes 3 and 4 are irreducible, which means that "blood test" and "fever with temperature higher than 38°C" are the irreducible attributes.

MATLAB graphical user interface for rough set reduction

Based on the rough set theory and its reduction methods, a MATLAB interface is written and included in the RSDA Toolbox. The command `rsdav3` can be given to start the interface, as shown in Figure 10.41. The Browse button can be used to import the decision table, and the column numbers for attributes in C and D can also be specified. Then the Redu button can be used to perform attribute reduction tasks, and the results are shown in the Results box. The free toolbox is provided on the companion CD.

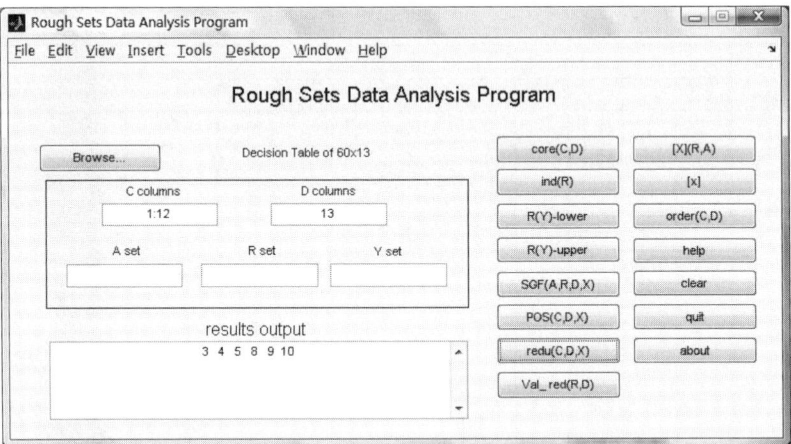

FIGURE 10.41: Graphical user interface for data analysis with rough sets

10.6 Fractional-Order Calculus

In Chapter 3, the calculus problems have been discussed, and the related materials are also shown in other chapters. It is quite interesting to note that before the invention of calculus, only algebra, geometry, and statics could be subjects of study. With calculus, lots of new subjects emerge such as dynamics, modeling, etc., which enable us to better characterize the world around us. It is known that the nth order derivative of a function $f(t)$ can be mathematically described by $\mathrm{d}^n y/\mathrm{d}x^n$, a notation invented by Leibniz not Newton who suggested dot notation. With Leibniz's notation, one may ask "What does $n = 1/2$ mean to the notation?" Actually, this is the question asked in a letter by the French mathematician Guillaume François Antoine L'Hôpital to one of the inventors of calculus, German mathematician Gottfried Wilhelm Leibniz more than 300 years ago. In answering the letter, Leibniz said, "It will lead to a paradox, from which one day useful consequences will be drawn." This marks the beginning of fractional-order calculus. However, earlier research concentrated on theoretical math issues. It is now being widely used in many areas. For instance, in the discipline of automatic control, fractional-order controller is a promising new topic. In this section, an easy-to-follow introduction to fractional-order calculus will be presented with how to get started using MATLAB in mind. Our treatment in this section, although not very mathematically rigorous, is practically useful since we presented a comprehensive introduction to numerical solutions to fractional-order filtering and fractional-order ordinary differential equations. Furthermore, from a programming point of view, the design and application of MATLAB classes and objects dedicated for fractional-order calculus demonstrated thoroughly in this section are of great value.

10.6.1 Definitions of fractional-order calculus

Definitions and properties of fractional-order calculus

Various definitions appeared in the development and studies of fractional-order calculus. Some of the definitions are directly extended from the conventional integer-order calculus. The commonly used definitions are summarized as follows:

(i) **Fractional-order Cauchy integral formula** The formula is extended from integer-order calculus

$$\mathscr{D}^{\alpha} f(t) = \frac{\Gamma(\alpha+1)}{\mathrm{j}2\pi} \int_{\mathrm{C}} \frac{f(\tau)}{(\tau - t)^{\alpha+1}} \,\mathrm{d}\tau \tag{10.17}$$

where C is the closed-path that encircles the poles of the function $f(t)$.

The integrals and derivatives for sinusoidal and cosine functions can be expressed by

$$\frac{\mathrm{d}^k}{\mathrm{d}t^k}\left[\sin at\right] = a^k \sin\left(at + \frac{k\pi}{2}\right), \quad \frac{\mathrm{d}^k}{\mathrm{d}t^k}\left[\cos at\right] = a^k \cos\left(at + \frac{k\pi}{2}\right). \tag{10.18}$$

It can also be shown with Cauchy's formula that if k is not an integer, the above formula is still valid.

(ii) **Grünwald-Letnikov definition** The fractional-order differentiation and integral can be defined in a unified way such that

$$ {}_a\mathscr{D}_t^{\alpha} f(t) = \lim_{h\to 0} \frac{1}{h^\alpha} \sum_{j=0}^{[(t-a)/h]} (-1)^j \binom{\alpha}{j} f(t - jh) \tag{10.19}$$

where $\binom{\alpha}{j}$ are the binomial coefficients; the subscripts to the left and right of $\mathscr{D}$ are the lower- and upper-bounds in the integral. The value α can be positive or negative, corresponding to differentiation and integration, respectively. α is non-integer.

(iii) **Riemann-Liouville definition** The fractional-order integral is defined as

$$ {}_a\mathscr{D}_t^{-\alpha} f(t) = \frac{1}{\Gamma(\alpha)} \int_a^t (t-\tau)^{\alpha-1} f(\tau)\,\mathrm{d}\tau \tag{10.20}$$

where $0 < \alpha < 1$, and a is the initial value. Let $a = 0$, the notation of integral can be simplified to $\mathscr{D}_t^{-\alpha} f(t)$. The Riemann-Liouville definition is a widely used definition for fractional-order differentiation and integral[59]. Similarly, fractional-order differentiation is defined as

$$ {}_a\mathscr{D}_t^{\beta} f(t) = \frac{\mathrm{d}^n}{\mathrm{d}t^n}\left[{}_a\mathscr{D}_t^{-(n-\beta)} f(t)\right] = \frac{1}{\Gamma(n-\beta)} \frac{\mathrm{d}^n}{\mathrm{d}t^n}\left[\int_a^t \frac{f(\tau)}{(t-\tau)^{\beta-n+1}}\,\mathrm{d}\tau\right] \tag{10.21}$$

where $n - 1 < \beta \leqslant n$.

(iv) **Caputo definition** The Caputo fractional-order differentiation is defined by

$$ {}_0\mathscr{D}_t^{\alpha} f(t) = \frac{1}{\Gamma(1-\alpha)} \int_0^t \frac{f^{(m+1)}(\tau)}{(t-\tau)^{\alpha}}\,\mathrm{d}\tau \tag{10.22}$$

where $\alpha = m + \gamma$, m is an integer and $0 < \gamma \leqslant 1$. Similarly, by Caputo's definition, the integral is described by

$$ {}_0\mathscr{D}_t^{-\gamma} f(t) = \frac{1}{\Gamma(\gamma)} \int_0^t \frac{f(\tau)}{(t-\tau)^{1-\gamma}}\,\mathrm{d}\tau, \quad \gamma > 0. \tag{10.23}$$

It can be shown that for great varieties of functions, the Grünwald-Letnikov and Riemann-Liouville definitions are equivalent[60]. In this section, we focus on the Grünwald-Letnikov's definition and its applications.

The properties of fractional-order calculus are summarized below[61]:

(i) The fractional-order differentiation ${}_a\mathscr{D}_t^\alpha f(t)$ of an analytic function $f(t)$ with respect to t is also analytic.

(ii) If $\alpha = n$, the fractional-order derivative is identical to integer-order derivative, and also ${}_a\mathscr{D}_t^0 f(t) = f(t)$.

(iii) The fractional-order differentiation is linear, i.e., for any constants c, d

$$ {}_a\mathscr{D}_t^\alpha \left[cf(t) + dg(t)\right] = c\ {}_a\mathscr{D}_t^\alpha f(t) + d\ {}_a\mathscr{D}_t^\alpha g(t). \tag{10.24} $$

(iv) Fractional-order differentiation operators satisfy commutative-law, i.e.

$$ {}_a\mathscr{D}_t^\alpha \left[{}_a\mathscr{D}_t^\beta f(t)\right] = {}_a\mathscr{D}_t^\beta \left[{}_a\mathscr{D}_t^\alpha f(t)\right] = {}_a\mathscr{D}_t^{\alpha+\beta} f(t). \tag{10.25} $$

Integral transform of fractional-order operator

The Laplace transform of a fractional-order integral can be expressed by

$$ \mathscr{L}\left[\mathscr{D}_t^{-\gamma} f(t)\right] = s^{-\gamma}\mathscr{L}[f(t)] \tag{10.26} $$

and the transform for a fractional-order derivative can be evaluated from

$$ \mathscr{L}\left[{}_a\mathscr{D}_t^\alpha f(t)\right] = s^\alpha \mathscr{L}[f(t)] - \sum_{k=1}^{n-1} s^k \left[{}_a\mathscr{D}_t^{\alpha-k-1} f(t)\right]_{t=a}. \tag{10.27} $$

In particular, if the derivatives of the function $f(t)$ at $t = a$ are all equal to 0, one simply has $\mathscr{L}[{}_a\mathscr{D}_t^\alpha f(t)] = s^\alpha \mathscr{L}[f(t)]$.

Consider the linear fractional-order differential equation given by

$$ \begin{aligned} a_1\mathscr{D}^{\eta_1}y(t) + a_2\mathscr{D}^{\eta_2}y(t) + \cdots + a_{n-1}\mathscr{D}^{\eta_{n-1}}y(t) + a_n\mathscr{D}^{\eta_n}y(t) \\ = b_1\mathscr{D}^{\gamma_1}u(t) + b_2\mathscr{D}^{\gamma_2}u(t) + \cdots + b_m\mathscr{D}^{\gamma_m}u(t). \end{aligned} \tag{10.28} $$

If all the initial values of the input and output are zero, Laplace transform can be applied such that the differential equation can be mapped into an algebraic equation, from which the fractional-order transfer function (FOTF) can be defined

$$ G(s) = \frac{\mathscr{L}[f(t)]}{\mathscr{L}[u(t)]} = \frac{b_1 s^{\gamma_1} + b_2 s^{\gamma_2} + \cdots + b_m s^{\gamma_m}}{a_1 s^{\eta_1} + a_2 s^{\eta_2} + \cdots + a_{n-1} s^{\eta_{n-1}} + a_n s^{\eta_n}}. \tag{10.29} $$

The Fourier transform for the fractional-order derivative and integral can be defined in a unified way as

$$ \mathscr{F}\left[{}_{-\infty}\mathscr{D}_t^\alpha f(t)\right] = (\mathrm{j}\omega)^\alpha \mathscr{F}[f(t)] \tag{10.30} $$

where α can be either a positive or negative real number.

FOTF object design

In MATLAB, object data structure is very important. For instance, the FOTF object can be designed to describe a fractional-order transfer function model in (10.29). A "`fotf`" class can be constructed by creating the `@fotf` directory and putting in this directory the following `fotf()` function

```
function G=fotf(a,na,b,nb)
if nargin==0,
   G.a=[]; G.na=[]; G.b=[]; G.nb=[]; G=class(G,'fotf');
elseif isa(a,'fotf'), G=a;
elseif nargin==1 & isa(a,'double'), G=fotf(1,0,a,0);
else,
   ii=find(abs(a)<eps); a(ii)=[]; na(ii)=[];
   ii=find(abs(b)<eps); b(ii)=[]; nb(ii)=[];
   G.a=a; G.na=na; G.b=b; G.nb=nb; G=class(G,'fotf');
end
```

The syntax of the function is `G=fotf(`$\boldsymbol{a}$,$\boldsymbol{\eta}$,$\boldsymbol{b}$,$\boldsymbol{\gamma}$`)`, where $\boldsymbol{a}$ and $\boldsymbol{b}$ are the coefficients of denominator and numerator, respectively, while $\boldsymbol{\eta}$ and $\boldsymbol{\gamma}$ are the order sequences in the denominator and numerator, respectively.

A display function should also be created for the `fotf` class and saved in the `@fotf` directory such that

```
function display(G)
strN=polydisp(G.b,G.nb); strD=polydisp(G.a,G.na);
nn=length(strN); nd=length(strD); nm=max([nn,nd]);
disp([char(' '*ones(1,floor((nm-nn)/2))) strN]),
disp(char('-'*ones(1,nm)));
disp([char(' '*ones(1,floor((nm-nd)/2))) strD])
function strP=polydisp(p,np)
P=''; [np,ii]=sort(np,'descend'); p=p(ii);
for i=1:length(p),
   P=[P,'+',num2str(p(i)),'s^{',num2str(np(i)),'}'];
end
P=P(2:end); P=strrep(P,'s^{0}',''); P=strrep(P,'+-','-');
P=strrep(P,'^{1}',''); P=strrep(P,'+1s','+s');
strP=strrep(P,'-1s','-s');
if length(strP)>=2 & strP(1:2)=='1s', strP=strP(2:end); end
```

Example 10.35 Suppose that the fractional-order transfer function is given by

$$G(s)=\frac{-2s^{0.63}-4}{2s^{3.501}+3.8s^{2.42}+2.6s^{1.798}+2.5s^{1.31}+1.5}.$$

With the following statement, the fractional-order transfer function can be entered into MATLAB environment, and the object can be suitably specified and displayed.

```
>> b=[-2,-4]; nb=[0.63,0]; a=[2 3.8 2.6 2.5 1.5];
   na=[3.501,2.42,1.798,1.31,0]; G=fotf(a,na,b,nb)
```

Based on the newly defined `fotf` class, the `plus()`,`mtimes()` and `feedback()` functions can be written to implement interconnections of FOTF objects.

(i) **Plus function** `plus()` for block parallel connections:

```
function G=plus(G1,G2)
a=kron(G1.a,G2.a); b=[kron(G1.a,G2.b), kron(G1.b,G2.a)];
na=[]; nb=[];
for i=1:length(G1.a),
    na=[na G1.na(i)+G2.na]; nb=[nb, G1.na(i)+G2.nb];
end
for i=1:length(G1.b), nb=[nb G1.nb(i)+G2.na]; end
G=unique(fotf(a,na,b,nb));
```

(ii) **Multiplication function** `mtimes()` for block series connections:

```
function G=mtimes(G1,G2)
G2=fotf(G2); a=kron(G1.a,G2.a);
b=kron(G1.b,G2.b); na=[]; nb=[];
for i=1:length(G1.na), na=[na,G1.na(i)+G2.na]; end
for i=1:length(G1.nb), nb=[nb,G1.nb(i)+G2.nb]; end
G=unique(fotf(a,na,b,nb));
```

(iii) **Feedback function** `feedback()` for block negative feedback connections:

```
function G=feedback(F,H)
H=fotf(H);
b=kron(F.b,H.a); a=[kron(F.b,H.b), kron(F.a,H.a)]; na=[]; nb=[];
for i=1:length(F.b),
   nb=[nb F.nb(i)+H.nb]; na=[na,F.nb(i)+H.nb];
end
for i=1:length(F.a), na=[na F.na(i)+H.na]; end
G=unique(fotf(a,na,b,nb));
```

(iv) **Simplification function** `unique()`:

```
function G=unique(G)
[a,n]=polyuniq(G.a,G.na); G.a=a; G.na=n;
[a,n]=polyuniq(G.b,G.nb); G.b=a; G.nb=n;
function [a,an]=polyuniq(a,an)
[an,ii]=sort(an,'descend'); a=a(ii); ax=diff(an); key=1;
for i=1:length(ax)
   if ax(i)==0,
      a(key)=a(key)+a(key+1); a(key+1)=[]; an(key+1)=[];
   else, key=key+1; end
end
```

Other functions should also be designed, such as the `minus()`, `uminus()`, `inv()`, and the files should be placed in the `@fotf` directory to overload the existing ones. The listings of these functions are not given in the text.

To compare Bode diagrams, an overload `bode()` function for FOTF objects is written and placed in the `@fotf` directory such that

```
function H=bode(G,w)
a=G.a; na=G.na; b=G.b; nb=G.nb; j=sqrt(-1);
if nargin==1, w=logspace(-4,4); end
for i=1:length(w)
   P=b*((j*w(i)).^nb.'); Q=a*((j*w(i)).^na.'); H1(i)=P/Q;
end
H1=frd(H1,w); if nargout==0, bode(H1); else, H=H1; end
```

Example 10.36 Suppose in the unity negative feedback system, the models are given by $G(s)=\dfrac{0.8s^{1.2}+2}{1.1s^{1.8}+0.8s^{1.3}+1.9s^{0.5}+0.4}$, $G_c(s)=\dfrac{1.2s^{0.72}+1.5s^{0.33}}{3s^{0.8}}$. Find the closed-loop model.

Solution The plant and controller can easily be entered and the closed-loop system can be directly obtained with the following commands

```
>> G=fotf([1.1,0.8 1.9 0.4],[1.8 1.3 0.5 0],[0.8 2],[1.2 0]);
   Gc=fotf([3],[0.8], [1.2 1.5],[0.72 0.33]); H=fotf(1,0,1,0);
   GG=feedback(G*Gc,H)
```

and the result is given by

$$G(s)=\frac{0.96s^{1.92}+1.2s^{1.53}+2.4s^{0.72}+3s^{0.33}}{3.3s^{2.6}+2.4s^{2.1}+0.96s^{1.92}+1.2s^{1.53}+5.7s^{1.3}+1.2s^{0.8}+2.4s^{0.72}+3s^{0.33}}.$$

It can be seen from the above illustrations that, although both the plant and controller models are relatively simple, extremely complicated closed-loop models may be obtained. This makes the analysis and design of a fractional-order system a difficult task.

10.6.2 Evaluating fractional-order differentiation

Computation via Grünwald-Letnikov definitions

The most straightforward way in evaluating the fractional-order derivatives numerically is to use the Grünwald-Letnikov definition which is given again

$${}_a\mathscr{D}_t^{\alpha}f(t)=\lim_{h\to 0}\frac{1}{h^{\alpha}}\sum_{j=0}^{[(t-a)/h]}(-1)^j\binom{\alpha}{j}f(t-jh)\approx\frac{1}{h^{\alpha}}\sum_{j=0}^{[(t-a)/h]}w_j^{(\alpha)}f(t-jh) \tag{10.31}$$

where the binomial coefficients $w_j^{(\alpha)}=(-1)^j\binom{\alpha}{j}$ can be evaluated recursively from

$$w_0^{(\alpha)}=1,\ \ w_j^{(\alpha)}=\left(1-\frac{\alpha+1}{j}\right)w_{j-1}^{(\alpha)},\ j=1,2,\cdots. \tag{10.32}$$

If the step-size h is small enough, (10.31) can be used directly to calculate approximately the values of the fractional-order derivatives, with an accuracy of $o(h)$[60]. The following MATLAB function can be written, with the syntax `y1=glfdiff(y,t,γ)`, where $\boldsymbol{y}, \boldsymbol{t}$ describe the original function using evenly distributed samples. The returned vector $\boldsymbol{y}_1$ is the γth order derivative.

```
function dy=glfdiff(y,t,gam)
h=t(2)-t(1); dy(1)=0; y=y(:); t=t(:);
w=1; for j=2:length(t), w(j)=w(j-1)*(1-(gam+1)/(j-1)); end
for i=2:length(t), dy(i)=w(1:i)*[y(i:-1:1)]/h^gam;  end
```

Example 10.37 For function $f(t) = \mathrm{e}^{-t}\sin(3t+1)$, $t \in (0, \pi)$, study the behaviors of the fractional-order derivatives.

Solution Select the step-sizes, $T = 0.01$ and 0.001, the 0.5th order derivative can be evaluated as shown in Figure 10.42 (a). It can be seen that, when t is close to 0, there are some differences. For larger t, the two solutions are very close. Normally speaking, selecting $T = 0.01$ may give accurate results.

```
>> t=0:0.001:pi; y=exp(-t).*sin(3*t+1); dy=glfdiff(y,t,0.5); plot(t,dy);
   t=0:0.01:pi; y=exp(-t).*sin(3*t+1); dy=glfdiff(y,t,0.5); line(t,dy)
```

For different selections of γ, the 3D surface representation of the fractional-order derivative can be obtained as shown in Figure 10.42 (b).

```
>> Z=[]; t=0:0.01:pi; y=exp(-t).*sin(3*t+1);
   for gam=0:0.1:1, Z=[Z; glfdiff(y,t,gam)]; end
   surf(t,0:0.1:1,Z); axis([0,pi,0,1,-1.2,6])
```

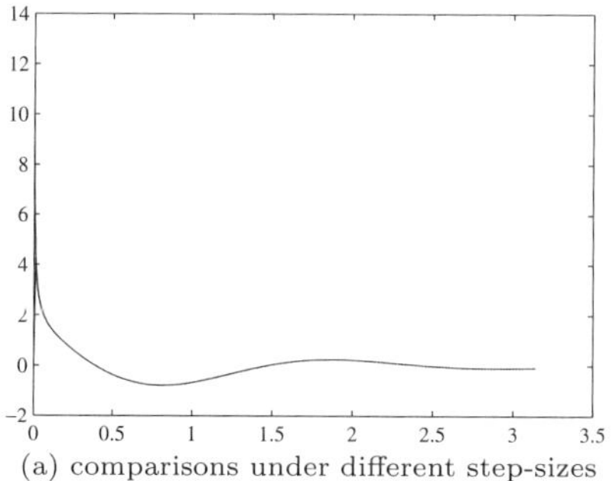

(a) comparisons under different step-sizes

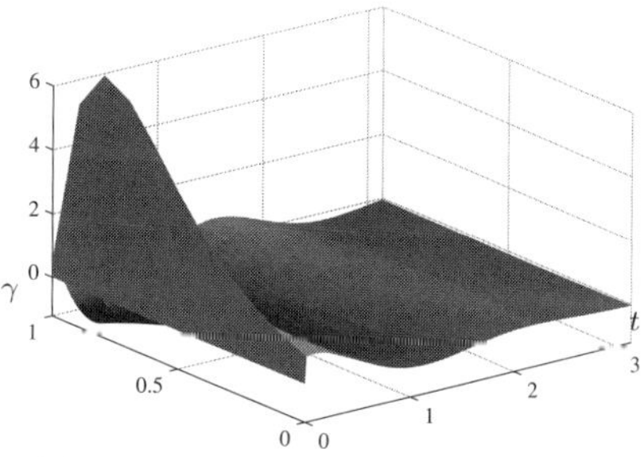

(b) fractional-order derivative surface

FIGURE 10.42: Fractional-order differentiation of the function

Example 10.38 Consider a sinusoidal function $f(t) = \sin(3t + 1)$. Find its 0.75th order derivative with Grünwald-Letnikov definition and Cauchy's formula, and compare the results.

Solution According to Cauchy's formula and (10.18), the 0.75th order derivative can be obtained as ${}_0\mathscr{D}_t^{0.75} f(t) = 3^{0.75}\sin(3t + 1 + 0.75\pi/2)$. With the Grünwald-Letnikov definition, the derivative can be evaluated with `glfdiff()` function. The two derivative curves are compared in Figure 10.43 (a).

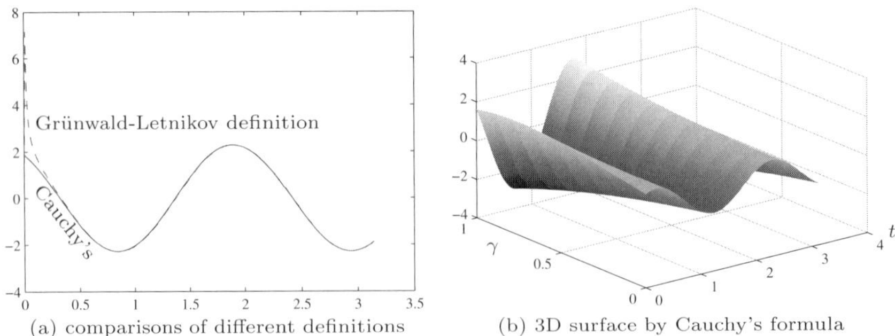

(a) comparisons of different definitions (b) 3D surface by Cauchy's formula

FIGURE 10.43: Comparisons and 3D surfaces

```
>> t=0:0.01:pi; y=sin(3*t+1); y1=3^0.75*sin(3*t+1+0.75*pi/2);
   y2=glfdiff(y,t,0.75); plot(t,y1,t,y2,'--')
```

It can be seen that when t is small, there exists significant difference between the two definitions. This is because, in these two definitions, the assumptions of the function at $t < 0$ are considered differently. In Cauchy's formula, it is assumed that the function can also be described by $f(t) = \sin(3t + 1)$ when $t < 0$, however, for Grünwald-Letnikov definition, it is assumed that $f(t) = 0$ for $t < 0$. Thus in the later case, there is a jump at $t = 0$, which causes the difference.

Again, the 3D surface can also be found for different orders γ, with the following statements, as shown in Figure 10.43 (b). It can be seen that the fractional-order derivatives provide information between the original function and its first-order derivative.

```
>> gam=[0:0.1:1]; Y=[]; t=0:0.01:pi; y=sin(3*t+1);
   for a=gam, Y=[Y; 3^a*sin(3*t+1+a*pi/2)]; end
   surf(t,gam,Y)
```

Using filtering algorithm to compute the fractional-order calculus

In the fractional-order derivative evaluation method discussed above, the function $f(t)$ or its samples are known. In many other applications, the signal $f(t)$ is generated dynamically. In this case, a filter can be designed to evaluate fractional-order derivative of $f(t)$ in real-time.

For the fractional-order derivative, its Laplace representation is s^γ, which exhibits straight lines in both Bode magnitude and phase plots. Thus it is not possible to find a finite order filter to fit the straight lines for all the frequencies. However, it is useful to fit the frequency responses over a frequency range of interest.

Different continuous filters have been studied in Reference [61], among which the Oustaloup's filter[62] has certain advantages. For the selected frequency range of interest, $(\omega_\mathrm{b}, \omega_\mathrm{h})$, the continuous filter can be written as

$$G_\mathrm{f}(s) = K \prod_{k=-N}^{N} \frac{s + \omega_k'}{s + \omega_k} \tag{10.33}$$

where, the poles, zeros and gain can be evaluated from

$$\omega'_k = \omega_b \left(\frac{\omega_h}{\omega_b}\right)^{\frac{k+N+\frac{1}{2}(1-\gamma)}{2N+1}}, \quad \omega_k = \omega_b \left(\frac{\omega_h}{\omega_b}\right)^{\frac{k+N+\frac{1}{2}(1+\gamma)}{2N+1}}, \quad K = \omega_h^{\gamma}. \quad (10.34)$$

Based on the above algorithm, the following function can be written

```
function G=ousta_fod(r,N,wb,wh)
mu=wh/wb; k=-N:N; w_kp=(mu).^((k+N+0.5-0.5*r)/(2*N+1))*wb;
w_k=(mu).^((k+N+0.5+0.5*r)/(2*N+1))*wb;
K=wh^r; G=tf(zpk(-w_kp',-w_k',K));
```

The continuous filter can be designed as `G=ousta_fod(`γ`,`N`,`ω_b`,`ω_h`)`, where γ is the order of derivative, and $2N+1$ is the order of the filter.

Example 10.39 Select the frequency range of interest as $\omega_b = 0.01, \omega_h = 100$ rad/sec, and design the continuous-time approximate fractional-order filters. For the function $f(t) = e^{-t}\sin(3t+1)$, calculate the 0.5th order derivative and verify the obtained results.

Solution The 5th order and 7th order filters can be designed

```
>> G1=ousta_fod(0.5,2,0.01,100),
   G2=ousta_fod(0.5,3,0.01,100), bode(G1,G2),
```

such that

$$G_1(s) = \frac{10s^5 + 298.5s^4 + 1218s^3 + 768.5s^2 + 74.97s + 1}{s^5 + 74.97s^4 + 768.5s^3 + 1218s^2 + 298.5s + 10};$$

$$G_2(s) = \frac{10s^7 + 509.4s^6 + 5487s^5 + 14988s^4 + 10786.7\times10^4 s^3 + 2045s^2 + 98.34s + 1}{s^7 + 98.34s^6 + 2045s^5 + 1.079\times10^4 s^4 + 1.499\times10^4 s^3 + 5487s^2 + 509.4s + 10}.$$

The Bode diagrams of the above approximate filter are shown in Figure 10.44 (a), superimposed by the theoretical straight lines. The filter output is shown in Figure 10.44 (b). Moreover, the 0.5th order derivative obtained through Grünwald-Letnikov definition can be obtained using the following MATLAB scripts and the obtained derivative curve is also shown in Figure 10.44 (b). Clearly, the filter output of this example is fairly accurate.

```
>> t=0:0.001:pi; y=exp(-t).*sin(3*t+1);
   y1=lsim(G1,y,t); y2=lsim(G2,y,t); y0=glfdiff(y,t,0.5);
   plot(t,y1,t,y2,t,y0)
```

Of course, if one is not satisfied with the filters, the frequency range of interest and the order of the filter can both be increased.

A modified Oustaloup filter

In practical applications, it is frequently found that the filter from using the `ousta_fod()` function cannot exactly fit the whole expected frequency range of interest. A new improved filter for a fractional-order derivative in

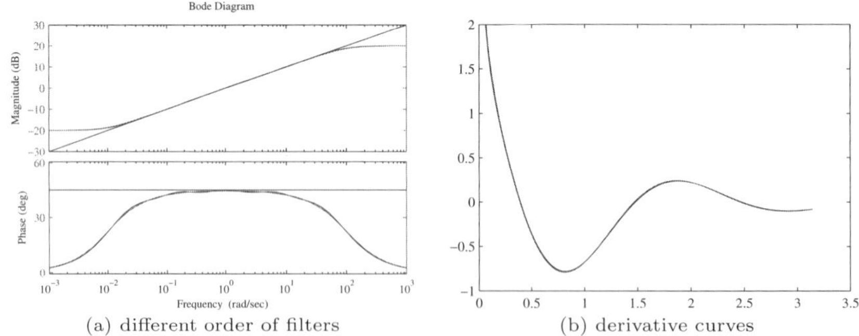

(a) different order of filters (b) derivative curves

FIGURE 10.44: Comparisons of approximate fractional-order filters

the frequency range of interest $[\omega_b, \omega_h]$, which is shown to perform better, is introduced in this subsection. The modified filter is[63]

$$s^\gamma \approx \left(\frac{d\omega_h}{b}\right)^\gamma \left(\frac{ds^2 + b\omega_h s}{d(1-\gamma)s^2 + b\omega_h s + d\gamma}\right) \prod_{k=-N}^{N} \frac{s+\omega_k'}{s+\omega_k} \tag{10.35}$$

where

$$\omega_k' = \omega_b \left(\frac{\omega_h}{\omega_b}\right)^{\frac{k+N+\frac{1}{2}(1-\gamma)}{2N+1}}, \quad \omega_k = \omega_b \left(\frac{\omega_h}{\omega_b}\right)^{\frac{k+N+\frac{1}{2}(1+\gamma)}{2N+1}}. \tag{10.36}$$

Through a number of experimentation confirmation and theoretic analyses, the modified filter achieves good approximation when $b = 10$ and $d = 9$. With the above algorithm, a MATLAB function `new_fod()` is written

```
function G=new_fod(r,N,wb,wh,b,d)
if nargin==4, b=10; d=9; end
mu=wh/wb; k=-N:N; w_kp=(mu).^((k+N+0.5-0.5*r)/(2*N+1))*wb;
w_k=(mu).^((k+N+0.5+0.5*r)/(2*N+1))*wb; K=(d*wh/b)^r;
G=zpk(-w_kp',-w_k',K)*tf([d,b*wh,0],[d*(1-r),b*wh,d*r]);
```

with the syntax G_f=`new_fod`$(\gamma, N, \omega_b, \omega_h, b, d)$.

Example 10.40 Consider a model $G(s) = \dfrac{s+1}{10s^{3.2} + 185s^{2.5} + 288s^{0.7} + 1}$, which is a fractional-order model. Compare the two Oustaloup filters.

Solution The exact Bode diagram can be obtained with the `bode()` function. The approximations to the 0.2th order derivative using the Oustaloup's filter and the modified Oustaloup's filter can be obtained as shown in Figure 10.45 (a). The frequency range of good fitting is larger with the improved filter. Also the approximation to the $G(s)$ model is shown in Figure 10.45 (b). It can be seen that the modified method provided a much better fit.

```
>> b=[1 1]; a=[10,185,288,1]; nb=[1 0]; na=[3.2,2.5,0.7,0];
   w=logspace(-4,4,200); G0=fotf(a,na,b,nb); H=bode(G0,w);
```

```
s=zpk('s'); N=4; w1=1e-3; w2=1e3; b=10; d=9;
g1=ousta_fod(0.2,N,w1,w2); g2=ousta_fod(0.5,N,w1,w2); a1=g1;
g3=ousta_fod(0.7,N,w1,w2); G1=(s+1)/(10*s^3*g1+185*s^2*g2+288*g3+1);
g1=new_fod(0.2,N,w1,w2,b,d); g2=new_fod(0.5,N,w1,w2,b,d);
g3=new_fod(0.7,N,w1,w2,b,d); bode(g1,a1); figure
G2=(s+1)/(10*s^3*g1+185*s^2*g2+288*g3+1); bode(H,G1,G2)
```

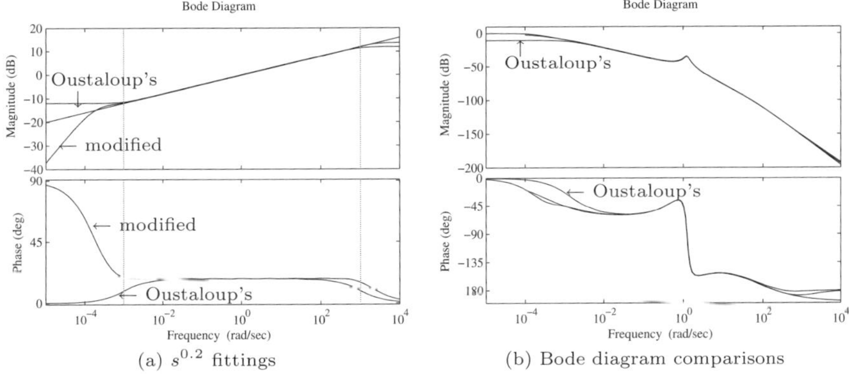

(a) $s^{0.2}$ fittings (b) Bode diagram comparisons

FIGURE 10.45: Bode diagram comparisons

10.6.3 Solving fractional-order differential equations

Solving fractional-order linear ordinary differential equations

A class of linear time-invariant (LTI) fractional-order differential equations (FODE) can be written as[60]

$$a_n \mathscr{D}_t^{\beta_n} y(t) + a_{n-1} \mathscr{D}_t^{\beta_{n-1}} y(t) + \cdots + a_1 \mathscr{D}_t^{\beta_1} y(t) + a_0 \mathscr{D}_t^{\beta_0} y(t) = u(t) \quad (10.37)$$

where $u(t)$ is the input signal, the simplified transfer function can be written as

$$G(s) = \frac{Y(s)}{U(s)} = \frac{1}{a_n s^{\beta_n} + a_{n-1} s^{\beta_{n-1}} + \cdots + a_1 s^{\beta_1} + a_0 s^{\beta_0}}. \quad (10.38)$$

Consider the Grünwald-Letnikov definition in (10.31), the modified discrete version can be written as

$${}_a\mathscr{D}_t^{\beta_i} y(t) \approx \frac{1}{h^{\beta_i}} \sum_{j=0}^{[(t-a)/h]} w_j^{(\beta_i)} y_{t-jh} = \frac{1}{h^{\beta_i}} \left[y_t + \sum_{j=1}^{[(t-a)/h]} w_j^{(\beta_i)} y_{t-jh} \right] \quad (10.39)$$

where $w_0^{(\beta_i)}$ can still be evaluated recursively

$$w_0^{(\beta_i)} = 1, \quad w_j^{(\beta_i)} = \left(1 - \frac{\beta_i + 1}{j}\right) w_{j-1}^{(\beta_i)}, \; j = 1, 2, \cdots. \quad (10.40)$$

Substituting the formula to (10.37), the closed-form numerical solution to the fractional-order differential equation can be obtained as

$$y_t = \frac{1}{\displaystyle\sum_{i=0}^{n} \frac{a_i}{h^{\beta_i}}} \left[u_t - \sum_{i=0}^{n} \frac{a_i}{h^{\beta_i}} \sum_{j=1}^{[(t-a)/h]} w_j^{(\beta_i)} y_{t-jh} \right]. \tag{10.41}$$

Now consider the general form of the fractional-order differential equation in (10.29), the right-hand side of the function can be evaluated and the equation can be converted into the form in (10.37). Based on this fact, an overloaded function `lsim()` can be written for `fotf` object, and placed in the `@fotf` folder.

```
function y=lsim(G,u,t)
a=G.a; eta=G.na; b=G.b; gamma=G.nb; nA=length(a); h=t(2)-t(1);
D=sum(a./[h.^eta]); W=[]; nT=length(t); vec=[eta gamma];
D1=b(:)./h.^gamma(:); y1=zeros(nT,1); W=ones(nT,length(vec));
for j=2:nT, W(j,:)=W(j-1,:).*(1-(vec+1)/(j-1)); end
for i=2:nT, A=[y1(i-1:-1:1)]'*W(2:i,1:nA);
   y1(i)=(u(i)-sum(A.*a./[h.^eta]))/D;
end
for i=2:nT, y(i)=(W(1:i,nA+1:end)*D1)'*[y1(i:-1:1)]; end
```

Example 10.41 Solve numerically the following fractional-order differential equation

$$\mathscr{D}_t^{3.5}y(t) + 8\mathscr{D}_t^{3.1}y(t) + 26\mathscr{D}_t^{2.3}y(t) + 73\mathscr{D}_t^{1.2}y(t) + 90\mathscr{D}_t^{0.5}y(t) = 90\sin(t^2).$$

Solution The FOTF object can be created first and then the overloaded `lsim()` function can be used to solve the FODE problem. There is not any existing method which ensures the accuracy of the solution. So, one should verify the numerical results obtained. The simplest way is to modify the control parameters in the solution. For instance, one can change the step-size from 0.002 to 0.001, and check whether they give the same results. For this example, it can be seen that the results are the same, as shown in Figure 10.46.

```
>> a=[1,8,26,73,90]; n=[3.5,3.1,2.3,1.2,0.5];
   G=fotf(a,n,1,0); t=0:0.002:10; u=90*sin(t.^2); y=lsim(G,u,t);
   subplot(211), plot(t,y); subplot(212), plot(t,u)
```

Analytical solutions of linear fractional-order equations

It is possible to solve the above fractional-order differential equation analytically by using Mittag-Leffler function in two parameters, which is a generalization of the exponential function e^z. The Mittag-Leffler function is defined as

$$\mathscr{E}_{\alpha,\beta}(z) = \sum_{k=0}^{\infty} \frac{z^k}{\Gamma(\alpha k + \beta)}, \quad (\alpha, \beta > 0). \tag{10.42}$$

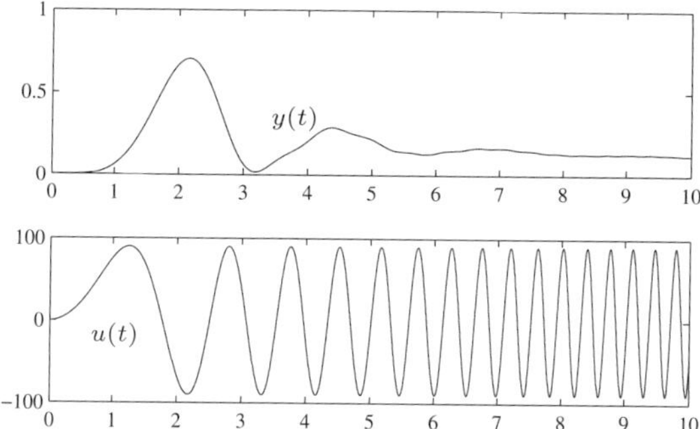

FIGURE 10.46: Input signal and solution of the equation

Clearly, e^z is a particular case of the Mittag-Leffler function[64]:

$$\mathscr{E}_{1,1}(z) = \sum_{k=0}^{\infty} \frac{z^k}{\Gamma(k+1)} = \sum_{k=0}^{\infty} \frac{z^k}{k!} = \mathrm{e}^z.$$

Furthermore, one can get more particular cases for the Mittag-Leffler function in two parameters, for example,

$$\mathscr{E}_{2,1}(z) = \cosh(\sqrt{z}), \quad \mathscr{E}_{1,2}(z) = \frac{\mathrm{e}^z - 1}{z}, \quad \mathscr{E}_{2,2}(z) = \frac{\sinh(\sqrt{z})}{\sqrt{z}}, \tag{10.43}$$

$$\mathscr{E}_{1/2,1}(\sqrt{z}) = \frac{2}{\sqrt{\pi}} \mathrm{e}^{-z} \mathrm{erfc}(-\sqrt{z}). \tag{10.44}$$

The analytical solution of the n-term fractional-order differential equation is given in general form[64] by

$$\begin{aligned} y(t) = & \frac{1}{a_n} \sum_{m=0}^{\infty} \frac{(-1)^m}{m!} \sum_{\substack{k_0+k_1+\cdots+k_{n-2}=m \\ k_0 \geqslant 0; \cdots, k_{n-2} \geqslant 0}} (m; k_0, k_1, \cdots, k_{n-2}) \\ & \prod_{i=0}^{n-2} \left(\frac{a_i}{a_n} \right)^{k_i} t^{(\beta_n - \beta_{n-1})m + \beta_n + \sum_{j=0}^{n-2} (\beta_{n-1} - \beta_j) k_j - 1} \\ & \mathscr{E}^{(m)}_{\beta_n - \beta_{n-1}, \beta_n + \sum_{j=0}^{n-2} (\beta_{n-1} - \beta_j) k_j} \left(-\frac{a_{n-1}}{a_n} t^{\beta_n - \beta_{n-1}} \right), \end{aligned} \tag{10.45}$$

where $\mathscr{E}_{\lambda,\mu}(z)$ is the Mittag-Leffler function defined in (10.42) and

$$\mathscr{E}^{(n)}_{\lambda,\mu}(y) \equiv \frac{\mathrm{d}^n}{\mathrm{d}y^n} \mathscr{E}_{\lambda,\mu}(y) = \sum_{j=0}^{\infty} \frac{(j+n)! y^j}{j! \Gamma(\lambda j + \lambda n + \mu)}, \text{ for } n = 0, 1, 2, \cdots. \tag{10.46}$$

A MATLAB function `ml_fun.m` can be written to evaluate the nth-order derivative of Mittag-Leffler function

```
function f=ml_fun(a,b,x,n,eps0)
```

```
if nargin<5, eps0=eps; end
if nargin<4, n=0; end, f=0; fa=1; j=0;
while norm(fa,1)>=eps0
   fa=gamma(j+n+1)/gamma(j+1)/gamma(a*j+a*n+b)*x.^j;
   f=f+fa; j=j+1;
end
```

The syntax of the function is `f=ml_fun(α,β,y,n,ε0)`, where ϵ_0 is the acceptable error tolerance. The returned argument $\boldsymbol{f}$ vector is the expected $\mathscr{E}_{\alpha,\beta}^{(n)}(\boldsymbol{y})$, where $\boldsymbol{y}$ is the vector of the independent variable. The default values of the variables $n = 0$ and ϵ_0 =`eps`.

Example 10.42 Draw the functions $\mathscr{E}_{1,1}^{(5)}(\boldsymbol{y})$ and $\mathscr{E}_{\sqrt{2},1.3}^{(2)}(\boldsymbol{y})$.

Solution From the discussions given earlier, the analytical solution of the former function is $\mathrm{e}^{\boldsymbol{y}}$, while the latter has no analytical solution. The functions of the two curves are obtained as shown in Figures 10.47 (a) and (b), respectively. It can be seen from the former curve, it is identical to that of the analytical solution.

```
>> x=0:0.001:2; y1=ml_fun(1,1,x,5); plot(x,y1,x,exp(x),'--')
   figure; y2=ml_fun(sqrt(2),1.3,x,2); plot(x,y2)
```

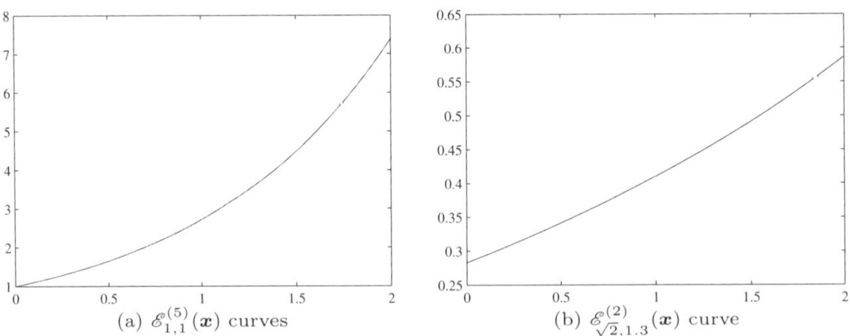

(a) $\mathscr{E}_{1,1}^{(5)}(\boldsymbol{x})$ curves (b) $\mathscr{E}_{\sqrt{2},1.3}^{(2)}(\boldsymbol{x})$ curve

FIGURE 10.47: Mittag-Leffler directive functions

The system response expression (10.45) sometimes is too complicated to use. Here a special form of the model $G(s) = \dfrac{1}{a_2 s^{\beta_2} + a_1 s^{\beta_1} + a_0}$ is studied. The step response of the system is written as[65]

$$y(t) = \frac{1}{a_2} \sum_{k=0}^{\infty} \frac{(-1)^k \hat{a}_0^k t^{-\hat{a}_1 + (k+1)\beta_2}}{k!} \mathscr{E}_{\beta_2-\beta_1, \beta_2+\beta_1 k+1}^{(k)} \left(-\hat{a}_1 t^{\beta_2-\beta_1}\right) \quad (10.47)$$

where $\hat{a}_0 = a_0/a_2, \hat{a}_1 = a_1/a_2$. Similar to the `ml_fun()` function, the step response solution function can be written as

```
function y=ml_step(a0,a1,a2,b1,b2,t,eps0)
y=0; k=0; ya=1; a0=a0/a2; a1=a1/a2;
```

```
if nargin==6, eps0=eps; end
while max(abs(ya))>=eps0
    ya=(-1)^k/gamma(k+1)*a0^k*t.^((k+1)*b2).*...
        ml_fun(b2-b1,b2+b1*k+1,-a1*t.^(b2-b1),k,eps0);
    y=y+ya; k=k+1;
end
y=y/a2;
```

whose syntax is `y=ml_step(`a_0`,`a_1`,`a_2`,`β_1`,`β_2`,t,`ϵ_0`)`.

Example 10.43 Consider the model $G(s) = \dfrac{5}{s^{0.8} + 0.75s^{0.4} + 0.9}$. Draw the step response of the system with different algorithms.

Solution It is obvious that $a_0 = 0.9, a_1 = 0.75, a_2 = 1, \beta_1 = 0.4, \beta_2 = 0.8$. The step response of the system, with the Mittag-Leffler function `ml_fun()` and `fotf` object, can also be obtained and it can be seen that they are almost exactly the same, as shown in Figure 10.48 (a). The difference of the two curves cannot be found, which means that the two methods are both accurate. The error plot can also be obtained as shown in Figure 10.48 (b).

```
>> t=0:0.001:5; y=ml_step(0.9,0.75,1,0.4,0.8,t);
   G=fotf([1 0.75 0.9],[0.8 0.4 0],5,0); y1=step(G,t);
   plot(t,5*y,t,y1,'--'), figure; plot(t,5*y-y1)
```

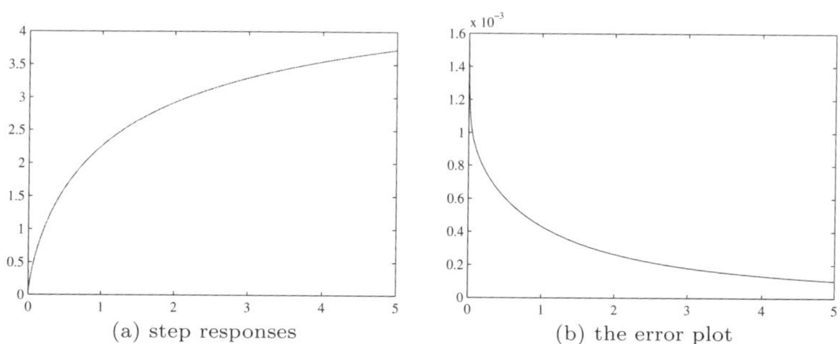

(a) step responses (b) the error plot

FIGURE 10.48: The step responses and error signal

From the above simulation example, it can be seen that both methods give correct results. However, the computation based on Grünwald-Letnikov's definition is much simpler than the Mittag-Leffler function-based method. Thus it will be used later in the book.

Approximate algorithm for nonlinear fractional-order ordinary differential equations

It can be seen that the modified Oustaloup filter given in the previous subsection is an effective way for evaluating fractional-order differentiations. Since the orders of the numerator and denominator in the ordinary Oustaloup filter are the same, it is likely to cause algebraic loops. Thus a low-pass filter should be appended to the filter. The block in Figure 10.49 (a) can be used for modeling fractional-order differentiators.

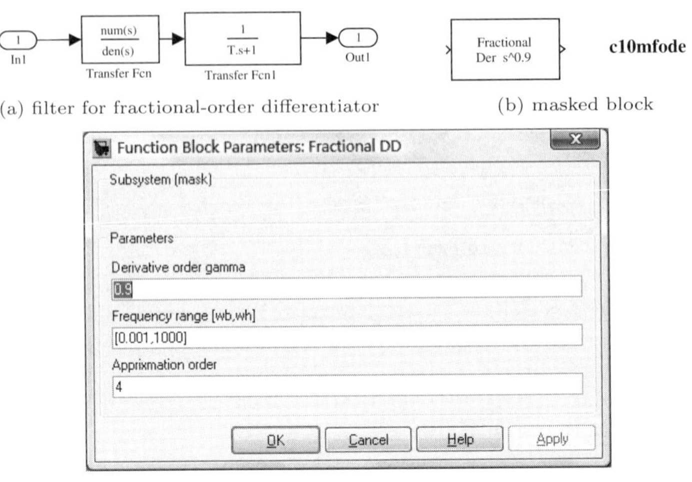

(a) filter for fractional-order differentiator (b) masked block

(c) parameter dialog box

FIGURE 10.49: Fractional-order differentiator block design

With the use of masking technique in Simulink[35], the designed block can be masked as shown in Figure 10.49 (b). Double click such a block and a dialog box appears as in Figure 10.49 (c). The corresponding parameters can be filled into the dialog box to complete the fractional-order differentiator block. The following code can be attached to the masked block.

```
wb=ww(1); wh=ww(2); G=new_fod(gam,n,wb,wh);
num=G.num{1}; den=G.den{1};  T=1/wh; str='Fractional\n';
if isnumeric(gam)
   if gam>0, str=[str, 'Der  s^' num2str(gam) ];
   else, str=[str, 'Int  s^{' num2str(gam) '}']; end
else, str=[str, 'Der  s^gam']; end
```

Example 10.44 Solve the linear fractional-order differential equations in Example 10.41, and compare the results with other methods.

Solution For linear fractional-order differential equations, the block diagram-based method is not as straightforward as the method used in Example 10.41. An auxiliary variable $z(t) = \mathscr{D}_t^{0.5}y(t)$ can be introduced, and the original differential equation can be rewritten as

$$z(t) = \sin(t^2) - \frac{1}{90}\left[\mathscr{D}_t^3 z(t) + 8\mathscr{D}_t^{2.6}z(t) + 26\mathscr{D}_t^{1.8}z(t) + 73\mathscr{D}_t^{0.7}z(t)\right].$$

The Simulink block diagram shown in Figure 10.50 can be established based on the new equation. With stiff ODE solvers, the numerical solution to the problem can be found and the results are exactly the same as the one in Figure 10.46.

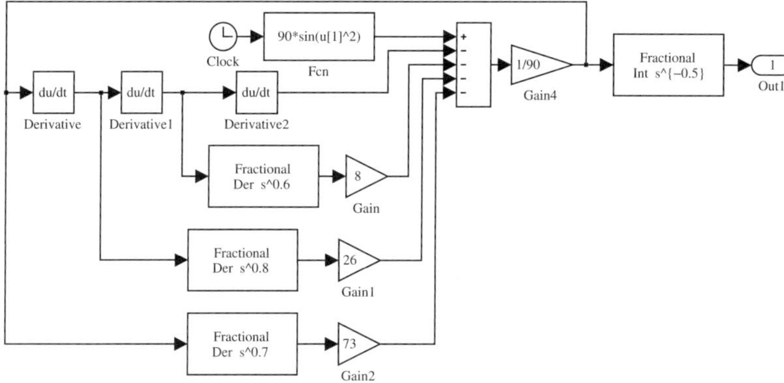

FIGURE 10.50: Simulink block diagram (file: c10mfod1.mdl)

Example 10.45 Solve the nonlinear fractional-order differential equation

$$\frac{3\mathscr{D}^{0.9}y(t)}{3+0.2\mathscr{D}^{0.8}y(t)+0.9\mathscr{D}^{0.2}y(t)}+\left|2\mathscr{D}^{0.7}y(t)\right|^{1.5}+\frac{4}{3}y(t)=5\sin(10t).$$

Solution From the given equation, the explicit form of $y(t)$ can be written as

$$y(t)=\frac{3}{4}\left[5\sin(10t)-\frac{3\mathscr{D}^{0.9}y(t)}{3+0.2\mathscr{D}^{0.8}y(t)+0.9\mathscr{D}^{0.2}y(t)}-\left|2\mathscr{D}^{0.7}y(t)\right|^{1.5}\right]$$

and from the explicit expression of $y(t)$, the block diagram in Simulink can be established in Figure 10.51 (a). From the simulation model, the simulation results can be obtained as shown in Figure 10.51 (b). The results are verified by different control parameters in the filter and they give consistent results.

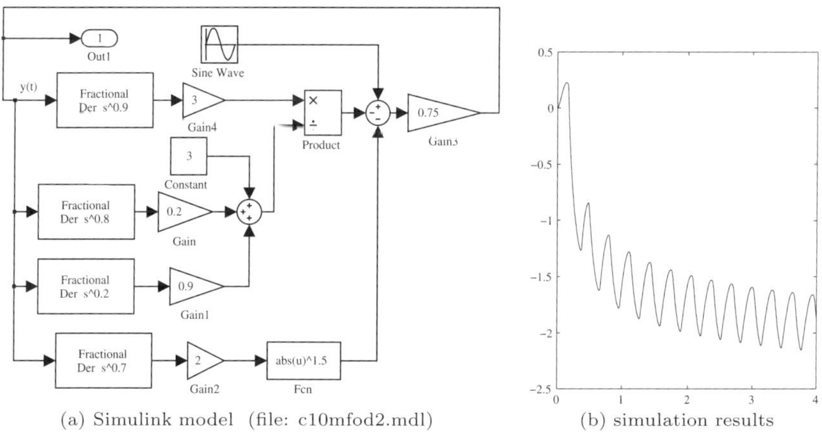

(a) Simulink model (file: c10mfod2.mdl) (b) simulation results

FIGURE 10.51: Simulink description and simulation results

Exercises

1. Consider a tipping problem in a restaurant[66]. Assume that the average rate for the tips is 15% of the consumption. The service level and food quality are used to calculate the tip. The service level can be written as "good," "average" and "poor," and the food quality can also be expressed as other fuzzy descriptions. Establish a fuzzy inference system for determining the tips.
2. Consider the sampled data set (x_i, y_i) given below. Construct a neural network model and plot the training curve in the interval $x \in (1, 10)$. Test different neural network structures and training algorithms. Compare the fitting results under different structures.

x_i	1	2	3	4	5	6	7	8	9	10
y_i	244.0	221.0	208.0	208.0	211.5	216.0	219.0	221.0	221.5	220.0

3. For the actual measured data given below, construct a neural network to fit the surface in the rectangular region $(0.1, 0.1) \sim (1.1, 1.1)$. Compare the results with data interpolation algorithms.

y_i	x_1	x_2	x_3	x_4	x_5	x_6	x_7	x_8	x_9	x_{10}	x_{11}
0	0.1	0.2	0.3	0.4	0.5	0.6	0.7	0.8	0.9	1	1.1
0.1	0.8304	0.8273	0.8241	0.8210	0.8182	0.8161	0.8148	0.8146	0.8158	0.8185	0.8230
0.2	0.8317	0.8325	0.8358	0.8420	0.8513	0.8638	0.8798	0.8994	0.9226	0.9496	0.9801
0.3	0.8359	0.8435	0.8563	0.8747	0.8987	0.9284	0.9638	1.0045	1.0502	1.1	1.1529
0.4	0.8429	0.8601	0.8854	0.9187	0.9599	1.0086	1.0642	1.1253	1.1904	1.257	1.3222
0.5	0.8527	0.8825	0.9229	0.9735	1.0336	1.1019	1.1764	1.254	1.3308	1.4017	1.4605
0.6	0.8653	0.9105	0.9685	1.0383	1.118	1.2046	1.2937	1.3793	1.4539	1.5086	1.5335
0.7	0.8808	0.9440	1.0217	1.1118	1.2102	1.311	1.4063	1.4859	1.5377	1.5484	1.5052
0.8	0.8990	0.9828	1.082	1.1922	1.3061	1.4138	1.5021	1.5555	1.5573	1.4915	1.346
0.9	0.9201	1.0266	1.1482	1.2768	1.4005	1.5034	1.5661	1.5678	1.4889	1.3156	1.0454
1	0.9438	1.0752	1.2191	1.3624	1.4866	1.5684	1.5821	1.5032	1.315	1.0155	0.6248
1.1	0.9702	1.1279	1.2929	1.4448	1.5564	1.5964	1.5341	1.3473	1.0321	0.6127	0.1476

4. Solve the constrained optimization problem with genetic algorithms and PSO methods and compare the results with traditional optimization algorithms.

$$\max_{\boldsymbol{x} \text{ s.t.} \begin{cases} 0.003079x_1^3x_2^3x_5-\cos^3 x_6\geqslant 0 \\ 0.1017x_3^3x_4^3-x_5^2\cos^3 x_6\geqslant 0 \\ 0.09939(1+x_5)x_1^3x_2^2-\cos^2 x_6\geqslant 0 \\ 0.1076(31.5+x_5)x_3^3x_4^2-x_5^2\cos^2 x_6\geqslant 0 \\ x_3x_4(x_5+31.5)-x_5[2(x_1+5)\cos x_6+x_1x_2x_5]\geqslant 0 \\ 0.2\leqslant x_1\leqslant 0.5, 14<\leqslant x_2\leqslant 22, 0.35\leqslant x_3\leqslant 0.6, \\ 16\leqslant x_4\leqslant 22, 5.8\leqslant x_5\leqslant 6.5, 0.14\leqslant x_6\leqslant 0.2618 \end{cases}} \frac{1}{2\cos x_6}\left[x_1x_2(1+x_5)+x_3x_4\left(1+\frac{31.5}{x_5}\right)\right]$$

5. Solve the benchmark problems in Exercise 6.7 using genetic algorithms and PSO methods.
6. Assume that the corrupted signal is generated from

```
>> t=0:0.005:5;y=15*exp(-t).*sin(2*t);r=0.3*randn(size(y));y1=y+r;
```

Perform de-noising tasks with wavelet transforms and compare the results with the filtering techniques in Exercise 8.8.

7. A series of experimental data is given in file c10rsdat.txt which is made up as a 60×13 table. Each column corresponds to an attribute and the last column is the decision attribute. Use rough set reduction technique to check which attribute is most important to the event in the decision.
8. For the signal $f(t) = \mathrm{e}^{-3t}\sin(t+\pi/3)+t^2+3t+2$, find the 0.2th order derivative and 0.7th order integral. Draw the relevant curves.
9. Design a filter for Exercise 10.8. The fractional-order derivatives and integrals can be obtained with the filter. Compare the results with the ones obtained with Grünwald-Letnikov method.
10. Consider a fractional-order linear differential equation described by[60]
$$0.8\mathscr{D}_t^{2.2}y(t) + 0.5\mathscr{D}_t^{0.9}y(t) + y(t) = 1,\ y(0) = y'(0) = y''(0) = 0.$$
Solve the solution using numerical method. If one changes the orders of 2.2 and 0.9 respectively to 2 and 1, an approximate integer-order differential equation can be obtained. Compare the accuracy of the integer-order approximation.
11. Evaluate and draw the following Mittag-Leffler functions and verify (10.43).
(i) $\mathscr{E}_{1,1}(z)$, (ii) $\mathscr{E}_{2,1}(z)$, (iii) $\mathscr{E}_{1,2}(z)$, (iv) $\mathscr{E}_{2,2}(z)$
12. Prove the following identities involving Mittag-Leffler functions and graphically visualize both sides of each identity for verification purposes.
(i) $\mathscr{E}_{\alpha,\beta}(x) + \mathscr{E}_{\alpha,\beta}(-x) = 2\mathscr{E}_{\alpha,\beta}(x^2)$ (ii) $\mathscr{E}_{\alpha,\beta}(x) - \mathscr{E}_{\alpha,\beta}(-x) = 2x\mathscr{E}_{\alpha,\alpha+\beta}(x^2)$
(iii) $\mathscr{E}_{\alpha,\beta}(x) = \dfrac{1}{\Gamma(\beta)} + \mathscr{E}_{\alpha,\alpha+\beta}(x)$ (iv) $\mathscr{E}_{\alpha,\beta}(x) = \beta\mathscr{E}_{\alpha,\beta+1}(x) + \alpha x\dfrac{\mathrm{d}}{\mathrm{d}x}\mathscr{E}_{\alpha,\beta+1}(x)$
13. Find the closed-loop model from the typical negative feedback structure.
(i) $G(s) = \dfrac{211.87s + 317.64}{(s+20)(s+94.34)(s+0.17)}$, $G_c(s) = \dfrac{169.6s+400}{s(s+4)}$, $H(s) = \dfrac{1}{0.01s+1}$
(ii) $G(s) = \dfrac{s^{0.4}+5}{s^{3.1}+2.8s^{2.2}+1.5s^{0.8}+4}$, $G_c(s) = 3+2.5s^{-0.5}+1.4s^{0.8}$, $H(s)=1$.
14. Consider the linear fractional-order differential equation given by
$$\mathscr{D}x(t) + \left(\frac{9}{1+2\lambda}\right)^{\alpha}\mathscr{D}^{\alpha}x(t) + x(t) = 1,\ 0 < \alpha < 1$$
where $\lambda = 0.5$, $\alpha = 0.25$. Solve the equation numerically.
15. Find a good approximation for the modified Oustaloup's filter, to $s^{0.7}$ and see which N can best fit the fractional-order differentiator.
16. Solve the following nonlinear fractional-order differential equation with the block diagram-based algorithm
$$\mathscr{D}^2x(t) + \mathscr{D}^{1.455}x(t) + \left[\mathscr{D}^{0.555}x(t)\right]^2 + x^3(t) = f(t)$$
where $f(t)$ is a unit square wave input signal of 1 Hz and the system has zero initial conditions.

References and Bibliography

[1] Garbow B S, Boyle J M, Dongarra J J, et al. Matrix eigensystem routines — EISPACK guide extension, Lecture notes in computer sciences, volume 51. New York: Springer-Verlag, 1977

[2] Smith B T, Boyle J M, Dongarra J J, et al. Matrix eigensystem routines – EISPACK guide, Lecture notes in computer sciences, volume 6. New York: Springer-Verlag, second edition, 1976

[3] Dongarra J J, Bunsh J R, Moler C B. LINPACK user's guide. Philadelphia: Society of Industrial and Applied Mathematics, 1979

[4] Press W H, Flannery B P, Teukolsky S A, et al. Numerical recipes, the art of scientific computing. Cambridge: Cambridge University Press, 1986. Free textbook at `http://www.nrbook.com/a/bookcpdf.php`

[5] Lamport L. LaTeX: a document preparation system — user's guide and reference manual. Reading MA: Addision-Wesley Publishing Company, second edition, 1994

[6] Xue D. Analysis and computer aided design of nonlinear systems with Gaussian inputs. D.Phil. thesis, Sussex University, U.K., 1992

[7] Strang G. Calculus. Free textbook at `http://ocw.mit.edu/ans7870/resources/Strang/strangtext.htm`: Wellesley-Cambridge Press, 1991

[8] Dawkins P. Calculus I, II, & III. `http://tutorial.math.lamar.edu/pdf/CalcII/CalcI_Complete.pdf`; `http://tutorial.math.lamar.edu/pdf/CalcII/CalcII_Complete.pdf`; `http://tutorial.math.lamar.edu/pdf/CalcIII/CalcIII_Complete.pdf`, 2007

[9] Hefferon J. Linear algebra. Saint Michael's College, USA: Open source textbook at `http://joshua.smcvt.edu/linearalgebra/`, 2006

[10] Meyer C D. Matrix analysis and applied linear algebra. Philadelphia: Society for Industrial and Applied Mathematics. Free at `http://www.matrixanalysis.com/DownloadChapters.html`, 2001

[11] Dawkins P. Linear algebra. `http://tutorial.math.lamar.edu/pdf/LinAlg/LinAlg_Complete.pdf`, 2007

[12] Moler C B, Van Loan C F. Nineteen dubious ways to compute the exponential of a matrix. SIAM Review, 1979, 20:801–836

[13] Huang L. Linear algebra in systems and control theory. Beijing: Science Press, 1984 (in Chinese)

[14] Mauch S. Advanced mathematical methods for scientists and engineers. Open source textbook at `http://www.its.caltech.edu/~sean/ applied_math.pdf`, 2004

[15] Boyd S, Vandenberghe L. Convex optimization. Cambridge University Press, 2004. Free textbook at `http://www.stanford.edu/~boyd/cvxbook/bv_cvxbook.pdf`

[16] Boyd S, Ghaoui L El, Feron E, et al. Linear matrix inequalities in systems and control theory. Philadelphia: SIAM books, Volume 15 of Studies in Applied Mathematics. Free textbook at `http://www.stanford.edu/~boyd/lmibook/lmibook.pdf`, 1994

[17] Mittelmann H D. Decision tree for optimization software. `http://plato.asu.edu/guide.html`, 2007

[18] Nelder J A, Mead R. A simplex method for function minimization. Computer Journal, 1965, 7:308–313

[19] Willems J C. Least squares stationary optimal control and the algebraic Riccati equation. IEEE Transactions on Automatic Control, 1971, 16(6):621–634

[20] The MathWorks Inc. Robust control toolbox user's manual, 2007

[21] Löfberg J. YALMIP: a toolbox for modeling and optimization in MATLAB. Proceedings of IEEE International Symposium on Computer Aided Control Systems Design. Taipei, 2004, 284–289

[22] Chipperfield A, Fleming P. Genetic algorithm toolbox user's guide. Department of Automatic Control and Systems Engineering, University of Sheffield, 1994

[23] Ackley D H. A connectionist machine for genetic hillclimbing. Boston, USA: Kluwer Academic Publishers, 1987

[24] Goldberg D E. Genetic algorithms in search, optimzation and machine learning. Reading, MA: Addison-Wesley, 1989

[25] Henrion D. A review of the global optimization toolbox for Maple, 2006. `http://www.laas.fr/~henrion/Papers/mapleglobopt.pdf`

[26] Press W H, Teukolsky S A, Vetterling W T, et al. Numerical recipes in C, second edition. Cambridge University Press. Free textbook at `http://www.nrbook.com/a/bookcpdf.php`, 1992

[27] Dawkins P. Differential equations. `http://tutorial.math.lamar.edu/pdf/DE/DE_Complete.pdf`, 2007

[28] Fehlberg E. Low-order classical Runge-Kutta formulas with step size control and their application to some heat transfer problems. Technical Report 315, NASA, 1969

[29] Forsythe G E, Malcolm M A, Moler C B. Computer methods for mathematical computations. Englewood Cliffs: Prentice-Hall, 1977

[30] Bogdanov A. Optimal control of a double inverted pendulum on a cart. Technical Report CSE-04-006, Department of Computer Science & Electrical Engineering, OGI School of Science & Engineering, OHSU, 2004

[31] Shampine L F, Thompson S. Solving DDEs in MATLAB. Applied Numerical Mathematics, 2001, 37(4):441–458

[32] Shampine L F, Kierzenka J, Reichelt M W. Solving boundary value problems for ordinary differential equation problems in MATLAB with bvp4c, 2000

[33] The MathWorks Inc. SimuLAB, a program for simulating dynamic systems, user's guide, 1990

[34] The MathWorks Inc. Simulink user's manual, 2007

[35] Xue D, Chen Y Q. MATLAB/Simulink based system simulation techniques. Beijing: Tsinghua University Press, 2002 (in Chinese)

[36] Moler C B. Numerical computing with MATLAB. MathWorks Inc, 2004

[37] Wikipedia. List of numerical analysis topics. http://en.wikipedia.org/wiki/List_of_numerical_analysis_topics, 2008

[38] Lancaster L, Salkauskas K. Curve and surface fitting: an introduction. London: Academic Press, 1986

[39] Xue D. Model reduction techniques and applications. Shenyang, China: Lecture Notes of Northeastern University, 1996

[40] Bosley M J, Lees F P. A survey of transfer function derivations from higher-order state-variable models. Automatica, 1972, 8:765–775

[41] The MathWorks Inc. Signal processing user's guide, 2007

[42] Grinstead C M, Snell J L. Grinstead and Snell's introduction to probability. The CHANCE Project: Open source textbook at http://math.dartmouth.edu/~prob/prob/prob.pdf, 2006

[43] StatSoft Inc. Electronic statistics textbook. Tulsa, OK: StatSoft. Electronics textbook at http://www.statsoft.com/textbook/stathome.html, 2007

[44] Landau D P, Binder K. A guide to Monte Carlo simulations in statistical physics. Cambridge University Press, 2000

[45] Conover W J. Practical nonparametric statistics. New York: Wiley, 1980

[46] Lu X. Applied statistics. Beijing: Tsinghua University Press, 1999 (in Chinese)

[47] Cody R P, Smith J K. Applied statistics and the SAS programming language. Prentice Hall, fifth edition, 2006

[48] The MathWorks Inc. Statistics toolbox user's manual, 2007

[49] Weisstein E W. Goldbach conjecture. From MathWorld — A Wolfram Web Resource. http://mathworld.wolfram.com/GoldbachConjecture.html

[50] Zadeh L A. Fuzzy sets. Information and Control, 1965, 8:338–353

[51] Hagan M T, Demuth H B, Beale M H. Neural network design. PWS Publishing Company, 1995

[52] Houck C R, Joines J A, Kay M G. A genetic algorithm for function optimization: a MATLAB implementation, 1995

[53] Kennedy J, Eberhart R. Particle swarm optimization. Proceedings of IEEE International Conference on Neural Networks. Perth, Australia, 1995, 1942–1948

[54] Birge B K. PSOt, a particle swarm optimization toolbox for MATLAB. Proceedings of the 2003 IEEE Swarm Intelligence Symposium. Indianapolis, 2003, 182–186

[55] Trelea I C. The particle swarm optimization algorithm: convergence analysis and parameter selection. Information Processing Letters, 2003, 85(6):317–325

[56] Clerc M, Kennedy J. The particle swarm: explosion, stability, and convergence in a multidimensional complex space. IEEE Transactions on Evolutionary Computation, 2002, 6(1):58–73

[57] Pawlak Z. Rough sets — theoretical aspects of reasoning about data. Boston, USA: Kluwer Academic Publishers, 1991

[58] Zhang X F. Research and program development of rough set data analysis system. Master's thesis, Northeastern University, 2004 (in Chinese)

[59] Hilfer R. Applications of fractional calculus in physics. Singapore: World Scientific, 2000

[60] Podlubny I. Fractional differential equations. San Diego: Academic Press, 1999

[61] Petráš I, Podlubny I, O'Leary P. Analogue realization of fractional order controllers. Fakulta BERG, TU Košice, 2002

[62] Oustaloup A, Levron F, Nanot F, et al. Frequency band complex non integer differentiator: characterization and synthesis. IEEE Transactions on Circuits and Systems I: Fundamental Theory and Applications, 2000, 47(1):25–40

[63] Xue D, Zhao C N, Chen Y Q. A modified approximation method of fractional order system. Proceedings of IEEE Conference on Mechatronics and Automation. Luoyang, China, 2006, 1043–1048

[64] Podlubny I. The Laplace transform method for linear differential equations of the fractional order. Proceedings of the 9th International BERG Conference. Kosice, Slovak Republic (in Slovak), 1997, 119–119

[65] Podlubny I. Fractional-order systems and PI$^\lambda$D$^\mu$-controllers. IEEE Transactions on Automatic Control, 1999, 44(1):208–214

[66] The MathWorks Inc. Fuzzy logic toolbox user's manual, 2007

MATLAB Functions Index

Bold page numbers indicate where to find the syntax explanation of the function.

Index

$\mathscr{J}$

$\mathscr{K}$

$\mathscr{L}$